Foundry Technology

Foundry Technology

Second edition

Peter Beeley BMet, PhD, DMet, CEng, FIM, FIBF

Life Fellow and formerly Senior Lecturer in Metallurgy, University of Leeds

BUTTERWORTH
HEINEMANN

OXFORD   AUCKLAND   BOSTON   JOHANNESBURG   MELBOURNE   NEW DELHI

Butterworth-Heinemann
Linacre House, Jordan Hill, Oxford OX2 8DP
225 Wildwood Avenue, Woburn, MA 01801-2041
A division of Reed Educational and Professional Publishing Ltd

ᚱ A member of the Reed Elsevier plc group

First published 1972
Reprinted 1979, 1980, 1982
Second edition 2001

© Peter Beeley 1972, 2001

British Library Cataloguing in Publication Data
A catalogue record for this book is available from the British Library

Library of Congress Cataloguing in Publication Data
A catalogue record for this book is available from the Library of Congress

ISBN 0 7506 4567 9

Typeset by Laser Words, Madras, India

Printed and bound in Great Britain by Biddles Ltd
www.biddles.co.uk

Contents

Preface to the second edition

The production of a revised edition of Foundry Technology was encouraged by a sustained demand for the book despite many changes in processes and products since the appearance of the original edition, which has been out of print for several years. The intervening period has seen major transformations in the foundry industry and its products and supporting organizations.

In this new edition the aim has been to update the book without losing its identity. The original framework was found useful by many readers and has been retained, but every effort has been made to accommodate changes of emphasis in various aspects of metal casting. Much new material has been introduced, especially in those chapters dealing with production processes and product characteristics. This has not, however, been at the expense of the basic principles of moulding and other operations which underpin more recent progress.

One rapidly developing aspect, shared with other fields of activity, is the application of the computer in areas ranging from process modelling and prototyping to plant operations. The subject of modelling is reviewed from a potential user perspective, mainly in the context of solidification and feeding and using specific examples. The broader treatment of methoding retains the original emphasis on the still widely accepted Chworinov–Wlodawer approach to the understanding of casting soundness. Rapid prototyping too is surveyed as an important advance in the overall design to production sequence. The growing significance of environmental and health issues has been recognised with the addition of a separate new chapter.

In considering design matters attention has been given to modern approaches to materials selection. As in the original work frequent reference is made to British Standards, where major changes have occurred, including increased coordination with international standards: a fully revised appendix reflects the replacement of long familiar numbers with new designations and changes in content.

Additional references to the literature reflect the continued active roles of the Institute of British Foundrymen and the American Foundrymen's Society, the output of numerous conferences and working groups, and the emergence of new journals. Many of the original references have nevertheless been retained for their direct links with a classical and fruitful period of progress in work on casting problems.

The author acknowledges willing help from many sources with the practicalities of producing this revised edition. Special thanks are due to those individuals, firms and organizations who have provided illustrations, separately credited, demonstrating the outstanding capabilities of the industry. In this respect I am especially grateful to Dr. Michael Ashton, Chief Executive, for advice and help from the Castings Development Centre. I also thank Dr David Driver for authorizing use of particular aerospace examples.

I owe much to the interest and encouragement of former colleagues and research students at the University of Leeds over a prolonged period and also to many years of working contacts and discussions with fellow practitioners in the wider field of metal casting: Professor John Camphell, Cyril McCombe and the late Dr Voya Kondic are representative of many friends with whom I have had fruitful contacts in varied capacities and aspects of the subject. I am indebted to Dr Michael Willison, Environmental Management Consultant, for valuable comments on a draft of Chapter 11.

My family has been a source of encouragement thoughout the production of this edition. My special thanks are due to my son John of Avesta Sheffield for his perceptive comments and for practical help and advice beyond price. My wife Jean has been a constant support during production and a valuable helper in the assembly of the product.

Note on units. The dual system of units as employed in the original edition has been retained, but in reverse order to reflect the full adoption of the SI system: bracketted Imperial values are thus included where appropriate. Some original illustrations based on the latter system have been retained unchanged.

Preface to the
first edition _____

This book is intended to serve as a bridge between the study of the basic principles of metal founding and their application in the producing and user industries. It has been the aim to further the understanding of modern techniques of casting manufacture and of the characteristics of the product. The work is an outcome of periods spent in foundry management and subsequently at the University of Leeds, where the author has been able to maintain his contacts both with the industry and with a stimulating field of study and research.

The book was originally conceived as a text and work of reference for use in universities and other branches of higher education and for managerial and technical staff in the industry, but it is hoped that it will be useful to all metallurgists and foundrymen with an interest in the scientific background to the subject. A particular aim is to assist engineers and engineering students in appreciating the role of castings in design and materials selection.

In 1934 J. Laing and R. T. Rolfe wrote, in a preface to their Manual of Foundry Practice 'It cannot be denied that foundry practice has experienced, during the last few years, a fundamental transformation, changing from a basis of crude empiricism to one of scientific and systematic control'. That the same point can still be made is not surprising, since the change is a continuing one for which there can be no more appropriate theme than the motto of the Institute of British Foundrymen 'Science hand in hand with labour'. Science and technology are reducing the amount and severity of labour and at the same time upgrading the quality of the product.

The general approach used in the work is pragmatic: principles are examined mainly in relation to specific situations in the manufacture and use of castings. However, whilst the subject must be seen in an industrial context, it has been considered equally essential to include ideas arising from basic research in the field of casting: it is hoped that such a treatment will help to strengthen the links between the industry and the centres of education and research.

The book is arranged in three main sections. The first four chapters deal with topics central to founding – the phenomena involving metal and mould during the crucial stages of moulding and casting: the reader is assumed to be familiar with the elements of these operations. Solidification is treated from the viewpoints both of structure and feeding; the quantitative approach to the latter topic is examined in Chapter 3. The following three chapters are concerned mainly with the cast product, including defects, quality evaluation and aspects of design. An outline survey of non-destructive testing techniques is included in Chapter 6, whilst Chapter 7 contains comparative data summaries of important properties of cast alloys.

The last three chapters deal with the casting processes themselves. The long Chapter 8 is really three chapters in one but has been retained as a unit to embrace the main elements in the sand foundry production sequence; die casting and other special techniques are examined in the subsequent chapters.

In treating fringe topics it was considered insufficient to refer the reader to standard specialized works. Whilst such references have been given, essentials of the topics have been presented in so far as they impinge on metal founding. Concerning more central themes, although it has been the aim to cover a wide spectrum of foundry activity, reference to the specialized metallurgy and problems associated with individual alloys has been restricted to cases which illustrate a principle or in which general comparisons are assisted. Mention is frequently made of economic repercussions of matters discussed: industrial readers will need no reminder of the commercial background to all foundry activities.

As a guide to the literature, selected references are given to original sources, amongst which the valuable contributions to knowledge contained in the publications of the American Foundrymen's Society and the Institute of British Foundrymen deserve special mention. Illustrations are individually acknowledged in the text: whilst it has been the aim to do this in all cases it is hoped that this note will serve in case of inadvertent omission.

The author is indebted to many friends and colleagues who have given encouragement and practical help, including Professor Jack Nutting, now Chairman of the Houldsworth School of Applied Science, for his constant support in furthering the study of metal founding at Leeds University. Above all I must acknowledge profound debts to my wife for her unstinting help and to my father, the late A. J. Beeley of Edgar Allen & Co., Sheffield, for his early guidance in the field of industrial metallurgy.

P. R. B.

About the author _____

Peter Beeley was born in Sheffield and graduated in metallurgy at the University of Sheffield, where he later obtained a PhD and was awarded a Brunton Medal for research on steel castings. Metallurgical posts with David Brown and Sons Ltd and Hadfields Ltd were followed by six years as Foundry Manager with Sheepbridge Alloy Castings Ltd, producers of alloy steel and superalloy cast components for aircraft and general engineering. In his subsequent post at Leeds University he undertook castings research and produced publications which led to the award of the further degree of Doctor of Metallurgy by the University of Sheffield. At Leeds the author served as Dean of the Faculty of Applied Science and also as a director of University of Leeds Industrial Services Ltd. He was made a Life Fellow of the University in 1988 and in the same year became the first recipient of the Voya Kondic Medal for services to foundry education. He served for several years as Chairman of the Editorial Board of the journal Cast Metals, and is a member of committees concerned with publications of the Institute of Materials. He is also co-editor and part author of a specialized book on investment casting.

Introduction _____

Castings, the products of the metal founding industry, are manufactured in a single step from liquid metal without intermediate operations of mechanical working such as rolling or forging. Shaped castings are thus distinguished from ingots and other cast forms which are only at an intermediate stage of their metallurgical life.

The basic simplicity of this, the most direct of all metallurgical processes, has provided the foundation for the growth of a vast industry with a wide diversity of products. To obtain a true perspective of the casting process, however, its characteristics must be seen in relation to the whole range of processes available for the production of metal structures.

The principal methods of shaping metals may be classified in five groups:

1. *Casting*. The production of shaped articles by pouring molten metal into moulds.
2. *Mechanical working*. The shaping of metals in the solid state by plastic deformation above or below the recrystallization temperature – by hot or cold working. The starting point for this group of processes is the cast ingot or billet and the metal must possess the capacity for plastic deformation. Much output in this category is of standard primary or semi-finished shapes such as bars, plates, sheets and sections, produced by rolling and extrusion and providing the basic material for further shaping operations. Other mechanical working processes, for example forging, produce varied shapes more directly analogous to castings.
3. *Fabrication by joining*. The production of structural units by the joining of smaller components manufactured in other ways. The most notable method employed is welding, much of which is carried out using components cut from standard wrought materials. Weld fabrications compete directly with castings over a considerable weight range, but composite structures are also produced in which the two processes can be combined to mutual advantage. Welding is also extensively used for the assembly of very large monolithic structures; applied on this scale as a field joining process, however, welding is in competition not with founding but with rivetting, bolting and other fastening devices.
4. *Machining*. The production of shaped articles by cutting from plain or roughly shaped forms using machine tools. Whilst components are often shaped wholly by cutting from blanks, machining is also frequently

needed as a finishing operation to develop accurate final dimensions on components formed by other methods.

5. *Powder metallurgy.* The production of shaped parts by the die pressing and sintering of metal powders.

Apart from these main divisions several hybrid methods of shaping metals have crossed the traditional boundaries. Semi-solid processing and squeeze casting combine features of both cast and wrought product manufacture, whilst spray deposition contains elements of both casting and powder metallurgy.

Consideration of this brief outline shows that the main shaping processes are in many respects complementary. Over wide fields, however, their capabilities overlap, allowing alternative decisions and providing a basis for competition with respect to properties and cost. Castings must variously compete with forgings, weldments and sintered compacts, quite apart from plastics, ceramics and other non-metallic materials. These alternative products can offer a challenge to the founder in terms both of price and functional performance. Castings, on the other hand, can make inroads in fields traditionally supplied by other types of product, as in the replacement of certain sheet metal constructions by pressure die castings or of machined components by investment castings. The value analysis approach in engineering design means that materials and products are increasingly selected on rational grounds rather than by long established custom.

The full exploitation of the casting process requires careful study not only of its advantages but of potential difficulties and limitations. The following summary shows some of the major characteristics which determine the place of the casting in engineering.

Design versatility

Weight range. There is little weight or size limitation upon castings, the only restriction being imposed by the supply of molten metal and the means of lifting and handling the output. The range of products accordingly extends from castings represented by the heavy steel structure illustrated in Figure 0.1 down to the precision dental casting shown in Figure 0.2; many of the same fundamental principles are involved in their manufacture. Some idea of the logistics involved in the production and handling of the heaviest castings as in the example shown can be derived from Reference 1.

Shape and intricacy. No other process offers the same range of possibilities for the shaping of complex features, whether in respect of elaborate contours or of intricate detail which would be expensive or impossible to machine. Some examples of the capabilities of founding in these respects are illustrated in Figures 0.3–0.6. Although restrictions on shape are fewer than in any other process, components should ideally be designed with

Figure 0.1 Cast steel mill housing: delivered weight 280 tonnes, poured weight 467 tonnes (courtesy of River Don Castings, now part of Sheffield Forgemasters Engineering Ltd.)

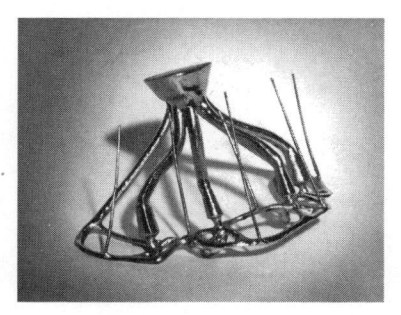

Figure 0.2 Investment casting for dental implant in cobalt–chromium alloy: finished weight 9 g

the manufacturing process in mind from the outset, a principle to be more fully examined in Chapter 7. Requirements for shape and accuracy can be catered for not only by the highly versatile process of sand casting but by a range of more specialized techniques such as die and investment casting.

It is perhaps fortunate that the geometric forms which present the greatest difficulties in casting, for example long and excessively thin sections, happen also to be those most cheaply and effectively produced by alternative methods, although continuous casting has itself proved suitable for some shapes of uniform cross section.

Figure 0.3 Cast iron gate assembly (courtesy of G. Bell & Sons Ltd., reprinted from Decorative Cast Iron Work, by Raymond Lister)

Figure 0.4 Railway crossing fabricated from manganese steel cast sections (courtesy of British Rail and Osborn-Hadfield Steel Founders Ltd.)

Figure 0.5 Set of steel castings for paper mill stockbreaker (courtesy of Osborn-Hadfield Steel Founders Ltd.)

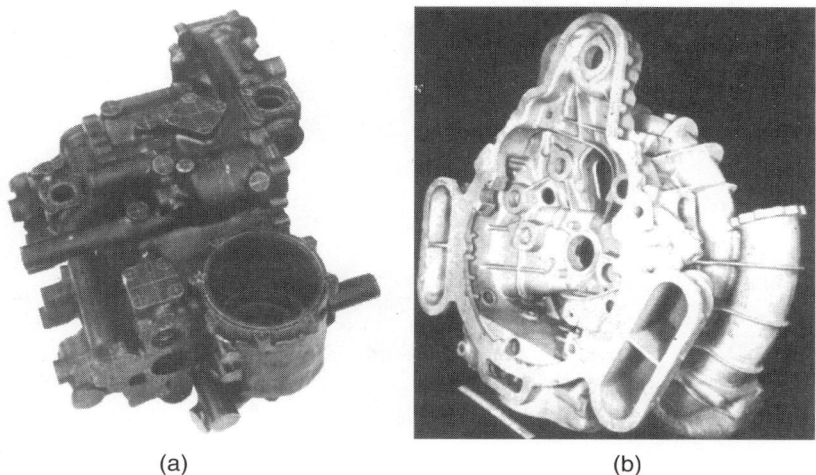

(a)   (b)

Figure 0.6 Magnesium alloy sand castings: (a) pump filter and valve, (b) super-charger casing (courtesy of Sterling Metals Ltd.)

Material composition.   Virtually all types of engineering alloy can be cast using the appropriate foundry techniques. In the past there were notable exceptions in the shape of highly reactive and refractory metals, but even these have yielded to special techniques developed to overcome difficulties in melting and casting. Certain important classes of engineering alloy can be formed by casting alone, the most notable example being the cast irons, around which the foundry industry originally grew and which now afford, within their own field, a wide selection of materials of diverse properties including some of the strongest alloys of all. The process also offers a special advantage for hard or tough alloys with poor machinability.

Some of the major groups of casting alloys are summarized in Table 0.1, which demonstrates the variety of available compositions. However, whilst

Table 0.1 Casting alloys

Alloy group	Main types of alloy	Principal British Standards for castings	Some alloying elements
Cast irons	Grey cast irons; malleable cast irons; spheroidal graphite cast irons; austempered ductile cast irons	BS EN 1561 BS EN 1562 BS EN 1563 BS EN 1564	C, Si, P, Ni, Cr
Steels	Carbon and low alloy steels; high alloy steels, including stainless and heat resisting alloys	BS 3100	C, Mn, Cr, Ni, Mo
Copper base alloys	Brasses, including high tensile brasses; miscellaneous bronzes and gunmetals, including aluminium and silicon bronzes	BS EN 1982	Zn, Sn, P Pb, Ni, Al, Fe, Mn, Si
Aluminium base alloys	Alloys for sand and die casting	BS EN 1706	Si, Cu, Mg, Mn, Zn, Ni
Magnesium base alloys	Alloys for sand and die casting	BS EN 1753	Al, Zn, Zr, Mn, rare earths
Zinc base alloys	Pressure die casting alloys	BS EN 12844	Al, Mg, Cu
Nickel base alloys	High temperature alloys; corrosion resisting alloys		Cr, Co, C, Ti, Al; Cu, Cr, Mo, Si
Miscellaneous	Lead and tin base alloys, especially for bearings; cobalt base alloys for heat, corrosion and wear resistance; permanent magnet alloys; titanium, chromium, molybdenum and other special materials for high temperature and corrosion resistance		Sn, Pb, Sb, Cu Cr, Ni, Mo, W, Nb Fe, Co, Ni, Al

almost any individual alloy composition can be cast, best results are obtained from casting specifications which have been specially developed to exploit the optimum combination of foundry and engineering properties.

Tooling costs

Casting is equally suitable for one-off or for quantity production. Since a foundry pattern usually takes the form of a positive replica, often of inexpensive material, its cost compares favourably with that of the form tools required for some alternative processes. This is especially true for jobbing and small quantity production, where the versatile techniques of moulding can be used to simplify the equipment required. The cost of the more elaborate and durable pattern assemblies needed for mass production can still compare favourably with that of the dies and fixtures used in other industries. The same advantage applies to the relatively simple dies used in gravity die casting, although pressure die casting requires much more costly equipment, and large volume production is needed for amortization of die costs.

Coupled with the low cost of patterns is the relative ease with which design changes or dimensional modifications can be introduced during the development and early production stages of a casting. Computer modelling and modern techniques for rapid prototyping can provide powerful assistance in further reducing costs and lead times during these stages.

The cast structure

The casting acquires many of its metallurgical characteristics during solidification. Even in cases where the main structural features are modified in further cooling or subsequent heat treatment, solidification may still exercise a lasting influence on structure and properties. The basic metallographic features, for example grain size and the form and distribution of micro-constituents, are themselves sensitive to casting conditions, whilst properties may be further influenced by segregation and microporosity.

In wrought metals mechanical deformation in manufacture not only confers shape but produces grain refinement, closure of voids and redistribution of segregates; it may even, in the presence of severely defective conditions, lead to failure during manufacture. These features have often been regarded as providing an element of insurance which is absent in castings. Such insurance can nevertheless be provided by close control of the cast structure and by non-destructive testing, now developed to a high standard of reliability. These issues are fully examined in later chapters, where it will be demonstrated that under controlled conditions valid comparisons can indeed be made between cast and wrought materials. The fibrous nature of the wrought structure is only advantageous if the stress pattern is such as to utilize the maximum properties associated with the direction of working; in other cases the more isotropic properties of castings may well be preferred, although it will be shown that not all cast structures fall into this category.

In some cases specific features of cast structure are turned to useful account; notable examples include the columnar and single crystal structures exploited in gas turbine rotor blades, the cored structure in bearing bronzes, and the graphite flake structure which gives rise to the high damping capacity of grey cast iron.

Flexibility of the process

Castings may be manufactured with relatively small capital investment when compared with the mills, presses and similar heavy plant required in some other fields. An appreciable proportion of output has traditionally been produced by small foundries catering for limited or local needs; the wide geographical distribution of foundries illustrates this feature. The basic elements of foundry operations, on the other hand, lend themselves readily to mechanization and automation, so that in larger foundries heavy capital investment can be deployed for quantity production and low unit costs. Recent years have seen a general trend towards the concentration of production in these larger and more efficient plants.

Historical development and the modern industry

Metal founding is one of the oldest of all industries, both ancient and medieval history offering examples of the manufacture and use of castings. From simple axeheads poured from copper in open moulds some 5000 years ago, founding in the pre-Christian world developed to a point at which elaborate bronze statuary could be produced in two-piece and cored moulds[2]. The investment casting process too is now recognized as having been in operation during that period of history[3]. By the end of the medieval period, decorated bronze and pewter castings had begun to be used in European church and domestic life, whilst cast iron made a more sombre appearance in the shapes of cannon shot and grave slabs. In the sixteenth century Biringuccio wrote a detailed account of metal founding, giving an impression of working conditions and the emotions of the foundryman which would not have been out of place even in recent times. Moulding boxes and sand were in use by this time[4].

Amongst the early scientists the work of Réaumur (1683–1757) is of outstanding interest to founders. In the course of wide ranging activities he became interested in cast iron: not only did he produce malleable iron but he showed a clear perception of the range of cast iron structures and of the factors influencing the production of white, grey and mottled irons[5]. The widespread adoption of cast iron as an engineering material awaited the success of Abraham Darby in 1709 in smelting in the coke blast furnace; this paved the way for the massive use of cast iron in construction during the years following the industrial revolution, a position which was only yielded when bulk supplies of cheap steel became available after Bessemer's development of the converter in 1856.

Despite the interest of a few early scientists, metal casting was looked upon as an art and this view persisted until well into the twentieth century. Many foundries sprang up after the industrial revolution, the vast majority being for the manufacture of the cast iron then being used as a structural material. The organized industry grew around the skills of the moulder and the patternmaker, craftsmen in the direct tradition of the earlier masters of the art of casting. Their skill enabled complex forms to be produced but the enhancement of metallurgical quality and soundness awaited later scientific understanding and the advent of non-destructive testing.

The quantity production of iron castings in the nineteenth century was not, therefore, matched by a universal advance in quality, and the engineering use of the products encountered risks that were the more serious in a non-ductile material. After the collapse of the Tay Bridge in 1879, its cast iron columns, already of doubtful design, were found to have been excessively porous and to have been indiscriminately doctored with a putty filler. This episode illustrated the potential weakness of the structural castings of that time, although the qualities of cast iron when properly manufactured and used are equally demonstrated in the survival of many magnificent bridges such as those of Telford and the famous example at Ironbridge. Despite the skill of the moulder in producing complex forms, there was little change in the metallurgical and engineering situation until the modern era brought a better understanding of the factors determining quality, with improved means for its evaluation. With modern techniques of process control the rudimentary judgement of the operator could give way to objective measurements of metal temperature, moulding material properties and other production variables; their effects too could be more precisely determined. These improvements have been applied not only to cast iron but to a wide range of cast alloys.

A further feature of the development of the modern industry has been the shift in the responsibilities carried by operators and supervisors alike. It was common practice for the moulder to receive the pattern, determine the casting method and carry out his own sand preparation, moulding and pouring; he might even melt his own metal and fettle his own castings. In certain foundries the operator was virtually a sub-contractor selling his product to the company; the only technical control was that exercised by the craftsman. Later development increased the role of the shop floor supervisor in determining manufacturing techniques: this type of organization promoted the best craftsman to guide the work of his fellows.

The further development of the industry, however, and particularly the advent of mass production of many types of casting, brought its corollary, specialization. Not only is the tendency for an operator to be confined to a limited task but he, and frequently his immediate supervisor, may be removed from direct responsibility for the method of production; this is often concentrated in the hands of specialist technical staff whose sole duty is to establish and record methods and casting techniques. This task requires increasing expertise and access to data sources, both written and in the form of software. The pre-engineering of casting methods is accompanied by metallurgical control based on modern laboratory facilities for the determination of metal composition, metallographic structure and properties: scientific controls can be applied at all stages from raw materials to finished product. These changes alter the demands on the skills of the operators since the major decisions are made before the pattern equipment reaches the shop. The primary reliance placed upon craft has thus given way to technology and a new outlet must be found to utilize the initiative and abilities of the man on the shop floor.

The structure of the modern foundry industry is complex. Directly related to the traditional industry are the jobbing foundries with their capacity for undertaking work involving a wide range of sizes and designs. Quantity requirements are usually small and there is still some dependence on manual operations even though much of the heavy labour is removed by mechanical aids. At the opposite end of the scale are the specialized foundries, with their emphasis either on the mass production of a limited range of articles or on the use of a single special casting process. Many such foundries are captive to engineering organizations which incorporate castings in their own finished products.

The jobbing foundry is constantly presented with new problems in the moulding of individual design features and in the determination of casting methods which will ensure a sound product at the first attempt. Whilst some minor design variations can be accommodated by recourse to the skill of the moulder, the casting method must either be systematically evolved from an understanding of the underlying principles or must incorporate wide margins of safety at the risk of uneconomic production. This is where the introduction of computer simulation can save both costs and time by validating the intended casting method before any molten metal is actually poured.

In the mass production foundry, by contrast, the emphasis is upon close process control to maintain consistency in materials and procedures. Sophisticated pattern equipment eliminates the need for a high degree of skill in moulding, whilst there is opportunity for progressive development of the casting method to reduce margins and achieve the most economic production.

This picture of the industry is necessarily simplified since many companies operate in several fields, with jobbing and mass production, conventional and highly specialized processes operating in parallel. Similarly, although most foundries base their activities on a limited range of alloys, for example grey cast iron or steel, copper base or die casting alloys, other firms produce several of these materials side by side.

The technology of founding

The processes employed in the industry include gravity and pressure die casting, investment casting and centrifugal casting, but the general characteristics of founding can be conveniently introduced in relation to a central theme, the production of sand castings by conventional moulding. The basic steps in the sequence from design to finished product are summarized in the flow diagram in Figure 0.7. A similar sequence applies with minor modifications to a wide range of casting processes. The major exception is die casting, in which the principle of a pattern and an expendable mould is replaced by that of a permanent mould containing a negative impression. The individual features of both these systems will be examined in detail in the main body of the book.

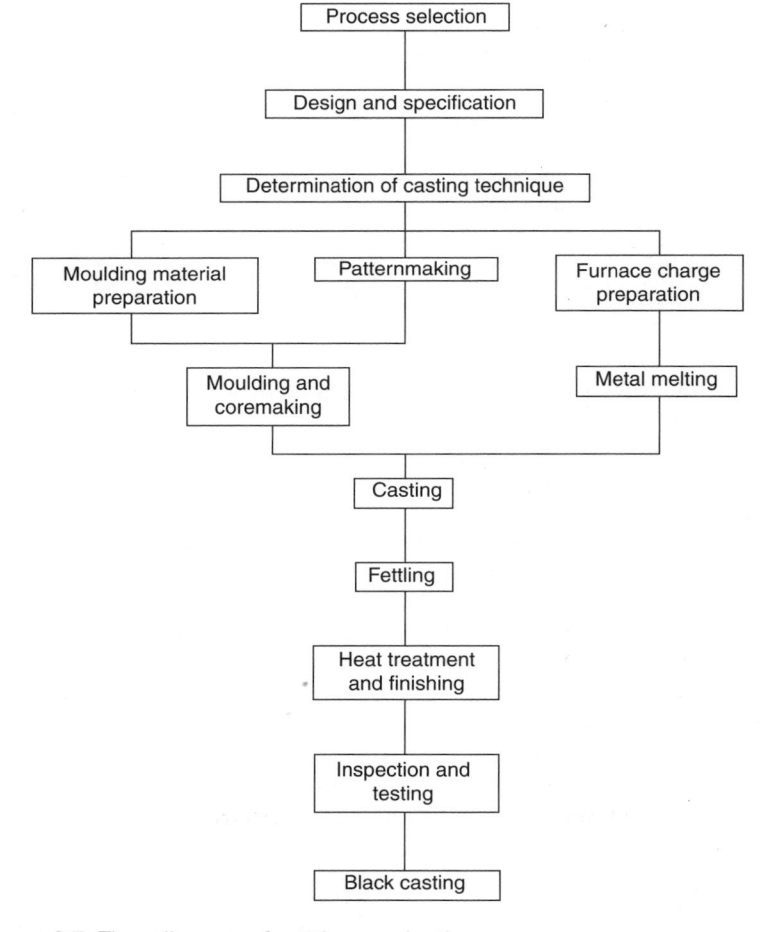

Figure 0.7 Flow diagram of casting production

Table 0.2 Melting points of some metals

Metal	Melting point °C	Metal	Melting point °C
Iron	1536	Magnesium	649
Nickel	1453	Zinc	420
Copper	1083	Lead	327
Aluminium	660	Tin	232

The variety of technological problems encountered in metal casting arises largely from differences in the physical properties and constitution of the alloys. This is exemplified in the wide range of process temperatures: Table 0.2 lists the melting points of parent metals of important groups of casting alloys. Other significant differences include chemical activity and solubility for gases, mode of solidification, and contraction characteristics. These affect the problems of metal flow, mould stability, feeding, stress-free cooling and the structure and properties of the cast material: individual techniques are thus required to meet conditions peculiar to the type of alloy.

Once the essential foundry technique has been established the quality of a casting is influenced by numerous process variables. Some of these can only be adjusted within narrow limits. Metal composition, for example, is often restricted by specification, although in the broader sense the needs of the casting process have influenced the ranges adopted in cast alloy specifications: this accounts for the frequent use of alloys close to eutectic composition, which show favourable characteristics with respect to fluidity, feeding and freedom from hot tearing. However, inoculation with very small amounts of certain elements can exercise potent effects both on foundry characteristics and casting structure, whilst control can also be exercised through moulding material properties, melting practice, casting temperature and pouring speed. These variables, together with the techniques of gating, risering and chilling, can be used to control the entire pattern of cooling: this is the main key to the structure and quality of the finished casting.

Control is exercised through many types of observation and measurement. In some cases direct determinations of temperature, time, composition or properties are useful, whilst other empirical tests bear specific relationship to aspects of foundry behaviour, for example metal fluidity, mould gas evolution or scabbing tendency. These techniques, whether used in the laboratory or the foundry, form part of the framework for the scientific control of metal founding. This also depends upon systematic evaluation of the cast product by non-destructive testing and metallurgical examination, and its correlation with production variables: statistical process and quality control systems can play an important role in this respect.

The application of science to founding is also seen in research into the fundamentals of casting and cast materials. This involves basic studies of the phenomena of liquid metal flow and solidification, high temperature stress–strain relationships in cast alloys and mould materials, and the origins of various forms of defect. Considerable effort is also being devoted to the computer modelling of many aspects of the casting process, including solidification with its repercussions on shape design and feeding technique, mould filling, stress development and microstructure.

Perhaps the most crucial areas for investigation are the process–structure– properties relationships in cast materials. Modern engineering requirements involving high stresses and extreme service conditions must be met by more exacting standards of quality and reliability. Apart from the development of new casting alloys and improved techniques of non-destructive examination, the capacity to control structure can provide more positive correlation between the true properties within castings and the nominal or test bar values. This approach is seen in the concept of 'premium' castings, with their high and guaranteed levels of internal properties, although assured product performance has since become a more general expectation in the design and purchase of castings. Modern procedures, involving the use of recognized standards in quality management and assurance, are now established as important features in both manufacture and procurement.

Other major activities lie in the fields of process development and production engineering. Casting processes have undergone radical changes during the modern era, some affecting the basic sand casting system and others involving the introduction of new processes with special characteristics. In sand casting the major development has been the wide application of hard sand technology, built upon the availability of chemical binders, greatly increasing the strength and rigidity of mould and core components and permitting precise assembly. Greensand practice has been similarly transformed by high pressure compaction in both box and flaskless moulding systems.

These developments, together with the various specialized casting processes, have refined the foundry operation and have contributed to the much discussed 'near net shape' objective. Stress on the greater accuracy, soundness and general quality of the cast component is in keeping with the broad quest for higher added value. This has arisen in part from a relative decline in the traditional demand for castings for machine tools, railways and heavy engineering, in favour of chemical industry, aerospace, offshore and automotive applications: these involve reduced overall tonnages but more exacting requirements. The dramatic growth in investment and die casting, the introduction of uphill casting, filtration, molten metal treatments and other process advances owe much to these industrial and market trends. Facilities to undertake the supply of fully machined castings are also often seen as a prerequisite.

The challenge in production engineering is to combine the philosophy of mass production with the flexibility of the casting process for moderate quantities of components. There is a continued shift towards capital intensive facilities: automatic plant, robots and sophisticated controls are increasingly featured in moulding and pouring operations, whilst computer aids are sought throughout the process from design to finished casting. Such developments improve not only productivity but quality, arising from greater consistency and reproducibility of casting conditions.

The nature of production problems in casting is closely associated with design features, so that the task of the foundry industry can be greatly simplified by cooperation during the preliminary stages between designer and producer. Apart from possibilities for personal contact, much has been accomplished in recent years to establish principles for use by the engineer in the design of castings. Proper communication can ensure that the most economic and technically sound use is made of foundry processes and that the casting is correctly assessed in relation to the whole range of metal products. The modern casting can emerge from such assessments as a versatile and reliable unit.

References

1 Nicholls, I. *Foundrym.* 92, 309 (1999)
2 Singer, C. J., Holmyard, E. J. and Hall, A. R. *A History of Technology*, Clarendon Press, Oxford, (1954–8)
3 Taylor, P. R. Metals and Materials 2, 11,705 (1986)
4 Biringuccio, V. *de la pyrotechnia*, Venetia (1559). Translation C. S. Smith and M. T. Gnudi, New York, *A.I.M.M.E.* (1942)
5 De Réaumur, R. A. F. *Memoirs on Steel and Iron*, Paris (1722). Translation A. G. Sisco, Chicago, University of Chicago Press (1956)
6 Aitchison, L. *A History of Metals*, Macdonald and Evans, London (1960)
7 Agricola, G. *de re metallica*, Basel (1556). Translation H. C. Hoover and L. H. Hoover, New York, Dover Publications (1950)
8 Roxburgh, W. *General Foundry Practice*, Constable, London (1910)
9 Coghlan, H. H. *Notes on the Prehistoric Metallurgy of Copper and Bronze in the Old World*, Pitt Rivers Museum, Oxford (1951)
10 Gale, W. K. V. 'Some notes on the history of ironfounding', *Br. Foundrym.* 56, 361 (1963)

1

Liquid metals and the gating of castings

The pouring of molten metal into the mould is one of the critical steps in founding, since the behaviour of the liquid and its subsequent solidification and cooling determine whether the cast shape will be properly formed, internally sound and free from defects. Campbell[1] has advanced the view that the great majority of scrap castings acquire that status during the first few seconds of pouring. Few would disagree with that assessment.

The success of the pouring operation depends partly upon certain qualities of the metal itself, for example its composition and temperature, which influence flow, and partly upon properties and design of the mould, including the nature of the moulding material and the gating technique used to introduce the metal into the mould cavity. Whilst the metal is in the liquid state the foundryman is also concerned with forces acting upon the mould and with volume contraction occurring during cooling to the solidification temperature. These aspects will be considered separately, beginning with the flow properties of the liquid metal under foundry conditions.

Fluidity of liquid metals

Although other terms such as *castability* have been used to describe certain aspects of flow behaviour, the term *fluidity* is most widely recognized. In the broad sense it can be defined as that quality of the liquid metal which enables it to flow through mould passages and to fill all the interstices of the mould, providing sharp outlines and faithful reproduction of design details. It follows that inadequate fluidity may be a factor in short run castings or in poor definition of surface features. It can at once be appreciated that fluidity is not a single physical property in the same sense as density or viscosity, but a complex characteristic related to behaviour under specific conditions within a foundry mould.

In considering the factors influencing flow, viscosity might be expected to predominate. Viscosity is defined as the force required to move a surface of unit area at unit velocity past an equivalent parallel surface at unit distance: it is thus a measure of the capacity of a liquid to transmit a dynamic stress in shear. When liquid is flowing in an enclosed passage, its viscosity will determine the extent to which the drag imposed by the passage wall is transmitted to the bulk of the liquid: it will therefore influence the rate of flow, which is found to bear a simple reciprocal relation to the viscosity. More directly related to the capacity of a liquid to flow under its own pressure head is the kinematic viscosity, that is the absolute viscosity divided by the density.

Further consideration indicates that these properties will not be decisive in determining the relative mould filling capacities of metals under foundry conditions. One of the fundamental characteristics of the liquid state is the ability of any liquid, however viscous, to conform in time to the shape of its container. This would occur rapidly in the case of liquid metal held at constant temperature since viscosities of liquid metals are very low. Under casting conditions failure to fill the mould cavity results not from high viscosity but from premature solidification. Thermal conditions and mode of solidification are thus the critical factors with respect to cessation of flow. The concept of fluidity takes these aspects into account.

The measurement of fluidity

Since fluidity cannot be assessed from individual physical properties, empirical tests have been devised to measure the overall characteristics. These are based on conditions analogous to the casting of metals in the foundry and measure fluidity as the total distance covered by molten metal in standardized systems of enclosed channels before cessation of flow. A further parameter in such tests is the flow time or *fluid life*.

Much of the earlier experimental work on fluidity was the subject of detailed reviews by Clark[2,3] and Krynitsky[4]. Early uses of a straight flow channel, with its disadvantages of excessive length and sensitivity to angle, were discontinued in favour of the spiral test, of which numerous variations have been used (e.g. References 5–7). A typical spiral fluidity test is illustrated in Figure 1.1. Variations in the spiral test have been mainly concerned with the problem of obtaining truly standard conditions of flow. This problem has been approached through various designs of reservoir system to regulate the pressure head, and constant speed pouring devices to ensure a uniform rate of metal delivery to the system[8]. Since fluidity measurements are also sensitive to small changes in thermal properties and surface characteristics of the mould, graphite and metal moulds were used by some investigators in attempts to minimize variation in these factors[9].

The closest approach to complete standardization, however, is achieved in the vacuum fluidity test devised by Ragone, Adams and Taylor[10]. Using this apparatus, illustrated in Figure 1.2, the metal flows through a smooth glass tube under suction induced by a partial vacuum; the pressure head is thus accurately known and the human factor in pouring eliminated.

These refinements of technique approach the ideal of excluding mould variables and measuring fluidity as a property of the metal alone. Using these and other techniques the major factors in fluidity were established.

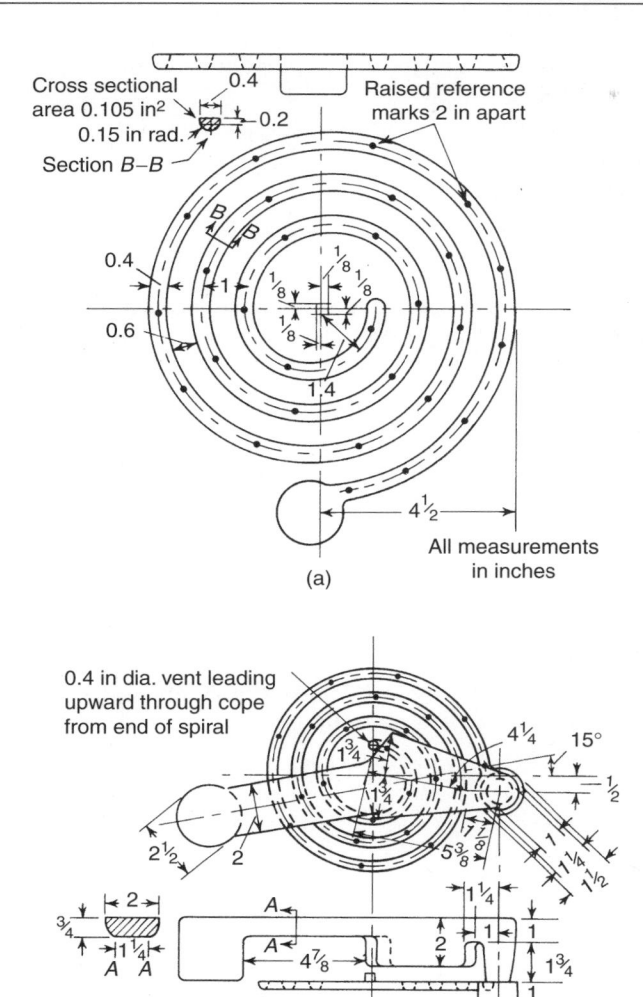

Cross sectional area 0.105 in² 0.15 in rad.

Section B–B

Raised reference marks 2 in apart

All measurements in inches

(a)

0.4 in dia. vent leading upward through cope from end of spiral

(b)

Figure 1.1 Spiral fluidity test casting. (a) Standard fluidity spiral, (b) arrangement of down-gate and pouring basin for standard fluidity spiral (courtesy of American Foundrymen's Society)

Variables influencing fluidity

Temperature

The initial temperature of the metal is found to be the predominant factor, several investigators having shown the fluidity of a given alloy to be directly related to the superheat[2,3,8,9,11,12]. This would be expected from the fundamental effect of solidification in controlling the duration of flow, since the

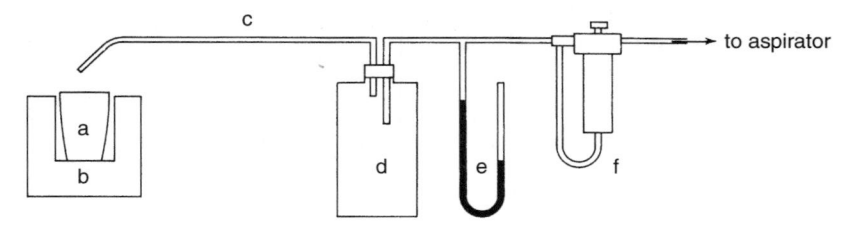

Figure 1.2 Vacuum fluidity test apparatus (from Ragone *et al*[10]). (a) Crucible of metal; (b) electric resistance furnace; (c) fluidity test channel; (d) pressure reservoir; (e) manometer; (f) cartesian manostat (courtesy of American Foundrymen's Society)

superheat determines the quantity of heat to be dissipated before the onset of solidification. Typical fluidity–superheat relationships are illustrated in Figure 1.3.

Composition

The other major factor is metal composition. Valid comparisons of the fluidities of various alloys can only be made at constant superheat but under these conditions a marked relationship emerges between alloy constitution

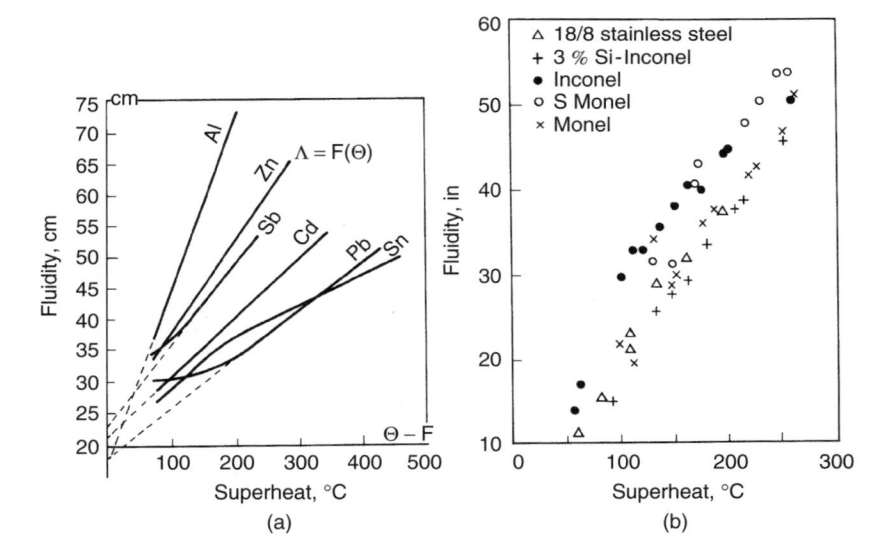

Figure 1.3 Influence of superheat on fluidity: (a) pure metals (from Portevin and Bastien[11]) (courtesy of Institute of Metals) (b) alloys (from Wood and Gregg[12]) (courtesy of Institute of British Foundrymen)

and fluidity. High fluidity is commonly found to be associated with pure metals and with alloys of eutectic composition; alloys forming solid solutions, especially those with long freezing range, tend to show poor fluidity. Portevin and Bastien[6] established an inverse relationship between fluidity and solidification range. This was later confirmed by other workers, for instance by Floreen and Ragone in their work with aluminium alloys[13]. The relationship between composition and fluidity for one alloy system is illustrated in Figure 1.4.

Differences in the behaviour of various types of alloy can be attributed primarily to their characteristic modes of freezing. In the case of alloys in which solidification occurs by progressive advance of a plane interface from the mould wall, flow can continue until the channel is finally choked; this is found to occur near to the point of entry[10,14]. Pure metals solidify in this manner and show appreciable fluidity even when poured at the liquidus temperature, flow continuing during evolution of the latent heat of crystallization.

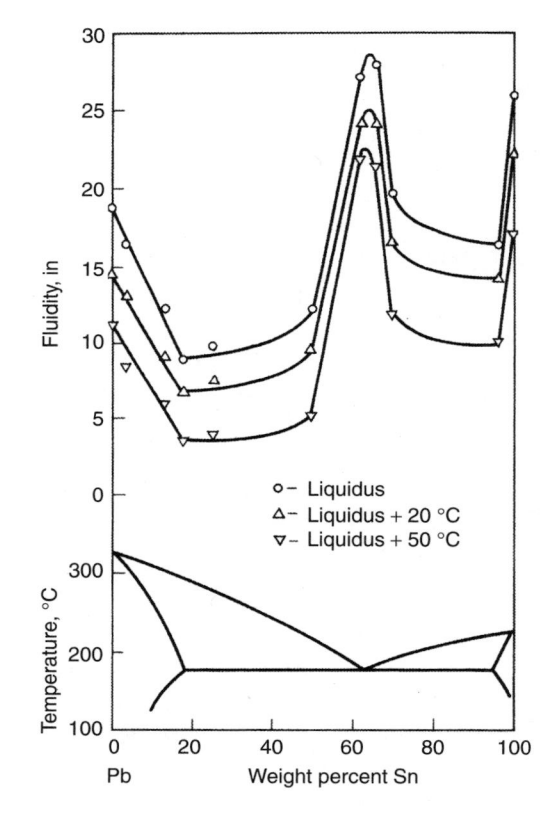

Figure 1.4 Relationship between composition and fluidity of lead–tin alloys (from Ragone *et al*[10]) (courtesy of American Foundrymen's Society)

In alloys in which constitutional undercooling and other phenomena produce independent crystallization in the main mass of liquid, flow is arrested by the presence of free crystals in the liquid at the tip of the advancing stream. These alternative modes of freezing are illustrated in Figure 1.5, together with the intermediate situation involving a dendritic interface (q.v. Chapter 2).

Although constitution and mode of freezing are of major significance in accounting for differences between alloys, fluidity comparisons depend upon additional factors. Even in the case of alloys exhibiting similar modes of freezing the fluidity–superheat relationships are not identical: the time to cool to the freezing temperature depends upon heat content and thermal properties rather than upon temperature alone. The distance of flow is thus affected by the volume specific heat, latent heat of fusion and thermal conductivity of the alloy. The influence of thermal properties is exemplified in the aluminium–silicon system, in which maximum fluidity does not occur at the eutectic composition as in many other alloy systems. In this case the hypereutectic alloys show greater fluidity due to the high heat of fusion of the primary silicon, although at least a part of this increase has been attributed to a shift to a non-equilibrium composition with a higher than normal silicon content[15].

Apart from basic composition, other characteristics affecting fluidity include the presence of dissolved gases and non-metallic inclusions in the liquid.

Other factors in fluidity

Although the spiral and vacuum fluidity tests have achieved a high degree of refinement for fundamental work, other tests have been employed in attempts to obtain a more comprehensive representation of conditions in foundry moulds, especially those incorporating a wide range of passage sizes. Ragone *et al.*[10] used the vacuum test to investigate the flow of molten tin in channels of various diameters down to 3 mm and established a simple relationship between channel diameter and observed fluidity, a finding subsequently extended by Barlow[20], down to 0.5 mm channels in various alloys. It is not clear, however, to what extent such relationships would hold good in extremely small mould passages for alloys and conditions susceptible to the growth of surface films. Under these circumstances, which are relevant to the reproduction of sharp corners and fine detail, surface phenomena must assume greater significance. The importance of surface tension with respect to flow in small passages was demonstrated by Hoar and Atterton[16], who found a direct relationship between surface tension and the pressure required to produce penetration of liquid metals into surface voids in sand compacts.

The distinction between this aspect of fluidity and that measured, for example, by the spiral test was drawn by Wood and Gregg[12], who expressed

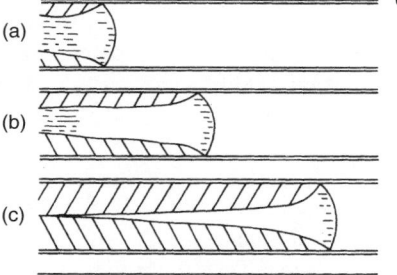
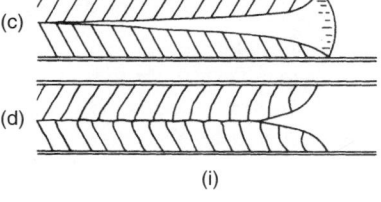
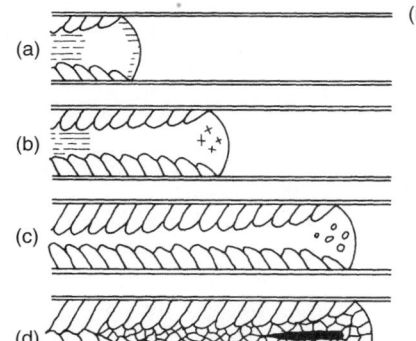

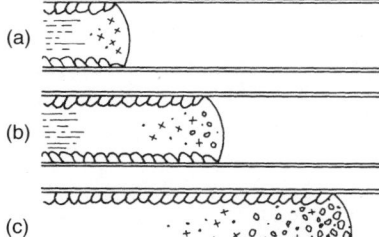
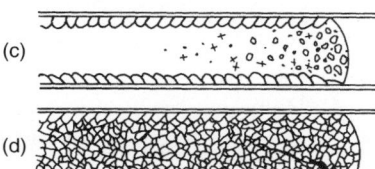

(i) Plane interface:
   (a) Liquid enters flow channel and columnar grain formation with smooth liquid–solid interface begins,

   (b) Columnar grains continue growing in upstream direction,

   (c) Choking-off occurs,

   (d) Remainder of casting solidifies with rapid grain growth and formation of shrinkage pipe;

(ii) Jagged interface:
   (a) Liquid enters flow channel and columnar grain formation with jagged liquid–solid interface begins,

   (b) Columnar grains continue to grow, fine grains nucleating at tip,

   (c) Choking-off occurs at entrance to flow channel, though cross-section is not completely solid,

   (d) Remainder of casting solidifies with equi-axed grains and formation of shrinkage cavity near tip;

(iii) Independent crystallization:
   (a) Liquid enters flow channel, columnar grain formation begins, and fine grains nucleate,

   (b) Fine grains grow rapidly as flow progresses,

   (c) Flow ceases when critical concentration of fine grains in tip is reached,

   (d) Remainder of casting solidifies with equi-axed grains and distributed microshrinkage;

Figure 1.5 Modes of solidification in flow channel. (from Feliu *et al*[14]) (courtesy of Institute of British Foundrymen)

the capability of the metal to conform closely to the mould surface as 'casting quality', determined by making a separate measurement of the length of spiral perfectly formed and expressing this as a percentage of total length. Comparisons of values for casting quality and fluidity revealed significant differences between the two according to alloy type.

There is wide agreement that the surface tension factor becomes significant in the channel size range 0.5–5 mm[17,18]. Under the conditions prevailing in casting, the surface tension of the metal itself, which may have a value as high as 1.5 N/m, is modified by the influence of the surface films existing on metals in normal atmospheres. Evidence of these films has long been available: in some cases, as in the aluminium bronzes, they may be visible, whilst in other cases they may be detected by their marked effect upon the emissivity of the metal surface[19]. Liquid aluminium for instance, carries surface films which increase surface tension by a factor of three[21], whilst the reduced fluidity of steels containing aluminium has also been explained by the presence of an alumina film, preferentially formed because of the high oxygen affinity of the element[8]; titanium produces similar oxide films on stabilized stainless steels and nickel base alloys. Chromium is yet another element known to produce strong surface films; Wood and Gregg used this fact in partial explanation of poor casting quality in some of their alloys. Baruch and Rengstorff[17] made direct determinations of the influence of surface films on the surface tension of liquid cast iron and found increases of up to 0.5 N/m, considered high enough to be significant with respect to flow in small passages.

The presence of a restrictive oxide film may mean, therefore, that the film rather than the mould will determine the final outline of the casting in confined corners. Alloys carrying the films are particularly susceptible to poor definition and to the formation of surface laps and wrinkles: they therefore need particular care in gating. Such defects will be further discussed in Chapter 5. The effect of oxide films is not, however, universally restrictive. A liquid film may exert the opposite effect and the influence of phosphorus in increasing the fluidity of copper alloys, for example, may be partly explained by such a film.

Flow behaviour in very narrow channels has been incorporated in certain further techniques for fluidity measurement. In multiple channel systems, wider representation of casting conditions is sought by integrating flow distances obtained in channels of greatly differing thicknesses[22,23]. Two such tests are illustrated in Figure 1.6. In a further test designed by Kondic[24] a mould cavity with a large surface area to volume ratio is used to provide an analogous mould filling problem to that encountered in investment castings of thin section. In this case the area of specimen produced is the test criterion. These types of test provide a wide range of conditions for the exercise of the solidification and surface influences upon flow distance. They thus offer some parallel with a similar range of actual mould conditions.

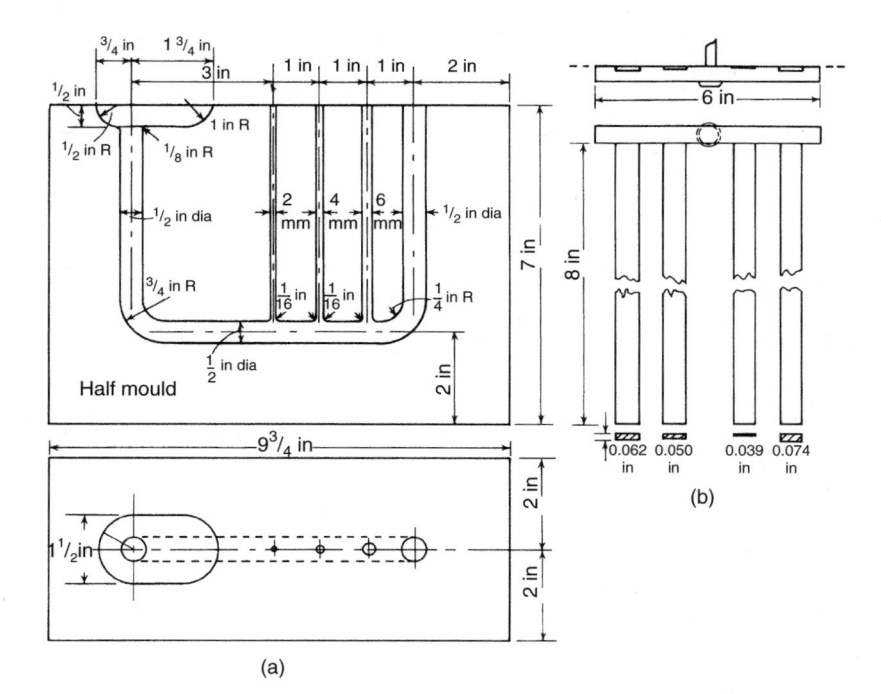

Figure 1.6 Multiple channel fluidity test castings. (a) U test (from Reference 23), (b) Strip test (from Cooksey *et al*[22]) (courtesy of Institute of British Foundrymen)

Campbell[15] subsequently carried out a close analysis and interpretation of the large body of fluidity results obtained by earlier investigators: it was concluded that the findings from the widely varying test methods employed could all be reconciled and brought to a common basis in terms of the fundamental influences on flow. Also included was a discussion of the separate but significant concept of 'continuous fluidity', representing flow behaviour through short channel conditions, in which flow continues indefinitely without the arrest which determines normal fluidity test results.

Mould factors in metal flow

Although fluidity should be clearly isolated as a property of the metal alone, the flow of metal under a given pressure head is also strongly influenced by the nature of the mould. Since metal flow is arrested through solidification, mould conditions can affect flow duration either directly through thermal properties or indirectly through flow velocity: if a restraining force reduces this velocity, increased time for heat loss per unit length of passage wall causes the final arrest to occur in a shorter distance.

Thermal properties

The rate of cooling to the temperature at which flow ceases is largely determined by the heat diffusivity* of the moulding material

$$D = (kc\rho)^{1/2}$$

where $k =$ thermal conductivity,
$\quad c =$ specific heat,
$\quad \rho =$ density.

In certain cases the cooling rate may be further influenced by absorption of the latent heat of vaporization of volatile mould constituents.

Rapid freezing thus results from the use of mould refractories of high heat diffusivity, or, in the extreme case, of metal chills or moulds[18,25]. Cooling is also accelerated by evaporation of water from the surface layers of green-sand moulds: this effect is found to retard metal flow in thin sections, hence the preference for drysand techniques for extremely thin castings. In investment casting the rate of heat removal is further reduced by the use of a high mould preheat temperature.

Mould surface effects

Flow down a mould channel is restrained by frictional forces dependent on the roughness of the mould surface. This roughness is related to the grain size of the moulding material, which explains the preference for fine grained sands and for metal or glass flow channels for fluidity testing.

Apparent fluidity as represented by flow distance in a test mould is found to be greatly increased by mould coatings[25–27]. These may be used simply to reduce friction by providing a smooth surface; increased flow has also been obtained by the use of reactive coatings designed to influence wetting characteristics at the metal–mould interface. Hexachloroethane, which generates active chlorine, has been found to be especially effective in increasing flow distance in aluminium alloys; it is suggested that the restrictive oxide film is reduced by chlorine[27].

Air pressure effects

As metal enters the mould it must displace a rapidly expanding mass of air from the mould cavity. This is accomplished through vents and open risers and through the permeable structure of the moulding material. If these channels are inadequate, back pressure of air is liable to retard flow and produce an apparent loss of fluidity.

---

* The significance of the term *heat diffusivity* is discussed by Ruddle (Ref. 29, Chapt. 3).

The various mould effects necessitate rigid standardization of mould conditions in fluidity testing.

To summarize, the flow of metal in a given mould cavity is determined by both metal and mould characteristics. The metal property of fluidity was shown to be primarily a function of composition and superheat. Since composition is normally governed by the alloy specification and can only be varied within narrow limits, casting temperature remains as the metal variable most readily open to practical control in the foundry. Such control must, however, be exercised with many other factors in mind, including feeding technique, metallurgical structure and the suppression of defects such as hot tears and metal penetration. Since casting temperature can be readily selected and measured, the fluidity test has not acquired universal status as a routine melt quality test for shop control purposes.

Mould conditions affecting flow are seen to be mainly inherent in the moulding process and its associated materials. Successful mould filling thus depends primarily on the use of a gating technique suited to the particular combination of alloy and moulding material.

The gating of castings

The gating system is that part of the mould cavity through which the metal is poured to fill the casting impression; its design is the principal means by which the foundryman can control the pattern of metal flow within the mould. The importance of gating technique arises from its fourfold purpose:

1. The rate and direction of metal flow must be such as to ensure complete filling of the mould before freezing.
2. Flow should be smooth and uniform, with minimum turbulence. Entrapment of air, metal oxidation and mould erosion are thus avoided.
3. The technique should promote the ideal temperature distribution within the completely filled mould cavity, so that the pattern of subsequent cooling is favourable to feeding.
4. The system can incorporate traps or filters for the separation of non-metallic inclusions, whether dislodged in the gating system or introduced with the metal.

Gating technique must be designed to take account of the weight and shape of the individual casting, the fluidity of the metal and its relative susceptibility to oxidation. Although techniques vary widely according to these conditions, the basic objectives must be achieved at minimum cost in moulding and fettling time and in metal consumption.

Essential features of gating systems

In the simplest of all gating systems the metal is poured down an open
feeder head situated at the top or the side of the casting. This technique is
often followed for small castings, in which the provision of an additional
system of mould passages would greatly lower the yield of useful metal; at
the same time the temperature distribution is favourable and the distance
through which the liquid metal must fall is short.

In most cases, however, a separate gating system is used, the metal being
distributed through passages into selected parts of the mould cavity. The
essential features of a typical system are illustrated in Figure 1.7. The metal
is poured into a bush or basin, whence a downrunner or sprue descends
to the required level. From the base of the sprue further runners conduct
the metal to the ingates, through which it enters the casting impression.
Entry may be directly into the casting or into one or more feeder heads;
the practice of gating through feeder heads will be examined more fully
in Chapter 3. The gating system may consist of a single passage or the
metal stream may be subdivided and directed into the casting at widely
dispersed points. For very large castings more than one sprue may be
required, the individual sprues then being filled either from separate ladles
or from a common tundish or launder. In addition to these basic features the

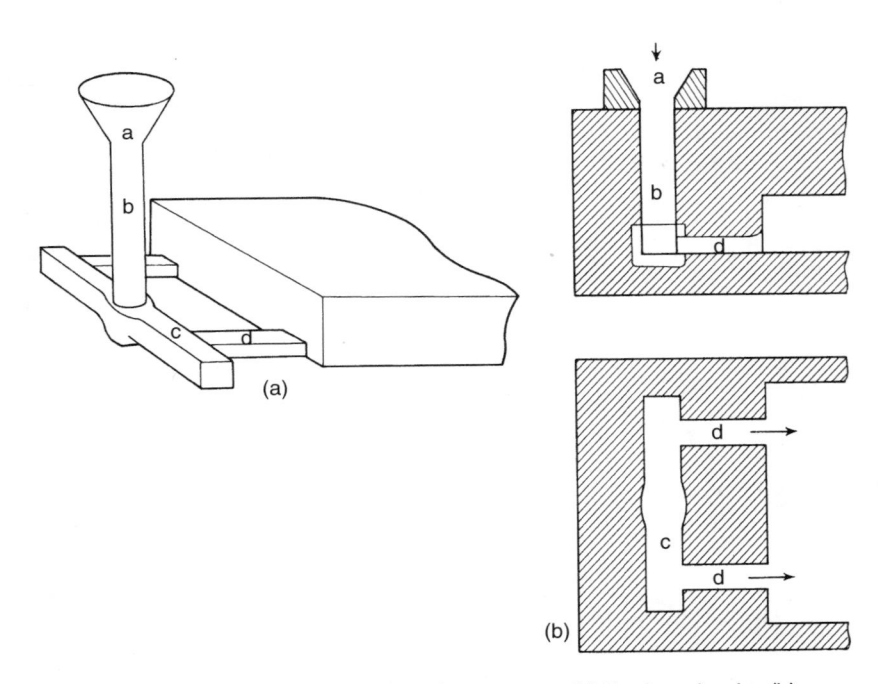

Figure 1.7 Principal features of a gating system. (a) Bush or basin, (b) sprue
or downrunner, (c) runner, (d) ingates

gating system often incorporates devices for the separation of non-metallic inclusions.

Before examining gating systems in more detail, consideration should be given to the general nature of metal flow in foundry moulds.

The mould erosion tendency during casting is accentuated by turbulent as distinct from smooth or laminar flow of the molten metal. In turbulent flow, although the mass of a liquid may have a resultant direction, there are wide local variations of direction and velocity within the stream. Laminar flow gives way to turbulence when the mean velocity reaches a critical value for the particular combination of metal properties and mould dimensions.

The balance of conditions determining the type of flow is represented in the Reynolds Number.

$$(Re) = \frac{Vd}{v}$$

where V = mean velocity,
$\quad d$ = linear dimension of the mould channel section,
$\quad v$ = kinematic viscosity of the liquid.

Turbulent flow is associated with high values of (Re) and therefore with high velocity, large flow channels and low kinematic viscosity. The value of (Re) at which flow becomes turbulent depends upon the geometry of the system but normally lies within the range 2000–4000. Turbulence will occur at lower values if smooth flow is disrupted by sudden changes in the dimensions and direction of the passages. Under most foundry conditions, given the minimum passage dimensions and flow velocities needed to avoid premature freezing, turbulent flow is encountered; a primary function of the gating system is to minimize its violence by shaping a smooth path for the liquid and by using the lowest flow rates compatible with mould filling. Streamlined passages are thus ideal.

Design of gating systems

Most modern studies of gating systems have been based upon consideration of two laws of fluid dynamics. The first of these, the Equation of Continuity, states that the volume rate of flow is constant throughout a system and is expressed by

$$Q = A_1 V_1 = A_2 V_2 \text{ etc.} \tag{1.1}$$

where Q = volume rate of flow,
$\quad A$ = cross-sectional area of flow passage,
$\quad V$ = linear velocity of flow.

The linear velocity of flow in a system is related to other factors in Bernoulli's Theorem, which states that the total energy of unit weight of fluid is constant throughout a system:

$$\frac{V_1^2}{2g} + h_1 + \frac{P_1}{\rho} = \frac{V_2^2}{2g} + h_2 + \frac{P_2}{\rho} \text{ etc.} \qquad (1.2)$$

where V = linear velocity of flow,

h = height above the datum plane,

P = pressure,

ρ = density.

The successive terms in the equation represent the kinetic, potential and pressure energies respectively. Flow calculations based upon these laws are subject to errors arising from friction, from sudden changes in cross-section, and from sharp changes in direction at bends and junctions. Corrections can, however, be applied using experimentally determined loss coefficients. Research has indicated that such losses are not so high as was at one time thought and that they are in many cases negligible[28].

These principles may be used to estimate the velocities of flow in individual systems. The behaviour of metal in the downrunner, for example, may be deduced from consideration of the linear velocity attained in a metal stream falling from an initial point of rest in the absence of frictional resistance. In this case, under atmospheric pressure, the potential energy at rest may be equated directly with the kinetic energy at any point in the moving stream. Thus, using the point in the stream as datum, the Bernoulli equation 1.2 becomes

$$h = \frac{V^2}{2g} \qquad (1.3)$$

Thus, $V = (2gh)^{1/2}$ and the velocity is simply related to the distance of fall from the initial point of rest.

In free fall or in a parallel downrunner the metal, in gaining velocity during descent, must contract inwards or away from the walls: this follows from the law of continuity, equation 1.1. Thus, in a long downrunner, constriction of the stream must cause aspiration of air through the permeable walls of the mould[29]. This reasoning was used as a basis for the adoption of down-runners progressively tapered so that diminishing cross-sectional area accompanies the increasing linear velocity of the metal to provide a constant volume rate of flow[30]. The dimensions of the down-runner theoretically needed to meet this requirement are defined by the relation

$$\frac{A_2}{A_1} = \left(\frac{h_1}{h_2}\right)^{1/2}$$

derived from equations 1.1 and 1.3, where A_1 and A_2 represent the cross-sectional areas, and h_1 and h_2 the depths of metal, at the entry to the down-runner and at any point within it.

Detailed treatments of the application of the laws of hydraulics to gating systems were included in a classical monograph by Ruddle[31], and in other reviews and papers dealing with particular systems[32–36]. Studies of flow behaviour have been carried out both by direct observation and by the analogy of water in transparent plastic moulds, whilst quantitative data for metals has been obtained by techniques including weighing, observation of trajectories of emergent streams, and pressure measurements at points within the system. A technique of considerable potential for such work is X-ray fluoroscopy, but its extensive application awaited the development of portable high energy sources and safe observation and recording systems[37,38].

Further contributions have employed mathematical modelling and simulation. This aspect is embodied in some of the commercial software designed to provide comprehensive pictures of events occurring in the mould throughout the casting process, a topic to be further considered in Chapter 3. The problem is a complex one, involving both hydraulic behaviour and heat transfer throughout the process. Some computer assisted practical investigations and simulation studies of gating practice were examined in References 39–41, whilst a further general treatment of gating by Webster was included in Reference 42.

The most important characteristics of an individual gating system are the shape and dimensions of the passages, which determine the rate and type of flow, and the position at which the metal enters the mould cavity in relation to the casting design.

The dimensional characteristics of any gating system can be generally expressed in terms of the gating ratio:

$$a : b : c$$

where a = cross-sectional area of downrunner,
b = total cross-sectional area of runners,
c = total cross-sectional area of ingates.

This ratio conveniently emphasizes an important general feature of any system, namely whether the total cross-section diminishes towards the casting, providing a choke effect which pressurizes the metal in the system (high ratio), or whether the area increases so that the passages are incompletely filled, the stream behaving largely in accordance with its acquired momentum (low ratio).

In pressurized systems, although the rate and distribution of flow are more predictable, the metal tends to enter the casting at high velocity, producing a jet effect, whereas in unpressurized systems irregular flow and aspiration of air are the predominant tendencies. These factors need to be considered in the light of the particular metal being cast, but an ideal system for all purposes would be one in which pressures were barely sufficient to maintain all passages full yet just enough to avoid aspiration.

The position and proportions of the elements of the gating system are not of importance solely with respect to minimising turbulence. For thin sectioned castings the mould filling objective predominates, so that the system must be designed to flush metal rapidly but progressively through sections otherwise susceptible to laps and misruns. Care is needed to avoid cold shuts due to the meeting of thin streams at points remote from the ingates.

In castings of thicker section there is greater preoccupation with turbulence, associated with mould erosion, and with the temperature gradients which so markedly influence the subsequent feeding of the casting. The latter aspect of gating will be referred to more fully in Chapter 3. Both these aspects will, however, be briefly considered here in relation to some of the main possibilities for variation in gating technique. The principal variations of gating position are illustrated schematically in Figure 1.8.

Top gating

When metal is poured through a top gate or directly into an open feeder head, the stream impinges against the bottom of the mould cavity until a pool is formed; this is kept in a state of agitation until the mould is filled. The erosive effect of the unconfined stream can be severe, whilst the associated splashing gives opportunity for oxidation. The mould surface can, however, be protected at the point of impact by preformed refractory tiles; the intensity of erosion can also be reduced, where fluidity of the metal permits, by the use of pencil gates. This method and others which divide the metal stream are unsuitable for alloys which are sluggish or prone to rapid oxidation, but are used with success, for example, in the gating of cast iron. Some variations in top gating are illustrated in Figure 1.9.

The principal advantages of top gating are its simplicity for moulding, its low consumption of additional metal and, above all, the generation of temperature gradients favourable to feeding from top heads; this arises from

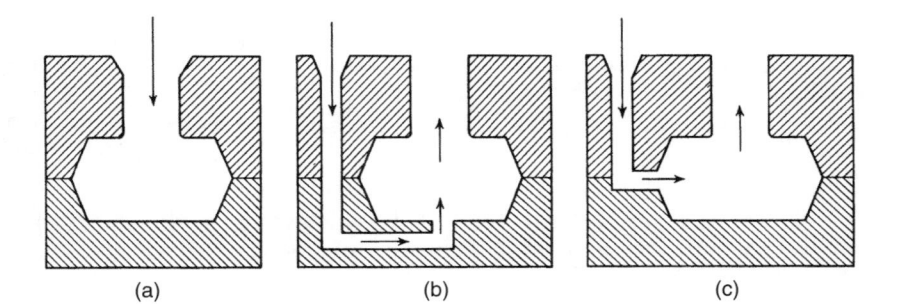

Figure 1.8 Variations in gating position. (a) Top gating, (b) bottom gating, (c) side gating

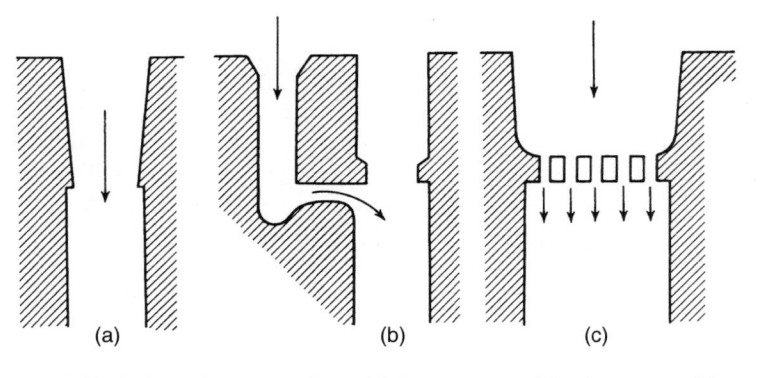

Figure 1.9 Variations in top gating. (a) Open pour, (b) edge gate, (c) pencil gates

the proportionately rapid cooling of the first metal poured, followed by the progressive accumulation of metal from above until the mould is full.

Bottom gating

Quiet entry of metal into the mould cavity is best achieved by its introduction at the lowest level. Using this method the metal rises steadily through the mould, splashing is eliminated and dislodged moulding material tends to be carried to the surface.

If bottom gates are used with top feeder heads the resulting temperature gradients are opposed to feeding, but various measures are available to mitigate this effect: these will be detailed in Chapter 3. Despite the greater complexity of moulding the method is much used for heavy castings.

Side gating

Moulding can be simplified by the discharge of metal into the side of the mould cavity through ingates moulded along a parting plane; this practice frequently offers the best compromise between moulding convenience and the ideal gating arrangement. Using side gating, progressive mould filling can be achieved by tilting the mould towards the ingates to provide uphill casting conditions.

Many variations of bottom and side gating find practical application. Apart from the multiple systems to be separately considered, the horn gate is one widely applied form of bottom gate; normal and reversed horn gates are illustrated in Figure 1.10. This type of gate, with its smooth curves and progressive change of dimensions, is designed to minimize erosion and oxidation. Since the normal horn gate is prone to jet effects[43], the reversed horn gate has frequently been adopted for those alloys such as

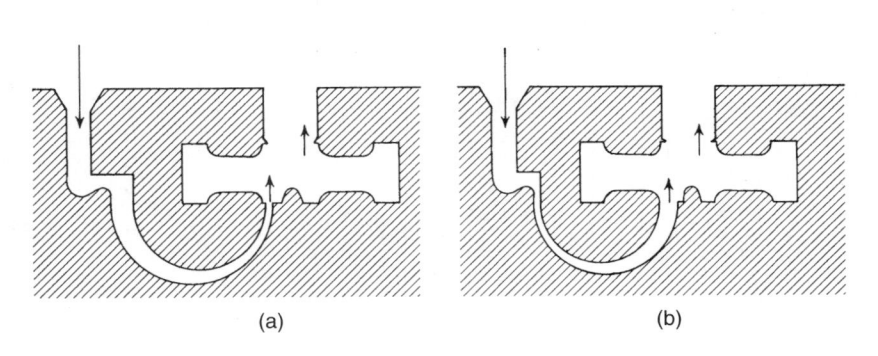

Figure 1.10 (a) Normal and (b) reverse horn gates

the aluminium bronzes which are especially susceptible to skin formation, although opinion is not unanimous as to its effectiveness[44].

These fundamental differences in running and gating practice have been the subject of such attention in the production of aluminium alloy castings, given the now well-recognized dangers of insidious oxide film defects arising from surface turbulence and air entrainment (q.v. Chapter 5). Observations by Runyoro, Boutorabi and Campbell[45,46], based in part on video recordings of molten alloy emerging from ingates, indicated a critical flow velocity of around 0.5 m/s, above which the restraining meniscus carrying oxide film becomes unstable, the surface breaking up and engulfing the oxide. Below this velocity the metal emerged by simple flooding. The conclusion was that internal oxide defects were inevitable under conditions of top gating. Bottom gating was suitable given carefully designed systems, but the results were in keeping with the developing use of upward displacement filling methods for castings of the highest quality.

Campbell later demonstrated the effects of metal filling velocity on the strength of castings using statistical analysis[47]: properties were again highly sensitive to the critical velocity of 0.5 m/s. Sutton[48] described the implementation of this concept in conjunction with other quality measures: the poorest results were invariably associated with top poured castings.

Multiple gating systems

Whilst small castings are commonly either top poured or gated through a single sprue and ingate, the latter method becomes increasingly unsuitable for castings of larger dimensions, because of the danger of overheating the mould adjacent to the point of entry and because of the long flow distances within the mould cavity.

The solution lies in the use of more complex gating systems in which the metal is directed through separate gating elements to different parts of the mould. Multiple systems can be used either to introduce the metal at widely separated points in the same horizontal plane, or to obtain flow at progressively higher levels as the mould fills (step gating). The choice or combination of these methods depends upon the shape and orientation of the casting.

In multiple gating it is necessary to control the distribution of metal between the separate elements of the system. Extensive studies of the behaviour of liquid metal in such systems were carried out by Johnson, Baker and Pellini[49], who demonstrated the failure of common designs of finger and step gates in unpressurized systems to function in the manner intended.

A horizontally disposed system embodying finger gates is shown in Figure 1.11a and b. Such an arrangement finds use for extended castings of plate form, whilst this system is also represented in other types of arrangement exemplified in Figure 1.11c. The tendency in such systems is for the greater flow to take place through the ingates farthest from the downrunner: the metal continues to flow along a straight path unless an obstruction creates a back pressure which initiates flow into other elements of the system.

These findings were reinforced by later quantitative studies carried out by Johnson, Bishop and Pellini[50]. Work on horizontal systems of widely differing dimensions and gating ratios showed flow to depend strongly upon the relative dimensions of the ingates, but flow in the end gates always predominated except when using the highest ratios. Figure 1.12 illustrates the principal patterns of flow behaviour obtained.

Uniformity of flow in such systems may be accomplished by measures to induce back pressure in the runners, or, in unchoked systems, to arrest the

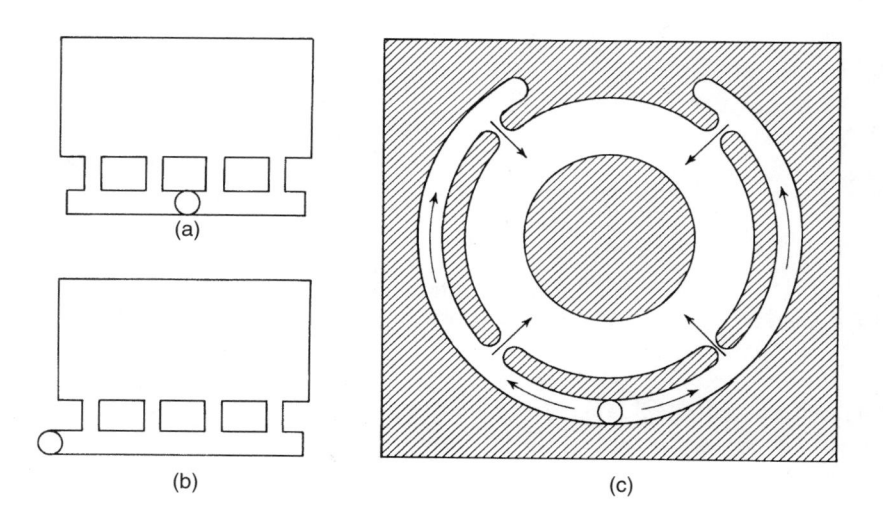

Figure 1.11 Multiple ingate systems. (a) and (b) parallel finger ingates, (c) circumferentially placed ingates

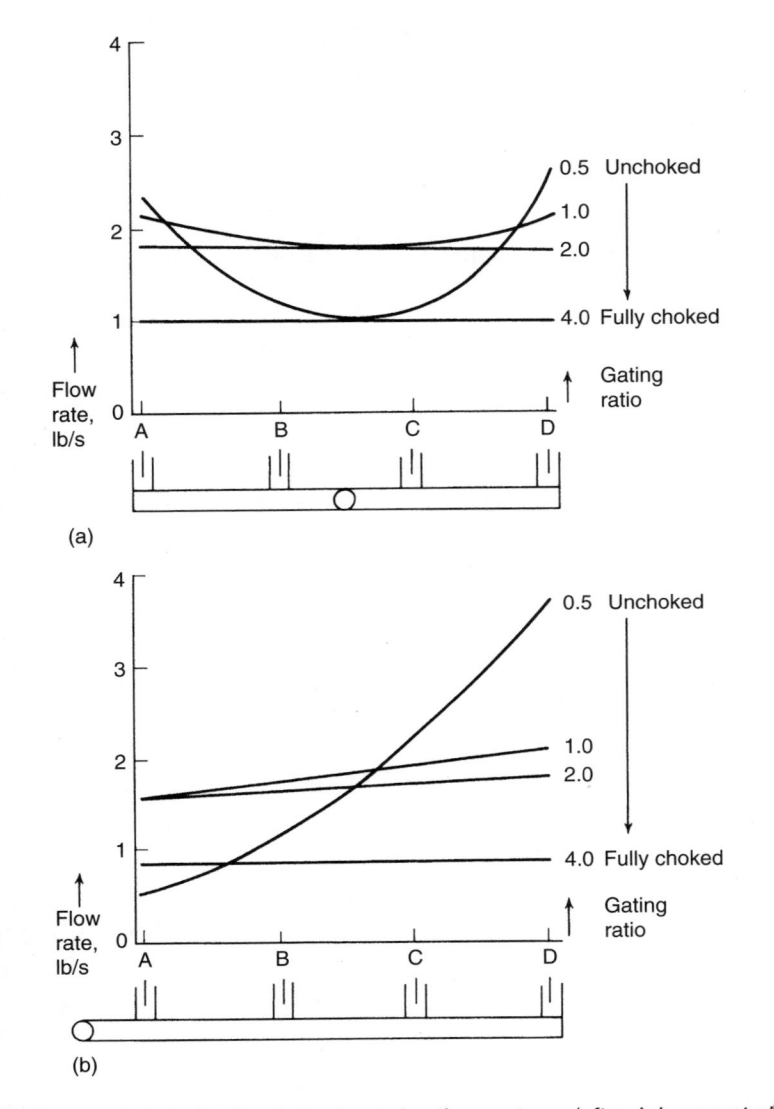

Figure 1.12 Flow behaviour in horizontal gating systems (after Johnson *et al*[50]). (a) Central sprue (b) end sprue (courtesy of American Foundrymen's Society)

momentum of the liquid before its entry to the casting by introducing pools into the system. Examples of these measures are illustrated in Figure 1.13.

The most complete approach to the design of horizontal multiple gating systems, however, was initiated by Grube and Eastwood[35] who obtained equal flow from six ingates by progressive diminution in the cross-section of the runner to maintain a constant linear velocity of flow throughout

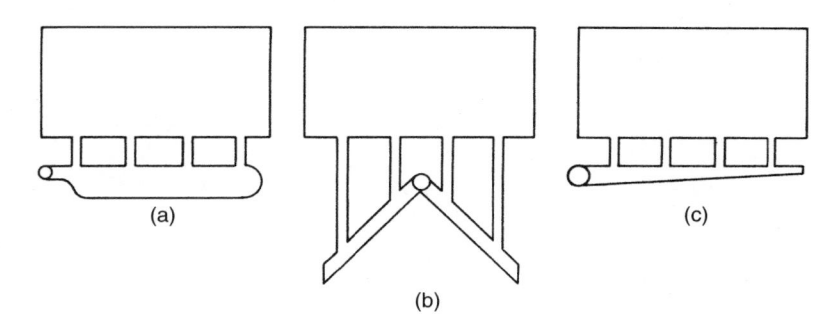

Figure 1.13 Multiple finger ingate systems designed to induce uniform flow. (a) Pool in system, (b) backswept runner, (c) tapered runner (from Johnson *et al*[50]) (courtesy of American Foundrymen's Society)

the runner and ingate system. This system, illustrated in Figure 1.14a, is also carefully streamlined to minimize turbulence and aspiration. The effectiveness of this principle was confirmed in work carried out by the Institute of British Foundrymen[32]; the work also demonstrated that precisely balanced flow rates from multiple gates could be achieved by using a parallel runner with angled ingates (Figure 1.14b). The respective angles were calculated to give identical discharge velocities after taking account of the diminishing velocity along the runner from the sprue.

The other basic type of multiple ingate system is that embodying the *step gate* for the introduction of metal at progressively higher levels in the mould. It offers a means of changing the adverse temperature distribution obtained in bottom gating whilst retaining many of the advantages of that system. It can, for example, be used to introduce hot metal directly into top feeder heads without the interrupted pouring otherwise required.

As with other multiple systems the step gate does not always function as predicted. In a simple system such as that illustrated in Figure 1.15a, filling of the mould cavity occurs principally through the bottom ingate and the upper ingates may even be subject to flow in the reverse direction. If each ingate is to function during its own intended stage of operation, measures are again required to arrest the initial pattern of flow. Figure 1.15 illustrates some methods by which this may be accomplished; tapered downrunners, inclined ingates and U-tube downrunners can be employed.

Other methods of gating

When an ingate at a single point is inadequate, wider distribution of metal may be sought by other methods besides multiple gating. A further group of methods has in common the use of a single long, continuous ingate of comparatively small breadth, connected to the casting through a narrow slit. Such an arrangement is seen in various types of knife, flash, web and

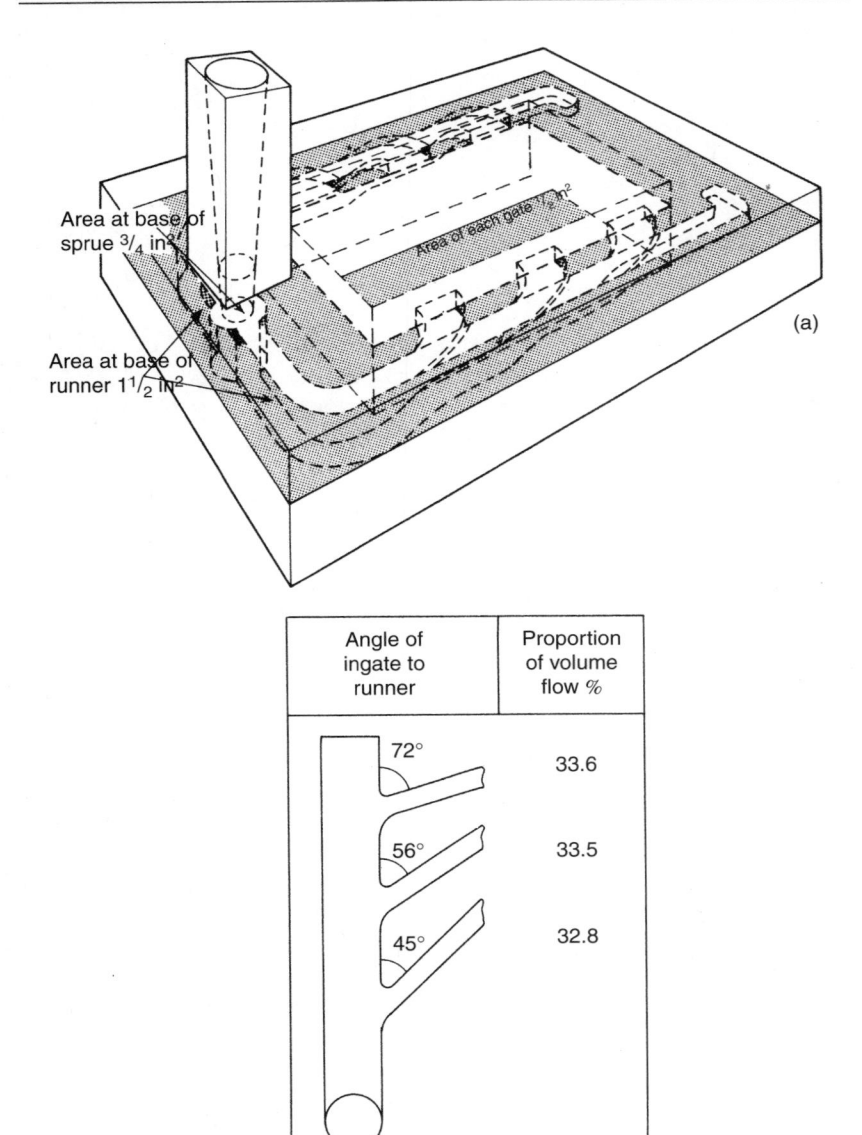

Figure 1.14 Horizontal gating systems. (a) Streamlined system with progressively diminishing cross-sectional area of passages (after Grube and Eastwood[35]) (courtesy of American Foundrymen's Society), (b) system using parallel runner with angled ingates (after Reference 32) (courtesy of Institute of British Foundrymen)

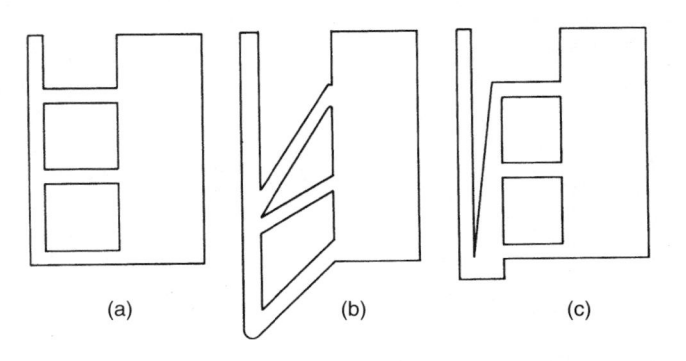

Figure 1.15 Step gating systems. (a) Simple system, (b) inclined steps with common junction, (c) reversed downrunner (from Johnson *et al*[50]) (courtesy of American Foundrymen's Society)

wedge gates and in the Connor runner. In the latter system a horizontal runner bar of square section is allowed to overlap the edge of the casting for a width of approximately 1–2 mm, permitting flow over an extended portion of its length. This technique has also been described as kiss gating and applied to the casting of brasses[51,52]. An example of a thin vertical web gate applied to aluminium alloys is that designed by Grube and Kura[53].

Although the method of metal distribution may sometimes be used to develop temperature gradients as an aid to feeding by directional solidification, it may in other cases be designed with the opposite purpose of minimising temperature gradients within the casting. This practice may be used, for example, as an aid to pressure tightness in castings prone to leakage at hot spots. In such cases, usually associated with alloys freezing in the pasty manner, multiple gating or gating into thin sections is employed to promote uniformity of temperature and so to equalize freezing rates: this topic will be further examined in Chapter 3.

Traps

Devices for the separation of non-metallic inclusions form an important feature of gating systems. Inclusions may be introduced into the mould with the liquid metal, in the form of slag or dross from the furnace, or may consist of dislodged particles of mould refractory. Methods of separation depend either upon differences in density or upon filtration.

The most common type of trap utilizes a simple float chamber formed by local expansion of the runner, within which non-metallic inclusions lighter than the metal can rise and remain at the top (Figure 1.16a). Many variations on this principle are possible, including those in which cross runners moulded in the top part are combined with ingates below the joint line, so that the runner itself serves as the float chamber (Figure 1.16b).

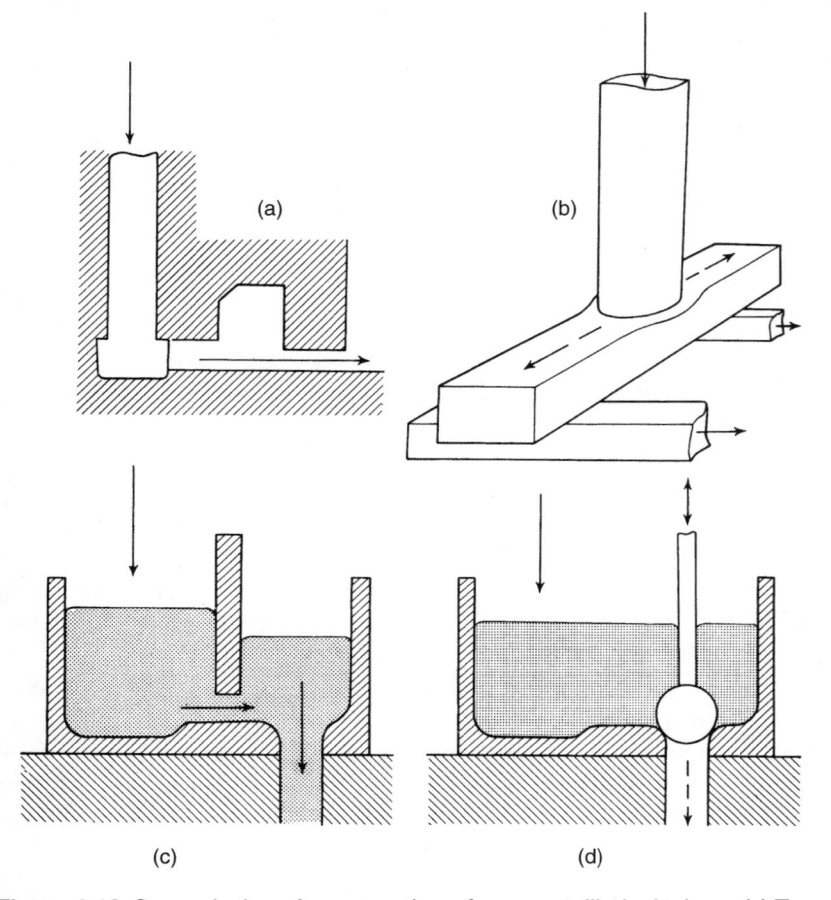

Figure 1.16 Some devices for separation of non-metallic inclusions. (a) Trap in gate member, (b) underslung gates, (c) baffle core, (d) ball plug

A similar principle operates in the case of baffle or skimmer cores placed across the runner basin as shown in Figure 1.16c. For separation by flotation in an open basin, opportunity must be given for a head of metal to accumulate; for this purpose, and to secure steady flow into the sprue, ball plug devices such as that illustrated in Figure 1.16d are sometimes used.

In readily oxidized alloys a dross trap is commonly provided by extending the runner beyond the ingates as illustrated in Figures 1.11 and 1.16b. Oxides carried with the first metal to enter the system are retained in the cul-de-sac and ingate flow occurs on back pressure.

The different densities of metal and inclusions also provide the basis for the design of the whirlgate or spinner. This principle is illustrated in Figure 3.30, Chapter 3, where it is seen in combination with an atmospheric feeder head. Tangential entry and rotation of the liquid produce centripetal forces which drive inclusions towards the centre whilst the metal passes into the casting.

Reference to the relative densities of some substances concerned (Table 1.1) shows that methods based upon gravitational separation are likely to be unsuitable for light alloys except in relation to certain chloride flux inclusions.

Filtration

The greatest advance in countering the problem of inclusions came with the development of modern filters for molten metals. For light alloys there has been long use of techniques depending upon the filtration principle for the separation of major inclusions; some of these are also used for the more fluid alloys based on the heavier metals. The commonest of the earlier devices is the strainer core, a perforated disc of sand or refractory placed in the runner bush or across the gate (Figure 1.17a), with the aim of arresting major inclusions and reducing turbulence in the system.

In the casting of light alloys thin perforated sheet metal screens were seen in the gating system developed by Flemings and Poirier[54] for magnesium alloys (Figure 1.17b). Elliott and Mezoff[55] described a further system in which the metal is filtered through a column of steel wool retained in a cylindrical tinned steel screen, again highly effective in trapping oxide dross. Various forms of wire mesh and filter cloth became widely used for the same purpose.

Rapid extension of the principle followed the introduction of ceramic filters with controlled pore characteristics. The most widely adopted type is produced by the impregration of low-density plastic foam with ceramic slurry containing a bonding agent. Drying and high temperature firing eliminates the plastic to leave a solid, three-dimensional ceramic mesh with a reticulated cellular structure of interconnected pores. The filters are made as rectangular or circular blocks in a range of pore sizes. An alternative type is produced by extrusion and firing of a ceramic mix to form a block

Table 1.1 Approximate densities of some metals and non-metallics

Metal	Relative density	Non-metallic inclusion	Relative density
Copper	8.0	Zinc oxide	5.6
Iron	7.0	Alumina	4.0
Aluminium	2.4	Magnesia	3.7
Magnesium	1.5	Quartz	2.7
		Fireclay	2.0
		Ferrous slag	$\sim$4.2
		Chloride flux	$\sim$1.6

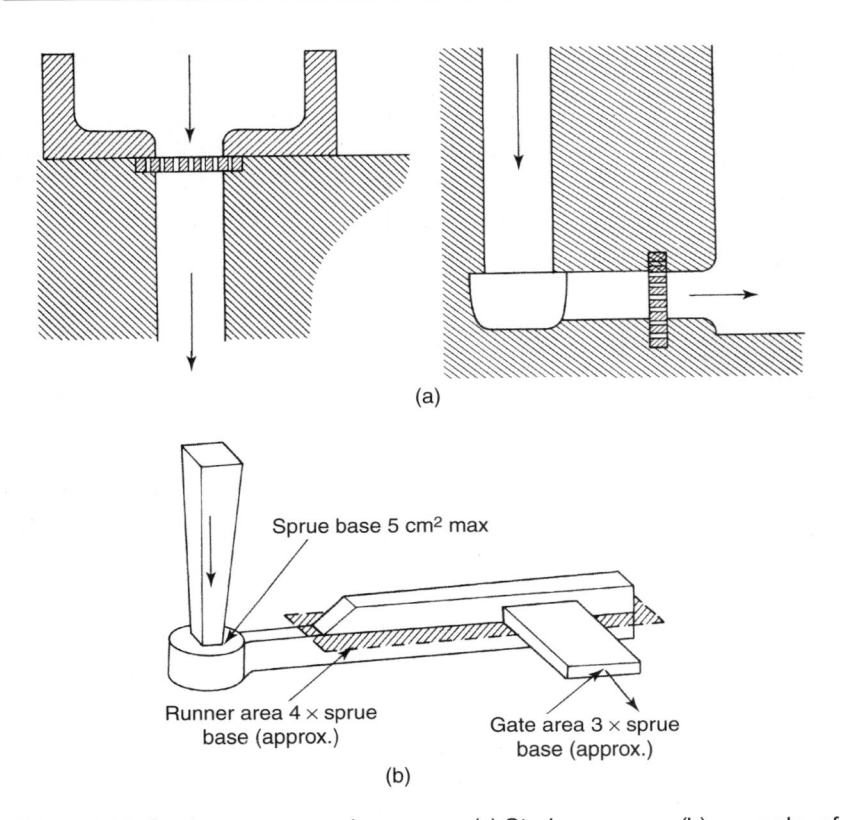

(a)

Sprue base 5 cm² max

Runner area 4 × sprue
base (approx.)

Gate area 3 × sprue
base (approx.)

(b)

Figure 1.17 Strainer cores and screens. (a) Strainer cores, (b) example of system using metal screen (Flemings and Poirier[54]) (courtesy of Foundry (Penton Publications))

or disc with a regular array of parallel holes, or cells. Examples of both types are shown in Figure 1.18. The filters are usually inserted in the gating system, located in prints analogous to coreprints and similarly embodied in the pattern equipment: some of these are custom-designed and made by suppliers of the filters.

The role of filters has been widely researched, the operating parameters established, and numerous studies made of their effects on casting quality. Apart from the removal of inclusions, with proven benefit to mechanical properties, a filter controls metal flow, reducing turbulence and ensuring smooth entry into the mould cavity. Yield is improved, not only through a reduction in scrap but by simplification of the gating system.

Inclusions are removed in two ways. Larger inclusions become trapped at or near the upstream face of the filter, but smaller inclusions, especially the finer indigenous types, are retained within the mesh, being deposited on the ceramic surfaces due to the low-velocity metal flow though the confined

Figure 1.18 Typical ceramic filters (from Hack *et al*[57]) (courtesy of Institute of British Foundrymen)

passages of the filter. The particles adhere to the ceramic; together with the material held at the face they eventually impede flow and the filter becomes blocked. Thus behaviour is predictable, and filters can be selected with known pore sizes and flow capacities to meet the particular casting requirement.

The major reduction in flow velocity though a filter requires it to have a substantial working area to maintain an adequate bulk flow rate. This is normally achieved by local expansion of the runner bar as discussed in References 56 and 57 and exemplified in Figure 1.19. The magnitude of this expansion is usually expressed as a multiple of the choke area, and factors of 3 and upwards are customary; in the case of steel, values of 4.5 and much higher have been quoted. The actual ratios required are naturally dependent on the pore size characteristics of the filter employed, an aspect examined by Devaux *et al*[58] in work with AlSi 7 Mg alloys.

The resistance to flow introduced by a filter is highest at the start of pouring and needs to be overcome by the build-up of a hydraulic head. This 'priming' stage is crucial in the case of steel and was extensively investigated by Wieser and Dutta[59]. Failure was found to occur if the local temperature drop during the heating of the filter exceeded the superheat, causing blockage by premature freezing. Increased pouring temperature is thus advisable in the presence of a filter.

Filters have become widely employed in the production of aluminium alloy sand and die castings and also for nodular cast irons and other alloys

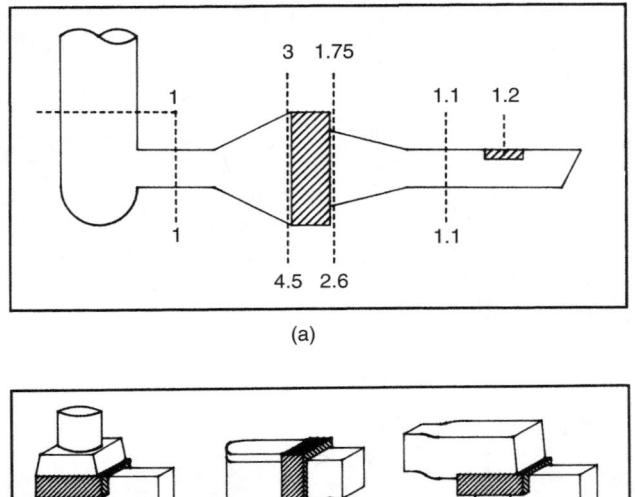

(a)

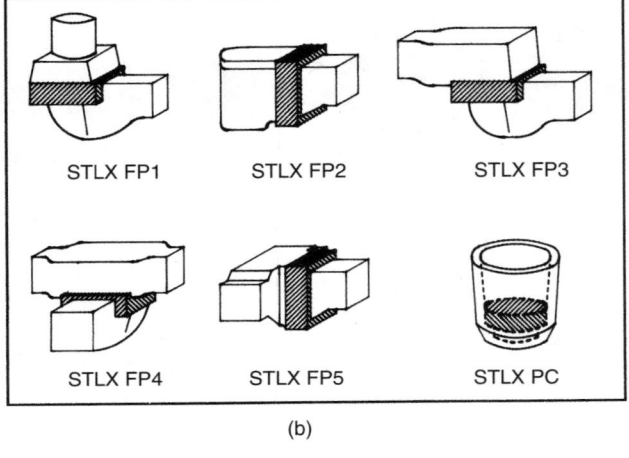

(b)

Figure 1.19 (a) Schematic illustration of filter position and running system area ratios; (b) Variants on filter arrangement shown in (a) (from Hack *et al*[57]) (courtesy of Institute of British Foundrymen)

prone either to oxide film or to dross inclusions. The suppression of such defects also enhances machinability by removing sources of excessive tool wear. The benefits of filtration in controlling flow velocity have been demonstrated by Halvae and Campbell[60] in their work with aluminum bronzes, which are particularly prone to internal oxide film defects. The authors again referred to the concept of a critical entry velocity, for the avoidance of a hydraulic jump effect associated with surface film fold-in, and the role of filters in achieving the required conditions was established: premature jet flow along the bottom surface of the runner was checked by the filter.

Despite initial difficulties, filter use became well established in the manufacture of steel castings, using special refractories with high strength and resistance to thermal shock. Hack *et al*[57] have reviewed practice in this field.

Filtration is also an integral feature of the investment casting of superalloy components.

Perhaps the most unorthodox development in the use of filters came with the introduction by Foseco of the direct pour unit Dypur®, as described by Sandford and Kendrick in Reference 61. This consists of a ceramic foam filter contained in an insulating sleeve, which also serves as a pouring cup and feeder head (Figure 1.20). This unit has been applied successfully to the production of a variety of complex aluminium alloy castings, with a dramatic increase in yield in the absence of a conventional gating system as used in bottom pouring. It offers the thermal advantage of top pouring and the main requirement is to minimize the free-fall distance, which might give opportunities for mould erosion and re-oxidation of the broken metal stream. This principle too has been adopted for other cast alloys.

An extensive illustrated account of filtration practice will be found in Reference 62, which embodies recommended systems for the use of one major supplier's range of filters and accessories, for various specific groups of alloys.

General aspects of gating practice

A number of general principles should be applied in gating, irrespective of the technique being adopted.

The disposition and direction of the ingates should always be designed to avoid direct impingement of the liquid against mould or core surfaces. It is usually possible for an ingate to converge with a limb of the casting: for annular castings, for example, tangential ingates provide ideal conditions for minimising erosion. General streamlining, both of casting design and of the gating system itself, eliminates sharp corners of sand liable to be dislodged by the metal and reduces the danger of *vena contracta*, a local constriction of the metal stream which can give rise to aspiration and turbulence.

The runner bush or basin has been shown to be a vital feature of the gating system: its purpose is to suppress fluctuations in pouring so as to secure steady flow and filling of the downrunner. The qualities of castings produced with bushes ranging from simple cups and funnels to basins offset from the downrunner were investigated by Richins and Wetmore[30] who conclusively demonstrated the advantages of basins of the offset pattern, similar to that shown in Figure 1.16d, for the suppression of turbulence and air entrainment. These advantages were confirmed in work carried out by the Institute of British Foundrymen[43]. Later work by the Institute[32] illustrated the importance of providing a radius at the entry from the runner basin into the sprue. In work with water in transparent plastic moulds, discharge coefficients* of 0.854 and 0.655 were obtained for radiused and

---

* The discharge coefficient D of a system is the ratio of the volume flow rate actually obtained to that theoretically predicted.

(a)

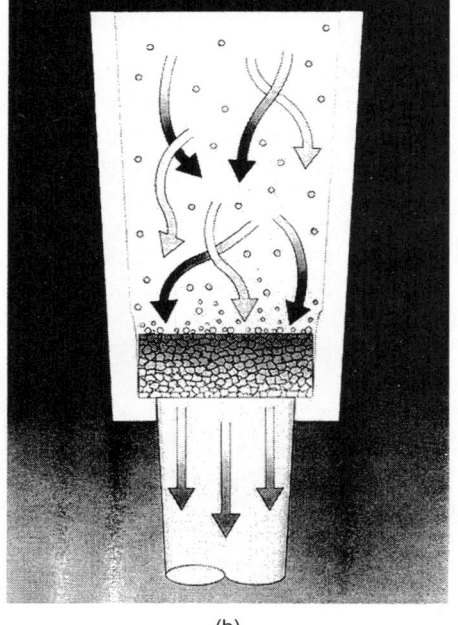

(b)

Figure 1.20 (a) Direct pour unit embodying ceramic filter and insulating sleeve; (b) Role of direct pour unit in reducing turbulence (from Sandford and Kendrick[61]) (courtesy of Institute of British Foundrymen)

sharp entries respectively, confirming the earlier findings of Richins and Wetmore with aluminium and of Swift *et al* with water[29,30].

The runner basin may be formed as an integral feature of the upper mould surface, or may be made separately and superimposed on the downrunner.

Moulding factors

Every precaution is needed to prevent the gating system itself from becoming a source of inclusions. Since the whole volume of liquid metal must pass through the system, breakdown of the surface within the flow channels leads to rapid and progressive erosion, the resulting inclusions being distributed throughout the casting. This can be avoided by ensuring that the whole gating system is formed as an integral part of the moulding operation. Too frequently the completion of the gating system, especially in the jobbing foundry, requires that the hard rammed mould surface be cut away by the moulder, exposing less dense layers behind the mould face. The provision of shaped pattern pieces, on the other hand, not only enables the hard rammed passage surface to be retained but ensures correct reproduction of its intended dimensions. Superimposed runner bushes or basins should be similarly made: these are frequently coated with refractory dressings to provide hard, smooth surfaces.

Gating systems may be reinforced against erosion by the use of pre-moulded refractory sleeves and tiles, or by separately prepared cores. A useful development in this field was the introduction by SCRATA of the Contirun® system, consisting of assemblies of interlocking pre-fired refractory gating components, the whole being held together in cardboard tubes for incorporation in the mould, so enabling heavy castings to be poured without danger of sand erosion from the system. Figure 1.21 shows an inverted pattern for a steel casting with such an assembly in position ready for moulding. It is essential for all forms of insert, whether sand or refractory, to be carefully dried to avoid gas evolution on casting.

To achieve the ideally shaped gating system moulding difficulties need to be surmounted. Horizontally disposed multiple systems can often be accommodated by utilizing an existing parting at or above the bottom of the mould cavity, but in other cases an additional joint or runner core may be needed. Multiple step gates in the vertical plane are more difficult to accommodate, especially if the ingates are to be sloped to provide the ideal flow sequence. Specially designed moulding boxes, or core assembly moulding using vertical joints, are only limited solutions to this problem. The most effective method of forming elaborate gating features is by the use of runner cores containing part or the whole of the system. These simplify mould construction, introduce well consolidated passage walls and ensure close control of design and dimensions of the system.

Figure 1.21 Pattern with gating system assembled from pre-fired refractory units and cardboard tubes (Contirun® system) (courtesy of CDC)

Metal factors

The characteristics of the individual alloy, especially its fluidity and susceptibility to oxidation, influence the type of gating system used. Alloys prone to oxidation and to the formation of strong surface films demand methods ensuring quiet, progressive mould filling with minimum division of the metal stream.

The case of aluminium alloys and the importance of low metal velocities has already been mentioned. Alloys particularly prone to surface film formation include the aluminium bronzes and high tensile brasses. Bottom gating, reversed horn gates and similar techniques are preferred for these metals, highly pressurized systems being avoided.

A comparative study of gating systems using two contrasting types of copper alloy[63] indicated that harmful effects resulting from ingate jets in high ratio systems were not confined to the oxide skin forming alloys, although defects from aspiration in low ratio systems were mostly encountered in this group, confirming the overriding importance of gating technique for this type of alloy. Grey cast irons and the fluid copper alloys such as the phosphor-bronzes, bronzes and gunmetals may be gated by a greater variety of methods, including top gating. A major factor in the case of steel is its greater tendency to premature freezing, combined with its powerful erosive action on the mould: this creates a strong preference for gating to points low in the mould, and for the use of refractory lined, unpressurized low ratio systems of the type referred to previously.

Very specific gating design recommendations for ductile cast irons were advanced by Karsay and Anderson[64-66], including proposed pouring times for castings of various weights. The general aim in this case is rapid filling, to minimize the opportunity for dross formation. Pouring times are themselves determined by the cross-sectional area of the choke incorporated in the system, using charts designed to simplify the process. The choke can be located between runner and ingate or between sprue and runner. The former system is often used but the latter is needed in cases where multiple castings are gated from a common runner.

Despite the general points discussed above, there remains no universal agreement as to the design of gating systems, only a series of useful principles. Valid practical points on gating can still be found in the classic work by Dwyer[67], although this antedated the widespread availability of filtration and much knowledge arising from modern studies. These are the subject of wide-ranging discussions in the much later work by Campbell[1]. This addresses the whole subject of fluid dynamics as affecting the behaviour of molten metal in the mould. Further notable contributions can be found in References 68 and 69.

Forces acting on the mould

As the mould fills it becomes exposed to high metallostatic pressures which tend to displace or distort the mould sections and cores. These forces can be accurately predicted and contained by foundry measures.

The first need is for a dense, rigid mould, since the pressure tends to dilate the mould cavity, especially in greensand practice. Rigidity of the mould parts can be increased by using box bars or cover plates to reinforce the sand mass; these measures become increasingly necessary with moulds of large area.

Assuming a rigid mould, the next concern is with the force tending to separate the mould parts. The upward force acting on a flat mould surface is equal to $9.81\ \rho h A$ N, where

$$\rho = \text{density of the metal, kg/m}^3$$

$$h = \text{head of metal, m}$$

$$A = \text{superficial area, m}^2$$

A completely flat mould surface gives rise to the maximum lifting force: calculations for other shapes can thus be safely based upon their projected areas. The force is resisted by using box clamps and arrangements of plates and tie-bars to hold the mould parts together and by weighting the top part. In the latter case the minimum weight required is $\rho h A$ kg including the weight of the cope itself.

Cores too are exposed to an upthrust, equivalent in this case to the weight of metal displaced; the net upward force is thus considerable except in the casting of light alloys. The force is countered by high mechanical strength and rigidity in core construction, enhanced by reinforcing grids and irons.

Cores must be firmly supported against movement in the mould. A core relying upon a single coreprint in the mould bottom, for example, tends to float out of position and must be anchored with wires or sprigs. Cores of large dimensions may require several points of support: if these cannot all be provided in the form of coreprints, studs or chaplets may be needed to prevent movement on casting.

Liquid contraction

Shrinkage of casting alloys in the liquid state is significant principally in bringing about loss of metallostatic head during cooling to the solidification temperature. The liquid shrinkage of most alloys lies in the range 1.0–2.5% per 100°C of superheat. In the case of slender top risers and sprues this small volume change in the body of the casting may be sufficient to cause appreciable loss of the liquid head. The liquid shrinkage for the whole mass of the casting must therefore be considered in relation to the dimensions of the elements at the top of the mould cavity. The implications of liquid shrinkage in risers will be further examined in the following chapter.

References

1 Campbell, J., *Castings*, 2nd Ed Butterworth-Heinemann Oxford (1993)
2 Clark, K. L., *Proc. Inst. Br. Foundrym.*, **39**, A52 (1945–6)
3 Clark, K. L., *Trans. Am. Fndrym. Ass.*, **54**, 37 (1946)
4 Krynitsky, A. I., *Trans. Am. Fndrym. Soc.*, **61**, 399 (1953)
5 Saeger Jr., C. M. and Krynitsky, A. I., *Trans. Am. Fndrym. Ass.*, **39**, 513 (1931)
6 Portevin, A. M. and Bastien, P., *C.R. Acad. Sc. (Paris)*, **194**, 80 (1932)
7 Taylor, H. F., Rominski, E. and Briggs, C. W., *Trans. Am. Fndrym. Ass.*, **49**, 1 (1941)
8 Worthington, J., *Proc. Inst. Br. Foundrym.*, **43**, A144 (1950)
9 Kondic, V. and Kozlowski, H. J., *J. Inst. Metals*, **75**, 665 (1949)
10 Ragone, D. V., Adams, C. M. and Taylor, H. F., *Trans. Am. Fndrym. Soc.* **64**, 640 and 653 (1956)
11 Portevin, A. and Bastien, P., *J. Inst. Metals*, **54**, 45 (1934)

12 Wood, D. R. and Gregg, J. F., *Br. Foundrym.*, **50**, 2 (1957)
13 Floreen, S. and Ragone, D. V., *Trans. Am. Fndrym. Soc.*, **65**, 391 (1957)
14 Feliu, S., Flemings, M. C. and Taylor, H. F., *Br. Foundrym.*, **53**, 413 (1960)
15 Campbell J. Cast Metals **7**, 4, 227 (1995)
16 Hoar, T. P. and Atterton, D. V., *J. Iron Steel Inst.*, **166**, 1 (1950)
17 Baruch, T. A. and Rengstorff, G. W. P., *Trans. Am. Fndrym. Soc.*, **71**, 595 (1963)
18 Flemings, M. C., Mollard, F. R., Niiyama, E. F. and Taylor, H. F., *Trans. Am. Fndrym. Soc.*, **70**, 1029 (1962)
19 Knowles, D. and Sarjant, R. J., *J. Iron Steel Inst.*, **155**, 577 (1947)
20 Barlow, G. and Beeley, P. R., *Br. Foundrym.*, **63**, 61 (1970)
21 Portevin, A. M. and Bastien, P., *Proc. Inst. Br. Foundrym.*, **29**, 88 (1935–6)
22 Cooksey, C. J., Kondic, V. and Wilcock, J., *Proc. Inst. Br. Foundrym.*, **52**, 381 (1959)
23 Report of Sub-Committee T.S.6., *Proc. Inst. Br. Foundrym.*, **39**, A28 (1945–6)
24 Oliff, I. D., Lumby, R. J. and Kondic, V., *Fndry Trade J.*, **119**, 469 (1965)
25 Flemings, M. C., Mollard, F. R. and Taylor, H. F., *Trans. Am. Fndrym. Soc.*, **69**, 566 (1961)
26 Betts, B. P. and Kondic, V., *Br. Foundrym.*, **54**, 1 (1961)
27 Flemings, M. C., Conrad, H. F. and Taylor, H. F., *Trans. Am. Fndrym. Soc.*, **67**, 496 (1959)
28 Webster, P. D. *et al.*, *Br. Foundrym.*, **57**, 524 (1964)
29 Swift, R. E., Jackson, J. H. and Eastwood, L. W., *Trans. Am. Fndrym. Soc.*, **57**, 76 (1949)
30 Richins, D. S. and Wetmore, W. O., Ref. 34, p. 1
31 Ruddle, R. W., *The Running and Gating of Sand Castings*, Inst. Metals, London (1956)
32 Report of Sub-Committee T.S.54., *Br. Foundrym.*, **58**, 183 (1965)
33 Report of Sub-Committee T.S.24., *Proc. Inst. Br. Foundrym.*, **48**, A306 (1955)
34 Symposium on Principles of Gating, *Am. Fndrym. Soc.* (1951)
35 Grube, K. and Eastwood, L. W., *Trans. Am. Fndrym. Soc.*, **58**, 76 (1950)
36 Jeancolas, M., De Lara, G. C. and Harf, H., *Trans. Am. Fndrym. Soc.*, **70**, 503 (1962)
37 Hall, H. T., *Cast Metals Res. J.*, **4**, 164 (1968)
38 Hall, H. T., Lavender, J. D. and Ball, J., *Br. Foundrym.*, **62**, 296 (1969)
39 Webster, P. D. and Young, J. M., *Br. Foundrym.*, **79**, 276 (1986)
40 Xu, Z. A. and Mampaey, F., *Trans. Am. Fndrym. Soc.*, **104**, 155 (1996)
41 Runyoro, J., *Cast Metals*, **5**, 4, 224 (1998)
42 Webster, P. D., (Ed) *Fundamentals of Foundry Technology*, Portcullis Press, Redhill (1980)
43 Report of Sub-Committee T.S.35., *Br. Foundrym.*, **53**, 15 (1960)
44 Johnson, W. H. and Baker, W. O., *Trans. Am. Fndrym. Soc.*, **56**, 389 (1948)
45 Runyoro, J. and Campbell, J. *Foundrym.*, **85**, 117 (1992)
46 Runyoro, J., Boutorabi, S. M. A. and Campbell, J., *Trans. Am. Fndrym. Soc.*, **100**, 225 (1992)
47 Campbell, J., *Foundrym.*, **92**, 313 (1999)
48 Sutton, T., *Foundrym.*, **90**, 217 (1997)
49 Johnson, W. H., Baker, W. O. and Pellini, W. S., *Trans. Am. Fndrym. Soc.*, **58**, 661 (1950)

50 Johnson, W. H., Bishop, H. F. and Pellini, W. S., *Trans. Am. Fndrym. Soc.*, **61**, 439 (1953)
51 Measures, J. F., *Proc. Inst. Br. Foundrym.*, **43**, B62 (1950)
52 Ward, C. W. and Jacobs, T. C., *Trans. Am. Fndrym. Soc.*, **70**, 865 (1962)
53 Grube, K. R. and Kura, J. G., *Trans. Am. Fndrym. Soc.*, **62**, 33 (1955)
54 Flemings, M. C. and Poirier, E. J., *Foundry*, **91**, October, 71 (1963)
55 Elliott, H. E. and Mezoff, J. G., *Trans. Am. Fndrym. Soc.*, **55**, 241 (1947)
56 Sandford, P. and Sibley, S. R., *Trans. Am. Fndrym. Soc.*, **104**, 1063 (1996)
57 Hack, J. A., Clark, H. and Child, N., *Foundrym.*, **83**, 183 (1990)
58 Devaux, H., Hiebel, D., Jacob, S. and Richard, M., *Cast Metals*, **3**, 2, 91 (1990)
59 Wieser, P. F. and Dutta I., *Trans. Am. Fndrym. Soc.*, **94**, 85 (1986)
60 Halvae, A. and Campbell, J., *Trans. Am. Fndrym. Soc.*, **105**, 35 (1997)
61 Sandford, P. and Kendrick, R., *Foundrym.*, **83**, 262 (1990)
62 Brown, J. R., ed., *Foseco Foundryman's Handbook*, 10th Edition, Butterworth-Heinemann, Oxford (1994)
63 Glick, W. W., Jackson, R. S. and Ruddle, R. W., *Br. Foundrym.*, **52**, 90 (1959)
64 Karsay, S. I., *Ductile Iron Production Practices*. Am. Fndrym. Soc., Des Plaines, Ill. (1975)
65 Anderson, J. V. and Karsay, S. I., *Br. Foundrym.*, **78**, 492 (1985)
66 Karsay, S. I., *Trans. Am. Fndrym. Soc.*, **105**, vii (1997)
67 Dwyer, P., *Gates and Risers for Castings*, Cleveland, Penton (1949)
68 Wukovich, N. and Metevelis, G., *Trans. Am. Fndrym. Soc.*, **97**, 285 (1989)
69 Hill, J. L., Berry, J. T. and Guleyupoglu, S., *Trans. Am. Fndrym. Soc.*, **99**, 91 (1991)

2

Solidification 1 Crystallization and the development of cast structure

The pouring of metal into a relatively cool mould initiates the processes of solidification, during which stage the cast form develops cohesion and acquires lasting structural characteristics.

The mode of freezing exercises a twofold influence upon the final properties of a casting. The normal metallographic structure determines many of the properties inherently available from the cast metal. This structure – the grain size, shape and orientation and the distribution of alloying elements as well as the underlying crystal structure and its imperfections – is largely determined during crystallization from the melt. Even in those cases where the as-cast structure is modified by subsequent treatments it still exerts a residual influence upon the final structure.

The properties and service performance of an individual casting are, however, also a function of its soundness – the degree of true metallic continuity. This too is established during solidification, since the volume shrinkage accompanying the change of state must be fully compensated by liquid feed if internal voids are to be prevented.

Structure and soundness, being dependent upon the mechanism of solidification, are influenced by many factors, including the constitution and physical properties of the alloy. Steels, bronzes, and cast irons, for example, exhibit wholly different feeding characteristics. Other important factors are the pouring and mould conditions and it is with the manipulation of these conditions to achieve full control of the pattern of freezing that much of the technique of founding is concerned.

In the present chapter the fundamental mechanisms of solidification will be examined with particular reference to the characteristic metallographic structures developed in castings, whilst the question of soundness and the techniques of feeding will be separately treated in the following chapter. Macroscopic defects will be examined in Chapter 5.

Crystallization from the melt

Matter in the crystalline state exhibits long range order, being characterized by regularity of atomic spacing over considerable distances. Metallic crystals deposited from the melt, although possessing internal symmetry, are commonly of irregular external form as a result of uneven growth rates and mutual constraint in the last stages of freezing. The allotriomorphic grain is thus the normal crystalline unit in the cast structure, although in the case of the eutectic grain* or cell two or more separate crystal structures are associated through the simultaneous growth of separate phases.

Since the crystal lattice usually represents a more closely packed state of matter than the liquid, freezing is almost invariably associated with volume contraction. The reduction in molecular motion is also accompanied by liberation of energy in the form of latent heat of crystallization, which exerts a marked effect upon the rate and mode of crystal growth. Latent and specific heat values for some common metals are given in Table 2.1.

Crystallization from a metal melt involves the successive stages of nucleation and growth. The location and relative rates of these two phenomena within the liquid determine the final structure of the solid and establish the extent to which freezing is directional or occurs in a discrete manner throughout the liquid.

Nucleation

Nucleation is the appearance at points in the liquid of centres upon which further atoms can be deposited for the growth of solid crystals. Such nuclei can be produced in two ways.

Table 2.1 Specific heats and latent heats of fusion of some metals (after Smithells[1])

Metal	Specific heat J/kg K	Latent heat of fusion kJ/g-atom or mol
Iron	456	15.2
Nickel	452	17.2
Copper	386	13.0
Aluminium	917	10.5
Magnesium	1038	8.8
Zinc	394	7.3
Lead	130	4.8
Tin	226	7.1

---

* The term *grain* is preferred in this context since the term *cell* has become graphically associated with specific substructures in cast alloys. The former term offers a better analogy with other cast structures even though the unit contains two phases with separate crystal structures. The description 'cellular' can then be confined to the distinctive hexagonal substructures common both to single phase and eutectic grains: the separate term *colony* for the latter substructure is then unnecessary.

Homogeneous nucleation is the occurrence, during the general movement within the liquid, of ordered groups of atoms forming small zones of higher than average density. These embryonic crystals are ephemeral and unstable, but some reach a critical size at which they become stable and grow.

General nucleation theory explains this in terms of the change in free energy resulting from the precipitation of a particle. This change is compounded of changes in both volume and interfacial free energies.

The formation of solid from liquid results in a negative free energy change directly proportional to the volume transformed. Thus, for a spherical particle

$$\Delta G = -\tfrac{4}{3}\pi r^3 \Delta G_v$$
(volume)

where r is the radius of the particle and ΔG_v the bulk free energy change per unit volume.

The creation of a new interface, however, requires a local gain in free energy proportional to the surface area of the particle. In this case, for the spherical particle,

$$\Delta G = 4\pi r^2 \gamma$$
(interface)

where γ is the interfacial free energy per unit area for a spherical surface.

The sum of the respective free energy changes is found to be positive for very small values of r but becomes negative at higher values of r (Figure 2.1). The peak positive value of ΔG corresponds to the critical radius above which growth can take place with a decrease in free energy. These conditions constitute an energy barrier inhibiting nucleation: however, the critical size of the nucleus diminishes with falling temperature and increases the probability of homogeneous nucleation occurring.

In *heterogeneous nucleation* the initial growth interface is provided by a foreign particle included or formed in the melt. For a second phase to act as a nucleus in this way it must be capable of being wetted by the metal, forming a low contact angle, and must possess some structural affinity with the crystalline solid.

The nucleation and growth of crystals occurs below the equilibrium freezing temperature. Because conditions are initially unfavourable to the survival of small nuclei, considerable undercooling is necessary to produce homogeneous nucleation[2,3]. In practice, however, some foreign nuclei are always present in the form of impurities having their origin in the metal, the walls of the mould and the atmosphere. Further foreign nuclei may be deliberately added to encourage a particular mode of crystallization. Solidification under foundry conditions thus occurs with little undercooling.

Once the initial nuclei are established, two possibilities exist for further crystallization. More solid may be deposited upon the first nuclei, or fresh nucleation of the same or of a different phase may occur in the liquid.

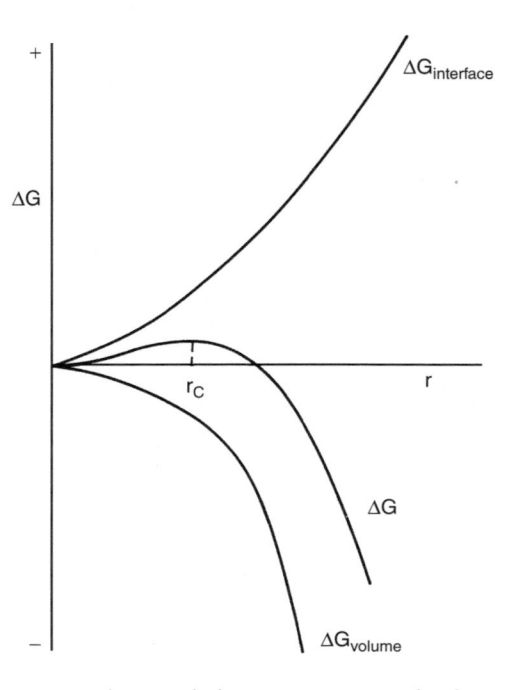

Figure 2.1 Free energy changes in homogeneous nucleation

The most effective site for the precipitation of a phase being an existing mass of the same phase, growth might be expected to predominate over further nucleation. In practice, however, barriers to growth result from the evolution of the latent heat of crystallization and, in alloys, from the change in composition of the adjacent liquid through differential freezing. The influences of these phenomena upon growth will now be further considered.

Growth

The growth processes which follow nucleation determine the final crystallographic structure of the solid. The mode of growth, both of individual grains and of the general mass of solid, depends upon thermal conditions in the solidification zone and the constitution of the alloy.

The freezing of a pure metal presents the most straightforward case. If the temperature were capable of being lowered in a completely uniform manner, random nucleation might be expected throughout the liquid. Under practical conditions of heat flow, however, temperature gradients produce initial nucleation at the relatively cool mould surface, growth proceeding directionally towards the centre of the casting (Figure 2.2). Such growth tends to be associated with a preferred crystallographic direction: favourably oriented crystals grow more rapidly than their neighbours in a temperature gradient.

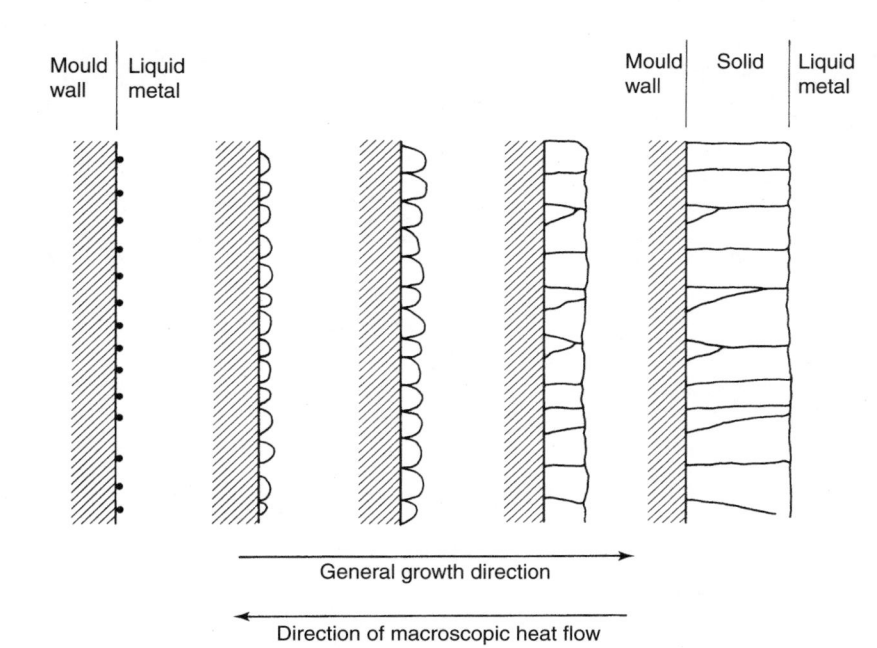

General growth direction

Direction of macroscopic heat flow

Figure 2.2 Development of columnar grain structure (schematic)

At a stage depending upon the number of effective nuclei and the initial growth rate, lateral growth becomes restricted by the mutual constraint of neighbouring crystals. An approximately plane interface then advances by the progressive deposition of atoms to join the lattice of one or other of the existing crystals. This growth actually takes place by the edgewise extension of close packed atomic planes, producing a minute terraced structure on the growth interface, but this interface is macroscopically smooth and, in most metals, non-crystallographic. Such is a typical sequence leading to the production of the characteristic *columnar* structure often observed in ingots and castings, with the individual grains elongated in the general direction of heat flow.

One set of thermal conditions producing such a structure is illustrated in Figure 2.3. This case assumes the existence of a positive temperature gradient at the solid–liquid interface, the interface advancing progressively as the growth temperature T_G is reached at points deeper inside the casting. Freezing can occur in this manner when the latent heat of crystallization is insufficient to reverse the direction of the heat flow in the liquid adjacent to the interface. It is only under these conditions that the latter approximates to a flat plane, a mode of growth promoted by slow cooling and a steep temperature gradient.

In many conditions, however, although heat transfer from the casting to the cooler mould produces a positive gradient on a macroscopic scale, local

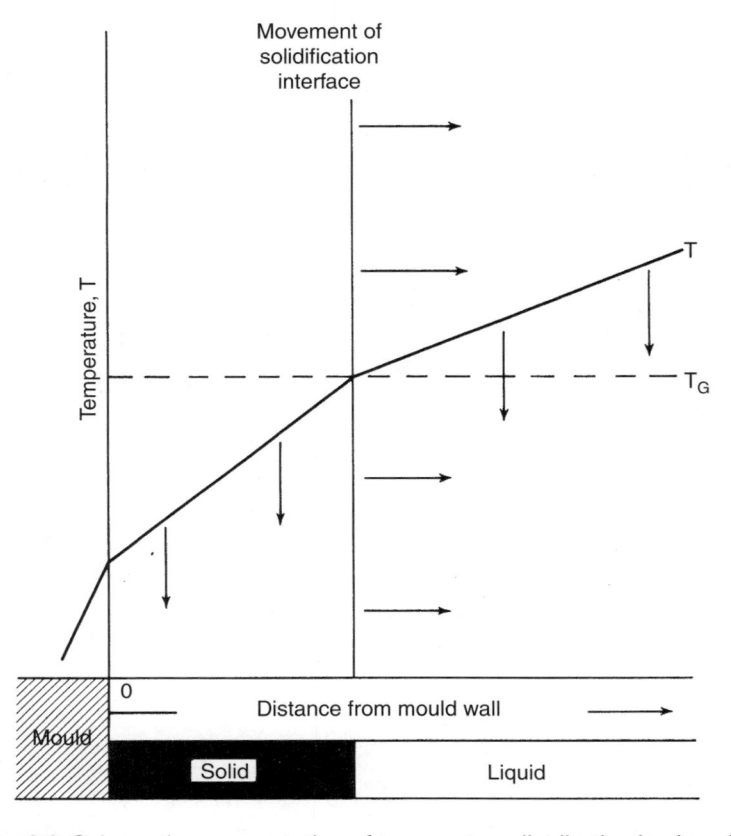

Figure 2.3 Schematic representation of temperature distribution in plane front solidification

evolution of latent heat is sufficient to reverse the temperature gradient at the interface. The thermal conditions in this case are represented in Figure 2.4. Since the minimum temperature in the liquid is no longer adjacent to the interface, growth by the general advance of a smooth solidification front gives way to other modes of growth in which deposition can occur in regions of greater undercooling. Microscopic heat flow can thus be a major factor in the formation of cast structures.

Before proceeding to examine these alternative modes of growth, brief consideration will be given to the alloys which freeze to form solid solutions. These exhibit differential freezing, which again promotes growth in a manner other than by the advance of a smooth interface.

In the binary alloy system illustrated in Figure 2.5, the solid initially deposited from the alloy C_0 is of composition C_1, so that the residual liquid becomes slightly enriched in the solute element B. Solid formed

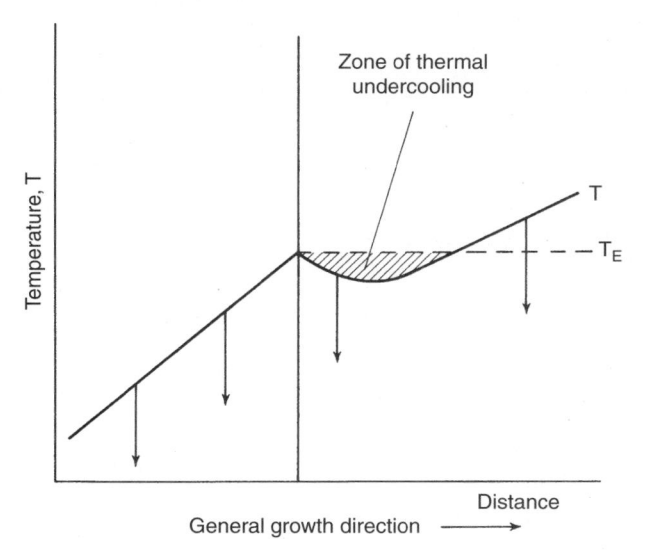

Figure 2.4 Thermal conditions with reversal of temperature gradient in liquid adjoining interface

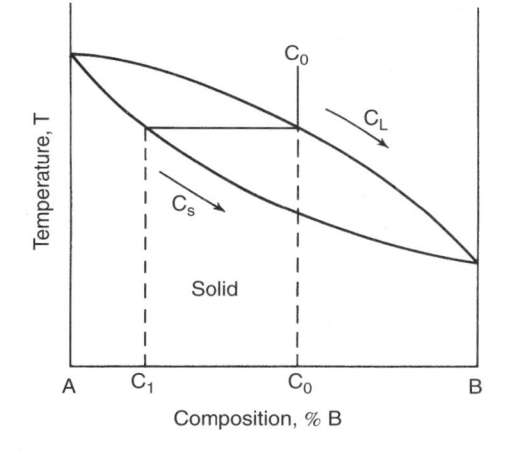

Figure 2.5 Thermal equilibrium diagram for two metals forming a continuous range of solid solutions

during further cooling being similarly deficient in element *B* relative to its own parent liquid, this element is continuously rejected into the liquid phase throughout freezing. The unidirectional freezing of a small element of such an alloy can now be considered. Were the system allowed to reach equilibrium at any intermediate stage, the situation could be defined by a

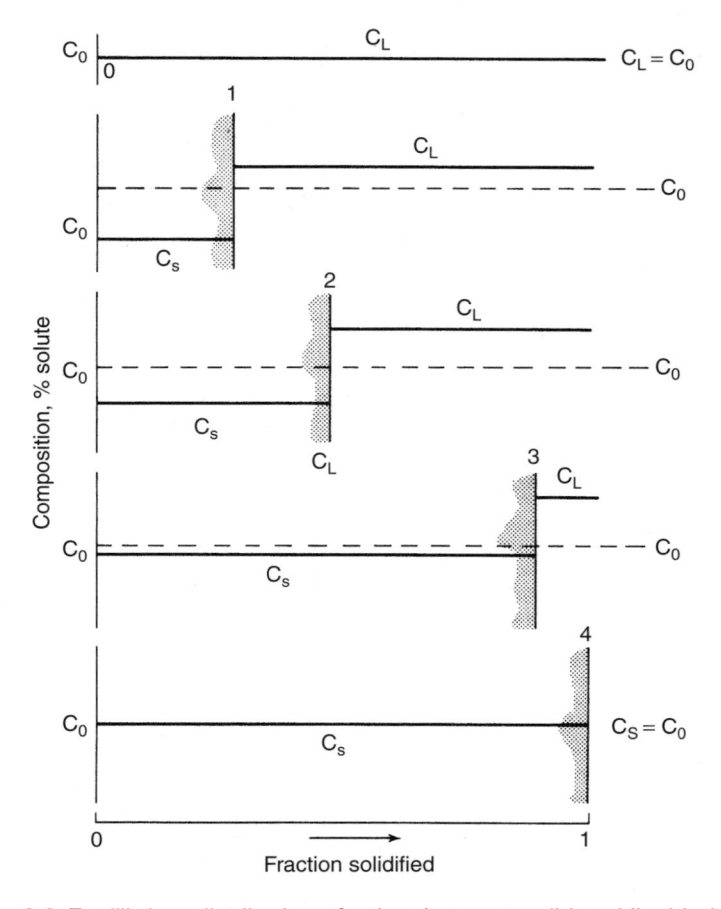

Figure 2.6 Equilibrium distribution of solute between solid and liquid phases at successive stages of unidirectional freezing: C_0 = mean composition before and after freezing, C_S and C_L = compositions of liquid and solid fractions at successive static positions of interface

diagram of the type shown in Figure 2.6, with complete homogeneity of the solid and liquid phases. Under dynamic conditions, however, a non-equilibrium situation prevails. The change in composition of the whole of the liquid is brought about by solute rejection at the interface, so that a compositional gradient must exist in the liquid, its profile being determined by the rate of solute transport relative to the rate of crystallization.

This compositional variation, illustrated in Figure 2.7a, introduces a corresponding variation in freezing temperature as shown in Figure 2.7b. Each composition on the solute distribution curve has its own characteristic equilibrium freezing temperature, below which a particular amount of undercooling will produce further crystallization.

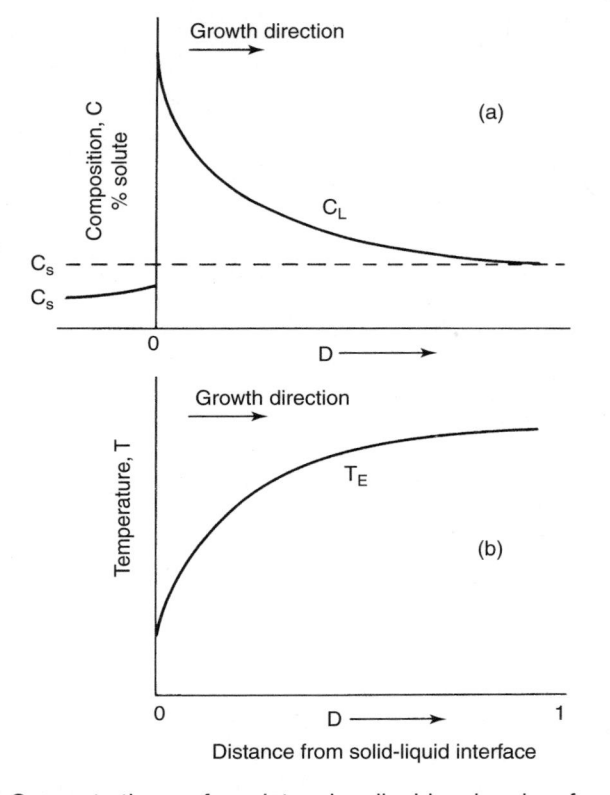

Figure 2.7 Concentration of solute in liquid ahead of advancing interface – non-equilibrium conditions. (a) Solute distribution C_L, (b) equilibrium liquidus temperature T_E corresponding to solute content at distance D from solidification interface

If the freezing temperature curve is related to the actual temperature gradient existing in the liquid, as shown in Figure 2.8, it can be seen that before the general temperature level falls sufficiently to produce fresh deposition of solid at the interface, considerable undercooling may exist at points further within the liquid, i.e. there may be a band of liquid within which conditions are more favourable to freezing than at the interface. This condition is frequently termed *constitutional undercooling*; its importance was appreciated many years ago by Hultgren, Genders and Bailey, Northcott and others and its influence upon the mode of crystallization was later subject to detailed investigation, particularly in the period dating from the 1960's[4,5–10].

The interaction of the temperature and compositional gradients in the liquid can thus be regarded as the most important single factor influencing the structure of a casting. Both grain and substructure depend upon this factor.

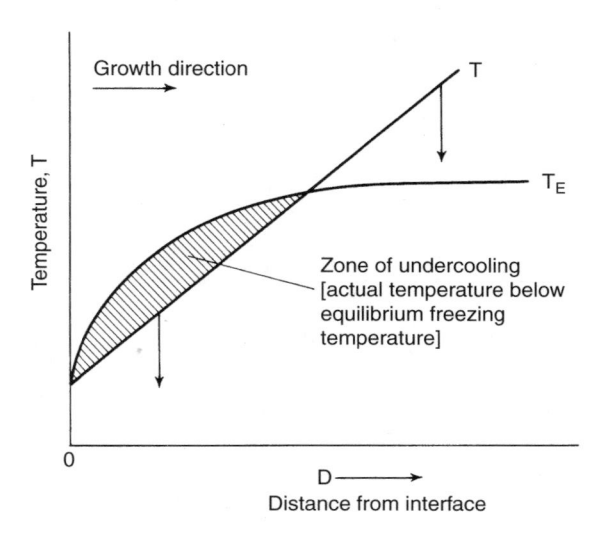

Figure 2.8 Relation of temperature gradient in liquid to equilibrium freezing temperature profile

Dendritic growth

When undercooling occurs in the band of liquid adjoining the interface, any existing protuberance on the solid face tends to become stable and to act as a centre for preferential growth. Whilst general advance of the interface is retarded by the latent heat or solute barrier, such local growth centres can probe further into the zone of undercooling.

These are favourable conditions for dendritic growth, with its characteristic tree-like form. This is the type of growth most commonly encountered in the freezing of commercial casting alloys forming solid solutions. The primary axis of the dendrite is the result of preferred growth at an edge or corner of an existing crystallite. The projection develops into a needle and subsequently a plate following the general direction of heat flow; this growth direction is usually associated with a particular crystallographic direction as exemplified in the literature[8,11].

Lateral growth of the primary needle or plate is restricted by the same latent heat or solute accumulation as inhibited general growth at the original interface, but secondary and tertiary branches can develop by a similar mechanism to that which led to the growth of the primary stem (Figure 2.9).

The factors determining the structure of an individual dendrite, including the spacing of the dendrite arms, were examined by Spear and Gardner[12], Alexander and Rhines[13], Flemings[14] and others. The secondary spacing, which decreases with increasing cooling rate, was originally seen as resulting from normal growth of the dendrite arms into undercooled liquid, but it was later established that the overriding influence on the final

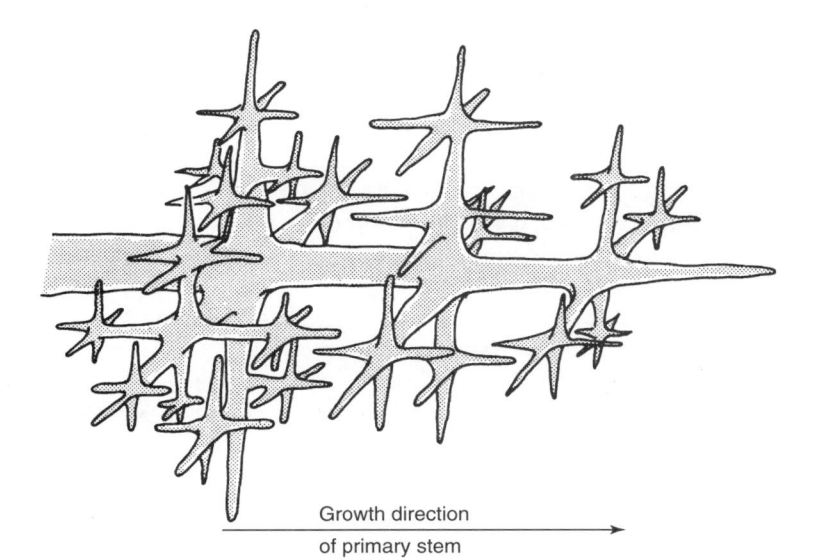

Growth direction
of primary stem
(a)

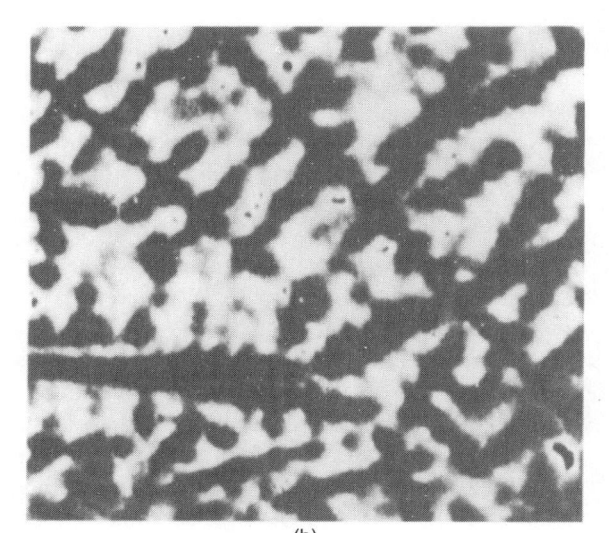

(b)

Figure 2.9 Dendritic growth. (a) Classical concept of a dendrite, (b) dendritic microstructure in carbon–chromium steel. ×5 approx.

spacing is the mechanism of Ostwald ripening, which accounted for certain discrepancies between theoretical and measured spacings. This and other aspects of dendrite arm spacing were examined in a later paper by Flemings *et al*[15], which also treated earlier findings[16–18]. Following establishment of the initial spacing by the previously discussed branching mechanism, the structure undergoes progressive coarsening during the remaining period of solidification. This occurs through the influence of interface curvature on the surface energy, so that the smaller arms remelt and disappear, whilst the neighbouring larger arms continue to grow and increase their surface area. The final spacing is thus much greater than the original.

It will later be shown that the features of the dendrite substructure are more significant than grain size in relation to the properties of many cast alloys, although some refining treatments produce copious nucleation of small grains without dendritic features.

Whilst this picture of dendrite development has been presented in terms of unidirectional growth from a major interface – this would produce columnar-dendritic grains – dendritic growth may equally be associated with a crystal growing independently within the melt. In this case the growth interface is the whole periphery of the crystal and the fully developed grain is approximately equiaxed.

Dendritic growth in a pure metal can only be detected by interrupted freezing and decantation, but is evident in alloys through the persistence of compositional differences, revealed on etching as the characteristic *cored* structure. Coring results from the previously mentioned differential freezing process, which leaves the centres of the dendrites deficient in solute relative to the interdendritic zones. The resulting compositional gradient can only be eliminated by high temperature diffusion in the solid state, time for which is not normally available during cooling from the casting temperature.

Prolonged annealing or homogenization can bring about complete diffusion of the solute, but even in this case visible evidence of the original coring may persist as a pattern of segregated impurities. The actual time for homogenization is less when grain size and the spacings of the dendrite substructure are small, since the diffusion distances are then shorter. Coring will be further referred to in Chapter 5.

Dendritic growth in alloys is preceded, with less marked undercooling, by the formation of a highly distinctive *cellular substructure*, direct evidence for which was originally obtained by the examination of interfaces from which the liquid had been decanted. This structure (see Figure 2.10) is produced as a cluster of hexagonal rods which grow into the liquid and reject solute to their boundaries. On attaining a certain level of undercooling the cellular gives way to the dendritic mode of growth by the preferential development of a limited number of cells. Intermediate rod-like structures have been described by Flemings[14] as 'fibrous dendrites'. The successive steps in the evolution of the substructure from a plane interface to the full dendritic condition were examined in detail by Biloni and Chalmers[19,20].

A substantial review of much of the modern work or dendritic structure, including detailed aspects of the growth processes operating under the special conditions of directional and rapid solidification, is that by Trivedi and Kurz in Reference 21.

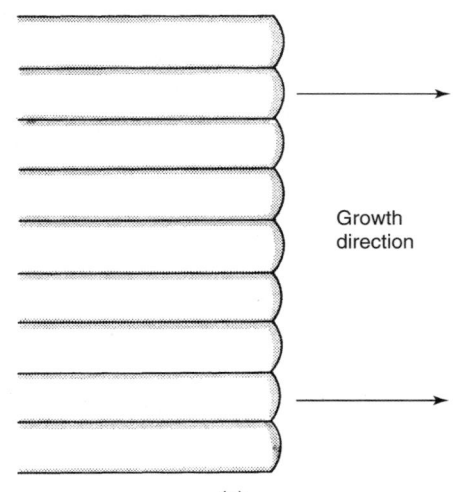

(a)

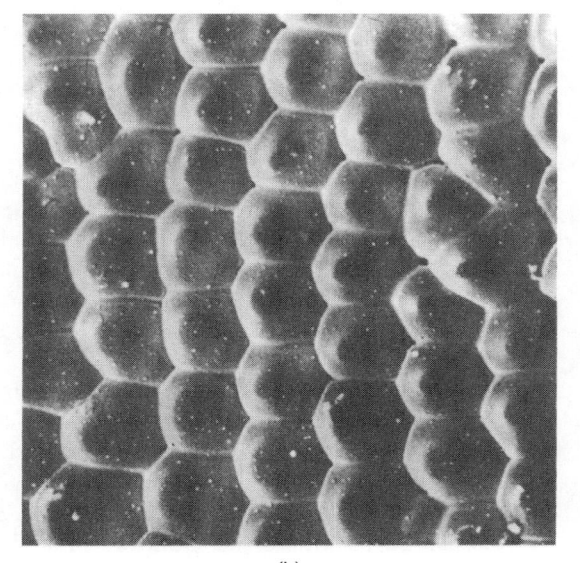

(b)

Figure 2.10 Cellular substructure formed by undercooling. (a) Structure in growth direction, (b) hexagonal cells on growth interface (nickel base alloy, ×120) (courtesy of Dr E. Grundy)

Independent nucleation

When freezing occurs under a shallow temperature gradient or at a very rapid rate, the undercooling may be sufficient to promote fresh nucleation at points distant from the main interface. Under these conditions, further

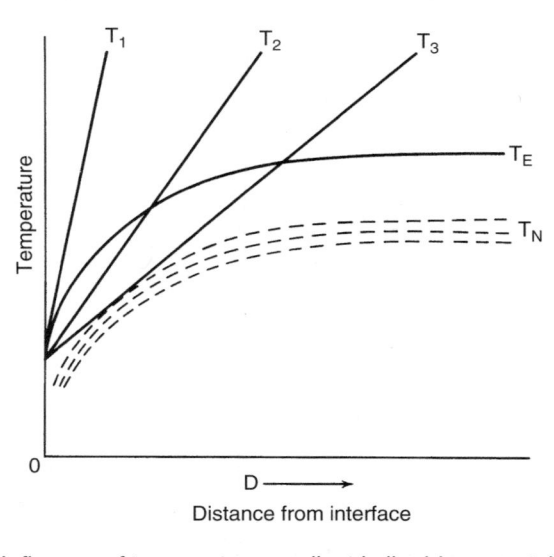

Figure 2.11 Influence of temperature gradient in liquid on crystallization. $T_E =$ equilibrium freezing temperature, $T_N =$ nucleation temperature, depending on nature of heterogeneous nuclei, T_1, T_2, and T_3 = temperature gradients producing increasing undercooling and associated changes in morphology: planar → cellular → dendritic → independently nucleated

solidification produces equiaxed grains, such as are frequently present in the central zone of both castings and ingots. The degree of undercooling at which this will occur will depend upon the potency of the nuclei present and will later be shown to be affected by other conditions. Figure 2.11 summarizes the effect of the increased undercooling produced by varying the temperature gradient, upon the mode of growth.

In sand casting, although the tendency is towards an extremely steep temperature gradient in the mould, the temperature gradient in the casting is comparatively shallow: the entire structure is consequently often equiaxed. This is especially the case in alloys of high thermal conductivity, in which a high degree of temperature uniformity is maintained during freezing.

To this purely thermal explanation of the equiaxed structure, which has so far implied that all such grains originate in heterogeneous nucleation *in situ*, must be added other important mechanisms to be examined in a subsequent section dealing with the structure of castings.

Eutectic freezing

Many casting alloys are either of approximately eutectic composition or contain appreciable amounts of eutectic constituent. Typical conditions for eutectic crystallization are represented by the binary equilibrium diagram in Figure 2.12, in which an alloy of composition E freezes at a single

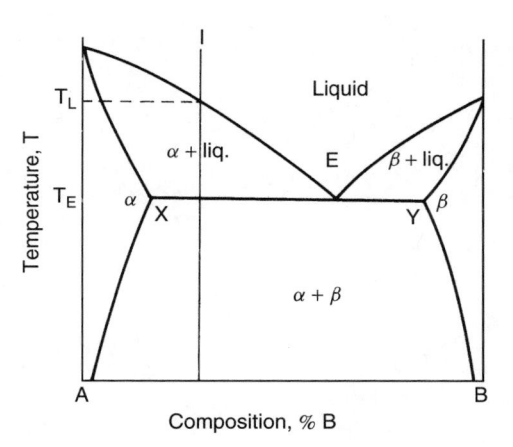

Figure 2.12 Thermal equilibrium diagram for two metals forming a eutectic

temperature to precipitate a mixture of two phases, α and β, of compositions X and Y.

Unlike the dendritic growth of grains of solid solution, each eutectic 'grain' is formed by the simultaneous growth of two or more separate phases in close association. The resulting structure may consist of alternate lamellae of the two phases, or may feature rod-like or globular elements of one of the phases in a matrix of the second.

The growth of a lamellar eutectic commences with the nucleation of one of its phases. The solute concentration in the liquid increases until nucleation of the second phase occurs at the interface, after which the growth of both phases proceeds with an orientation relationship between the two. It was originally postulated by Tiller[22] that additional lamellae could be produced either by fresh nucleation or by a process of bridging.

Growth of the eutectic grain as a whole takes place by the simultaneous edgewise extension of the plates of both phases. It was later appreciated that each eutectic grain in the final structure is the product of associated growth from a single centre to form a microscopically distinct unit. Such units may be equi-axed in form, analogous to the equi-axed grains in single phase systems (see Figure 2.13), although it is also possible to grow columnar eutectic structures by controlled unidirectional freezing: this special case will be further considered in a later section. As an example of radial growth from a single centre, consideration of the three-dimensional structure of the austenite–graphite eutectic in cast iron produced the schematic picture of a eutectic grain illustrated in Figure 2.14a[23]; an equivalent two-dimensional photomicrograph is also shown.

Major differences in the morphology of eutectics have been related to the concept of a 'coupled zone' below the eutectic temperature, within which the two phases can grow at equal velocities at a common interface,

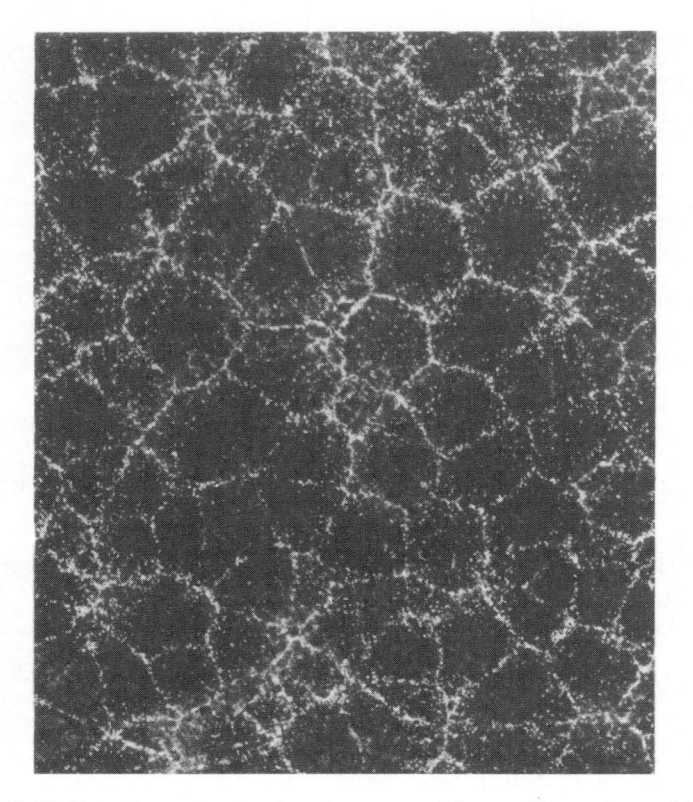

Figure 2.13 Eutectic grain structure in grey cast iron. ×7 (courtesy of I. C. H. Hughes)

producing regular structures as in lamellar eutectics. Growth outside the coupled zone occurs at different rates and produces irregular structures.

Lamellar or rod-like eutectic structures tend to be obtained when the volume fractions of the two phases are similar and when the eutectic composition lies roughly midway between those of the constituent phases, whilst the less regular structures are associated with the opposite conditions. In the latter case the second phase is subject to heterogeneous nucleation and there is no consistent orientation relationship between the two phases. This condition has been frequently described as *anomalous* in contrast with the 'normal' conditions described earlier.

Many eutectic phases previously thought to be discontinuous are now known to be continuous throughout a eutectic grain, notable examples being flake graphite and the silicon phase in the aluminium–silicon eutectic. The final sizes of such constituents are thus determined by the growth phenomena of branching and growing out rather than by repeated nucleation. Growth mechanisms and the relationships between the phases have been the subject of an extensive literature, detailed at the end of this section,

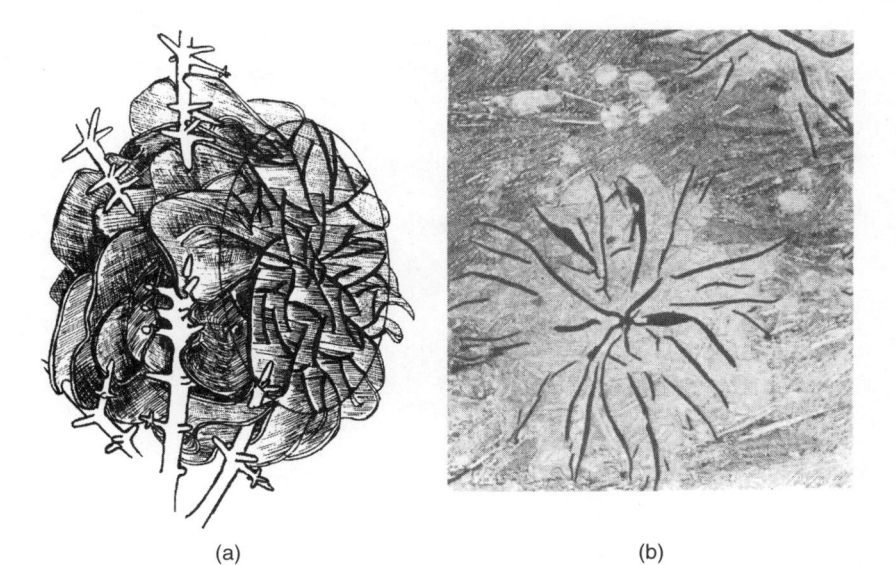

(a)   (b)

Figure 2.14 (a) Schematic picture of graphite structure in a eutectic grain (from Morrogh[23]) (courtesy of Institute of British Foundrymen), (b) microstructure illustrating radial nature of graphite flake growth. ×170 approx. (from Hultgren et al[24]) (courtesy of Iron and Steel Institute)

and specific examples encountered in important commercial eutectics will be further examined in relation to the structure of castings.

A general theory was developed by Hunt and Jackson[25] based on studies of transparent organic compound systems analogous to metals. The principal differences in mode of growth and structure were attributed to the entropies of fusion of the constituent phases[26]. Materials with high entropies of fusion, which include most non-metals, solidify with crystallographically faceted solid–liquid interfaces. Most metals, by contrast, possess low entropies of fusion and grow with non-crystallographic interfaces of planar, cellular, or dendritic form according to the local thermal conditions as previously described. Lamellar or rod-like eutectics are found to occur principally in systems where both phases have low entropies of fusion, as for example in the lead–tin eutectic, whilst irregular or complex-regular structures of the types encountered in the aluminium–silicon and magnesium–zinc systems are produced where one phase has a high and the other a low entropy of fusion. Where both phases have high entropies of fusion the structures are always irregular, with independent growth of the two sets of crystals.

Apart from the state of dispersion of the phases, impure eutectic grains are also distinguished by a colony* or cellular substructure analogous to that

---

* see footnote to p. 52

occurring in solid solutions and developing by transition from a planar to a cellular growth interface under conditions of constitutional undercooling. This type of cellular structure in eutectics is exemplified in Figure 2.15.

The morphology of eutectics and the alternative mechanisms of eutectic crystallization have been the subject of numerous reviews and specialized

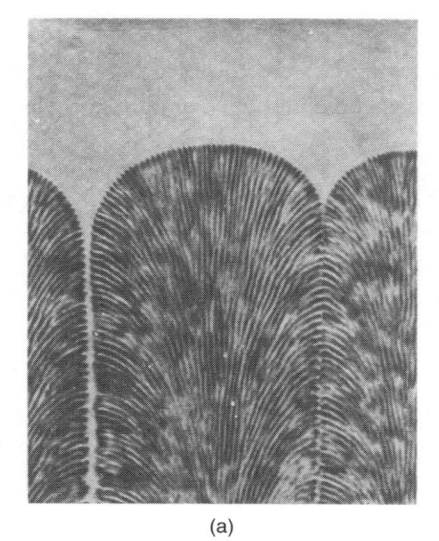

(a)

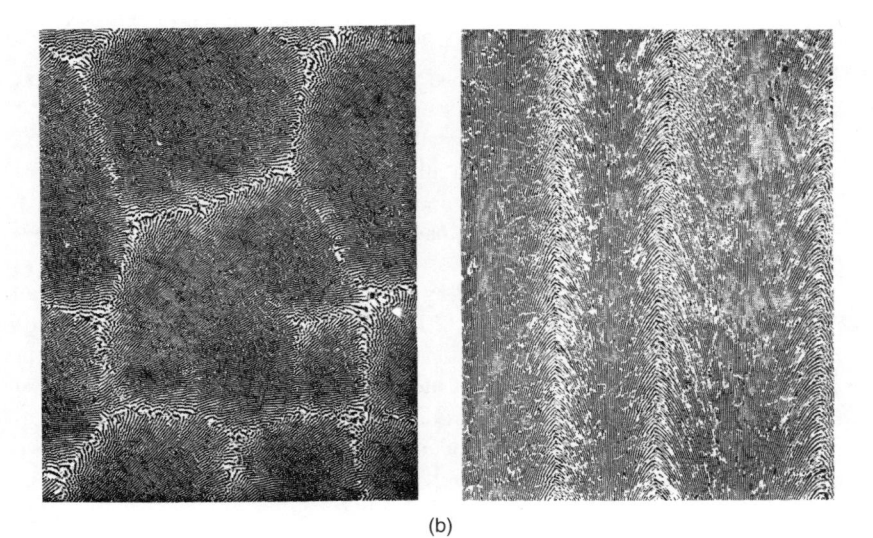

(b)

Figure 2.15 Cellular structures in impure eutectic grains. (a) Transparent organic system: camphor–succinonitrile. ×410 (from Hunt and Jackson[25]) (courtesy of Metallurgical Society of A.I.M.E.), (b) Al–Zn alloy. ×130 approx. (from Kraft[27]) (courtesy of Metallurgical Society of A.I.M.E., Journal of Metals)

works[27-33] and have also been extensively treated in more general publications on solidification and casting[7,9,10,34-40].

Peritectic reactions

Solidification involving peritectic reactions is represented in the portion of a binary equilibrium diagram shown in Figure 2.16. In this case the equilibrium freezing of alloys of compositions between X and P involves the precipitation of primary crystals of solid solution α and their subsequent reaction with residual liquid at temperature T_P to form a second solid solution β of composition Y. For alloy compositions to the right of point Y the whole of the α of composition X reacts with a proportion of the liquid of composition P to form β and freezing of the remaining liquid is completed at lower temperatures by the precipitation of additional β. In alloys to the left of point Y the whole of the liquid is consumed in reaction with part of the α and the structure on completion of freezing consists of two solid phases.

Peritectic reactions do not proceed readily to equilibrium since the initial reaction product forms a layer surrounding the primary solid phase; the subsequent reaction rate is controlled by diffusion through this layer and a compositional gradient persists in the final solid. Non-equilibrium residues of primary α thus tend to be retained even in alloys to the right of point Y. In experiments with silver–zinc alloys, Uhlmann and Chadwick[41] confirmed the presence of primary α dendrites together with the β phase in all alloys between the extreme points of the peritectic horizontal. The two phases under these conditions are precipitated over separate and successive ranges of temperature.

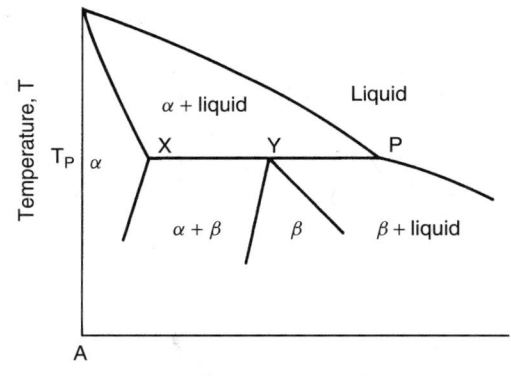

Figure 2.16 Section of thermal equilibrium diagram involving a peritectic reaction

The structure of castings

In the foregoing section the processes of nucleation and growth have been considered in relation to the principal alternative features of crystal structure in cast metals. So far this structure can be seen as an outcome of three major influences:

1. *Alloy constitution.* Metal composition governs the basic mode of crystallization and determines whether the structure will consist of single phase or eutectic grains or both. The alloy composition is also characterized by particular distribution and diffusion coefficients for solute in liquid and solid phases and thus establishes the relative tendency to constitutional undercooling.
2. *Thermal conditions.* The temperature distribution and rate of cooling in a casting are derived from the initial temperature conditions and the thermal properties of metal and mould.
3. *Inherent nucleation and growth conditions in the liquid.* The relative possibilities for nucleation and growth depend upon foreign particles or solutes present in the liquid, whether as trace impurities or as deliberate additions.

Since wide variations in thermal conditions can occur at various stages during the cooling of a single casting, its overall structure may consist of separate zones with widely different characteristics. The nature and extent of these zones depend partly upon factors yet to be discussed, but it will be useful at the present stage to summarize the effects of thermal conditions alone.

Considering firstly the alloys forming a continuous range of solid solutions, it has been shown that the mode of crystallization is governed by the interaction of temperature and compositional gradients in the liquid. Columnar growth from the mould wall towards the centre of the casting is favoured by the existence of a steep temperature gradient, since this increases the probability that renewed growth at the interface will occur before undercooling is sufficient to bring about further nucleation elsewhere (Figure 2.17a). Columnar growth is also promoted by slow cooling, which allows more time for solute transport in the liquid, diminishing the concentration gradient at the interface (Figure 2.17b). Slow cooling is, moreover, associated with an inherently low rate of nucleation relative to rate of growth in all processes of crystallization[42].

The influence of foundry variables upon structure must accordingly be determined by their effects upon the temperature gradient G and the rate of freezing R, and it follows that the ratio G/R is a significant parameter with respect both to mode of growth and to final grain structure in solid solution alloys. The effect of this ratio may now be summarized. Progressive change in the parameter from a high to a low value is accompanied

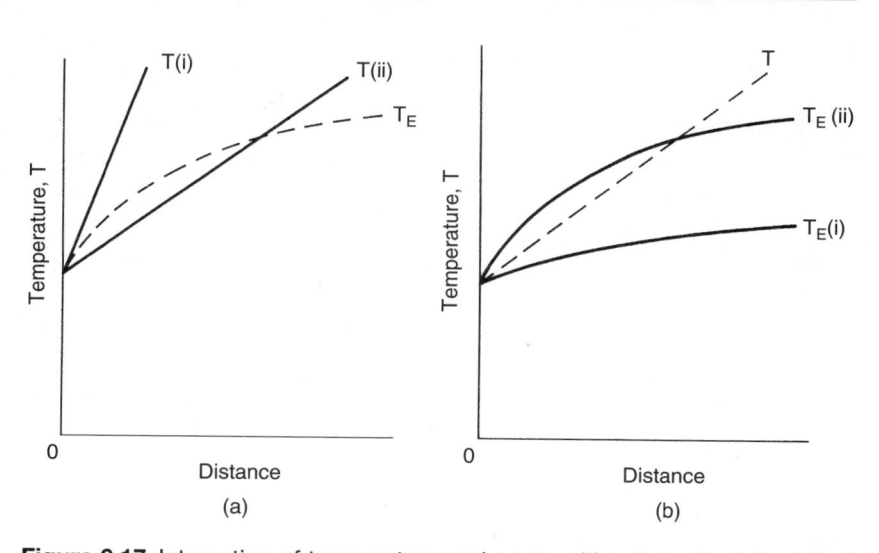

Figure 2.17 Interaction of temperature and compositional gradients in determining structure. (a) Influence of temperature gradient (T), (b) influence of liquidus temperature profile (T_E), (i) condition favouring plane front solidification), (ii) condition producing undercooling

by successive transitions in the mode of crystallization as the effect of undercooling becomes more pronounced. These successive stages are illustrated schematically in Figure 2.18 and the associated thermal conditions in Figure 2.19. With a very high ratio, columnar growth takes place with the advance of a plane interface, which first gives way to the cellular growth form previously described. Diminishing values of G/R bring about the cell → dendrite transition, with columnar growth now occurring on probes of solid some distance ahead of the main interface. At still lower values of G/R, heterogeneous nucleation in the undercooled zone brings about the growth of new grains in positions remote from the existing interface; these grains may themselves undergo dendritic growth, although there is some evidence[14] that the radial growth of independent grains begins with spherical morphology analogous to the plane interface condition in unidirectional freezing.

These successive transitions to cellular, to dendritic and thereafter to independently nucleated crystal growth, can be quantitatively represented in the manner of Figure 2.20, derived from experimental observations under controlled thermal conditions. From these types of relationship it can also be seen that alloys containing more solute become progressively more susceptible to growth forms associated with undercooling. The actual relationships between G, R and C_0 for the successive transitions are not identical. The planar → cellular transition is characterized by a clear linear relationship between G/R and C_0, as in Figure 2.20a, but evidence

Growth direction

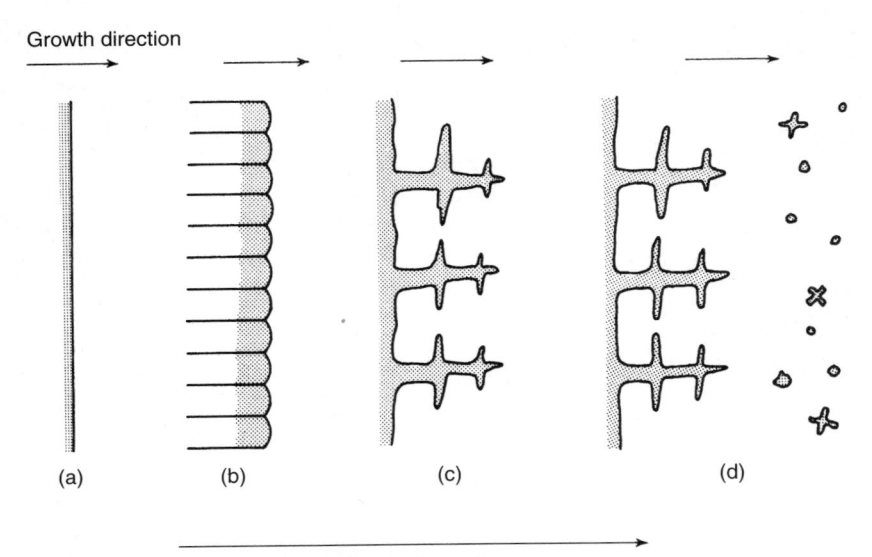

(a)   (b)   (c)   (d)

Increasing undercooling
decreasing value of G/R ratio

Figure 2.18 Influence of undercooling on interface morphology and mode of growth. (a) Planar interface, (b) cellular interface, (c) dendritic growth, (d) independent nucleation

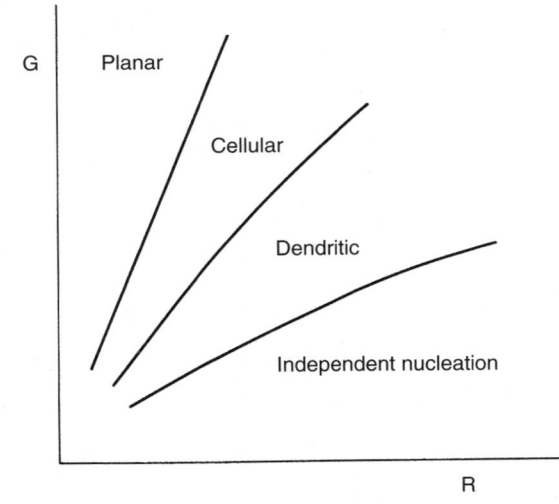

Figure 2.19 Influence of temperature gradient G and freezing rate R on solidification morphology of a given alloy

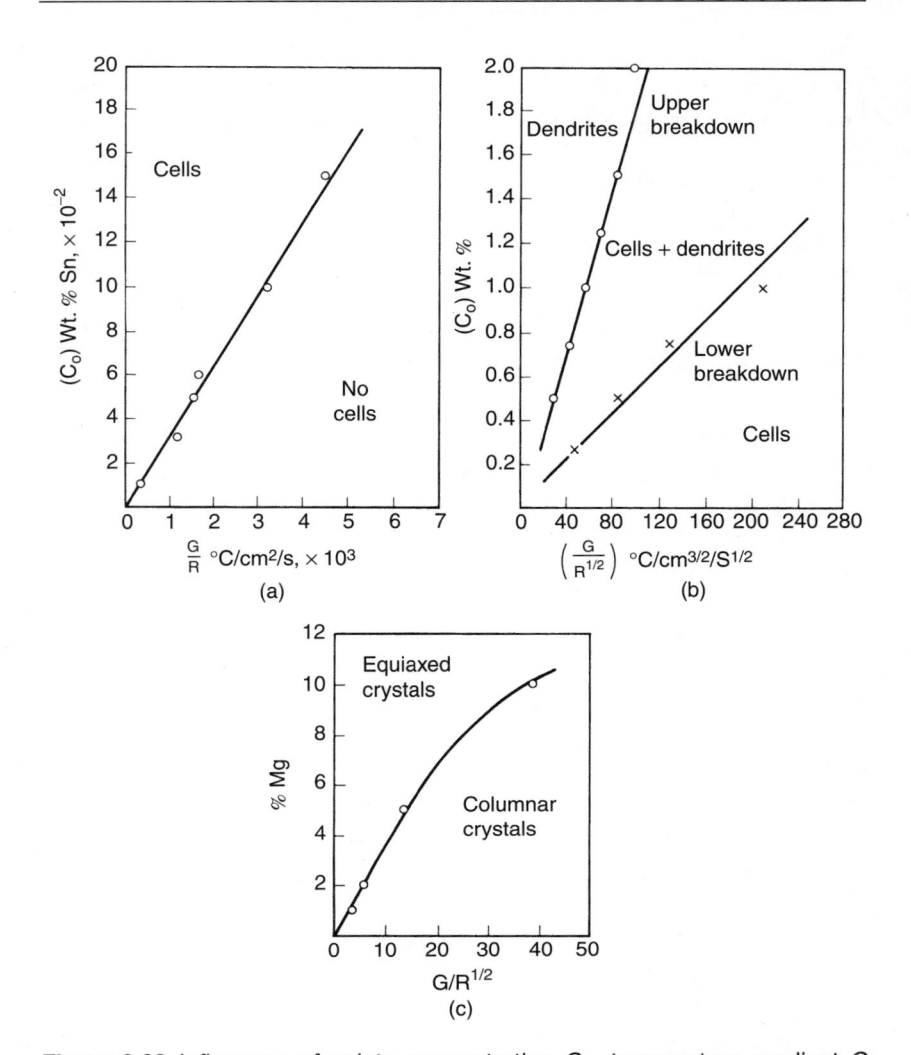

Figure 2.20 Influences of solute concentration C_0, temperature gradient G and freezing rate R on solidification phenomena in unidirectional freezing. (a) Formation of cellular interface, (b) cellular–dendritic transition in alloys of tin in lead (from Rutter[43] after Tiller and Rutter[44]), (c) onset of equiaxed growth in aluminium magnesium alloys (after Plaskett and Winegard[45]) (courtesy of American Society for Metals)

suggests[10] that there is no simple relationship between G/R and C_0 for either cellular → dendritic or columnar → equiaxed transition. Under most conditions, columnar and equiaxed dendritic structures predominate in production castings in this type of alloy system. Dendrite morphology is thus of great practical importance.

Because of the changing thermal conditions during freezing, therefore, the separate structural zones encountered in castings might be explained solely on the basis of critical changes in the G/R ratio. Initial solidification occurs under a marked temperature gradient, which frequently suffices to bring about columnar dendritic growth in the outermost zone. In the central zone, and in some cases throughout the casting, the temperature gradient is shallow and the zone of undercooling extensive, so that solidification proceeds by the widespread nucleation and growth of equiaxed grains. This explanation is portrayed in Figure 2.21. A typical duplex structure is shown in Figure 2.22. As will shortly be seen, however, other factors strongly influence this zonal transition and the local thermal conditions only provide a trend.

Although the foregoing discussion refers principally to alloys forming solid solutions, analogous changes occur in alloys subject to eutectic freezing. The colony or cellular growth structure within a eutectic grain develops when the G/R ratio falls below a value representing a critical

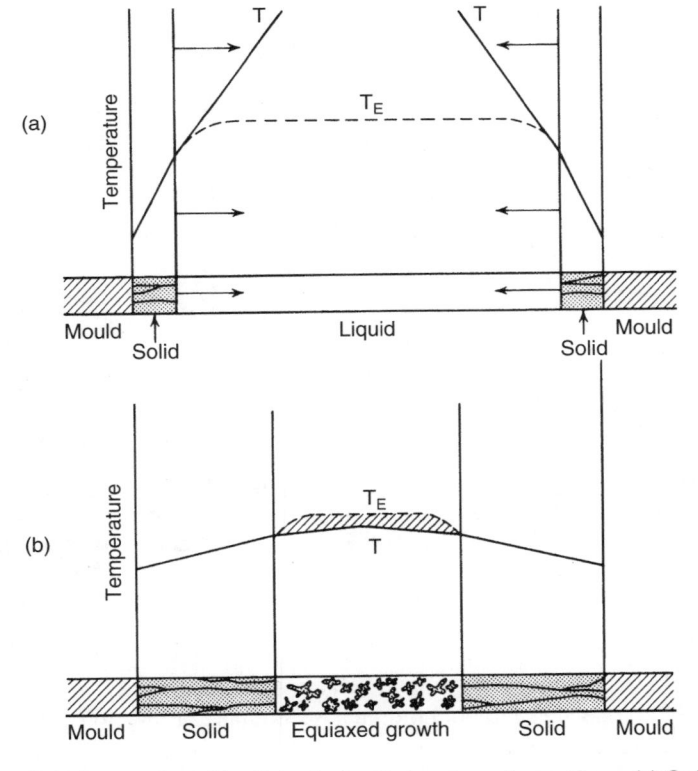

Figure 2.21 Thermal explanation of mixed structures in castings. (a) Columnar growth stage, (b) central equiaxed region

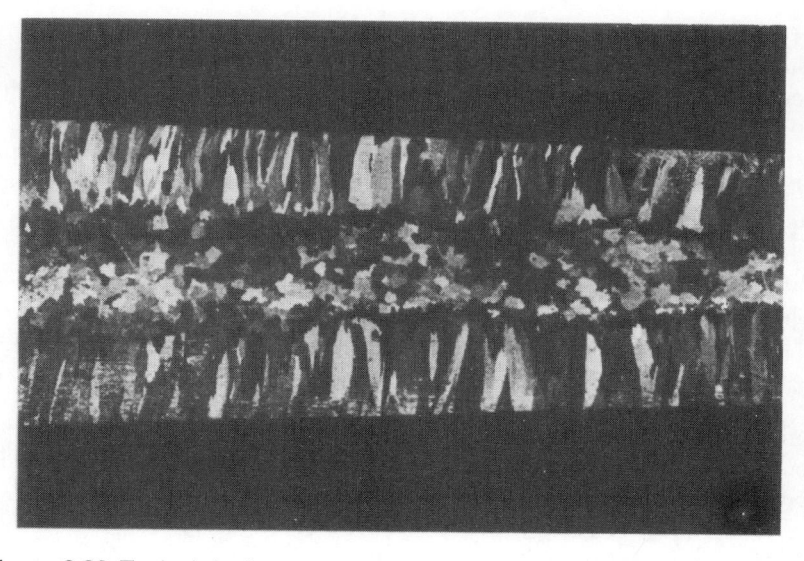

Figure 2.22 Typical duplex macrostructure in cast metals showing columnar and equiaxed zones. Actual size

degree of constitutional undercooling with respect to an impurity. With further undercooling the cellular gives way to irregular growth forms which could be regarded as analogous to the dendritic and independent nucleation stages in other types of alloy. The similarities between successive modes of freezing in the two types of alloy can be fully appreciated by reference to an examination by Hogan and Lakeland of solidification phenomena in cast iron[46].

Eutectic freezing commonly occurs as a second stage following initial dendritic crystallization of a primary phase. In an alloy of type I in Figure 2.12, the temperature range $T_L–T_E$ is marked by dendritic growth of crystals of α solid solution and the accumulation of component B in the residual liquid. On reaching temperature T_E the remaining liquid freezes to form eutectic grains in the manner already outlined. Freezing in such a case may be envisaged as the passage of successive waves of dendritic and eutectic crystallization from the outside to the centre of the casting. Hypo-eutectic grey cast iron offers the best example of this pattern of freezing.

Additional influences upon structure

The structural zones in castings have so far been explained largely in terms of thermal and constitutional influences upon nucleation and growth conditions at particular locations. Significant effects also arise, however, from crystal multiplication and from movement within the liquid of individual crystals or blocks of grains, away from the sites of their original nucleation. Such movement can result from turbulence originating during pouring, from mass feeding, from thermal convection, or from gravitational separation due to the difference in density between liquid and solid phases.

Direct evidence of dendrite fragmentation as an important mechanism in the formation of the equi-axed zone in castings was obtained by Jackson and Hunt and their collaborators in studies of solidification in transparent organic compounds analogous to metals[47]. Dendrite arms in the columnar zone become detached by local recalescence due to thermal fluctuations and changes in growth rate. These are carried by convective stirring and turbulence into the central region, where they grow independently in undercooled liquid. This phenomenon was confirmed by Tiller and O'Hara[48].

Fragmentation is not the only source of such centres for equi-axed growth. Contact with the cool mould surface on pouring initiates the nucleation of many crystallites, which become widely distributed by pouring turbulence. In castings poured with little superheat these continue to grow rapidly to form a wholly equiaxed structure: this has been termed 'big-bang' nucleation. With higher superheat the initial crystals may remelt completely or may survive in sufficient numbers to participate in forming the equi-axed zone. Crystallites can also become detached and fall through the melt from the upper metal–mould interface or from the exposed and chilled free surface of the liquid, a phenomenon long recognized in ingot solidification.

To summarize, therefore, three distinct phenomena can contribute to the formation of the equi-axed region, 1. heterogeneous nucleation *in situ*, 2. crystal multiplication and 3. transport of crystallites by gravity or by mass movement of liquid.

The significance and practical control of cast structure

The principal factors governing the final metallographic structure of a casting may now be listed. They are:

1. Constitution and thermal properties of the alloy;
2. Casting design and dimensions;
3. Thermal properties of the mould;
4. Superheat and final casting temperature;
5. Conditions for heterogeneous nucleation;
6. Conditions affecting motion during solidification;
7. Subsequent heat treatment.

Thus, although the structure of a casting is in the first instance a function of alloy composition and casting geometry, it is also sensitive to measures taken in founding. These include preliminary treatment of the liquid metal and variation of cooling rate within the mould. It would, however, be true to say that manipulation of the freezing process in castings has been more usually directed at the problems of feeding than at structure.

The importance of structure in cast alloys lies mainly in the structure-sensitive properties which can be utilized in engineering (q.v. Chapter 7). These properties are determined primarily by the influence of the microstructure on the behaviour of dislocations in the lattices of the individual crystals. Unlike wrought materials, in which further opportunities exist for changing both structure and dislocation density, the initial microstructure is frequently the main vehicle for the control of properties, although subsequent heat treatment plays this role in some cast alloys.

In examining the critical characteristics of cast structures, it will be convenient to consider two types of metallographic feature:

1. *Grain structure* (size, shape and orientation). The boundaries of primary grains formed during freezing are normally visible on macroscopic or microscopic examination, although their detection may be complicated by solid state transformations occurring on further cooling to shop temperature. In a structure consisting largely of eutectic grains, however, boundaries may be more difficult to distinguish unless highlighted by the segregation of impurities (see Figure 2.13). The determination of grain size in castings has been the subject of a special group of three papers dealing respectively with non-ferrous alloys, steels and cast irons[49–51].

2. *Substructure*. Important features include the distribution and state of division of alloy components and microconstituents, including solutes, second phase networks or particles, and the separate phases within the eutectic grain. Dendrite substructure too falls within this category.

The normal cast structure can be complicated by the presence of shrinkage cavities or by the macroscopic segregation of alloying elements, but these features will be examined more fully in Chapter 5.

Control of cast structure may take either of two directions. Most widely established are measures directed towards refinement of grain and constituent size, and special treatments for these purposes are a well established feature of the production of light alloys and cast irons. Some of these treatments produce more drastic changes in the microstructure, providing beneficial influences on the morphology of individual phases. The second form of structure control involves the manipulation of grain shape and orientation to provide anisotropy of structure and properties by design, a principle developed with great success in the field of high temperature superalloys.

Although structure control is primarily of interest for its effect upon properties of the alloy, this aspect cannot, in practice, be wholly divorced from the task of producing a sound casting. Thus refinement or columnar growth, for example, may be sought, not necessarily as a step towards superior mechanical or physical properties, but for the suppression of hot tears or for benefits to be derived in feeding. Both aspects must be considered when assessing the most appropriate structure. This partial interdependence of structure and soundness also means that the structure sensitivity of properties varies both with alloy and with solidification conditions. This will later be shown to be reflected in varying degrees of section sensitivity, and in some cases of anisotropy, in different cases.

Grain shape and orientation

Grain shape and orientation in castings are significant in relation both to material properties and to feeding. The size aspect of structure was the first characteristic to be assessed and subjected to controls, and at that stage the isotropy of cast material was tacitly implied. It remains true that in equiaxed regions, whilst each grain possesses its own crystallographic orientation, the statistical effect of a mass of grains produces isotropy of properties. In an extensive columnar zone this is not so: boundary alignment and common orientations can cause marked anisotropy.

The influence of directionality upon mechanical properties of single crystals and wrought materials is well documented, but information on cast metals had been mainly confined to straight comparisons between longitudinal properties of columnar structures and those of equi-axed or randomly structured material. The potential effect on the properties of polycrystalline cast metal is demonstrated in Figure 2.23, which illustrates the results of tensile tests taken at various angles to a columnar structure produced by unidirectional freezing of super-purity aluminium. This example of a pure metal emphasizes the need for the proper control of structure in relation to design stresses in castings, although in many commercial casting alloys directionality is much less marked, either because dendrite or eutectic

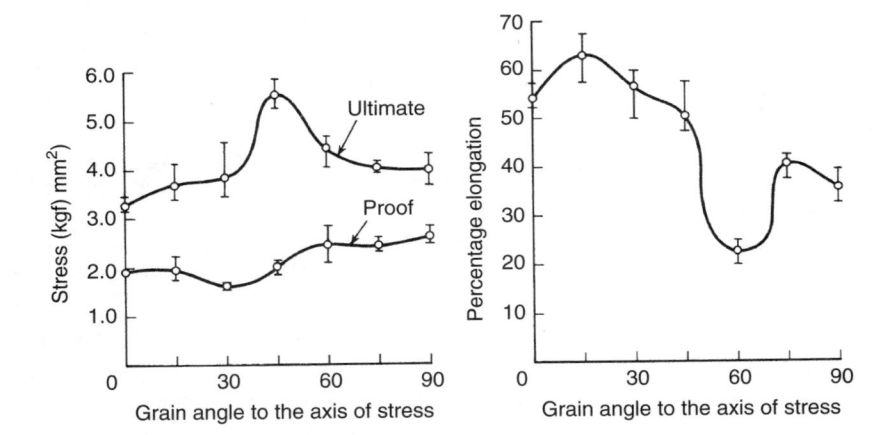

Figure 2.23 Example of influence of structural alignment on tensile properties of cast material. Super-purity aluminium (courtesy of Dr P. Thakur)

substructures dominate the properties or because solid state transformations modify the influence of the as-cast structure. In steel, for example, work by Flemings[52], using the principle of unidirectional freezing in the production of premium quality castings in low alloy steel, demonstrated how improvements resulting from refinement and increased soundness could be achieved without marked directionality. Thus the degree of anisotropy must depend both on alloy type and on structural characteristics. It can in some cases be exploited with powerful effect, as will later be shown.

In relation to feeding, columnar crystallization is often preferred for its elimination of the pasty mode of freezing and for its consequent association with greater freedom from discrete porosity. It was shown by Cibula[53] and by Ames and Kahn[54] that the columnar structure is beneficial to pressure tightness in bronzes and gunmetals and can be obtained by the use of relatively high pouring temperatures. Columnar structures having similarly been found to give superior soundness and properties in steel, the technique recommended by two groups of American workers was to employ a high casting temperature with rapid and directional freezing[55-57]. This combines a fine dendrite arm spacing with the exceptional freedom from microporosity which results from solidification under a steep temperature gradient. In certain other cases, however, the equiaxed structure is preferred for feeding: in aluminium alloys, for example, a suspension of fine equiaxed grains flows much more readily during the mass feeding stage[58].

The significance of the temperature gradient G and the rate of freezing R in determining the mode of crystallization has been previously discussed. Whilst a high G/R ratio has been stressed as a prerequisite to columnar growth, G and R both react in the same sense to changes in thermal conditions, so that the influence of such variables as casting temperature and mould properties cannot be automatically foreseen. Freezing under a maximum temperature gradient is associated with a particular degree of superheat, that at which solidification begins without delay at the mould wall. Excessive superheat tends, by mould preheating, to produce shallower temperature gradients during freezing: this might, independently of other factors, be expected to promote an equiaxed structure. However, a high pouring temperature is also conducive to slow cooling, which reduces the nucleation rate and minimizes the concentration gradient in the liquid; there is also an opportunity for the remelting of crystallites transported from the surface region. In practice, these effects usually predominate and columnar structures are favoured by high pouring temperature, as in the specific examples cited above.

Equiaxed structures, on the other hand, are encouraged by low pouring temperatures, which not only increase the nucleation rate but amplify the role of initial pouring turbulence in distributing crystallites formed at the earliest stage. Suppression of the columnar zone is also one of the first functions of inoculants added for grain refinement.

Even castings showing extensive columnar regions may exhibit a narrow zone of fine equiaxed grains immediately adjoining the mould surface. The instantaneous contact with the cold surface, especially in the case of a chill or metal mould, produces extreme undercooling and the formation of many crystals. Those most favourably oriented for growth may then develop preferentially to form a columnar zone as the rate of cooling diminishes. This separate zone of chill crystals is not always evident; the crystals may remelt due to heat flow from the mass of the casting, or may be transported to the interior to form potential centres for equi-axed growth. This mechanism has already been mentioned as being operative in the formation of wholly equi-axed structures in castings poured with little superheat.

Directional crystallization for development of special properties. The development of columnar or of mixed macrostructures in castings is in many cases a fortuitous result of casting geometry and of foundry conditions, especially moulding materials and casting methods. Longitudinally aligned structures can, however, be produced by techniques of controlled directional solidification within the mould. Heat flow in a single direction is maintained by casting against a water cooled chill surface forming one end of the mould cavity. Solidification from the remaining mould surfaces or within the main body of liquid ahead of the interface is suppressed by maintaining a steep temperature gradient throughout the freezing process.

Early work on metallurgical aspects of unidirectional solidification was carried out by Northcott[59], and advantages of columnar structures as perceived by Flemings[57] and others have already been mentioned. The principle came to be adopted for the production of anisotropic cast magnets, the all-columnar structures being obtained by selective chilling in combination with exothermic and insulating mould wall materials.

Fuller control of the directional solidification process came with the introduction of specially designed furnaces of the type shown schematically in Figure 2.24, in which a refractory mould unit is withdrawn progressively from a surrounding heat source. Nucleation occurs on the chill surface and columnar grains grow vertically, at a rate determined by the motor driven withdrawal system, until the cast component is fully solidified. The long axis of the columnar grains is normally associated with a particular crystallographic direction: together with the absence of transverse grain boundaries this can provide anisotropic characteristics favourable to the pattern of stresses in the component. This applies particularly to high temperature creep and stress rupture properties and to oxidation resistance. Conditions are intrinsically favourable to feeding, so that microporosity is suppressed, with further benefit to mechanical properties. Vacuum melting and processing are also usual in DS casting.

The significance of aligned structures of this type in relation to gas turbine development are more fully discussed in Chapter 9 in the context of the investment casting process, including the alloys used in gas turbine

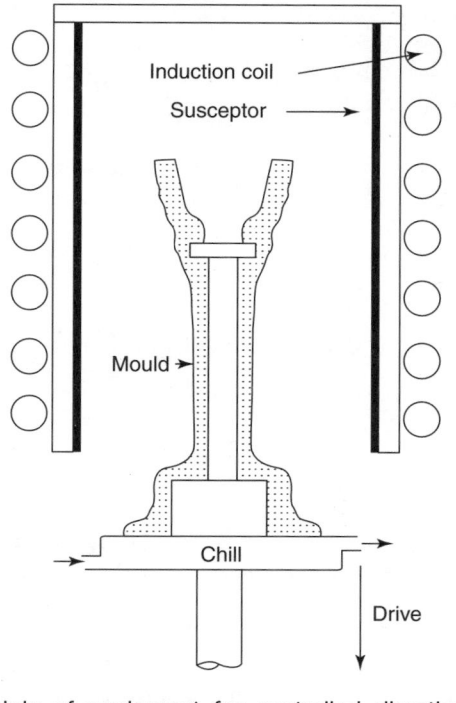

Figure 2.24 Principle of equipment for controlled directional solidification, employing electrical heating, water cooling and withdrawal of mould unit from hot zone

blading; a typical all-columnar structure is shown in Figure 9.26. Early developments in the directional solidification of high temperature superalloys were treated in References 60 and 61 and were fully reviewed in Reference 62. The subject was further examined by Versnyder[63] and in a comprehensive specialized work by McLean[64].

A similar directional solidification technique is employed for the production of single crystal castings, using specially designed moulds embodying constrictions and changes in the local growth direction. A polycrystalline columnar zone is formed on initial contact with the chill surface, and coarsening occurs by selective elimination of grains less favourably oriented for growth within the constricted passage, until only a single crystal survives at the point of entry to the casting itself. A notable example of this type of system can be seen in Figure 9.27.

The rationale for the use of DS techniques in the production of high temperative alloys can be readily appreciated from the comparison of stress rupture properties of conventional, DS columnar and DS single crystal materials as shown in Figure 2.25.

A further possibility offered by controlled directional solidification is the production of aligned structures of dissimilar phases, with the general

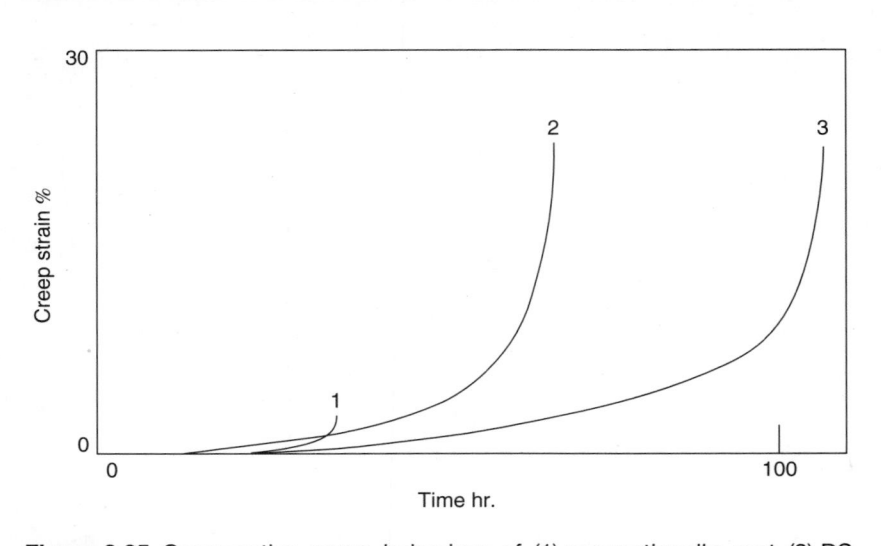

Figure 2.25 Comparative creep behaviour of (1) conventionally cast (2) DS columnar and (3) single crystal forms of nickel base alloy Mar M200 at 980°C approx. and 207 MPa (*after Versnyder and Shank*[62])

characteristics of composite materials. These are necessarily limited to alloy systems offering combinations of appropriate properties, for example for the reinforcement of a soft, ductile matrix by strong but relatively brittle filaments. An example of a eutectic structure of this type is shown in Figure 2.26. Gas turbine blades with aligned eutectic composites in Ni–Ta–C and Co–Ta–C alloys have been produced[65] and demonstrated some suitability, although subject to degradation at very high temperatures.

Grain size

The primary grain size is one of the most significant features emerging during solidification. Apart from its intrinsic effect, grain size influences the distribution of solute and of second phase constituents, whether produced during freezing, further cooling or subsequent heat treatment.

   Direct effects of grain size are expressed in well known relationships arising from the function of grain boundaries as obstacles to dislocation movement. These relationships, originally established by Hall and Petch, are between grain size and yield, flow and fracture stresses, taking the typical form:

$$\sigma_f = \sigma_o + kd^{-1/2}$$

where d = grain diameter,
       σ_o = frictional stress,
       k = constant

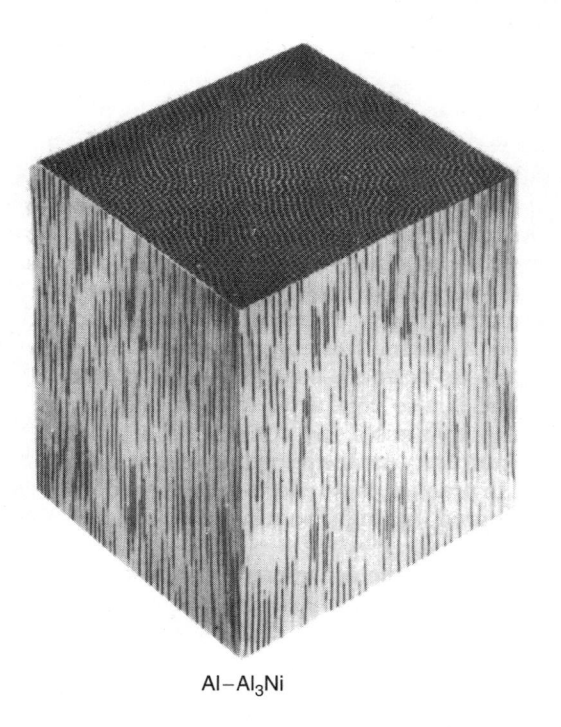

Al–Al₃Ni

Figure 2.26 Aligned eutectic structure in aluminium–nickel system, produced by unidirectional solidification. Al–Al₃Ni (courtesy of United Aircraft Research Laboratories, Connecticut)

This strengthening influence of grain boundaries, determined on wrought and recrystallized material, is also present in cast alloys, but the most important effects of grain size in castings arise from the influence of impurities and alloy phases in grain boundaries, particularly upon ductility. Refinement is commonly found to improve ductility by dissipating the embrittling influence of grain boundary constituents, including dispersed phases and non-metallic inclusions. In coarse grained material these tend to form a semi-continuous network providing a path for crack propagation; refinement produces a large increase in total grain boundary area, promoting better dispersion and a more controlled strengthening effect. (Brittle constituents affect tensile strength and ductility rather than yield stress since the latter property is essentially determined by the behaviour of the matrix.)

This effect is paralleled in the case of microporosity, also less serious when widely dispersed; the actual amounts of porosity may also be reduced through more effective mass feeding. The sensitivity of properties to porosity will be further examined in Chapter 5. In a typical investigation of the influence of structure on the properties of an 85/5/5/5 leaded gunmetal, Kura and Eastwood[66] found the critical factors to be the amount and distribution of microporosity, lead particles and local shrinkage, all influenced by the grain size of the primary phase. The direct and secondary influences of grain size upon the structure, hydraulic soundness and tensile properties of bronzes and gunmetals were also demonstrated by Cibula[53,67].

Although these comments are mainly devoted to primary grain size, it will be shown that dendrite substructure is in some cases the more significant characteristic: in alloys with marked dendritic features it is the elements of the substructure which determine the periodic spacing of composition and of second phases. Grain size in eutectics will be referred to elsewhere.

Other benefits of refinement include reduced susceptibility to hot tearing and to hot cracking in welding, improvements associated with the greater number of sites for deformation and the reduced degree of strain concentration in the solid–liquid brittle temperature range.

Yet another advantage of fine grained as-cast structures is the reduced scatter of properties obtained. Coupled with the greater uniformity is a reduction in solution or homogenization heat treatment times due to the smaller diffusion distances involved. This is the principal reason why shorter times are associated with castings of small mass.

To summarize, therefore, fine grain size is usually beneficial both to mechanical and to foundry properties. Since grain boundaries represent a source of weakness at high temperatures, however, the converse is the case for high temperature creep resistance.

Sub-grain characteristics

Within the individual grains, both dendrite and eutectic substructure are significant and often outweigh even grain size in importance. In dendritic structures, dendrite cell size and secondary arm spacing frequently determine the form and distribution of second phase constituents and the pitch of the compositional fluctuations due to microsegregation, so exercising the function fulfilled in other cases by primary grain size. This is illustrated in Figures 2.27 and 5.33 and the typical effect upon mechanical properties is shown in Figure 2.28.

The importance of the morphology of the phases within the eutectic grain to the properties of a cast alloy is exemplified by the effects of graphite dispersion in cast iron and especially by the contrast between spheroidal and flake forms. With the change to the spheroidal form the limitation on mechanical properties is shifted from the graphite to the matrix, giving both ductility and increased strength. These considerations do not just apply to mechanical properties: in grey cast iron, for example, the damping capacity is found to increase both with cooling rate and with the carbon equivalent of the iron[68].

The refinement and modification of cast structures

Three factors have been shown to be significant in determining the metallographic structure of castings, viz. nucleation, growth behaviour and crystal

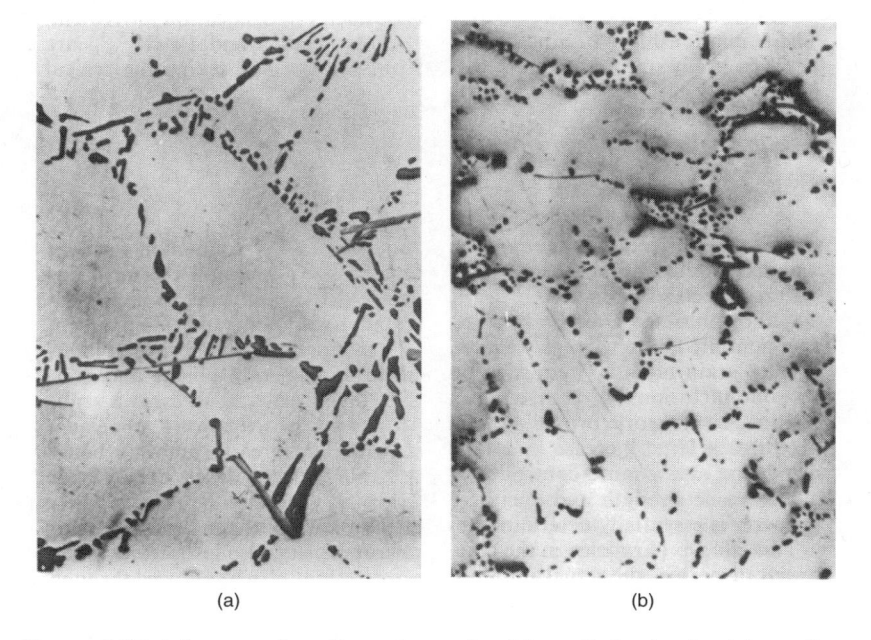

(a)   (b)

Figure 2.27 Influence of cooling rate on dendrite cell size in aluminium alloy, showing effect of primary structure on dispersion of secondary constituents. (a) $2\frac{1}{2}$ in plate, (b) $\frac{3}{8}$ in plate. LM8. $\times 120$ (courtesy of N. Aras)

multiplication. The most common aim in structure control is the pursuit of refinement through one or some of these possibilities. The first effect of most refining treatments is the reduction or elimination of the columnar zone, which is followed by a progressive increase in the number of equiaxed grains. Treatments may also be expected to influence growth substructures by their effects on the degree of undercooling: freezing rate, for example, is important in both respects. Grain refining treatments have progressed to an advanced level in the field of aluminium and magnesium alloy castings, where marked improvements are obtained in feeding and in tear resistance as well as in mechanical properties. In grey cast iron, inoculation is regularly employed to improve the graphite dispersion by increasing the nucleation of eutectic grains, whilst more radical treatments are exemplified in the production of spheroidal graphite cast iron and in the modification of aluminium–silicon alloys, the object being to optimize the size and shape of the individual phases.

Practical measures for size control and modification can be grouped as follows:

1. Variation of cooling rate;
2. Chemical treatment of the liquid metal;
3. Manipulation of melt superheat;
4. Agitation during freezing.

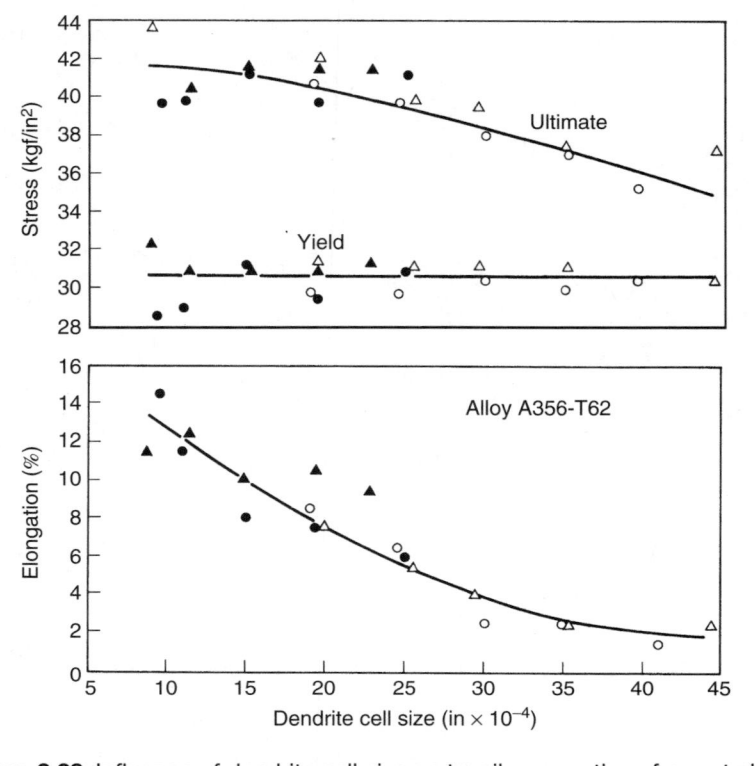

Figure 2.28 Influence of dendrite cell size on tensile properties of a cast alloy (after Spear and Gardner[12]) (courtesy of American Foundrymen's Society)

Cooling rate. The association of rapid cooling with fine grain size arises from the influence of undercooling on the comparative rates of nucleation and growth. The undercooling of a melt to a lower temperature increases the number of effective nuclei relative to the growth rate, the latter being restricted by the rate at which the latent heat of crystallization can be dissipated. Slow cooling, conversely, favours growth from few nuclei and produces coarse grain structures. This direct influence of undercooling on the freezing process was illustrated by Tammann in the manner shown in Figure 2.29.

The refining effect of increased cooling rate applies both to primary grain size and to substructure but in the latter case the effect is upon the growth processes rather than nucleation. The influence upon secondary arm spacing and dendrite cell size, essentially the same parameter, is typified in Figures 2.30 and 2.31 (the local solidification time plotted in the latter case is the liquidus–solidus interval and is inversely proportional to the cooling rate). Researches over a wide range of cooling rates[15,69] have shown that the arm spacing conforms to the relation $dr^a = c$, where d is the spacing, r the cooling rate and a and c are constants; the value of a in the cases quoted

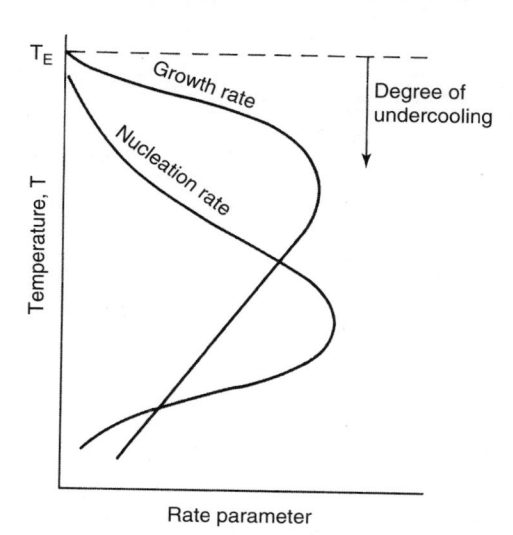

Figure 2.29 Relative rates of nucleation and growth as functions of under-cooling (after Tamman[42])

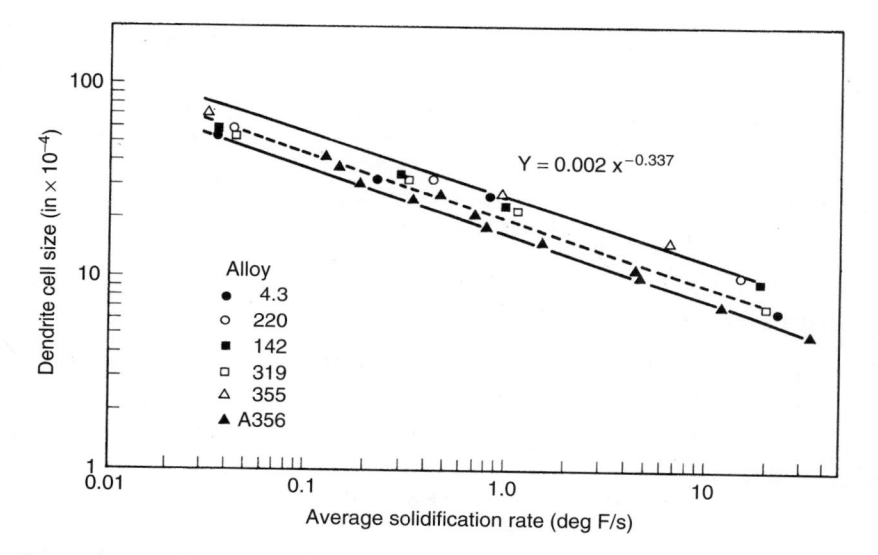

$$Y = 0.002 \, x^{-0.337}$$

Alloy
* 4.3
○ 220
■ 142
□ 319
△ 355
▲ A356

Figure 2.30 Influence of freezing rate on dendrite substructure (from Spear and Gardner[12]) (courtesy of American Foundrymen's Society)

was 0.32–0.33, although values approaching 0.4 have been obtained in other work. It will be noted from Figure 2.31 that the relationship remained valid over many orders of magnitude, covering the whole range commonly met in commercial castings but also including exceptional values such as those obtained in experimental splat cooled material.

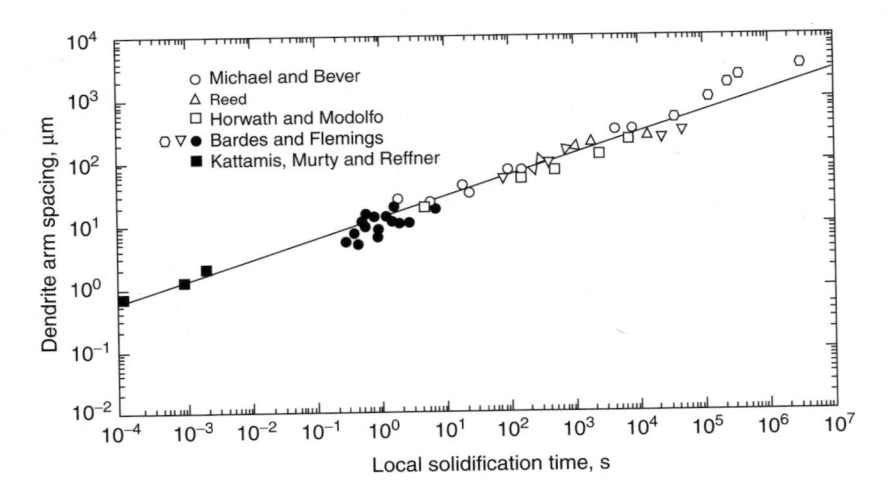

Figure 2.31 Relation between dendrite arm spacing and local solidification time for Al-4.5% Cu alloy (from Flemings *et al*[15]) (courtesy of American Foundrymen's Society)

In the case of eutectics the lamellar spacing is associated with the local rate of freezing in accordance with the relationship $\lambda r^n = $ constant, where λ is the spacing, r the rate of freezing and $n \approx 0.5$: the relationship was derived by Tiller[5] and has since been confirmed in most experimental work. The lower value of the exponent a obtained for the dendritic structure is considered to arise from the more complex mechanisms, particularly the ripening, involved in this case, an aspect pursued in detail by Flemings *et al*[15].

The combined effects of increased cooling rate upon both grain size and substructure in cast iron can be summarized in the following diagram, after Morrogh[23].

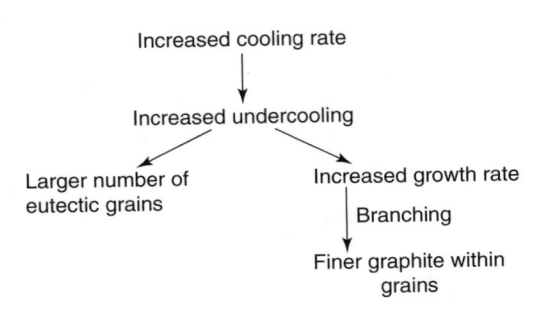

The graphite flakes thus become progressively finer with more rapid cooling, and with pronounced undercooling assume the characteristic finely divided, curled form with a ferritic matrix shown in Figure 2.32c and commonly referred to as Type D: this structure gives inferior mechanical properties and wear resistance. With still faster cooling and depending on the balance of composition, graphite precipitation gives way to the metastable Fe_3C constituent and the eventual production of wholly white cast iron structures.

The cooling rate in castings is to a large extent governed by the design and thermal properties of the casting itself. Cooling rate is, however, affected by variation in the mould material, the highest heat diffusivity being

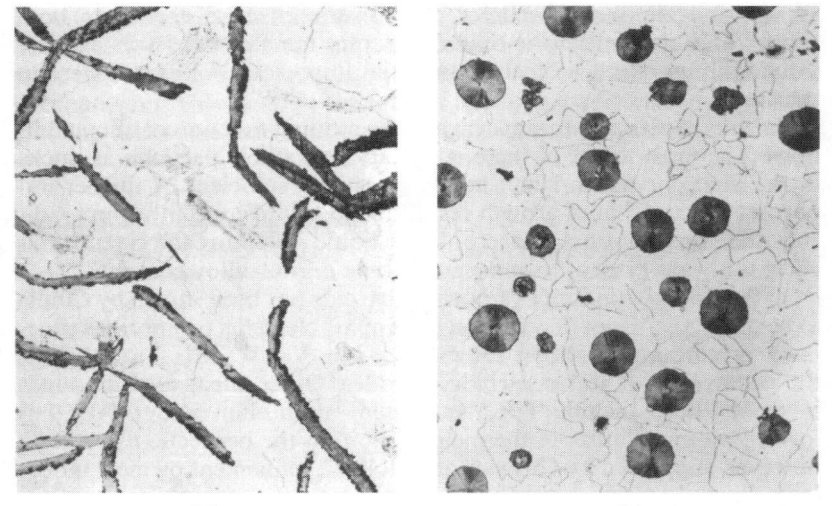

(a)   (b)

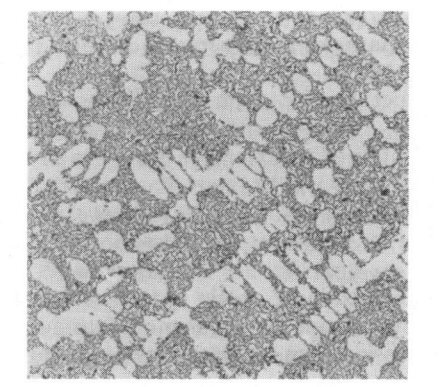

(c)

Figure 2.32 Flake and spheroidal graphite forms in cast iron. (a) Normal flake. ×200, (b) spheroidal. ×120 approx. (*a* and *b* courtesy of Dr I. Codd), (c) undercooled flake. ×70 approx. (courtesy of BCIRA and I. C. H. Hughes)

obtained in practice with the metal moulds used in gravity die casting. This influence of cooling rate is mainly responsible for the superior properties often associated with chill castings, thin sections and surface zones. The extensive use of chills both to refine the microstructure and to assist feeding by steepening the temperature gradients, is an important feature of the production of premium castings, with their high and guaranteed levels of mechanical properties[52,70].

The influence of very high cooling rates in refining the structure of cast materials has been further pursued using techniques such as the splat cooling of molten droplets projected against a chill surface. It was established that the relationship between dendrite arm spacing and freezing rate as portrayed in Figure 2.30 is maintained to cooling rates several orders of magnitude higher than those encountered in normal casting, but also that radically different types of structure can be obtained. Some of the developments in this greatly expanded field of work are reviewed in Chapter 10.

Amongst normal casting variables the most useful influence upon cooling rate is that of pouring temperature, an increase in which diminishes the freezing rate by preheating the mould, so reducing the rate of heat transfer during crystallization. Maximum undercooling for grain refinement thus requires a low casting temperature.

However, control of cooling rate through casting temperature and mould properties offers little scope for the decisive manipulation of structure in normal casting processes, since narrow limits are imposed by other foundry requirements such as fluidity and feeding. More specific techniques must therefore be sought.

Chemical treatment. The decrease in grain size brought about by increasing normal alloy content is comparatively small. However, highly effective grain refinement can be accomplished by inoculation – the addition to the melt of small amounts of substances designed to promote nucleation – although in certain cases the function may be to modify the growth rather than the nucleation process. Such small additions exercise a disproportionately large refining effect.

The basic requirement for heterogeneous nucleation lies in the ability of the liquid metal to wet the foreign particle. The factors determining the effectiveness of an inoculant are not, however, wholly understood: some structural affinity such as similarity of symmetry and lattice parameter is one criterion, hence the efficacy of a seed crystal in initiating solidification in an undercooled melt.

The nucleant must be capable of survival in superheated liquid and must be in a sufficiently fine state of division to remain as a widely dispersed suspension: Nelson[71] referred to the grain coarsening effect produced by the gravity segregation of nuclei in magnesium alloys.

Extensive investigations of the identity and effectiveness of various substances as nucleants in non-ferrous alloys were reported by Cibula[72–74],

Emley[75] and Kondic[76] and were the subject of a comprehensive review by Hughes[77]. Effective grain refinement in aluminium alloys is obtained from small additions of titanium and titanium–boron alloys, whilst for magnesium alloys carbon and zirconium, and for copper, iron are effective agents. In both copper and aluminium the oxides of the parent metals themselves exercise some refining effect, indicating the potential importance of atmosphere to structure.

Various explanations are advanced for the refining function of alloy additions. The most direct of these is the formation of stable particles as nuclei in the melt, for which small additions are usually sufficient. A further important effect is that of growth restriction due to solute concentration gradients and constitutional undercooling in the liquid adjoining the crystal: this effect is always present to some extent from the normal alloy content.

In the case of titanium as a nucleant, the carbide was shown by Cibula to be one active agent; a common feature of many elements promoting refinement in aluminium alloys, for example Zr, Nb, Va, W, Mo, Ta and B, is the occurrence of stable carbides and nitrides. Other effective compounds in these alloys include TiB_2. In magnesium–aluminium alloys, Al_4C_3 is formed on inoculation of the melt with carbon. In the peritectic hypothesis, however, supported by Crossley and Mondolfo[78], refinement by most inoculants in aluminium was attributed to peritectic reactions common to their respective systems with the parent metal; such a reaction also occurs in the magnesium–zirconium system, notable for its very fine cast structures. There is much evidence to suggest that both mechanisms are operative in certain circumstances.

Work by Turnbull et al on the cast structure of carbon steels[79] showed grain refinement to be effected by titanium, niobium, and titanium carbide additions; the paper reporting this work contained an extensive bibliography on many aspects of grain refinement. Other work on steel included investigations by Church et al[80] and by Hall and Jackson[81]. The former again demonstrated the marked influence of titanium; amongst the effects observed was a significant decrease in dendrite arm spacing. In the latter work, investigations into the grain refinement of austenitic steels indicated that nitrogen plays an important role; calcium cyanamide as well as more familiar inoculants proved effective, although it was in all cases necessary to avoid excessive superheat.

In cast iron the purpose of inoculation is to promote increased nucleation of eutectic grains for a given rate of cooling. This actually lessens the degree of undercooling, producing a slower growth rate and reduced branching of the graphite flakes, but the overall effect is to generate a uniform and random distribution of flake graphite, that commonly referred to as Type A. This can be accomplished by the addition of ferrosilicon or calcium silicide at the pouring stage. The graphite structure is, however, influenced by many other elements. Sulphur, for example, exercises a powerful effect in nucleating more eutectic grains and coarsening the graphite flakes; the influence of sulphur is neutralized in the presence of titanium, which acts as a scavenging element and thus as a graphite flake refiner[82].

The most drastic type of structural change, however, occurs with the modification of the eutectic structures in both cast iron and aluminium–silicon systems. In the former case the spheroidal graphite or nodular structure, typified in Figure 2.32b, is formed by the addition of certain elements, notably magnesium and cerium, which suppress the nucleation of flake graphite eutectic grains and bring about a condition of marked undercooling; heterogeneous nucleation of graphite then occurs on a copious scale and the characteristic ductile spheroidal structure is developed. Each spherulite forms the centre of an individual eutectic grain or cell, in which it becomes encased in a layer of austenite. The flake and nodular morphologies represent the extremes of a series in which intermediate structures, including compacted flake and fibrous or vermicular forms, can also occur.

The various graphite forms are associated with particular growth conditions in the melt. The main operating factors are the cooling rate, determining the degree of undercooling, and the presence of impurities which influence crystallographic aspects of the growth process. Graphite crystallizes in the hexagonal system, in which the basic unit is characterized by basal plane and prism faces as shown in Figure 2.33, which also indicates the potential a and c growth directions. Growth normally occurs most readily in the a direction, and the typical flake structure develops by the edgewise extension of the basal plane. This is encouraged by preferential absorption of sulphur and other surface active impurities on to the prism faces, reducing the interfacial energy with the melt. Curvature of the graphite flakes is attributed to faults or mismatch of lattice layers and local branching.

The introduction of magnesium or cerium removes the impurities, inhibiting the associated pattern of growth on the prism faces by increasing the interfacial energy. Growth occurs at higher undercoolings in the c direction and nodules are formed, in which the basal planes are oriented in a broadly circumferential manner. The complex processes involved in the development of spheroidal graphite and intermediate forms were examined and illustrated in detail by Lux, Minkoff, Hellawell and other authors in

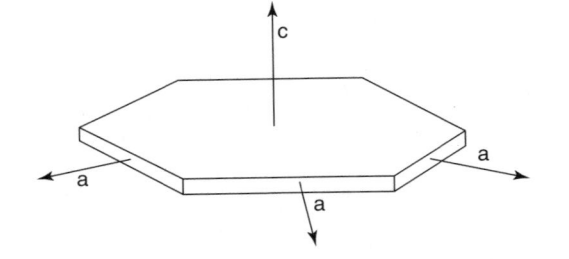

Figure 2.33 Basic unit of graphite crystal showing principal growth directions a and c, from prism faces and basal plane.

Reference 31. The many factors influencing graphite nucleation and the nature and role of inoculants in ductile ions have also been the subject of a substantial review by Harding et al[83].

Formation of the austenite layer surrounding the graphite spherulite in each eutectic cell follows local carbon depletion of the adjacent melt, and further graphite growth occurs either by the nucleation of new nodules or by diffusion of carbon though the austenite layer[84].

The available range of properties of the ductile nodular irons was further enhanced by the development of the ADI group of materials, based on the application of austempering heat treatments to the castings.

The major role of chemical treatments of the melt is similarly seen in the modification of the aluminium–silicon eutectic by small additions of sodium, strontium or other elements: the structure undergoes changes which offer some parallels with those occurring in the cast irons. The faceted plate morphology of the silicon in the unmodified eutectic, an inherently brittle distribution of the phase, can in this case be modified by undercooling through chilling alone, but chemical modifiers are required to accomplish the change at the cooling rates encountered under most foundry conditions. The resulting structure, once assumed to be discontinuous, is characterized by fibrous silicon, produced by continuous growth and branching from a single nucleus, in the aluminium matrix (Figure 2.34). The crystallographic basis for this was summarized by Elliott in Reference 85.

The modified structure provides superior mechanical properties, as indicated by the values shown in Table 2.2.

Although most chemical treatments involve additions to the melt, either within the furnace or during pouring, mould washes can also be used as a medium for the introduction of nucleants; this technique has been used to refine small investment castings. Control of grain size in such castings by a combination of pouring temperature and additives to the primary coating slurry is described in Reference 87.

Superheat. A further influence upon nucleation may be exercised by superheating the metal to a higher temperature than that used for pouring, whether in the normal course of production or as a deliberate preliminary treatment. This effect lends added significance to the pouring temperature itself and is exemplified in the decrease in the number of eutectic grains obtained in grey cast iron with increased pouring temperature and holding time[23,88]. Excessive superheat may even suppress graphite nucleation altogether and produce a chilled structure.

The influence of superheat is held to be associated with the behaviour of heterogeneous nuclei during thermal cycling. The most usual effect is to produce coarsening of the grain structure: this has been attributed to the elimination, by solution, either of the foreign nuclei themselves or of small amounts of solid parent metal which remain stable above the liquidus

(a)

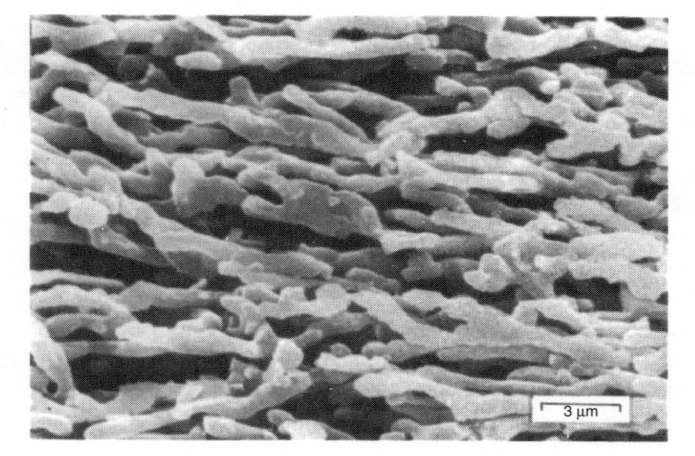

(b)

Figure 2.34 SEM micrographs of (a) An unmodified Al-Si eutectic alloy and (b) a strontium modified alloy grown at the same rate (*from Elliott*[85]) (*courtesy of CDC*)

Table 2.2 Influences of cooling rate and modification on mechanical properties of an Al-13% Si alloy (after Reference 86)

Condition	UTS MPa	Elongation %	Hardness HB
Normal (sand cast)	124	2	50
Modified (sand cast)	193	13	58
Normal (chill cast)	193	3.6	63
Modified (chill cast)	221	8	72

temperature in the surface pores of foreign particles or the walls of the container.

In certain cases, however, superheat produces the opposite effect of grain refinement, a notable example being seen in the treatment applied to magnesium–aluminium alloys. Detailed accounts of this and other refining techniques for magnesium alloys were given by Emley[75] and by Nelson[71]. The mechanism of refining by superheat has been explained as the solution of relatively large included particles of nucleating compounds and their precipitation in more finely divided and crystallographically favourable forms on cooling towards the liquidus temperature.

Thus the entire history of thermal conditions in the molten state can affect the structure produced on solidification.

Agitation.   It has long been realized that nucleation could be brought about by physical disturbance of undercooled liquid, for example by stirring or gas evolution. This may be attributed to the widespread distribution of nuclei originally produced at the surface, or to the fragmentation of already growing crystals. Vibration too is well known as a method of structural refinement. Its beneficial influence on the structure and properties of cast aluminium and copper alloys was established in early studies by Friedman and Wallace[89] and a reduction in graphite flake size in cast irons was also reported[90]. Subsequent studies by Southin[91] established the effect of vibration on crystal multiplication as the main determining factor. A later comprehensive review of effects of vibration in the solidification of castings was undertaken by Campbell[92], as was a valuable summary of the practical factors involved[93].

The use of high energy ultrasonic treatment during solidification as a means of structural refinement has been described in References 94 and 95. It was established that the application of ultrasonic energy induces cavitation, involving the creation, pulsation and collapse of bubbles in the melt, with associated shock waves contributing to mixing and fragmentation of growing crystals. Refined structures were obtained in various light alloys, and observed benefits also included a degassing effect.

Other practical possibilities include electromagnetic stirring with the aid of induction coils. This has been shown in certain cases to produce grain refinement as a shower effect at some critical value of the magnetic field[48,96]. It is not, however, always the case that stirring will produce crystal multiplication. In some cases the predominant effect is the elimination of the solute barrier and the remelting of equiaxed grains, encouraging columnar growth. The effect of stirring must therefore depend upon the particular thermal conditions in the system[97].

The structures encountered in castings, particularly large sand castings, can be comparatively coarse in the absence of the additional refining effects of mechanical working and recrystallization as encountered in wrought products. This lends point to the efforts to achieve close control of grain structure by any means available. Since the previously mentioned classical review by Hughes[77] there has been a steady output of further papers on the subject of grain refinement treatments, particularly in aluminium alloys. These have been a prominent feature in the series of international conferences on the solidification theme which followed the pioneering event recorded in Reference 10. The proceedings of the later conferences have been listed as References 35–38. These have shown an increasingly applied emphasis, dealing with new processes, products and metal treatment techniques, some of which are reviewed in other chapters. Other works on solidification, apart from those cited earlier in the chapter, include those by Davies[98] and Chadwick[99].

Further examples of the structural features referred to in the foregoing sections can be found in the volume of typical microstructures of cast alloys, published by the Institute of British Foundrymen[100], and in the wider literature dealing with metallographic topics.

References

1 Smithells, C. J., *Metals Reference Book*, 7th Edn. Brandes, E. A. and Brook, G. B., (Eds.), Butterworth-Heinemann, Oxford (1998)

2 Hollomon, J. H. and Turnbull, D., The Solidification of Metals and Alloys: 1950 Symposium, *A.I.M.M.E.*, New York (1951)

3 Hollomon, J. H. and Turnbull, D., *Prog. Metal Phys.*, **4**, 333 (1953)

4 Tiller, W. A., Jackson, K. A., Rutter, J. W. and Chalmers, B., *Acta. Met.*, **1**, 428 (1953)

5 Liquid Metals and Solidification, *Am. Soc. Metals*, Cleveland (1958)

6 Pfann, W. G., *Zone Melting, 2nd Ed.*, Wiley, New York (1966)

7 Winegard, W. C., An Introduction to the Solidification of Metals: Monograph No. 29, *Inst. Metals*, London (1964)

8 Winegard, W. C., *Metall. Rev.*, **6**, No. 53, 57 (1961)

9 Chalmers, B., *Principles of Solidification*, Wiley, New York (1964)

10 The Solidification of Metals. Proceedings of 1967 joint conference at Brighton. Sp. Publication 110, *Iron Steel Inst.*, London (1968)

11 Barrett, C. S. and Massalski, T. B., *Structure of Metals, 3rd Ed.*, McGraw Hill, New York (1966)

12 Spear, R. E. and Gardner, G. R., *Trans. Am. Fndrym. Soc.*, **71**, 209 (1963)

13 Alexander, B. H. and Rhines, F. N., *Trans. A.I.M.E.J. Metals*, **188**, 1267 (1950)

14 Flemings, M. C., *Ref.* 10, 277

15 Flemings, M. C., Kattamis, T. Z. and Bardes, B. P., *Trans. Am. Fndrym. Soc.*, **99**, 501 (1991)

16 Kattamis, T. Z., Coughlin, J. C. and Flemings, M. C., *Trans. Met. Soc. A.I.M.E.*, **239**, 1504 (1967)

17 Kirkwood, D. H., *Mater. Sci. Eng.*, **73**, L1 (1985)

18 Mortensen, A., *Met. Trans.*, **22**, 569 (1991)

19 Biloni, H. and Chalmers, B., *Trans. metall. Soc. A.I.M.E.*, **233**, 373 (1965)

20 Biloni, H., *Ref.* 10, 74

21 Trivedi, R. and Kurz, W., *Int. Mater. Rev.*, **39**, 2, 49 (1994)

22 Tiller, W. A., *Ref.* 5, 276

23 Morrogh, H., *Br. Foundrym.*, **53**, 221 (1960)

24 Hultgren, A., Lindblom, Y. and Rudberg, E., *J. Iron Steel Inst.*, **176**, 365 (1954)

25 Hunt, J. D. and Jackson, K. A., *Trans. metall. Soc. A.I.M.E.*, **236**, 843 (1966)

26 Jackson, K. A. and Hunt, J. D., *Acta. Met.*, **13**, 1212 (1965)

27 Kraft, R. W., *J. Metals*, **18**(i), 192 (1966)

28 Chadwick, G. A., *Prog. Mater. Sci.*, **12**, 97 (1963)

29 Cooksey, D. J. S., Munson, D., Wilkinson, M. P. and Hellawell, A., *Phil. Mag.*, **10**, 745 (1964)

30 Kerr, H. W. and Winegard, W. C., *J. Metals*, **18**(i), 563 (1966)

31 The Metallurgy of Cast Iron. Lux, B., Minkoff, I. and Mollard, F., (Eds.) Proceedings of Second Int. Symposium Geneva, Georgi Publishing, Geneva (1974)

32 Elliott, R., *Int. Metall. Rev.*, **22**, 161 (1977)

33 Elliott, R., *Eutectic Solidification Processing*, Butterworths, London (1983)

34 Flemings, M. C. *Solidification Processing*, McGraw-Hill, New York (1974)

35 Solidification and Casting of Metals. Proceedings of Sheffield joint conference. Book 192. The Metals Society, London (1979)

36 Solidification in the Foundry and Casthouse. Proceedings of International Conference at Warwick, Book 273, The Metals Society, London (1983)

37 Solidification Processing 1987. Proceedings of Third International Conference at Sheffield, Book 421, The Institute of Metals, London (1988)

38 Solidification processing 1997. Proceeding of Fourth Decennial International Conference at Sheffield. Ed. Beech, J. and Jones, H. Dept of Engineering Materials, University of Sheffield (1997)

39 Campbell, J., *Castings*, Butterworth-Heinemann, Oxford (1991)

40 ASM Metals Handbook 9th Ed. Vol 15 Casting: Granger, D. A and Elliott, R., 159; Stefanescu, D. M., 168; Magrin, P. and Kurty, W., 119. ASM International, Metals Park Ohio (1988)

41 Uhlmann, D. R. and Chadwick, G. A., *Acta Met.*, **9**, 835 (1961)

42 Tamman, G., *A Textbook of Metallography.*, Translation Dean and Stevenson, Chemical Catalog Co., New York (1925)

43 Rutter, R. W., *Ref.* 5, 243

44 Tiller, W. A. and Rutter, J. W., *Canad. J. Phys.*, **34**, 96 (1956)

45 Plaskett, T. S. and Winegard, W. C., *Trans. Am. Soc. Metals*, **51**, 222 (1959)

46 Lakeland, K. D. and Hogan, L. M., *Ref.* 10, 213

47 Jackson, K. A., Hunt, J. D., Uhlmann, D. R. and Seward, T. P., *Trans. metall. Soc. A.I.M.E.*, **236**, 149 (1966)

48 Tiller, W. A. and O'Hara, S., *Ref.* 10, 27

49 Cibula, A., *Br. Foundrym.*, **51**, 581 (1958)

50 Jackson, W. J., *Br. Foundrym.*, **51**, 584 (1958)

51 Morrogh, H., *Br. Foundrym.*, **51**, 588 (1958)

52 Flemings, M. C., *Foundry*, **91**, March, 72 and April, 69 (1963)

53 Cibula, A., *Proc. Inst. Br. Foundrym.*, **48**, A73 (1955)

54 Ames, B. N. and Kahn, N. A., *Trans. Am. Fndrym. Soc.*, **58**, 229 (1950)

55 Harris, R. F. and Chandley, G. D., *Trans. Am. Fndrym. Soc.*, **70**, 1287 (1962)

56 Ahearn, P. J., Form, G. W. and Wallace, J. F., *Trans. Am. Fndrym. Soc.*, **70**, 1154 (1962)

57 Flemings, M. C., Barone, R. V., Uram, S. Z. and Taylor, H. F., *Trans. Am. Fndrym. Soc.*, **69**, 422 (1961)
58 Cibula, A., *Fndry Trade J.*, **93**, 695 (1952)
59 Northcott, H. L., *J. Inst. Metals*, **62**, 101 (1938)
60 Piearcey, B. J. and VerSnyder, F. L., *Metal Prog.*, **90**, Nov, 66 (1966)
61 Northwood, J. E. and Homewood, T., *Ninth Commonwealth Min. and Metall. Congress*, Instn. Mining and Metallurgy (1969)
62 VerSnyder, F. L. and Shank, M. E., *Mater. Sci. Eng.*, **6**, 213 (1970)
63 VerSnyder, F. L. In *High Temperature Alloys for Gas Turbines*, Brunetaud *et al* (Eds), D. Reidel, London (1982)
64 McLean, M., *Directionally Solidified Materials for High Temperature Service*, Metals Society, London (1983)
65 Fras, E., Guzic, E., Kapturckiewicz, W. and Lopez, H. F., *Trans. Am. Fndrym. Soc.*, **105**, 783 (1997)
66 Kura, J. G. and Eastwood, L. W., *Trans. Am. Fndrym. Ass.*, **55**, 575 (1947)
67 Cibula, A., *Proc. Inst. Br. Foundrym.*, **46**, B92 (1953)
68 Planard, E., *Trans. Am. Fndrym. Soc.*, **70**, 299 (1962)
69 Matyja, H., Giessen, H. C. and Grant, N. J., *J. Inst. Metals*, **96**, 30 (1968)
70 Flemings, M. C. and Poirier, E. J., *Foundry*, **91**, October, 71 (1963)
71 Nelson, C. E., *Trans. Am. Fndrym. Soc.*, **56**, 1 (1948)
72 Cibula, A., *J. Inst. Metals*, **76**, 321 (1949)
73 Cibula, A., *J. Inst. Metals*, **82**, 513 (1953–4)
74 Cibula, A., *J. Inst. Metals*, **80**, 1 (1951–2)
75 Emley, E. F., *Principles of Magnesium Technology*, Pergamon Press, London (1966)
76 Kondic, V., *Br. Foundrym.*, **52**, 542 (1959)
77 Hughes, I. C. H., *Progress in Cast Metals*, Institution of Metallurgists, London (1971)
78 Crossley, F. A. and Mondolfo, L. F., *Trans. A.I.M.E.J. Metals*, **191**, 1143 (1951)
79 Turnbull, G. K., Patton, D. M., Form, G. W. and Wallace, J. F., *Trans. Am. Fndrym. Soc.*, **69**, 792 (1961)
80 Church, N., Wieser, P. and Wallace, J. F., *Br. Foundrym.*, **59**, 349 (1966)
81 Hall, H. T. and Jackson, W. J., *Ref.* 10, 313
82 Hughes, I. C. H., *Ref* 10, 184
83 Harding, R. A., Campbell, J. and Saunders, N. J., *Ref.* 38, 489
84 Lux, B., Mollard, F. and Minkoff, I., *Ref.* 31. 371
85 Elliott, R., *Cast Metals*, **1**, 1, 29 (1988)
86 Thallard, B. M. and Chalmers, B., *J. Inst. Metals*, **77**, 79 (1950)
87 Rapoport, D. B., *Fndry Trade J.*, **116**, 169 (1964)
88 Morrogh, H. and Oldfield, W., *Iron Steel, London*, **32**, pp. 431 and 479 (1959)
89 Freedman, A. H. and Wallace, J. F., *Trans. Am. Fndrym. Soc.*, **65**, 578 (1957)
90 Dmitrovich, A. M. and Butsel, K. T., *Gases in Cast Metals*, Ed. Gulyaev, B. B. Translation Consultants Bureau, New York, 128 (1965)
91 Southin, R., *Ref.* 10, 305
92 Campbell, J., *Inter. Metall. Rev.*, **26**, 2, 71 (1981)
93 Campbell, J., Solidification in the Foundry and Casthouse. Proc. 1979 Conference at Warwick, Book 273. The Metals Society, London (1983)
94 Abramov, V. O., Abramov, O. V. and Sommer, F., Reference 38, 58
95 Eskin, G. I., *Ref.* 38, 62

96 Cole, C. S. and Bolling, G. F., *Ref.* 10, 323
97 Melford, D. A. and Granger, D. A., *Ref.* 10, 289
98 Davies, G. J., *Solidification and Casting*, Applied Science Publishers, London (1973)
99 Chadwick, G. A., *Metallography of Phase Transformations*, Butterworths, London (1972)
100 *Typical Microstructures of Cast Metals, 2nd Ed., Inst. Br. Foundrym.*, London (1966)

3

Solidification 2
The feeding of castings

The solidification of most alloys is accompanied by appreciable volume contraction. The magnitude of this contraction is indicated in Table 3.1, which gives values for several metals forming alloys of commercial importance.

The precise influence of this contraction upon the casting varies with freezing conditions. When the solidification process is characterised by sharp demarcation between solid and still liquid zones, as occurs in pure metals and alloys of narrow freezing range, the contraction is readily compensated by a fall in the free liquid surface in the system. Provided that enough liquid is present this process can continue until the casting itself is wholly solid, the supply of liquid being maintained by external feeder heads or risers. The purpose of a feeder head, therefore, is to provide a reservoir of liquid metal under a pressure head sufficient to maintain flow into the casting. Without feeding, the final casting will be subject to shrinkage defects in the form of major internal cavities, centre line or filamentary voids, or surface sinks and punctures: these forms of defect will be examined in more detail in Chapter 5.

In many cases the alloy constitution and thermal conditions are such that there is no clearly defined macroscopic interface between solid and still liquid regions. Solidification progresses simultaneously throughout a zone which may in certain cases comprise the entire casting. This zone contains crystals at various stages of growth, intimately mixed with residual liquid of lower melting point, and the metal is said to be in a pasty or mushy state.

The sites of contraction are now dispersed through an extensive region within the casting, so that the mechanism of feeding is more complex: three successive stages are encountered. At an early stage, whilst bodies of growing crystals are suspended out of mutual contact in the liquid, free movement of the whole can take place as before to provide feeding through a fall in the level of the free surface: this is termed the mass feeding stage and a feeder head provides compensating metal under the necessary pressure.

At a certain stage, however, the grains form a contiguous network of solid. Movement of liquid is then confined to the diminishing intergranular

Table 3.1 Volume contraction of metals on freezing (Reference 1)

Metal	Contraction on freezing %
Iron	3.5
Nickel	4.5
Copper	4.2
Aluminium	6.5
Magnesium	4.1
Zinc	4.7
Lead	3.5
Tin	2.3

or dendritic channels, where increasing frictional resistance is encountered. In the final stage of freezing, isolated pockets of liquid are solidifying independently and contraction can no longer be fed from external sources: large feeder heads may not, therefore, be wholly effective in ensuring soundness. These conditions are characteristic of alloys of long freezing range and of cooling under shallow temperature gradients. In such alloys shrinkage defects commonly take the form of scattered porosity distributed throughout the casting and sometimes extending to the surface. Intercrystalline movement of liquid also plays a major role in macrosegregation (q.v. Chapter 5).

Measures adopted for the feeding of castings frequently use the principle of *directional solidification*. Cooling is controlled, by designing the mould and by utilizing the intrinsic design of the casting, so that freezing begins in those parts of the mould furthest from the feeder heads and continues through the casting towards the feeder heads. These are designed to solidify last so as to supply liquid metal throughout freezing.

The practical approach to directional solidification is based on a variety of measures designed to steepen those temperature gradients which lie in favourable directions. These measures include control of pouring rate and temperature, differential cooling by chilling, and differential heating with the aid of exothermic materials.

The influence of the temperature gradient upon the mechanism of freezing and the extent of the pasty zone is illustrated in Figure 3.1. This diagram indicates how steepening of the gradient narrows the zone of crystallization and so minimizes the distance over which intercrystalline feeding is required. The downward movement of the temperature line produces corresponding travel of successive start and end waves of freezing towards the centre of the casting.

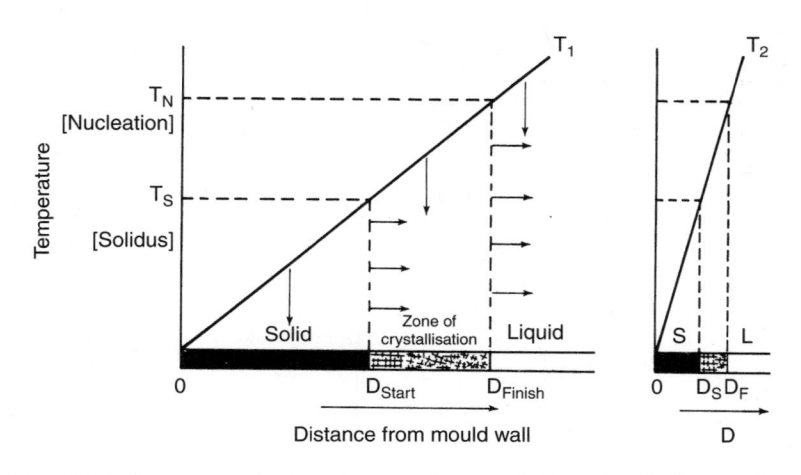

Figure 3.1 Influences of temperature gradient and alloy constitution on depth of pasty zone (schematic)

The feeding characteristics of alloys

The readiness with which progressive solidification can be used to promote feeding is partly dependent upon the constitution of the alloy, since this determines the freezing range and basic mode of crystallization. Further reference to Figure 3.1 demonstrates the importance of freezing range in establishing the extent of the pasty zone: this can be alternatively depicted, in the manner of Pellini and his colleagues, by distance–time curves representing the beginning and end of freezing. Figure 3.2 shows typical curves for a pure metal and a solid solution alloy of long freezing range, whilst the more complex solidification of a hypo-eutectic grey cast iron, involving successive stages of dendritic and eutectic crystallization, can be portrayed by the respective start and end waves as shown in Figure 3.2c. These and many other curves are available in the literature[2-4].

No two alloys have identical feeding characteristics, but it is possible to distinguish three groupings based upon the major contrasts in solidification behaviour emphasised above. These are:

1. Alloys freezing with marked skin formation;
2. Alloys showing extensive pasty zones during freezing;
3. Graphitic cast irons, which show an expansion stage during freezing.

Some important alloys in each of these categories are listed in Table 3.2, although it must be recognized that such a classification can only be approximate.

In the first group of alloys directional solidification can readily be achieved. A narrow wave of freezing proceeds from the surface towards the thermal axis of the casting and can be directed to terminate in the

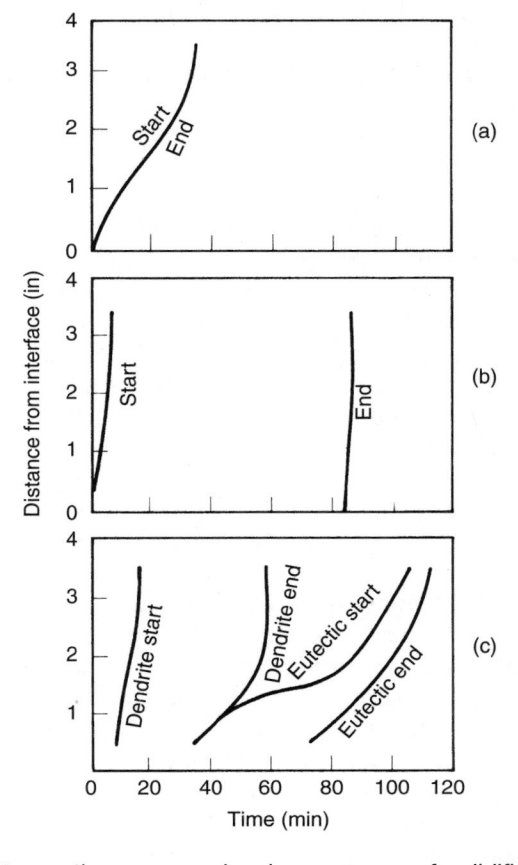

Figure 3.2 Distance–time curves showing progress of solidification in sand moulds. (a) Copper (99.8% Cu), (b) 88/10/2 gunmetal, (c) hypo-eutectic grey cast iron (after References 2 and 3) (courtesy of American Foundrymen's Society)

Table 3.2 Solidification characteristics of alloys

Group 1 Short freezing range	Group 2 Long freezing range	Group 3 Expansion stage
Low carbon steel	Medium and high carbon steels	Grey cast irons
Brasses	Nickel base alloys	Ductile cast irons
Aluminium bronzes	Phosphor bronzes	
High tensile brasses	Gunmetals	
Aluminium	Magnesium alloys	
Copper	Complex aluminium alloys	

feeder heads, where the final shrinkage usually takes the form of a major cavity or pipe. The metal itself tends to be extremely sound but low yield is often associated with heavily piped feeder heads. If feeding technique is unsatisfactory the resulting cavities tend to be concentrated in the region of the thermal centre.

The second group of alloys is characterised by long freezing range, so that solidification proceeds simultaneously in much of the casting. High thermal diffusivity* of the alloy also promotes this mode of freezing by reducing the temperature gradient for a given rate of cooling. Progressive freezing is much less pronounced in these alloys unless very steep temperature gradients can be induced, for example by heavy chilling. Flinn and Kunsman established that a minimum temperature gradient of 1.3 degC/mm is required for effective feeding in some alloys of this type[5]. Castings are consequently prone to widespread porosity, often resulting in low density and hydraulic leakage: density measurements and pressure testing may therefore be appropriate as tests of quality. A further factor opposing feeding in this type of alloy is gas pressure generated by diffusion of dissolved gases into cavities nucleated during freezing. The contributions of gas and shrinkage to pore nucleation in solidifying metals were theoretically treated by Campbell[6] and were also examined by Piwonka and Flemings[7]. Campbell also paid special attention to "burst feeding" occurring in this type of alloy, a phenomenon in which the dendritic mesh undergoes periodic collapse under the growing pressure difference created by shrinkage, atmosphere and metallostatic head[8,9]. The role of this phenomenon in porosity formation during the solidification of aluminium alloys has been investigated in detail by Dahle et al[10]; its particular significance in the absence of gas was emphasized. The general nature of the feeding processes in the long freezing range alloys was further examined by Flemings[11].

The casting technique used for this group of alloys is not invariably based on directional solidification. In some cases the existence of discrete porosity is accepted and effort directed to avoiding local concentrations of porosity by equalization of the cooling rate rather than by accentuating temperature gradients. If this approach is adopted, the casting is gated into its thinner sections and multiple gating used to avoid local hot spots during cooling.

Directional solidification, and especially the soundness of the surface layers can, nevertheless, be effectively promoted by chilling: this was illustrated by Johnson, Bishop and Pellini[12] for an 88/10/2 gunmetal typical of this series of alloys: Figure 3.3 shows the influence of chilling upon the start and end waves of freezing. Because of their tendency to widespread porosity, these alloys do not exhibit the heavy piping and local shrinkage effects associated with the previous group: casting yields are consequently

---

* The thermal diffusivity $\alpha = k/c\rho$ where k is the thermal conductivity, c the specific heat and ρ the density of the alloy.

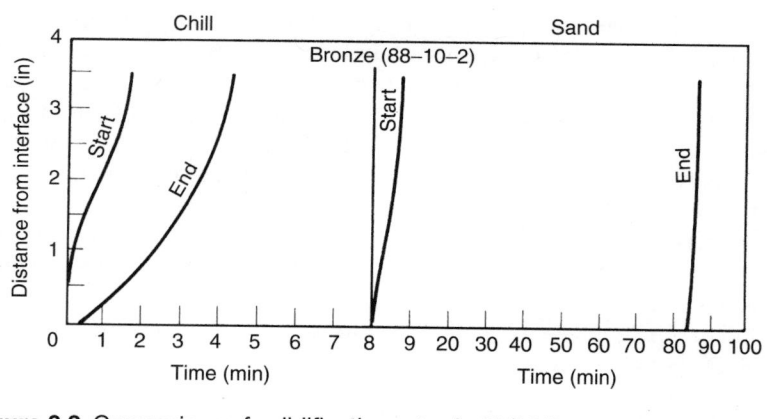

Figure 3.3 Comparison of solidification rates in 88/10/2 gunmetal in sand and chill moulds. (Note difference in time scales). (From Brandt et al[2]) (courtesy of American Foundrymen's Society)

high. It is, however, correspondingly difficult to achieve a really high standard of internal soundness and this is reflected in the greater tolerance of measures for salvage by impregnation. The contrasting problems posed by this type of alloy have been reviewed by Cibula[13].

Further investigations of feeding behaviour and techniques for various non-ferrous alloys, and especially for those of long freezing range, were the subject of References 14–19, whilst a substantial modern appraisal and review, based in part on their own earlier work, was conducted by Berry et al[20]. Those cast irons which form graphite on solidification fall into neither of the categories so far considered. A typical hypoeutectic grey iron undergoes two successive stages of freezing, previously illustrated in Figure 3.2; the second of these stages is wholly distinctive. Freezing proceeds initially with the growth of austenite dendrites. During this period contraction and feeding occur much as in the case of other alloys of appreciable freezing range, bulk or mass feeding being succeeded by an interdendritic feeding stage. With further fall in temperature, however, the eutectic stage of freezing begins, with precipitation of graphite following the carbon enrichment of the interdendritic liquid. Solidification of the austenite–graphite eutectic is accompanied by volume expansion, which creates a positive pressure within the system. In a completely rigid mould this expansion is sufficient to offset the contraction of the metallic phase, with the result that the casting becomes virtually self-feeding in the later stages of crystallization. A similar situation applies in the case of ductile cast irons.

Under practical conditions, however, the positive pressure induced by the graphite precipitation creates a strong tendency to mould wall movement before solidification is complete. This movement produces dimensional

growth and creates a corresponding demand for feed metal, giving susceptibility to internal porosity on further cooling. The importance of this
phenomenon was highlighted in systematic studies carried out by BCIRA
and it became clear that there was a direct relationship between the final
external dimensions of the casting and the amount of piping encountered
(see Figure 3.4) Findings were applicable both to experimental and to engineering and ingot mould castings[21–24]. A review of work on the factors
influencing the soundness of grey iron castings confirmed that, together
with the conditions for the nucleation of the eutectic, relative mould rigidity
is the most important factor[25].

Again, this influence is not confined to grey cast iron: nodular and other
graphitic irons behave in a similar manner, with even greater mould dilation
tendency. Much literature on mould wall movement was reviewed by Lee
and Volkmar in Reference 26. Feeding techniques used for these materials
also recognize this essential difference in solidification behaviour. These
will be the subject of more detailed review later in the present chapter.

Not surprisingly, given the above observations, it is found that hardened
sand moulds give superior results to green, although equivalent results can
be achieved in the latter case with the aid of high pressure moulding[21]. Work
was also carried out with additives and benefit was found to be derived from

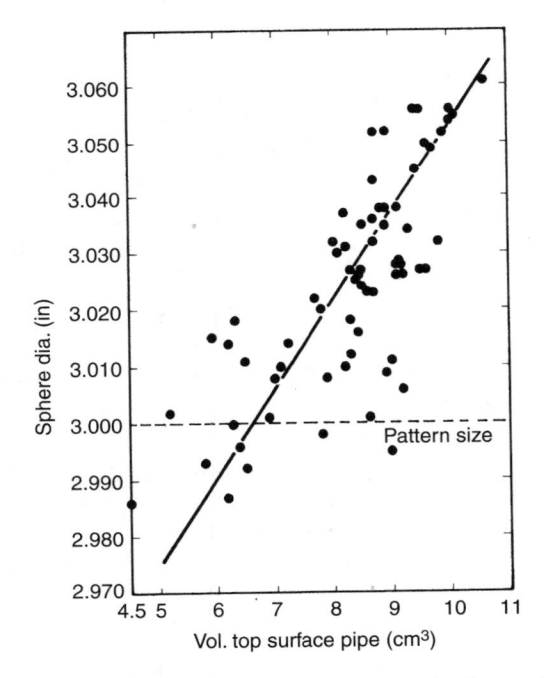

Figure 3.4 Relationship between casting size and pipe volume in cast
iron spheres (from Nicholas and Roberts[21]) (courtesy of Institute of British
Foundrymen)

pitch and coal dust; CO_2 process moulds were also shown to be effective in minimizing mould wall movement[27,28], a benefit associated with all types of modern moulding material with high strength in the hardened condition.

A further complication in the solidification of some cast irons is the occurrence of microporosity associated with the final freezing of small zones of phosphide eutectic at temperatures far below that of the normal solidus. This effect is, however, confined to high phosphorus irons and is not a general problem.

These comparisons of the feeding characteristics of alloys show that response to measures designed to bring about directional solidification is by no means uniform. This principle is nevertheless most widely adopted in feeding practice and will be further considered in some detail.

Geometric influences on solidification

It has already been shown that the soundness of a casting depends upon the relative progress of solidification through its different members, the pattern of freezing being influenced partly by externally imposed conditions and partly by the size and shape of the casting.

The freezing processes themselves are open to investigation in a number of ways, the crudest being the technique in which residual liquid is poured out from the casting after a given interval. The casting is subsequently sectioned to determine the thickness solidified, the internal contour of the solid shell approximating to an isotherm along the original plane of demarcation between liquid and solid. It follows that this technique can only be effective where a clear interface exists as in pure metals and alloys of short freezing range; it must be unreliable in alloys showing a pasty stage.

More accurate information can be derived from cooling curves determined from thermocouples positioned in the mould cavity. Using this technique a complete picture of temperature distribution and of the pattern of freezing can be deduced: the extensive information presented by Pellini and his colleagues was derived in this manner.

Solidification in castings has also been examined by mathematical analysis and by electrical analogue simulation of the processes of heat flow. These and other methods of investigation, together with many aspects of the solidification of castings, were examined in a comprehensive monograph by Ruddle[29].

The numerical modelling of solidification has since become a major field of activity and much information, including experimental data, has been incorporated in computer software for use in the practical design of feeding systems for casting. These developments will be further reviewed later in the chapter.

The temperature distribution within a casting can be portrayed either by isotherms approximately parallel to the mould surface or by thermal gradients in a direction normal to the surface. In general terms the rate of advance of a solidification front from a plane surface may be approximated by a parabolic relation

$$D = qt^{1/2} - c$$

where D = thickness of metal solidified,

         t = time elapsed,

         q and c are constants depending respectively upon thermal properties and the initial temperature conditions in the system.

Temperature distribution is, however, influenced by shape features which produce departures from the conditions at a plane surface; it is also affected by the initial temperature distribution arising from the method of introduction of the liquid into the mould. Since knowledge of solidification time provides a key to economic feeding technique, there has been much concern to establish the influence of variations in geometry and to provide a rational means of predicting freezing time irrespective of shape.

The total solidification time of a casting is always less than would be predicted from the above simple relationship for a plane surface since, considering the casting as a whole, freezing is occurring on a converging front. In respect of individual design features, however, this is not so: departure from the conditions of a plane surface may either accelerate or retard freezing locally, a fact readily illustrated by examples of typical isotherm patterns.

Figure 3.5 demonstrates the effects produced by external and re-entrant corners in the casting contour, both separately and in combination as at intersections. These effects were investigated by Briggs and Gezelius, Ruddle and Skinner, and Pellini[30–32]. The net effect of an intersection is to create a local thermal centre or hot spot, unless design or casting

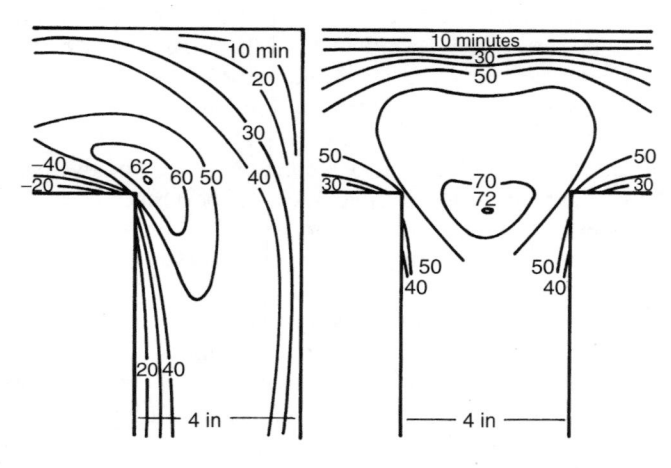

Figure 3.5 Effects of external and re-entrant corners on local rates of freezing: end of freeze waves after successive intervals (from Brandt *et al*[30]) (courtesy of American Foundrymen's Society)

techniques are used to suppress its influence. The relatively slow progress of solidification at a re-entrant angle is of particular interest, since not only is this behaviour utilized in atmospheric feeder head design but it is also responsible in some circumstances for surface defects such as skin puncture: these effects will be examined at a later stage.

The slow rate of freezing associated with a re-entrant feature reaches an extreme case when an internal core of small dimensions is surrounded by liquid metal. Such a core, if of small thermal capacity compared with that of the casting, can dissipate little heat; its surface temperature tends to approach that of the adjacent liquid so that solidification is delayed until a late stage. This effect is exploited in the Washburn core used for restricted neck feeding. The influences of core shape and dimensions upon the rate of freezing were quantitatively examined by Brandt, Bishop and Pellini[30].

Solidification time and the Chworinov rule

Despite local variations in the rate of freezing, the influence of casting geometry is chiefly of importance in determining the overall solidification time. Once this is known, feeder heads can be designed to maintain the supply of feed metal throughout freezing. The ideal is to be able to predict the solidification time irrespective of the shape of the casting.

A notable step in this direction was taken with the discovery of the Chworinov Rule, which postulates that the total freezing time of any casting is a direct function of the ratio of its volume to its surface area[33].

$$t = k(V/A)^2$$

This expression was developed theoretically and was found to correspond well with experimental determinations made on a wide variety of cast shapes ranging from plates to spheres. The data for steel castings in sand moulds is illustrated in Figure 3.6.

The overall solidification time for a given volume of metal is thus found to be greatest when the ratio V/A is a maximum, i.e. in the case of a sphere, becoming progressively less for cylinders, bars and plates.

Determination of the V/A ratio or of some analogous factor can thus be applied to the estimation of relative freezing times of feeder heads and the casting sections they are intended to feed, as well as giving guidance as to the best shape for a feeder head. It is the Chworinov rule which provides the basis for those later approaches to the feeding problem which use a shape factor or 'modulus' as the criterion of total freezing time. The application of these principles to feeder head design will be further discussed in a subsequent section.

Approaches to feeding based simply on casting geometry assume an initially uniform temperature distribution throughout the system. That this is not so can be readily demonstrated by thermocouple measurements, differences of more than 100 degC having been detected within a single casting

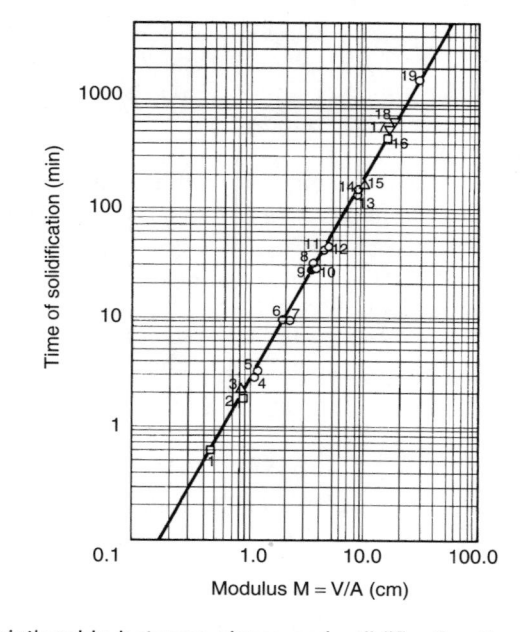

Figure 3.6 Relationship between shape and solidification time for steel castings (from Reference 34 after Chworinov[33]) (courtesy of Dr R. Wlodawer and Pergamon Press Ltd., reprinted from Directional Solidification of Steel Castings (1966))

shortly after pouring, by virtue of the gating technique employed[35,36]. Thus the relative freezing times of feeder heads and castings are not governed solely by their dimensions but also by initial temperature differentials. The relative importance of the initial temperature effect diminishes as the mass becomes greater and the freezing time long, but for small castings which freeze rapidly, initially adverse temperature gradients could invalidate feeder head calculations based on geometry alone. This is especially the case for plate sections: in such cases a margin of safety can be established by the use of casting methods which exploit the initial gradients as direct aids to feeding.

Methods for the feeding of castings

Since the basic objective is to produce a sound casting at minimum cost, every aspect of feeding technique must be considered from an economic as well as a technical viewpoint. The method of feeding may influence plant and labour productivity, but the cost factor most universally involved is that of metal consumption: this will be considered before further discussion of casting methods.

Cost and the concept of yield

An important item in foundry costs arises from metal consumption in excess of the weight finally sold as castings. The origin of these costs can be observed from a study of the pattern of metal utilization in a typical production cycle, as shown in Figure 3.7. This chart illustrates the path of the metal through the foundry and indicates the points at which the major losses occur.

For cost control purposes, the vital information concerning metal consumption may be conveniently expressed as percentage yield, but it is important to distinguish between the various meanings of this term as used in foundry management.

Yield may be used to denote the quantity of finished 'black' castings expressed as a percentage of the total weight of metal charged into the melting units. This can be referred to as the *overall yield* and provides a management ratio indicative of relative success in metal conservation at all points in the production cycle. This value is of particular importance in the case of intrinsically expensive alloys, in which metal cost is high in relation to operational costs.

Of more specific significance in relation to feeding methods is the weight of fettled castings expressed as a percentage of gross weight including feeder heads and gating systems. This ratio has the advantage of being

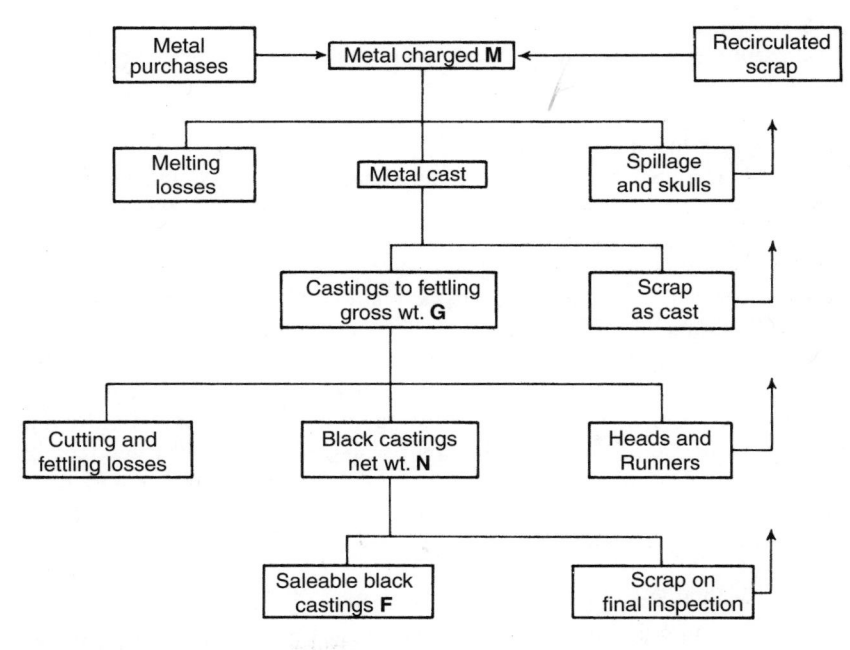

Figure 3.7 Metal utilization in the foundry

measurable for individual castings, either by prior calculation or from determined weights, and can equally be extended to the output as a whole.

This index, the *casting yield*, excludes melting practice and other production variables and so provides a direct measure of the relative success of individual casting methods in promoting metal economy. It must, however, be considered in conjunction with scrap analysis, since a proportion of rejects may themselves arise from faulty methods.

Referring to Figure 3.7,

$$\text{Overall yield} = (F/M) \times 100\%$$
$$\text{Casting yield} = (N/G) \times 100\%$$

The casting yield does influence the overall yield, since although the heads and runners are themselves recoverable, melting losses are incurred on the full gross weight including recirculated material. A high casting yield and low remelt ratio thus benefits metal costs by reducing overall losses.

Yield and policy

Any budget value for casting yield must be determined with due regard to its effect upon overall production costs. Although a low yield incurs additional costs in respect of immediate metallic losses, fuel consumption and handling, it may provide a margin of safety which reduces the costs of rectification or of scrap castings. Policy should accordingly be based upon careful analysis with the object of minimizing overall cost.

The highest levels of casting yield should be attainable under mass production conditions, where there is opportunity for early experiment at the risk of defective castings and where gross weight can be minimized by progressive paring of feeder head dimensions during production runs. Conversely a lower yield is to be expected in the jobbing foundry, where the numbers of castings are small and initial success of greater importance. In this type of production the cost of replacement castings must be measured not only in terms of immediate production costs but in dislocation of production planning and delays in delivery.

The level of yield attainable is also a function of quality as determined by the particular market for the castings: specialized components with exacting standards of inspection and a correspondingly high selling price tend to be manufactured by low yield methods involving wider safety margins. Other factors affecting yield include the feeding characteristics of the alloy and the design and dimensions of the individual casting. Values tend to be lower for small castings: these are more sensitive to minor fluctuations in production variables, requiring a greater margin for error, and the gating system dimensions cannot be scaled down proportionately to the mass of the casting.

In view of the many factors influencing the attainable yield it is not to be expected that a common budget value can be utilized for all types of castings produced in a particular foundry.

Measures available to induce feeding through directional solidification can be summarized under the following headings:

1. Choice of orientation of the casting to exploit the natural features of its design;
2. Design of the gating system;
3. Selection of casting temperature and pouring speed;
4. Design and location of feeder heads;
5. Use of supplementary techniques to increase feeder head efficiency;
6. Introduction of design modifications and padding to provide paths for feeding;
7. Differential cooling by local chilling, insulation and exothermic padding.

Thus the subject of feeder heads must be considered not in isolation but as part of an overall technique designed to produce the required pattern of solidification. Initial consideration will be given to the factors of orientation, gating and pouring technique, after which the design of feeder heads will be further examined.

Orientation

The orientation of the casting is important in relation to the general arrangement of the gating system and feeder heads, but must also be influenced by both moulding and design considerations: these factors are discussed in other chapters. The need to obtain a sound casting must weigh heavily in the choice.

Since the metallostatic pressure is an important aid to feeding, the most logical position for feeder heads is high in the mould: many simple shapes can be fed from a single superimposed head as typified in the casting of ingots. Since feeding is promoted by direct attachment of feeder heads to the heaviest sections of the casting, it is usually advantageous for these to be positioned uppermost in the mould.

Gating technique

The principal variations in gating technique were discussed in Chapter 1; their importance in feeding lies in the initial temperature gradients induced in the casting-feeder head system. The differences between gates adjacent to and remote from the feeder heads in this respect are illustrated in Figure 3.8, which shows schematic temperature gradients for the two cases in a side fed casting.

When gating through the feeder head the initial temperature distribution is favourable to feeding, so that a head of minimum size is adequate. In the opposite case the metal entering the feeder heads has passed progressively through the mould cavity, so that the initial temperature gradients

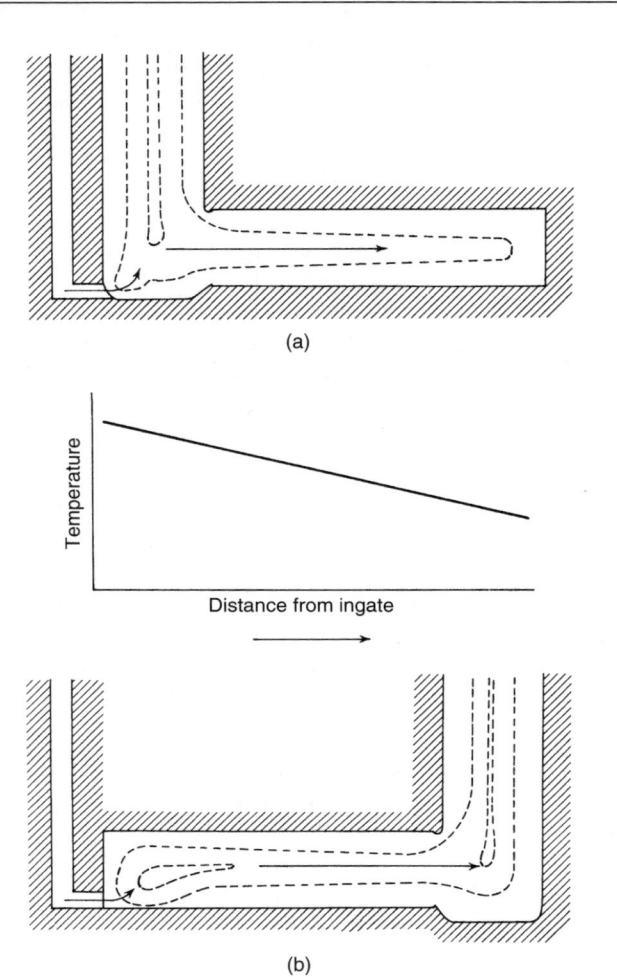

(a)

(b)

Figure 3.8 Direction of initial temperature gradient relative to ingate position. Feeder head (a) coincident with ingate (favourable gradient), (b) remote from ingate (adverse gradient). Isothermal surfaces during freezing suggested by dotted lines

are opposed to feeding, requiring the employment of larger heads than would otherwise be necessary. The conditions illustrated in Figure 3.8 are paralleled in the cases of top and bottom gating. The use of bottom gates in conjunction with top feeder heads provides quiescent mould filling and natural venting of the mould cavity but again suffers from the disadvantage of adverse gradients.

The influence of gating technique and initial temperature distribution upon feeding diminishes as the mass of metal increases. With time, the temperature distribution becomes progressively modified by local differences in cooling rate arising from the respective V/A ratios of feeder head and casting. An initially unfavourable temperature gradient is thus eventually reversed by the slower proportionate heat loss from the mass of metal in the head, as depicted in Figure 3.9: given a sufficiently long freezing time this can occur before the critical stage of freezing and satisfactory feeding is then achieved. Figure 3.10 illustrates this point using hypothetical cooling curves for points in a feeder head and an adjacent casting section. If freezing of the casting is completed before the temperature gradient can be reversed, the casting is likely to be unsound no matter how large the head: these conditions are prone to occur in castings of thin section and short freezing time, for example parallel walled plates.

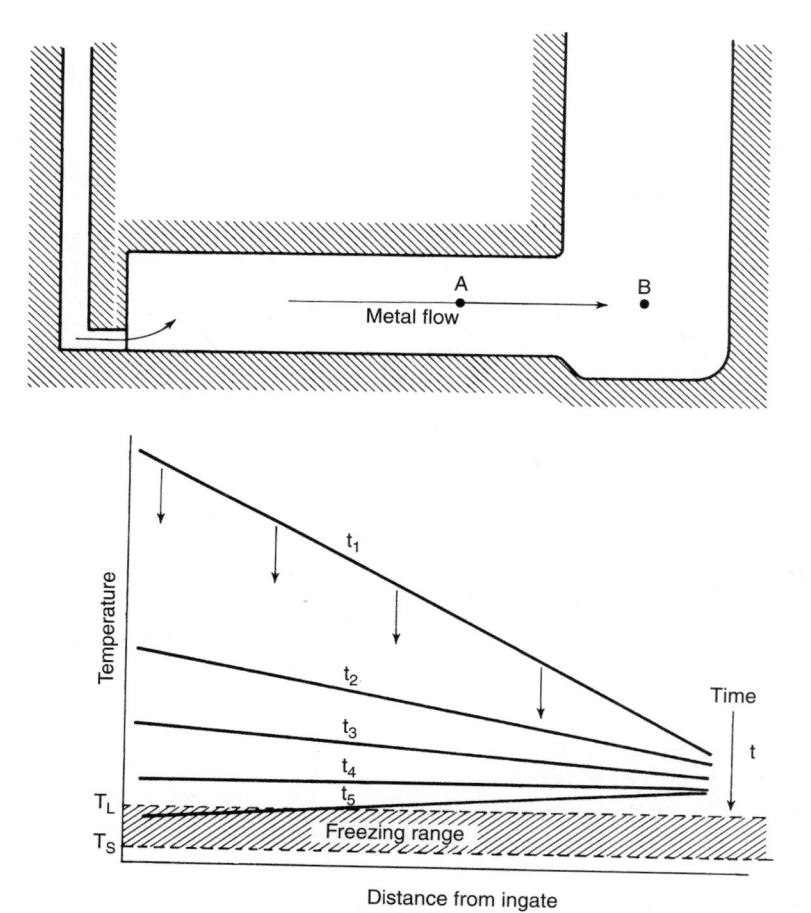

Figure 3.9 Reversal of adverse temperature gradient before freezing. Gradients at successive times t_1–t_5

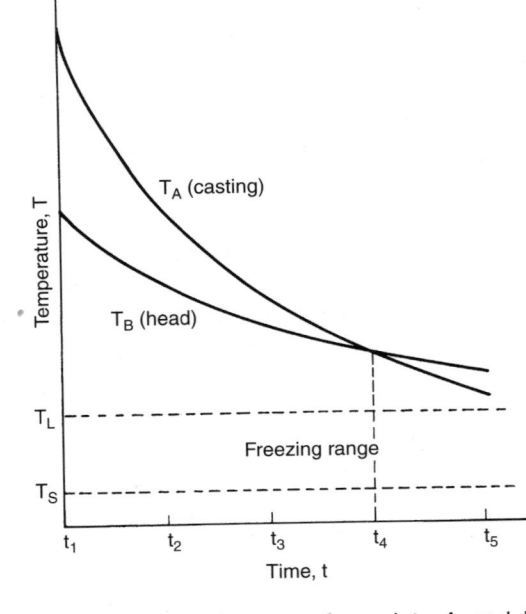

Figure 3.10 Hypothetical cooling curves for points A and B in Figure 3.9. Freezing under favourable conditions after critical time t_4

The part of a casting immediately adjacent to an isolated ingate is particularly susceptible to local shrinkage in the intense hot spot caused by the passage of a large volume of metal during pouring. This effect can be minimized by wider distribution of the metal through multiple ingate systems, which reduce the initial temperature differentials in the system.

The adverse temperature distribution resulting from thermally unfavourable ingate positions can be alleviated by a number of measures. One method is to interrupt the pouring operation when the metal level reaches the base of the feeder heads and to fill these directly with metal from the ladle, so ensuring that they contain metal at the highest temperature. The same object can be achieved by step gating directly into the base of the feeder head; additional step elements can be used to introduce hot metal at successively higher levels in the body of the casting, producing more favourable thermal conditions as illustrated in Chapter 1 (Figure 1.15).

The advantages of both top and bottom gating, or of both the methods of side gating illustrated in Figure 3.8, can be combined by using the technique of total or partial *mould reversal* after casting. The feeder heads are placed at the lower end of the mould cavity with directly attached ingates, pouring being carried out in this position. The mould is then reversed to achieve gravity feed from the heads, which operate under favourable temperature gradients. The partial reversal technique is illustrated in Figure 3.11. These methods have been fully described by Batty[35] and by Duma and Brinson[37].

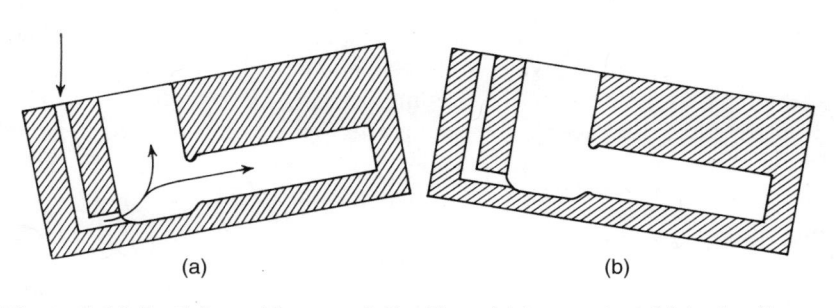

Figure 3.11 Partial mould reversal. Positions (a) for pouring, (b) for feeding

It is interesting to note that similar thermal conditions to those in mould reversal casting are achieved in low pressure upward fill casting techniques without the need for reversal: these developments are further examined in Chapters 9 and 10.

Casting temperature and pouring speed

The magnitude of the initial temperature gradients in the completely filled mould is influenced by both casting temperature and rate of pouring. The desirability of accentuating or of equalizing these gradients depends upon whether the gating system is inherently favourable to feeding.

Where the gating technique promotes favourable gradients, slow pouring is conducive to directional freezing under the steepest gradient. Where adverse gradients need to be minimized, however, as in the case of bottom gating in conjunction with top feeder heads, rapid pouring is to be preferred. The effect of pouring speed arises from the greater opportunity, in the case of slow pouring, for heat transfer from metal to mould during filling: regions remote from the ingates thus receive metal at a correspondingly lower temperature. Rapid pouring, on the other hand, promotes temperature uniformity.

In the case of pouring temperature a high degree of superheat provides opportunity for equalization of adverse gradients during the interval in which the entire casting is cooling to the solidification temperature. With a technique giving favourable temperature gradients, the maximum gradient will be attained when the pouring temperature is just low enough for freezing to begin without delay in the parts of the mould remote from the ingates. Still lower pouring temperatures reduce the temperature differential within the liquid and may also require higher pouring speeds for satisfactory mould filling. The optimum pouring temperature thus depends upon several circumstances.

The significance of pouring temperature in the production of sound castings was illustrated in investigations by Chursin and Kogan[38]. Leak-proof bronze castings were found to be associated with low pouring temperatures in cases where top gating and directional solidification were feasible, but bottom gating required high pouring temperatures for satisfactory results.

Both casting temperature and pouring speed must be optimized in relation to many additional factors, for example the possibilities of laps, misruns and hot tears.

The design and location of feeder heads

The design of the feeder head or riser system is of fundamental importance to the production of sound castings. As previously emphasized, however, an additional aim is to minimize metal consumption so as to achieve a high casting yield and minimum cost. These objectives lend point to efforts directed towards the systematic determination of feeder head dimensions, the subject of much work in recent years.

Although no simple principle exists for the design of feeding systems, extensive studies have enabled a quantitative approach to be developed to supplement the improvised techniques traditionally employed. These studies, mainly carried out in the field of steel castings, have coupled theory with the systematic collection of experimental data. Notable contributions include those of Heuvers, Chworinov, Briggs, Caine, Wlodawer and groups at M.I.T. and the United States Naval Research Laboratory.

In the design of a feeding system the shapes, sizes and positions of the heads must be established. Before considering these factors in turn, however, certain general principles can be emphasised.

The size and shape of each feeder head must satisfy two requirements. Firstly, the head must freeze sufficiently slowly to ensure that liquid metal will be available throughout the freezing of the section to be fed, so enabling directional solidification to continue from the casting into the head. Secondly, the head must be capable of supplying a sufficient volume of liquid to compensate for liquid and solidification shrinkage. Any resulting pipe must remain clear of the casting. These two requirements may be referred to respectively as the *freezing time* and *volume feed capacity* criteria.

The volume of metal required to compensate for shrinkage is itself quite small, being less than 7% for most casting alloys (see Table 3.1). If it were possible to maintain the head as a reservoir of completely liquid metal, the head volume need be of this order only and could be calculated simply from the volume of the casting. This ideal is most closely approached when solidification in the head is retarded by exothermic compounds or other means, or when freezing of the casting is accelerated by heavy chilling or by an extremely low V/A ratio as in thin plates: in these cases feeder head requirements are much closer to the actual volume of feed metal needed and very high yields are attainable.

In most cases, however, solidification is proceeding at similar rates in both feeder head and casting and the criterion for satisfactory feeding is that the solidification time for the head should exceed that for the casting section by a safe margin. The indirect comparison of solidification times by the use of appropriate geometric factors provides the quantitative basis for designing feeder heads to satisfy this criterion. Calculations based on the geometric principle may also incorporate correction factors for changes in freezing rate such as those introduced by local chilling or insulation; alternatively such aids may be used simply as a means of increasing the margin of safety which ensures a sound casting.

Other factors involved in the effective functioning of a feeder head are the need to maintain a pressure differential for feeding, and the joining of the head to the casting in a manner enabling flow of feed metal to continue throughout freezing.

Feeding of a compact and simply shaped casting may be accomplished by a single feeder head of dimensions calculated in accordance with the aforementioned principles. As the design becomes either more complicated or more extensive, however, additional heads may be required. Considering the first case, a frequent feature of design is the separation of heavy sections by lighter members, creating numerous thermal centres. In such cases much of the feeding occurs after the freezing of the intermediate thinner sections and the heavier masses then solidify after the manner of a series of separate castings. In these circumstances feeding is usually accomplished by several heads, each exerting its influence through a portion of the casting.

The second limitation to feeding from a single head arises in the case of extensive walls of parallel section. *Feeding range* then becomes a limiting factor and the task of supplying molten metal must again be divided between a number of heads in accordance with the effective range of each: this topic will be further examined in connection with feeder head positioning.

The converse case to multiple head feeding is the use of a single head to feed a number of separate castings. This technique offers important advantages which will be discussed in a later section.

In short, therefore, the three principal factors in the design of a feeding system are freezing time, feed volume and feeding range.

Considerations governing feeder head shape

In accordance with the previously discussed findings of Chworinov, solidification time is directly related to the ratio of volume to surface area. It follows that the shape of a feeder head of given mass which will remain liquid for the longest period is a sphere. By similar reasoning a cylindrical head is more efficient than a square or rectangular head of equivalent mass. This is illustrated in Table 3.3, which gives a comparison of V/A ratios for a number of simple shapes of identical volume, together with relative freezing times as calculated from the Chworinov relationship.

Spherical heads have not been widely adopted because of two disadvantages. They are inconvenient for moulding and they solidify towards a thermal centre: unlike the thermal axis of other types of head, this centre is inaccessible to external feeding aids, for example exothermic feeding compounds. Most feeder heads are accordingly made with parallel walls, the squat cylindrical form being widely adopted. The pattern pieces can in this case be withdrawn either through the upper surface of the mould or towards the parting plane.

The least satisfactory ratio of volume to surface area is obtained in the case of narrow rectangular heads. This type is also represented by the ring head frequently employed on large annular castings carrying a centre core. Despite its relatively low efficiency the ring head may nevertheless be preferred to a series of separate heads because of ease of removal by machine parting.

The shape of a feeder head is not, however, determined solely by the need to maximize freezing time. Other factors include the timing of the demand for feed metal, affecting the shape of the shrinkage cavity in the head, and the permissible area of junction with the casting: this should be as small as possible to minimize fettling costs. The whole question of feeder head dimensions will now be further considered.

Feeder head volume

Methods for determining feeder head size must be based upon meeting the separate requirements for freezing time and feed volume, so ensuring directional solidification and a sound casting. In each individual case either one or the other of these requirements will be the critical factor controlling the minimum size of head.

In the frequent instances where head size is governed by the freezing time criterion, the freezing times of head and casting need not themselves be estimated: except in the case of differential cooling through the use of chills or exothermic materials, a purely geometric comparison may be adopted to ensure that the head will solidify last.

In its simplest form this can be based on comparison of the ruling sections of casting and head, the essentially two dimensional approach of Heuvers[36,39] using the inscribed circle technique to be referred to again in considerations of casting design, but more comprehensive three dimensional techniques have followed the previously considered findings of Chworinov.

Caine's method[40,41] was based on an experimentally determined hyperbolic relationship between relative volumes and relative freezing rates of head and casting to produce castings free from shrinkage cavities. It was pointed out that if the freezing rates of head and casting were identical, the head volume required would theoretically be infinite, but that as the casting freezes increasingly rapidly relative to the head, the necessary head volume decreases towards a minimum represented by the shrinkage requirement alone, a condition obtained in the case of extended and thin walled castings of very low V/A ratio. This relationship is illustrated in Figure 3.12, for

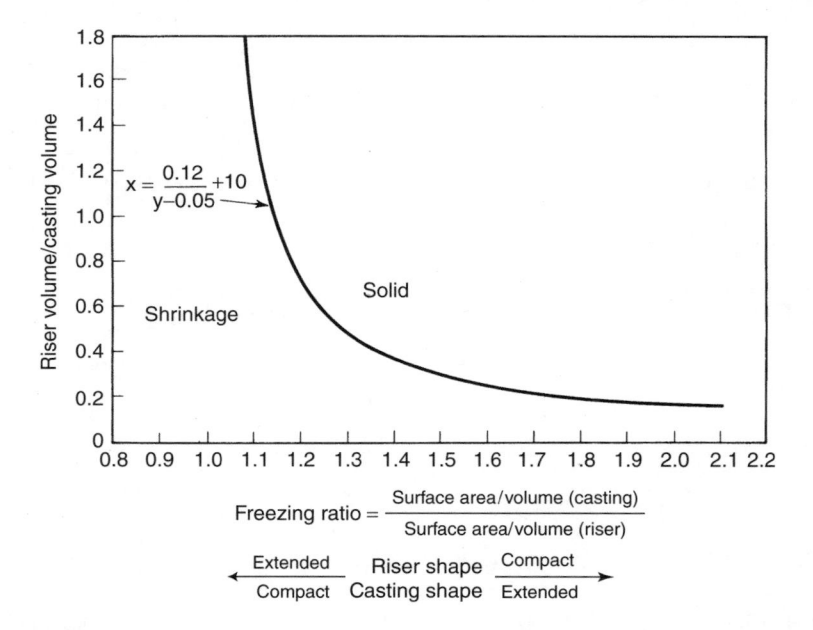

$$x = \frac{0.12}{y-0.05} + 10$$

Freezing ratio $= \dfrac{\text{Surface area/volume (casting)}}{\text{Surface area/volume (riser)}}$

$\longleftarrow$ Extended   Riser shape Compact $\longrightarrow$
Compact   Casting shape Extended

Figure 3.12 Caine's curve for minimum feeder head volume, based on relative freezing rates of casting and head, or freezing ratio (after Reference 41) Basic risering equation: $x = a/(y - b) + c$, where x = freezing ratio, y = riser volume/casting volume, a = freezing characteristics constant, b = liquid–solid solidification contraction, c = relative freezing rate of riser and casting (courtesy of American Foundrymen's Society)

steel castings. Adams and Taylor[42] subsequently indicated how corrections could be applied to values of surface area used in such calculations by adopting appropriate factors for local differences in cooling rate introduced by other mould materials: examples were a factor of 2.2 for chills used in the casting of steel and of 0.6 for gypsum insulation used in conjunction with bronzes.

A further and simplified development of the Caine approach was that of Bishop *et al*[43], who used the concept of a *shape factor* to replace the surface area to volume ratio used in the earlier relationship. The shape factor $S = (L + W)/T$, where L, W and T are the length, breadth and thickness of the section concerned. The relationship between this factor and feeder head requirements is illustrated in Figure 3.13, the empirical graph being essentially similar to that of Caine. With increasing values of S, accompanying transition towards thinner and more extensive shapes of casting, the feeder head requirement is again seen to diminish towards the limiting level at which the controlling influence is no longer the shape factor but the volume of feed metal required.

A further simplification as compared with the Caine graph is the omission of feeder head geometry as a variable: the plotted relationship is restricted to

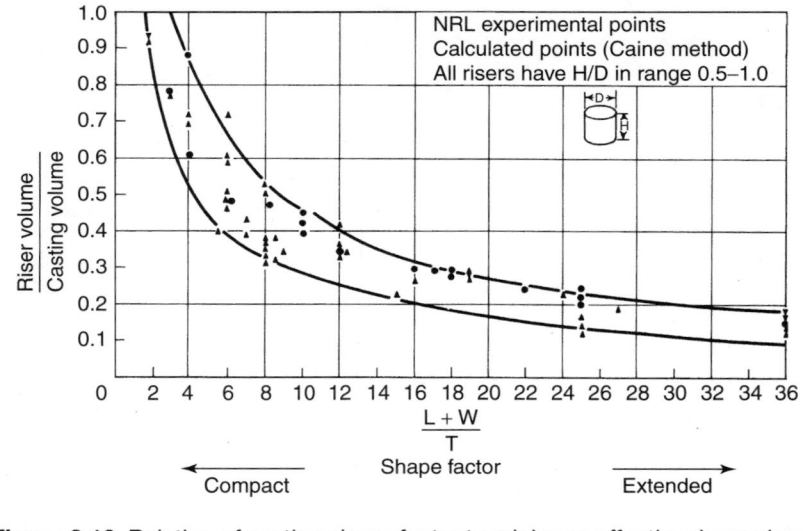

Figure 3.13 Relation of casting shape factor to minimum effective riser volume expressed as a fraction of casting volume. For cylindrical feeder heads with H/D ratios 0.5–1.0 (after Bishop *et al*[43]) (courtesy of American Foundrymen's Society)

cylindrical heads within an optimum range of V/A ratios, so permitting the direct derivation of feeder head volume from consideration of the casting geometry alone. This procedure involves three steps:

1. The shape factor $(L + W)/T$ for the casting is determined, using the dimensions of the main body of the casting or of the section being considered;
2. The ratio $V_{head}/V_{casting}$ is derived from the graph;
3. V_{head} is calculated from the volume of the casting or section. A further chart is provided to portray the various alternative height–diameter combinations for each feeder head volume.

Procedures were also evolved to allow for the effects of design features such as appendages or the presence of small cores. Despite the slight increase in mass, a thin rib or bracket freezes rapidly and may even accelerate the cooling of the casting by reducing the V/A ratio after the manner of a cooling fin: little additional feed metal is therefore required. Thicker projections, however, progressively lengthen the freezing time, so that the feeder head must be enlarged accordingly. The additional feeder head volume required to provide for appendages of various proportions relative to the parent section is derived as a separate step from empirical charts, the appendage being ignored in the main shape factor calculation.

Thin fins can be employed deliberately as an alternative to chills to accelerate local feeding, both as an aid to feeding and to reduce the risk of hot tearing. Wright and Campbell investigated the role of cooling fins in aluminium castings in some detail[44], including further consideration of the relative proportions of parent and appendage sections, an aspect treated in some detail in earlier works by Campbell and other authors.

In the neighbourhood of a small core, the cooling rate of a casting is retarded to a value equivalent to that of a thicker section containing no such core. The metal thickness used in computing the shape factor is therefore replaced by a value for effective thermal thickness, again derived from tables. For fuller details of these allowances and for the relevant charts and tables, reference should be made to the original paper. Detailed examples of the application of the above techniques to the feeding of a commercial steel casting, using various types of head, were described by Briggs[45].

The most comprehensive treatment of feeder head dimensioning is that of Wlodawer[34,46], based on the original approach of Chworinov and extended to include systematic consideration of exothermic materials, padding, chills and other aids to directional solidification. Theoretical studies were combined with experimental observations of his own and of earlier workers. The independently developed system described by Jeancolas[47] embodies similar principles.

Although Wlodawer's techniques have much in common with those of Caine and of Bishop *et al*, the latter was criticized on the grounds that the shape factor employed offered no real advantage that could not be attained with greater accuracy by direct use of the Chworinov ratio of volume to surface area, or *cooling modulus*, given simplified methods of deriving the values. In the Wlodawer system, the feeder head requirement is deduced from the cooling modulus for the casting, the latter being subdivided where necessary into basic shapes for each of which the V/A ratio can be readily determined. Only the cooling surfaces are included in the calculation, the imaginary contact surfaces being omitted. The feeder head is then selected on the principle that it should have a modulus value 1.2 times that for the casting or section concerned. As in previous cases, however, the feeder head requirement for a slender, extensive cast shape is governed not by its modulus but by the volume of feed metal required. A further check is therefore necessary to verify that the feed volume from the proposed head will be adequate in the particular circumstances.

Calculation of the casting modulus. To simplify the modulus calculation for more intricate castings, the principle of substitute bodies is employed, in which plain shapes are substituted for more elaborate shapes of equivalent modulus on the basis that two bodies of identical modulus will solidify in the same time. Angular and curved bodies of the same ruling dimension, for instance, have the same modulus, as shown by the sphere, cylinder and cube in Figure 3.14, which also indicates how the replacement of an irregularly

Shape		Volume V	Area A	Modulus V/A
Sphere		$\dfrac{\pi t^3}{6}$	πt^2	$\dfrac{t}{6}$
Cylinder $h = t$		$\dfrac{\pi t^3}{4}$	$\dfrac{3\pi t^2}{2}$	$\dfrac{t}{6}$
Cube		t^3	$6t^2$	$\dfrac{t}{6}$
Bar (square semi infinite)		$t^2 l$	$4tl$	$\dfrac{t}{4}$
Bar (Cylindrical semi infinite)		$\dfrac{\pi t^2 l}{4}$	$\pi t l$	$\dfrac{t}{4}$
Plate (semi infinite)		At	$2A$	$\dfrac{t}{2}$

(a)

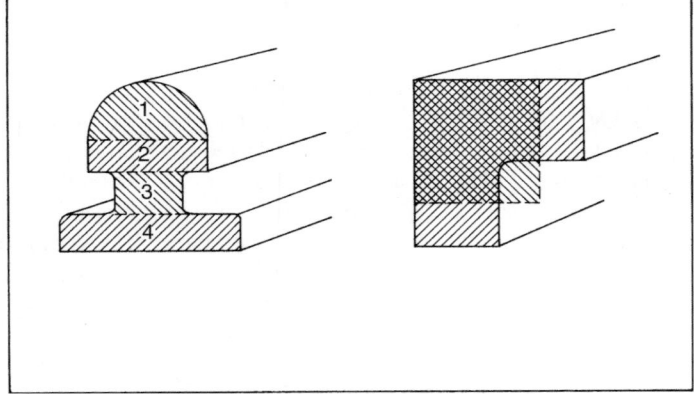

(b)

Figure 3.14 Derivation of cooling modulus values. (a) Simple shapes, (b) irregular shapes, (i) cross-section determined by subdivision into simple shapes, (ii) cross-section approximated by substitute shape (after Wlodawer[34])

shaped body by one of approximately equivalent mass can simplify the calculation. Similarly, rings and hollow cylinders can be treated as the bars or plates which they would form if opened out*.

For ready determination of the moduli of the simpler elements resulting from the above types of breakdown, use can be made of a number of generalized formulae and of tables and charts.

As shown in Figure 3.14, the modulus of the most compact bodies, namely a sphere, a cube and an equiv-axed cylinder, for a given ruling dimension t, is $t/6$. In the case of plates and bars of this same ruling dimension, t, the ratio of modulus to ruling dimension increases from the minimum represented by the cube or equi-axed cylinder towards a limiting value at which end effects become negligible: the modulus is then $t/2$ for a plate and $t/4$ for a square or cylindrical bar. This progressive change in the relationship of modulus to section thickness is illustrated in Figure 3.15. The condition in which edge or end effects can be ignored closely approximates to the true condition when assessing a plate or bar section as a member of a casting joined to other masses, or as a substitute body for a hollow cylinder or ring: the limiting values of $t/2$ and $t/4$ can then be employed.

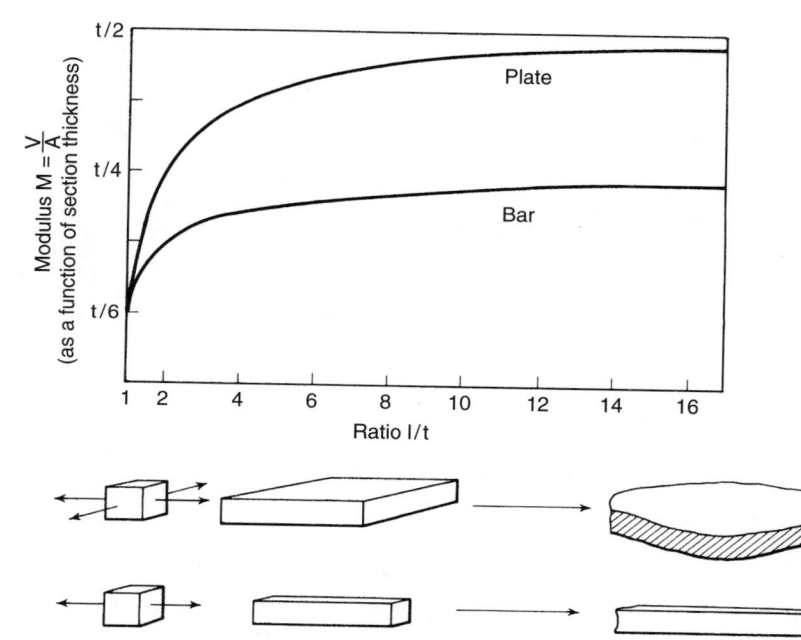

Figure 3.15 Change in modulus with increase in size for a constant thickness, t

---

* When the bore diameter is less than 27% of the outside diameter, the recommended approximation is to a solid body without the centre core; it will be noted, however, that this omits the Bishop allowance for delayed freezing in the presence of small cores.

In the absence of end effects the modulus of a bar of any section can be derived from the general formula $M = $ cross-sectional area/perimeter. The consequent expression $M = t^2/4t$ for a square bar, giving the above value $t/4$, gives way for bars of rectangular cross-section to the formula $M = ab/2(a + b)$. Values of M for rectangular bars of all dimensions can, however, be read directly from the chart illustrated in Figure 3.16. Values for bars of more complex cross-section can be derived directly from the above general formula $M = $ c.s.a./perimeter.

The modulus and feeding capacity of the head. The use of the factor 1.2 in deriving the modulus of the feeder head from that of the casting

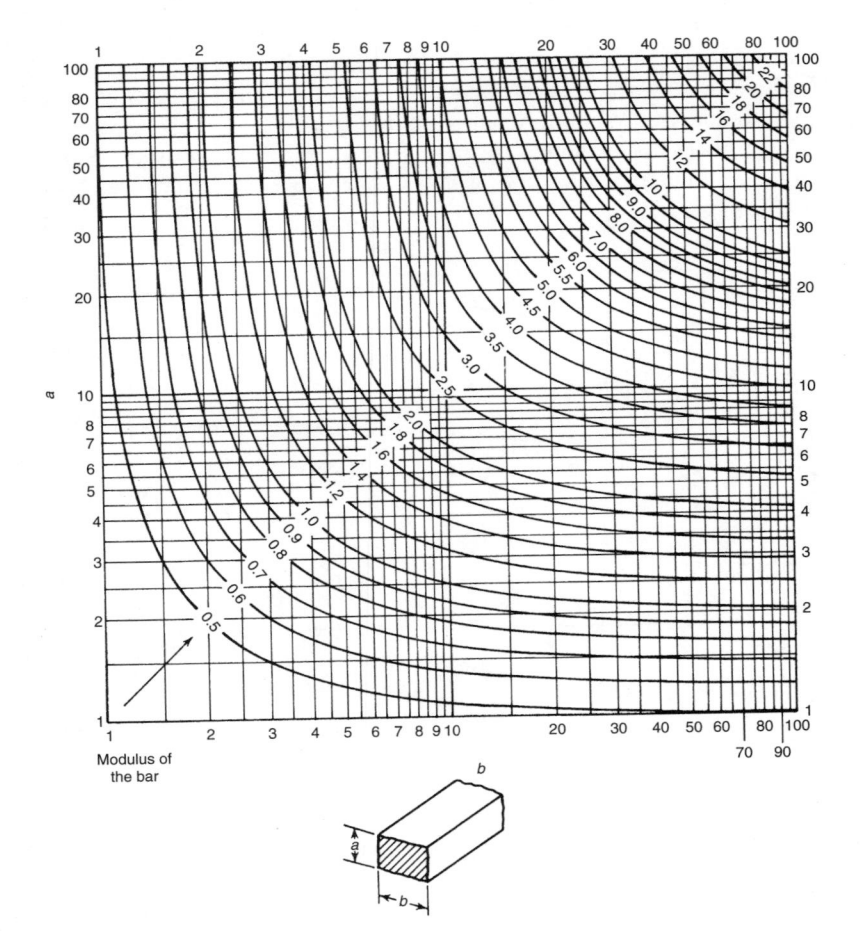

Figure 3.16 Determination of modulus for bars (from Wlodawer[34]) (courtesy of Dr R. Wlodawer and Pergamon Press Ltd., reprinted from Directional Solidification of Steel Castings (1966))

section is intended to allow for the unfavourable change in the feeder head modulus due to shrinkage during freezing. The shape of the shrinkage cavity also determines the maximum proportion of the initial feeder head volume actually available for feeding. For extensive castings the latter is the factor governing the minimum size of head that can be employed.

Considering a cylindrical head, the shrinkage cavity under normal cooling conditions takes up a general form approximating to that illustrated in Figure 3.17a. Although the shape of the cavity will vary to some extent with the timing of the demand for feed metal, it can be shown that the increased surface area reduces the final modulus to a value approximately 75% of that before the shrinkage occurred: the factor of 1.2 is used to offset the consequent reduction in freezing time.

The volume of the shrinkage cavity, representing the maximum volume of metal available for feeding, comprises roughly 14% of the original volume. Thus the maximum volume of casting which could be fed by a cylindrical head in the absence of exothermic feeding is given by

$$\underset{\text{(casting)}}{V_{\max}} = \underset{\text{(head)}}{V} \times \frac{14 - S}{S}$$

where S = specific shrinkage of the alloy (%). This expression also allows for the shrinkage of the metal in the feeder head itself.

Metal utilization is more efficient in a hemispherical head (Figure 3.17b), in which the normal shrinkage cavity represents 20% of the initial volume, hence

$$\underset{\text{(casting)}}{V_{\max}} = \underset{\text{(head)}}{V} \times \frac{20 - S}{S}$$

This shape of head is therefore advocated by Wlodawer for those cases where volume feed capacity rather than modulus is the factor governing the size of the head.

The most efficient metal utilization of all, however, is obtained with the use of exothermic sleeves. Since the metal surface falls uniformly and piping is virtually absent (Figure 3.17c), the residual metal need only comprise a small proportion of the original feeder head volume: a utilization of 67% is feasible in this instance, hence

$$\underset{\text{(casting)}}{V_{\max}} = \underset{\text{(head)}}{V} \times \frac{67 - S}{S}$$

The modulus value for an exothermic head is an effective value, derived from the geometric modulus using an experimentally determined conversion factor which depends on the nature and thickness of the exothermic material used. Figure 3.18 illustrates a range of values for this factor, obtained from experiments on the solidification time of cast spheres. For steel in sand moulds a typical value is 1.43, equivalent to a factor of 2 when

$$V_{max} = V \times \frac{67 - S}{S}$$

(Casting)   (Head)

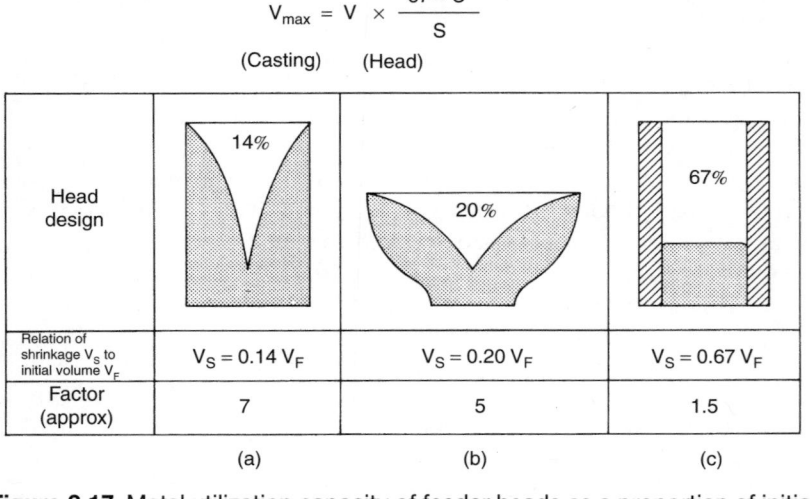

Head design	14%	20%	67%
Relation of shrinkage V_S to initial volume V_F	$V_S = 0.14 \, V_F$	$V_S = 0.20 \, V_F$	$V_S = 0.67 \, V_F$
Factor (approx)	7	5	1.5

    (a)   (b)   (c)

Figure 3.17 Metal utilization capacity of feeder heads as a proportion of initial volume. (a) and (b) cylindrical and hemispherical heads treated with normal feeding compounds, (c) cylindrical head lined with exothermic sleeve (after Wlodawer[34])

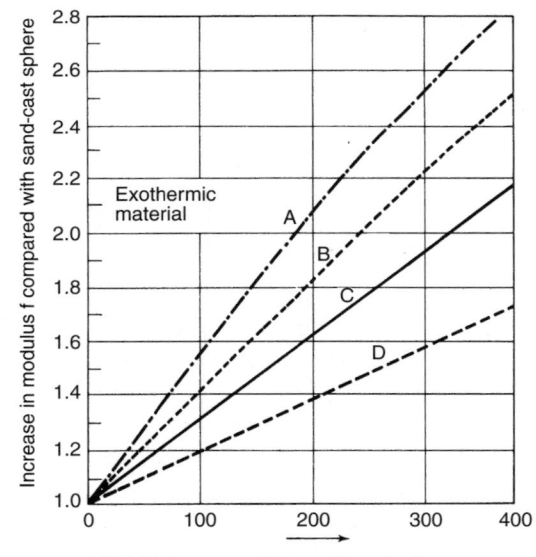

Increase in modulus f compared with sand-cast sphere

Exothermic material

Wall thickness w of the exothermic sleeve as %

of the geometrical modulus ($\frac{w \times 100}{M}$)

Figure 3.18 Increase in modulus of cast steel spheres due to exothermic mould lining. Sphere diameter 150 mm (from Wlodawer[34]) (courtesy of Dr R. Wlodawer and Pergamon Press Ltd., reprinted from Directional Solidification of Steel Castings (1966))

applied to freezing time: these values are applicable for the case of a typical commercial exothermic compound and for a sleeve thickness 20% of the feeder head diameter. Exothermic feeding will be further discussed at a later stage.

It will be seen that the feeding capacity of the head will determine the maximum casting yield attainable for a particular alloy. Based on the above values for metal utilization, an alloy with 4% shrinkage, for example, would give maximum yield values of 71, 80 and 94% with plain cylindrical, hemispherical and exothermic heads respectively.

Feeding criteria and the influence of casting shape. It has been emphasised in the foregoing sections that the dimensions of any feeder head must be established on the basis of satisfying the two separate criteria of modulus (representing freezing time) and volume feed capacity. The influence of casting geometry upon the relative importance of these criteria can be best appreciated by a study of the changing relationship between modulus and volume. Table 3.3 gives comparisons of moduli for variously shaped bodies of a given volume and for a series of plates of varying proportions; also included are the calculated freezing times for the case of steel in sand moulds. Table 3.4 provides the converse comparison of the respective volumes associated with a constant freezing time.

The influence of the proportions of a plate casting upon the volume associated with a particular freezing time (that for a modulus of 1) is illustrated in Figure 3.19. It could reasonably be assumed that any of the castings represented on this curve could be fed by a spherical head of modulus 1.2. Figure 3.20 (curve A) shows the volume of such a head expressed as a percentage of the volumes of the various plate castings. This hypothetical curve could be interpreted as a guide to the minimum theoretical feeder head requirement, up to some point at which the available feed volume replaces freezing time as the controlling factor. The minimum volume percentage of head required to meet the total demand for feed metal in each case is indicated separately in Figure 3.20, line B, having been calculated in this instance on the basis of a specific shrinkage of 3% and assuming 20% utilization of metal in the head. Under these particular conditions it is seen that the volume feed demand becomes critical when the l/t ratio for the plate attains a value of approximately 9.

An overall picture of feeder head requirement would, therefore, take the form of the schematic curve C, compounded from the two stages in which freezing time and feed capacity are successively critical. This curve is essentially a further form of that developed by Caine and of the similar shape factor curve used by Bishop *et al*: it again demonstrates the increased yield attainable in the case of thin walled castings.

The ranges of casting dimensions over which freezing time and volume feed capacity respectively control feeder head size depend upon the

Table 3.3 Moduli and freezing times of bodies of constant volume 1000 cm³

(a) Variously shaped bodies

Shape	Ruling dimension cm	Modulus $M = V/A$ cm	M^2 cm²	Freezing time $= 2.1\,M^2$ min	Freezing time as percentage of that of the sphere
Sphere	$D = 12.41$	2.068	4.277	9.0	100
Cylinder $H = D$	$D = 10.84$	1.806	3.26	6.8	76
Cube	$T = 10$	1.667	2.78	5.8	65
Cylinder $H = 10D$	$D = 5.03$	1.198	1.44	3.0	38
Square bar $L = 10T$	$T = 4.64$	1.101	1.23	2.6	29
Plate or slab $L = 10T$	$T = 2.15$	0.898	0.81	1.7	19

(b) Plates of varying proportions

	Ruling dimension cm	Modulus $M = V/A$ cm	M^2 cm²	Freezing time $= 2.1\,M^2$ min	Freezing time as percentage of that of the sphere
$L = T$(cube)	$T = 10$	1.667	2.78	5.8	65
$L = 2T$	$T = 6.30$	1.575	2.48	5.2	58
$L = 5T$	$T = 3.42$	1.221	1.49	3.1	35
$L = 10T$	$T = 2.15$	0.898	0.81	1.7	19
$L = 20T$	$T = 1.35$	0.617	0.38	0.8	9

specific shrinkage of the alloy. Volume feed capacity becomes increasingly significant with higher values of specific shrinkage, as is illustrated in Figures 3.21–3.23. Figure 3.21 shows the feeder head size needed to meet the volume feed criterion alone, expressed as a function of specific shrinkage. In Figure 3.22a, the volume feed requirements for various degrees of shrinkage are superimposed on the freezing time criterion curve for the plate castings fed by the spherical head; Figure 3.22b gives a similar comparison for the case of an exothermic head. In the former case it can be seen that at 10% shrinkage the volume feed capacity would determine the minimum feeder head size throughout the shape range of the casting.

The respective influences of the volume feed and freezing time criteria for various combinations of shrinkage and casting shape are summarized in

Table 3.4 Comparison of volumes of bodies of a given modulus $M = 1$ (equivalent to a constant freezing time of 2.1 min for steel cast in sand moulds)

(a) Variously shaped bodies

Shape	Ruling dimension cm	Volume V cm³	Ratio V/V equivalent sphere	Volume of equivalent sphere as a percentage of V
Sphere	$D = 6$	113	1	100
Cylinder $H = D$	$D = 6$	170	1.5	67
Cube	$T = 6$	216	1.91	52
Cylinder $H = 10D$	$D = 4.2$	582	5.15	19
Square bar $L = 10T$	$T = 4.2$	741	6.56	15
Plate or slab $L = 10T$	$T = 2.4$	1 382	12.23	8

(b) Plates of varying proportions

Shape	Ruling dimension cm	Volume V cm³	Ratio V/V equivalent sphere	Volume of equivalent sphere as a percentage of V
$L = T$(cube)	$T = 6$	216	1.91	52
$L = 2T$	$T = 4$	256	2.27	44
$L = 5T$	$T = 2.8$	549	4.86	21
$L = 10T$	$T = 2.4$	1 382	12.23	9
$L = 20T$	$T = 2.2$	4 259	37.3	3

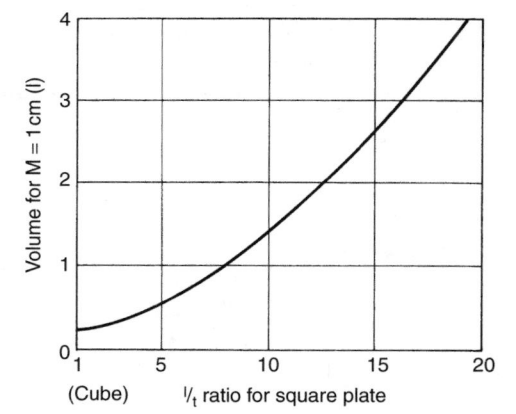

Figure 3.19 Influence of shape of plate castings upon volume associated with identical freezing times (values for M = 1 cm, equivalent to a freezing time of 2.1 min)

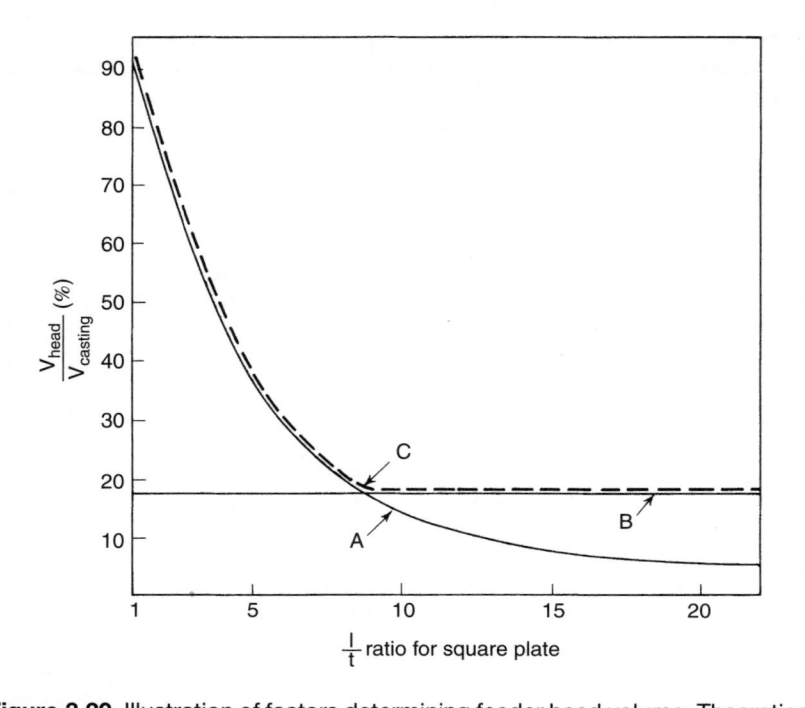

Figure 3.20 Illustration of factors determining feeder head volume. Theoretical feeder head requirements for plates of various proportions (spherical head, ignoring non-cooling interface). Curve A: Minimum head volume percentage to satisfy freezing time criterion (based on relation $M_F = 1.2M_C$). Curve B: Minimum head volume percentage to satisfy volume feed demand criterion (based on 3% specific shrinkage and 20% metal utilization). Curve C: Composite feeding curve

Figure 3.23. Although it is possible from such diagrams to predict which will be the factor governing minimum head size in particular cases, the safest procedure is to carry out independent calculations based on the separate criteria, and to establish feeder head volume on the higher of the two estimates, e.g.

$$M_f \geq 1.2M_c$$

$$V_f \geq V_c \times S/(U-S) \quad \text{(where } U \text{ is the percentage utilization of metal in the head).}$$

It will be seen from Figure 3.23 that for most practical cases the freezing times is the critical consideration.

Once the volume of the feeder head has been determined, its proportions can be designed to take account of three factors: the need for a high modulus during freezing; the need for the final shrinkage cavity to be kept clear of

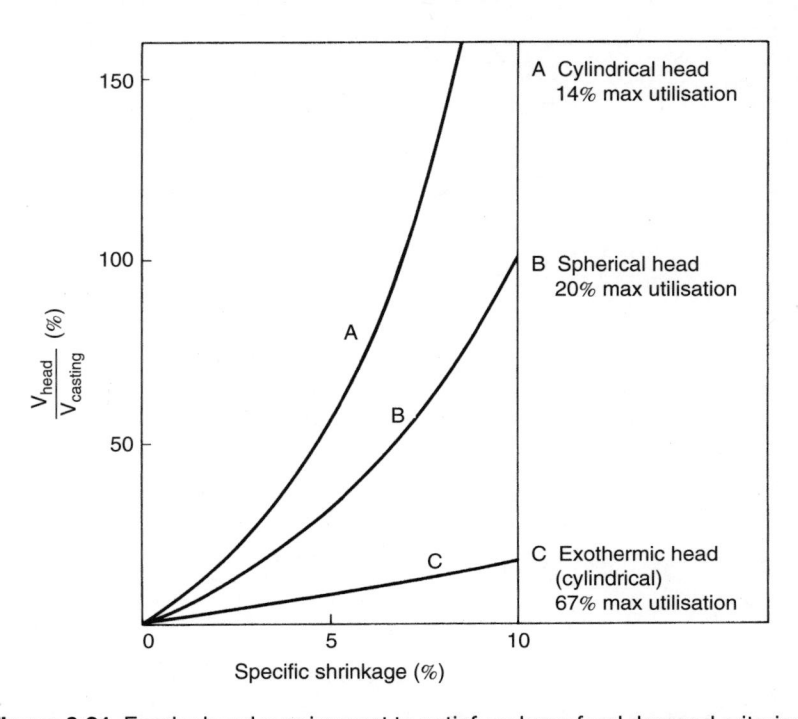

Figure 3.21 Feeder head requirement to satisfy volume feed demand criterion, as a function of specific shrinkage of alloy [from $V_{head} = V_{casting} \times S/(U–S)$]

the casting; and the need for the area of contact with the casting to be minimized for economy in fettling costs. In practice these considerations produce h/d ratios within the range 1–1.5 for cylindrical feeder heads: these proportions are recommended throughout the literature. The exact proportioning of feeder heads of various shapes was considered in detail by Wlodawer, who provided tables of dimensions for rapid selection. For these and for a comprehensive treatment of many further aspects of feeding, reference to the original work is recommended.

The junction of feeder head and casting, whilst being kept as small as possible, needs to be designed to allow feeding to be maintained throughout freezing: this aspect will be further discussed in a subsequent section.

Feeder head positioning

Feeder heads are normally placed in direct contact with the heavier sections of a casting, since this enables directional solidification to be maintained throughout freezing. Just as a compact mass can be fed by a single head, in many castings of complex design the shape divides itself into natural zones of feeding, each centred on a heavy section separated from the remainder of the casting by more constricted members. Each zone can in these cases be

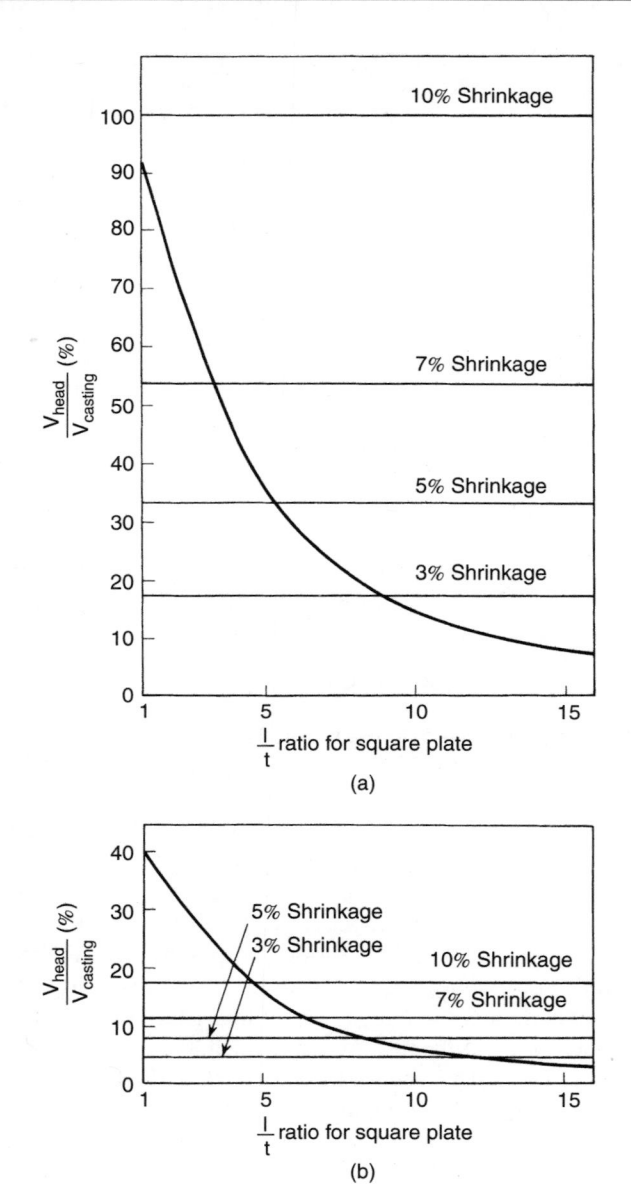

Figure 3.22 Feeder head requirement to satisfy volume feed demand criterion for various values of specific shrinkage. (a) For spherical head, (b) for exothermic head (cylindrical, h = 1.5d). Freezing time criterion curves for square plates superimposed. Exothermic curve calculated for residual geometric modulus of 0.7, equivalent to M = 1 without exothermic

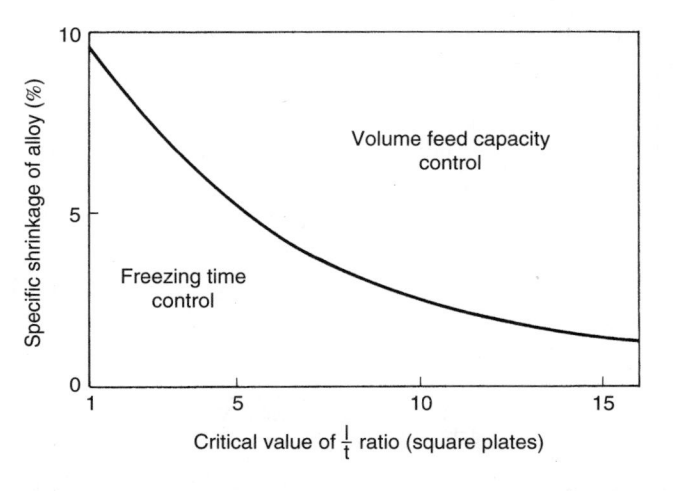

Figure 3.23 Diagram indicating controlling factor in feeder head size (for plate castings of various proportions, fed by spherical head)

fed by a separately calculated head, this being the main factor determining the number of heads required.

In the case of many extended castings, however, *feeding range* rather than constriction is the factor limiting the operation of each head. This is due to the fact that appreciable temperature gradients are needed for feeding and these are difficult to induce in parallel sections over long distances. It has been established, for instance, that to obtain castings free from dispersed porosity, gradients of ~0.2–0.4 degC/cm in plates and ~1.5–2.6 degC/cm in bars are necessary for steel, whilst values as high as ~5.5 degC/cm and even ~13 degC/cm have been quoted for some non-ferrous alloys of long freezing range[5,48,49].

The characteristic temperature distribution pattern within a parallel walled casting is exemplified in Figure 3.24: there is a general tendency towards steep gradients against end walls and shallow gradients adjacent to feeder heads, with isothermal plateaux in intermediate regions. Feeding can only take place over very short distances in such plateau zones and the presence of centre line shrinkage is thus accepted as inevitable in many castings incorporating long parallel sections.

The feeding distances obtained in parallel sections in steel castings were comprehensively investigated by Bishop, Myskowski and Pellini for a wide range of plate and bar castings and this data can be used to determine the spacing of feeder heads under similar conditions for large castings[50–53]. The data for plates is summarized in Figure 3.25 and that for bars in Figure 3.26: the respective contributions of end wall and feeder head effects to the cooling pattern are indicated in each case.

Of particular note is the powerful influence of chills in extending the feeding range of heads when placed at intermediate positions. The spacing

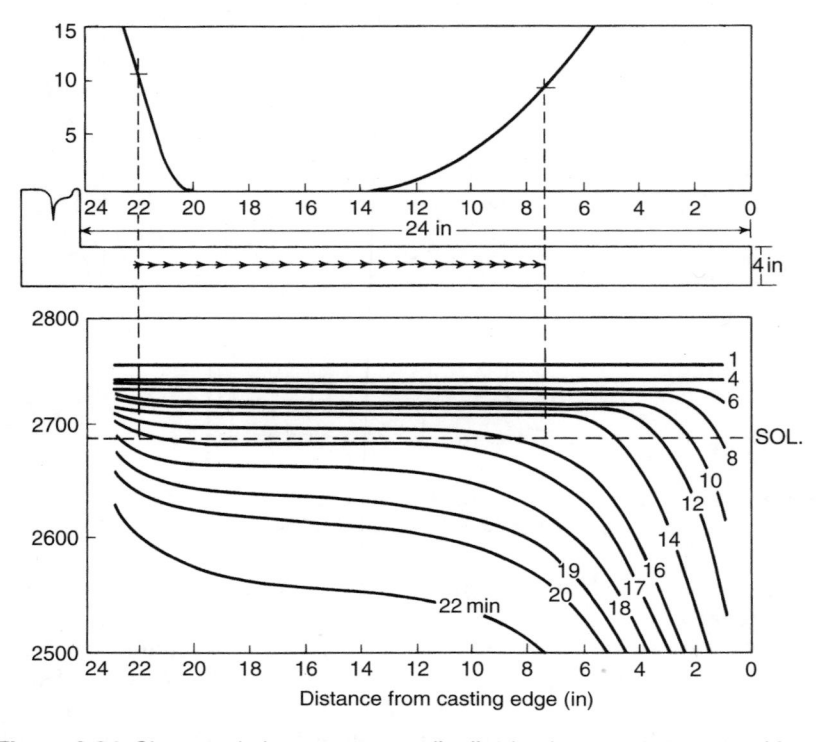

Figure 3.24 Characteristic temperature distribution in a parallel section (from Pellini[49]) (courtesy of American Foundrymen's Society)

between feeder heads can in this case be more than doubled and greatly increased yields thereby obtained. This effect is much as though the casting were severed and the new surfaces separately chilled.

Other aspects of feeding distance investigated by Bishop *et al* included the behaviour of plates joined at their edges to form stepped members: the principal finding in this work was that the conjunction of two plates differing in thickness by a factor exceeding 1.4 increases the feeding distance in the thinner or 'parasite' plate, when compared with that when the latter is cast alone[53]. The empirical data collected is summarized in the relationship portrayed in Figure 3.27.

Most of the systematic data on feeding distance was derived from steel containing 0.25–0.30% carbon, which has only a moderate freezing range and solidifies with marked skin formation. Subsequent investigations, however, confirmed that feeding distance diminishes with increasing freezing range: the steep gradients required for feeding in alloys freezing in the pasty manner have already been mentioned. In the extreme case of a gunmetal, satisfactory soundness was attained only with the aid of chills. In these alloys there is little purpose in attempting to eliminate porosity with

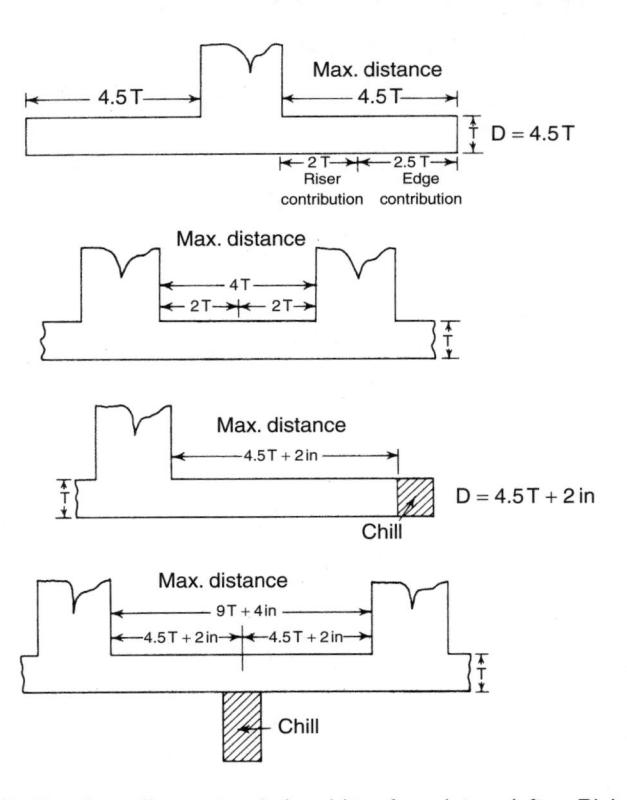

Figure 3.25 Feeding distance relationships for plates (after Bishop[50] and Myskowski[52]) (courtesy of American Foundrymen's Society)

feeder heads alone. Large feeder heads are self-defeating, since they delay freezing and so diminish the lateral temperature gradients which provide a relatively sound surface zone. The most suitable approach, therefore, lies in the combination of comparatively small feeder heads with the use of chills. A sound surface zone is also promoted by the use of moulding materials of higher heat diffusivity than normal moulding sands. Suitable base refractories are shown in Figure 3.41.

For comparison of feeding ranges in different alloys, Flinn[54] introduced the useful concept of a 'centre line resistance factor', developed from previously published beginning and end-of-freeze data referred to earlier in the present chapter[2–4].

$$\text{C.R.F.} = \frac{\text{time during which crystals are present at the centre line}}{\text{total solidification time}} \times 100$$

This index is really a measure of the periods during which metal flow for feeding through a central channel is successively free and obstructed by growing crystals. As expected (see Table 3.5) the highest values are

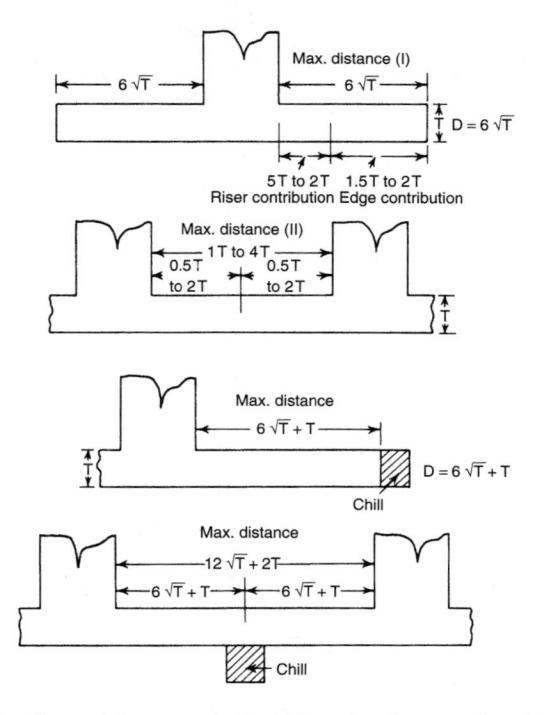

Figure 3.26 Feeding distance relationships for bars (after Bishop[52] and Myskowski[53]) (courtesy of American Foundrymen's Society)

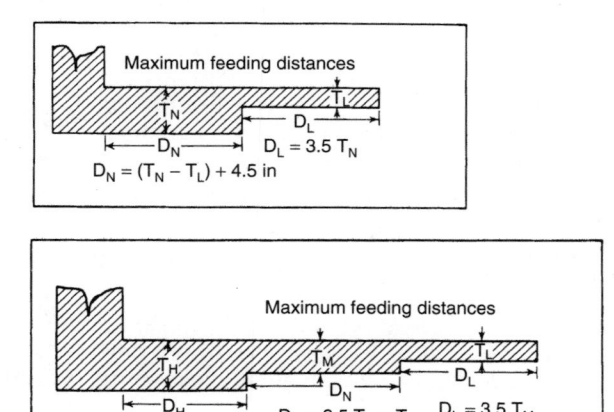

Figure 3.27 Feeding distance relationships for dual and multiple sections (from Pellini[49]) (courtesy of American Foundrymen's Society)

Table 3.5 Centre line resistance factors of typical alloys cast in sand moulds (after Flinn[54], courtesy of American Foundrymen's Society)

Alloy	Centre line resistance factor
Copper (99.8%)	<1
Lead (99.0%)	17
60–40 Brass	26
18.8 Steel (0.2% C)	35
0.6% C steel	54
Monel	64
88/10/2 gunmetal	95
Al 4% Cu	96

obtained in the long freezing range alloys typified by the gunmetal, whilst pure metals exhibit the lowest feeding resistance of all.

The great importance of feeding distance data lies in the ability to restrict the number of feeder heads to the minimum consistent with their respective ranges. From the point of view of the freezing time or cooling modulus criterion, metal is most advantageously employed in the form of a few large heads rather than as a greater number of small, since the collective V/A ratio is then at a maximum. The value of chills in contributing to this objective has already been demonstrated.

Apart from the questions of feeding range, a number of general factors need to be taken into account in positioning feeder heads, of which the most important is the generation of the pressure required for feeding. To achieve maximum metallostatic pressure, heads should be positioned as high as possible relative to the casting. This is particularly important in the case of alloys of long freezing range, for which taller heads than those recommended for steels are usually preferred. Metallostatic head is often the only wholly effective source of feed pressure in these alloys since, owing to the absence of an intact skin, the casting as well as the feeder head remains accessible to atmospheric pressure. In skin forming alloys, however, atmospheric pressure is the most important factor in feeding and heads can, subject to certain limitations, be placed in lower positions if required: the atmospheric feeding mechanism will be examined in greater detail in a later section.

Moulding convenience is served if head locations are consistent with existing parting surfaces so as to avoid the need for additional joints or cores. One potential means of surmounting moulding difficulties is provided by the full mould casting process, more fully described in Chapter 10. Using this technique a feeder head can be shaped in foamed plastic and rammed into the mould with the pattern; the plastic is displaced and volatilised by the molten metal on casting. The technique enables a feeder head to be positioned without regard to moulding considerations, giving an additional degree of freedom in establishing the feeding system. The successful use of this method was described in Reference 55.

Problems of feeder head removal and dressing must also be considered. Heads should be accessible for removal and should whenever possible be placed upon flat rather than contoured surfaces, avoiding the need for sculpture in the fettling shop.

Shared feeder heads. In discussing the positioning of feeder heads in relation to feeding distance, it has been emphasised that the number of heads should be minimized in order to obtain the maximum collective V/A ratio. A further step in this direction can be taken by the use of a single head to feed more than one casting. This practice is frequently adopted for intensive moulding box utilization, but it is not always appreciated that it provides a potential means of improving casting yield.

By sharing a feeder head the total volume of metal required is brought nearer to the minimum needed to satisfy the volume demand criterion. This point is best appreciated by considering a hypothetical example. Referring to Figure 3.20, a plate casting with a length–thickness ratio of 2 would require a feeder head representing 76% of the casting volume to meet the freezing time or cooling modulus criterion. Such a head would possess a large excess of volume feed capacity for the casting concerned: (it can be calculated that there would be only 7% utilization of the feeder head volume as against an attainable maximum of 20%). If the same feeder head were now shared between two castings, the relative freezing times would be unaffected, but the head volume would be reduced to 38% of the total casting volume, a level still, however, more than adequate to meet the volume feed requirement. The casting yield would thus be increased from 57% to 72%.

The same head would in fact be of sufficient theoretical capacity to feed four such castings, giving a yield of 84%, before reaching the limit imposed by availability of feed metal. It must, however, be appreciated that these advantages are confined to those cases in which the need to meet the freezing time criterion would otherwise provide surplus feeding capacity. These conditions are primarily associated with compact castings and low specific shrinkage.

Summary of principles in feeder head design

The foregoing account of the design of feeder head systems can be summarized as a series of general principles applicable to most types of casting. These principles will be stated before consideration of further aspects of feeding practice.

1. A casting needs firstly to be subdivided into zones for feeding by separate heads. For maximum yield the number of heads should be the minimum consistent with the availability of feed paths and with feeding distance. Further advantage may be derived from the sharing of heads between castings.
2. Heads should be positioned adjacent to thick sections and should exploit those factors favourable to feeding, namely temperature gradients, metallostatic pressure, or in certain cases atmospheric pressure. All available means, including gating technique, chills and other measures yet to be discussed, should be employed to induce favourable temperature gradients within the casting.
3. Using a method of calculation such as those developed by Caine, Bishop, Wlodawer and others, the freezing time of each head must be designed to exceed that of the casting or feeding zone by a safe margin. Feeder heads of compact form achieve this objective with minimum metal consumption.
4. The volume feed capacity of each head must be determined on the basis of the known pattern of feeder head behaviour during freezing. This capacity must be consistent with the volume contraction expected in the casting. The feeder head volume is thus fixed by the higher of the two estimates for freezing time and feed demand.

Variations of feeder head design

The simplest and most commonly used type of head is that placed at the top or the side of a casting with its upper surface open to atmosphere. A valuable feature of such heads is their function in venting the mould cavity during pouring.

The principal disadvantage of an open head is the heat loss to atmosphere by radiation: in the absence of insulation the exposed surface is susceptible to premature freezing, with corresponding loss in feeding efficiency, so that it is desirable to use feeding aids in the shape of insulating or exothermic compounds. Feeding techniques employing these materials, both as surface coverings and as feeder head lining sleeves, will be fully discussed at a subsequent stage. The practice of rod feeding is similarly used to combat the tendency for feeder heads to freeze over prematurely.

Open heads are subject to a further disadvantage in that their height is often determined not by feeding requirements but by the depth of moulding box employed. If several such heads are used, their common metal level is established by the requirements of that head placed at the highest point. In such circumstances blind heads can provide a considerable improvement in casting yield.

Blind heads. A blind or dummy head is domed off short of the upper mould surface and may be placed either at the top or relatively low down in the mould cavity. It must, however, be borne in mind that a plain blind head will be ineffective unless placed in such a position that metallostatic pressure is available for feeding. Plain heads of this type are used principally in the superimposed position to save metal which might otherwise be wasted in filling the mould to its upper surface. Blind heads required for the feeding of sections at lower levels are usually modified to exploit the atmospheric pressure effect for feeding.

The *atmospheric head* was introduced to overcome the limitations of a normal blind head. It depends for its effect upon the use of a permeable core to maintain access to the liquid metal in the head; its effective functioning also demands that freezing should occur by progressive inward growth of a hermetic solidified skin. The general arrangement of such a head is shown in Figure 3.28.

After casting, a solidified shell of metal forms round the periphery of the entire mould cavity, the gating system having become sealed off due to its relatively small cross-sectional area. As the remaining liquid continues to contract and a partial vacuum is created, the solid envelope punctures at its weakest point which, due to the slow rate of freezing at re-entrant angles in the casting surface, is at the end of the core insert. Atmospheric pressure is thus brought to bear on the liquid, creating the conditions for feeding to an appreciable height. Atmospheric pressure in a closed system of this nature is theoretically capable of exerting pressures equivalent to 1.30 m of steel,

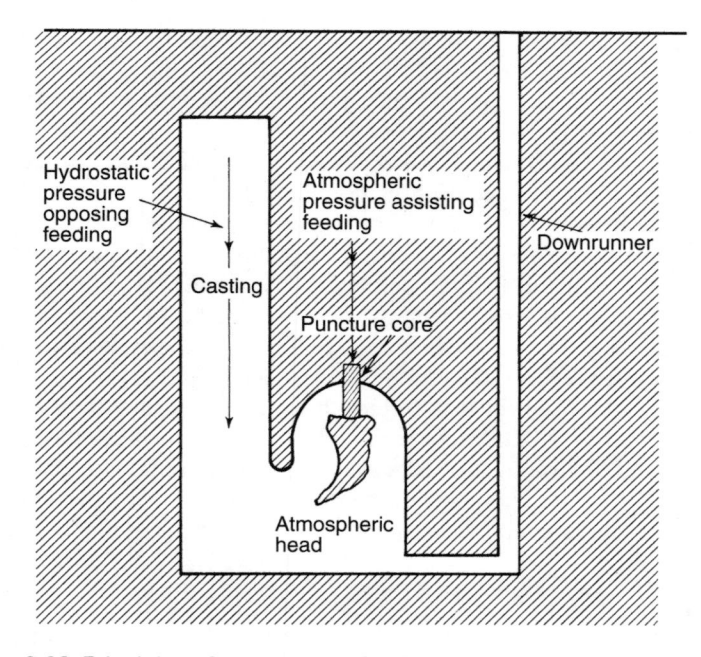

Figure 3.28 Principles of atmospheric feeder head

1.15 m of copper or 4.34 m of aluminium. A casting may therefore be fed to a height considerably greater than that of the head itself, where gravity is opposed to feeding. From the nature of the feeding mechanism, this design of head is only suitable for those alloys which form the necessary sealed layer of solid at the surface. In alloys of long freezing range liquid channels tend to be retained in the surface zone, so that the residual liquid in the casting remains directly accessible to atmosphere pressure until a late stage of freezing.

Atmospheric heads are frequently used to feed heavy sections situated low in the mould cavity, in the manner illustrated in Figure 3.29. In such cases the heads are frequently to be found in combination with open heads placed to feed the uppermost sections. For such multiple feeding systems to be effective, however, the heavy sections of the casting must be separated by lighter members as in Figure 3.29. The atmospheric head only begins to function when the intermediate section has solidified, isolating the lower heavy section from the remainder of the casting. Until this occurs

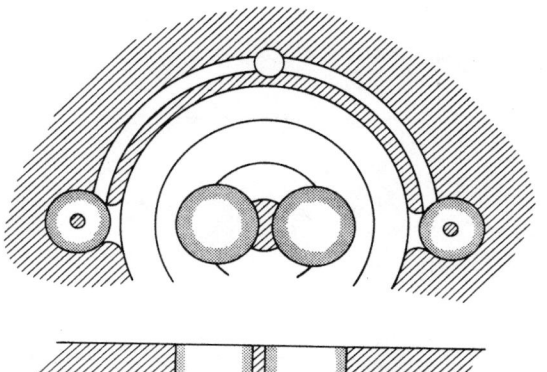

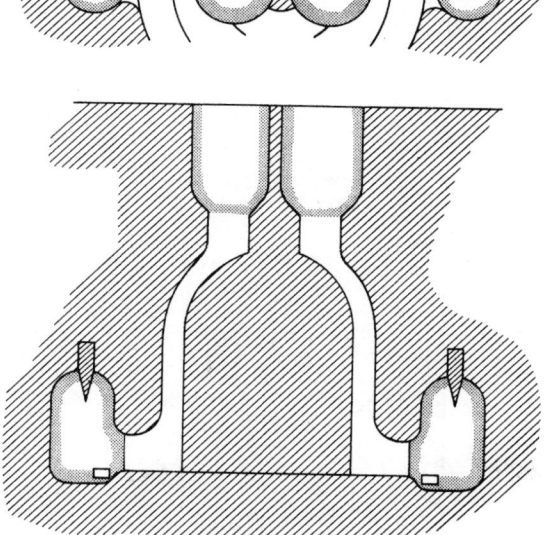

Figure 3.29 Atmospheric heads in combination with open heads

the top heads feed the shrinkage occurring in the entire system and must be proportioned accordingly. In the absence of the necessary constriction between the respective heads, an atmospheric head in such a position never develops the pressure differential required for feeding.

The atmospheric head thus facilitates the feeding of skin forming alloys at points which would be inaccessible or uneconomic for the use of open heads and where simple blind heads would be inoperative.

The *whirlgate* principle is frequently employed in conjunction with open or blind heads for lateral feeding. The metal enters a sump at the base of a cylindrical head through a tangentially placed ingate, producing a swirling effect which continues throughout pouring (see Figure 3.30). The mould surrounding the head and the neck between head and casting are effectively heated due to this circulation; the temperature differential between head and casting is thus accentuated, promoting directional solidification.

A whirlgate head has the added advantage of functioning as a trap for slag and sand inclusions: these are forced to the centre of the head and prevented from entering the casting.

All types of closed feeder head require venting to exhaust air displaced during mould filling: this is particularly important in respect of the air pocket in the upper section of the head.

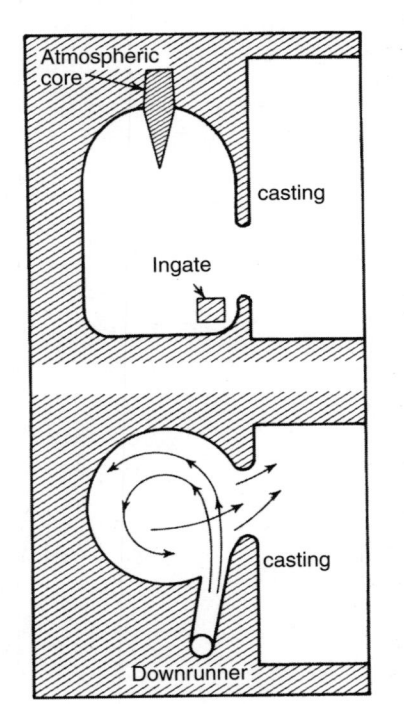

Figure 3.30 Whirlgate head

Further aids to feeder head efficiency

The efficiency of a feeder head may be defined as the amount of feed metal supplied to the casting in relation to the total weight of metal initially present in the head.

$$\text{Efficiency } U = [(I - F)/I] \times 100\%$$

where I = initial volume of metal in head;
F = final volume of metal in head.

The efficiency of a plain feeder head has been seen to be comparatively low, since solidification proceeds in the head at the same time as in the casting; it may however be increased by a variety of measures designed to delay the solidification of the metal or to assist its movement into the casting.

Topping up and rod feeding

The most direct method of extending the feeding period is by the further addition of superheated liquid metal after an interval in which some feeding has occurred. This renews the temperature differential and restores the metallostatic head. The practice is especially applicable to very heavy castings with long solidification times, and the initial size of the feeder head can be reduced to take account of the resulting increase in efficiency.

Rod feeding is used to maintain the influence of atmospheric pressure by mechanical disturbance of the partially crystallised metal in the upper region of the head.

Electric arc feeding

The solidification of an open feeder head may be delayed by applying an electric arc. This is accomplished by the incorporation of a steel electrode in the base of the head, the arc being struck on to the liquid metal surface from an external graphite electrode (see Figure 3.31). The intensive heating effect is maintained by intermittent arcing, heat being conserved by a refractory cover through which the graphite electrode can be adjusted. The technique is particularly suitable for feeder heads employed on very heavy steel castings.

A modification of this principle is seen in *electroslag feeding*, in which heat is supplied by the resistance melting of a consumable wire electrode fed through a molten slag layer on the metal surface. The successful use of this technique in the manufacture of large steel valve bodies and other castings was examined in detail in Reference 56. Induction heating coils or resistance windings could, in principle, be similarly employed, but the cost and complication of the procedures make their commercial application unlikely when simpler alternatives are available.

Electric feeding techniques were examined in Reference 57.

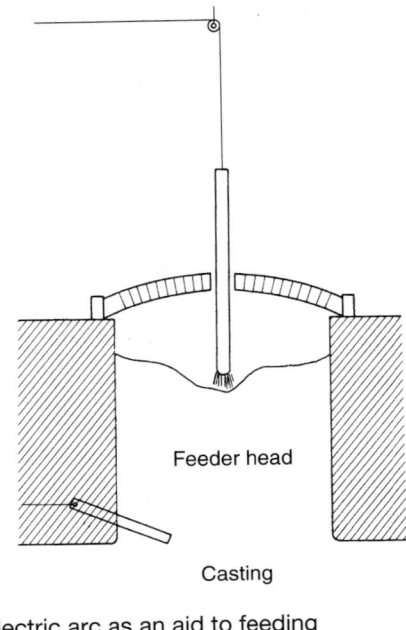

Figure 3.31 The electric arc as an aid to feeding

Insulation and exothermic feeding

Various substances may be used to delay solidification in the feeder heads, either by simple insulation or by the supply of additional heat from exothermic reaction. In certain cases a chemical effect is also used, the freezing temperature of the liquid metal being lowered by local change in its composition on contact with the material. Insulating materials or feeding compounds may be applied either as coverings applied to the surfaces of open heads or as linings for feeder head walls.

Insulation. The use of insulating materials to retard radiation losses from the exposed metal surface is a long established practice. More thorough insulation is now frequently sought by lining the heads with moulded sleeves to reduce conductive heat loss through the mould.

The profound importance of insulation in promoting efficient feeding was illustrated by Adams and Taylor[58], who computed the freezing times of standard cylindrical heads under conditions with and without top and side wall insulation. The results, summarized in Table 3.6, indicate that a combination of both types of insulation offers the best conditions for prolonged feeding.

In the case of steel, after a short initial period during which the cold walls of the feeder head extract heat rapidly from the metal, radiation losses predominate owing to the high casting temperatures involved: top insulation

Table 3.6 Feeder head insulation: freezing times for 4 in × 4 in cylindrical feeder heads (min) (Reference 58)

	No treatment	Top insulation only	Wall insulation only	Top and wall insulation
Steel	5.0	13.4	7.5	43.0
Copper	8.2	14.0	15.1	45.0
Aluminium	12.3	14.3	31.1	45.6

is therefore most important in this case. For aluminium, by contrast, the use of insulating sleeves is the most useful single measure, since conduction loss to the walls is proportionately greater at lower temperatures. In all cases, however, greatly lengthened freezing times can be expected from the combined use of both types of insulation.

The materials used derive their low heat diffusivity from porous or granular structures. Top coverings range from dry sand and powdered slag to organic substances such as chopped straw, which first char and then burn away to leave a bulky ash. Proprietary anti-piping compounds may contain a certain amount of exothermic material. Some compounds used for steel introduce carbonaceous matter which becomes absorbed by the metal and locally depresses its freezing temperature, prolonging feeding without directly affecting the temperature of the metal in the head.

For the lining of feeder head walls, insulating materials may be applied as pre-formed sleeves. One type of sleeve used for copper and light alloy castings is made from foamed gypsum plaster, the sleeves being pre-cast from a plaster mix to which a foaming agent has been added to develop high porosity. Numerous early accounts were given of the application of plaster sleeves to the casting of non-ferrous alloys[59–62]. Other materials suitable for sleeve production are diatomaceous silica and vermiculite, the former material, unlike plaster, being thermally stable to a temperature high enough to be used for steel castings. Even paper has proved effective as a feeder head lining material[58]. Valuable theoretical and practical data on the use of insulating sleeves for non-ferrous castings, particularly aluminium alloys, were collected and developed in References 63 and 64. The economics of these sleeves was examined in Reference 65, with particular attention to the wall thicknesses which are adequate for castings of different solidification times. The use of insulated risers for large steel castings was described in Reference 66.

Exothermic feeding. A further category of feeding compound is that in which considerable heat is generated by exothermic reaction; in some cases molten metal is also produced. The basis for exothermic feeding is expressed in the typical Thermit reaction between finely divided aluminium and suitable metallic oxides:

$$2Al + Fe_2O_3 \rightarrow Al_2O_3 + 2Fe \quad \Delta H_{298}(25°C) = -852.9\,kJ$$

$$8Al + 3Fe_3O_4 \rightarrow 4Al_2O_3 + 9Fe \quad \Delta H_{298}(25°C) = -3347\,kJ$$

Exothermic feeding compounds may be employed in either of two ways. In the first of these the material is added directly to the feeder head in relatively large quantities, control of feeding action being achieved by the amount and timing of the addition in accordance with head and casting dimensions. The composition of the superheated molten metal product of the reaction must be compatible with that of the casting. In certain cases alloying elements have been introduced so that the reaction products approximate to the normal composition of the cast alloy[67]. This type of practice subsequently declined, however, and the most significant advance in the use of exothermic compounds came with the introduction of mouldable materials capable of being preformed into sleeves for lining feeder heads. The use of exothermic materials in this way was advocated by Finch in 1947[68] and became the subject of extensive activity[34,46,69–72]. The wide adoption of both insulating and exothermic sleeves owed much to the pioneering development and commercial service provided by the supplier industry.

The exothermic reactants are blended with bonding materials and water to provide green and dried strength, permeability being obtained by mixing with relatively coarse sand. Substances are also added to delay the reaction and extend the period during which heat is generated. It is not the aim in this case to produce molten metal but rather to retain the moulded shape of the material throughout solidification. This diminishes the danger of contamination of the casting, since the exothermic reaction is confined to the mould wall and there is no direct association of the reaction products with the movement of liquid in the course of feeding.

The sleeves can be conventionally moulded and vented. Whilst mouldable powders remain available, there is widely established use of preformed sleeves manufactured by the specialist suppliers in varied sensitivities. Some typical properties as originally developed were as follows[73]:

Permeability	150
Green compression strength:	27.5 kPa (4 lbf/in^2)
Dry compression strength:	2.75 MPa (400 lbf/in^2)
Density	1.34 × 10^3 kg/m^3
Ignition temperature:	300°C
Effective calorific value	6.9 MJ/kg

The period during which the heat is generated is obviously of great importance for optimum feeding effect. If the compound is insufficiently sensitive the reaction may exert its major influence after solidification, especially when used in conjunction with a small feeder head. Conversely, an instantaneous reaction may liberate heat too soon for maximum feeding efficiency. Ignition and reaction rate are therefore controlled to provide the required delay. For low melting point alloys and for smaller feeder heads a sensitive compound is used while under the opposite conditions low sensitivity is preferred.

As observed earlier in the present chapter, the powerful local heating effect of exothermic compounds produces a fundamental change in the character of the feeder head shrinkage. The metal level in the head sinks in a relatively uniform manner across the section, with the elimination or drastic shortening of the pipe normally encountered (Figure 3.17). The level at which the shrinkage terminates can be predicted with greater certainty than when a narrow pipe is present: the utilization of metal in the feeder head can consequently be as high as 67% compared with 14% for a normal cylindrical head. The feeder head volume needed to meet the volume feed capacity criterion is thus reduced to $V_f \geq V_c \times S(67-S)$ where S is the specific shrinkage percentage of the alloy. In conjunction with the increase in the effective modulus, this can increase the casting yield out of all proportion to the amount of heat generated.

A systematic scheme for the design of exothermic heads for steel castings was described by Wlodawer[34,46]. Selection of head dimensions is facilitated by charts of the type shown in Figure 3.32, the appropriate head size being derived from the casting modulus and from the demand for feed metal as determined by casting volume and specific shrinkage. The feeder head proportions are designed on the principle that a compact residual cylinder $h/d \approx 1$ should remain after feeding. It is the effective modulus of this residual body, derived from the true modulus in accordance with the influence of the exothermic material, which determines the modulus of the largest casting section that can be fed.

The influence of exothermic compound on the effective modulus of a cast body was illustrated earlier in Figure 3.22. Considering the previously cited example of a conversion factor of 1.43, the freezing time criterion would be satisfied when the geometric modulus of the residual body

$$M_f \geq M_c/1.43 = 0.7M_c$$
$$\text{(residual)}$$

The major impact of the exothermic technique is again demonstrated in Table 3.7, where it is shown that in the case of a spherical head the exothermic compound would reduce the volume of metal associated with a particular freezing time by a factor approaching 3.

An alternative way of depicting the influence of the exothermic layer on the feeding of steel castings is its approximate equivalence to a layer of steel of identical thickness. Considering the case of a sphere surrounded by

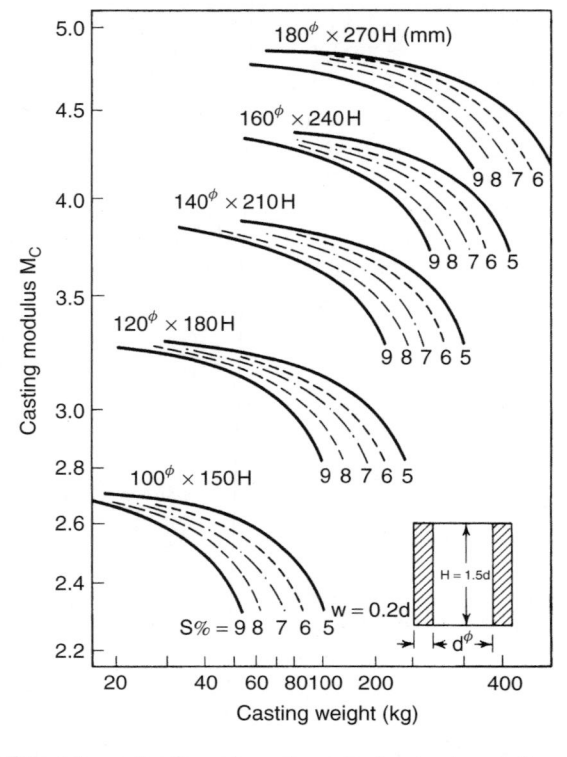

Figure 3.32 Chart for selection of exothermic feeder head dimensions (from Wlodawer[34]) (courtesy of Dr R. Wlodawer and Pergamon Press Ltd., reprinted from Directional Solidification of Steel Castings (1966))

Table 3.7 Exothermic feeder heads: comparison of volumes of spheres of a given modulus $M = 1$

Sphere	Actual diameter cm	Equivalent modulus M_E cm	Equivalent diameter cm	Equivalent volume cm³	Volume factor
Plain	6	1	6	113	1
Exothermic	6	1.43	8.58	330	2.92

a layer of exothermic material of thickness 20% of its diameter $2r$:

$$\text{Total volume of sphere and exothermic compound} = \frac{4}{3}\pi\left(\frac{7}{5}r\right)^3$$

$$\text{Volume of sphere alone} = \frac{4}{3}\pi r^3$$

Hence the volume factor is $(\frac{7}{5})^3 = 2.74$, a rough approximation to the value derived from the modulus factor 1.43.

Economic aspects of feeding compounds. Since the primary aim in using feeding compounds is the economic one of increasing casting yield, material and labour costs must be balanced against the associated metal and energy savings. It is evident that the greatest potential savings are to be made in the case of intrinsically expensive alloys, with their high costs of melting and of the metallic losses accompanying each melting cycle.

In the case of cheaper alloys top insulation is commonly employed, but the massive use of exothermic compounds to increase yield may in many cases be uneconomic. The compounds may nevertheless provide specific technical advantages by reducing defects in cases which do not readily respond to other methods. Cost savings of this type, together with fettling economies, must not be overlooked in establishing the overall cost balance for the materials.

A critical appraisal of insulation practice relative to the use of exothermics was carried out by Corbett[74], and a comprehensive review of this group of feeding aids was undertaken by Ruddle[75]. These authors emphasized the decline of highly exothermic metal-generating exothermic compounds in favour of the self-heating insulation principle, in which the exothermic reaction is used primarily to extend the time period during which the insulating properties of the sleeve or topping compound are most effective. Sleeve proportions had been examined and $H = D$ found to be suitable for most purposes.

Modern insulating and exothermics materials offer varying sensitivities, properties and modes of application, suitable for many different alloys and casting types. They consequently require careful selection based on the manufacturers' literature.

Pressure feeding

The pressure differential needed for effective feeding is normally provided either by gravity or by the atmospheric effect. The idea of a more direct application of feeding pressure is superficially attractive and has received some attention[76–80]. Two methods by which pressure can be artificially increased in a blind head are the use of pellets of gas forming compounds[77] and the direct application of compressed air[78]. It is interesting to note the parallel between such methods and the early practice in which foundrymen would cover an open feeder head with wet sand and seal the top with a weighted plate. Neither method has been widely adopted, largely because of difficulties of control and the associated dangers.

A weight of opinion suggests that little is to be gained in case of the skin forming alloys with their strong tendency to directional solidification[79,80].

That some benefit to soundness could be achieved with long freezing range alloys has been demonstrated in work on light alloys solidified under high pressures. The improvement can be attributed to suppression of the gas precipitation to which these alloys are particularly susceptible[81,82].

The use of closely controlled air pressure systems to assist feeding in long freezing range alloys would have considerable merit subject to appropriate safety precautions.

The junction of feeder head and casting

For the satisfactory completion of feeding, the neck joining the feeder head to the casting should be of sufficient section to solidify after the casting but before final solidification of the head. Should the neck freeze prematurely, the casting will contain shrinkage no matter how large the feeder head itself.

The necessary minimum dimensions vary with the particular casting design and method, but except in special cases the neck requires a cross-sectional area greater than that of the section which it is designed to feed. The importance of this principle was realized by Heuvers[36,39] who devised his inscribed circle method as an empirical test to be applied to sectional drawings to ensure conformity with the principle of progressive freezing. Any two dimensional approach must, however, necessarily be crude. Wlodawer later applied the modulus concept to the determination of neck dimensions on the basis that the moduli of casting, neck and feeder head should be in the ratio $1:1.1:1.2$[46]. The modulus of the neck can in this case be determined through its analogy to a semi-infinite bar, using the bar formula or the chart in Figure 3.16.

Neck size needs to be greatest when the metal enters the head through the casting cavity, but can be reduced when the surrounding mould is preheated by liquid metal, as for example in the whirlgate head. As in the case of feeder heads, therefore, neck dimensions must depend to some extent on gating methods and the initial temperature distribution. Wlodawer expressed the preheating effect in the modified relationship $M_{neck} = 1 - 1.03 M_{casting}$ according to the magnitude of the effect, representing from 15 to 21 per cent increase in the effective freezing time of the neck.

Restricted neck feeding

One method which can be adopted to restrict the neck without impairing feeding is based on the Washburn core. This is an extremely thin wafer with a central aperture, inserted into the mould across the junction of casting and feeder head (Figure 3.33). Provided that the core is sufficiently thin in relation to the surrounding mass of metal, its temperature rises rapidly because of its limited thermal capacity. Solidification thus proceeds extremely slowly adjacent to the core and the aperture remains open for feeding: the net effect is much as though the core were absent.

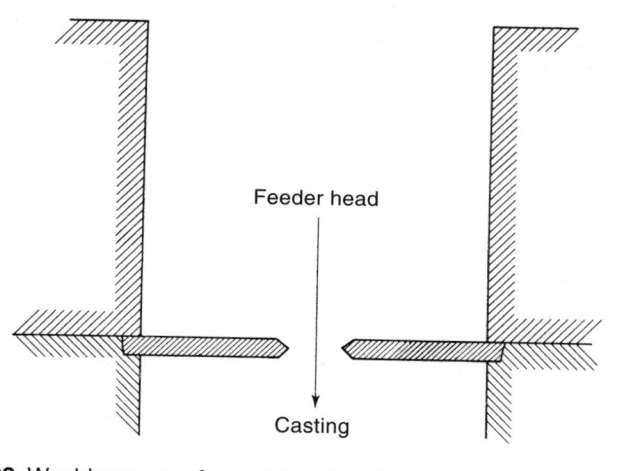

Figure 3.33 Washburn core for restricted neck feeding

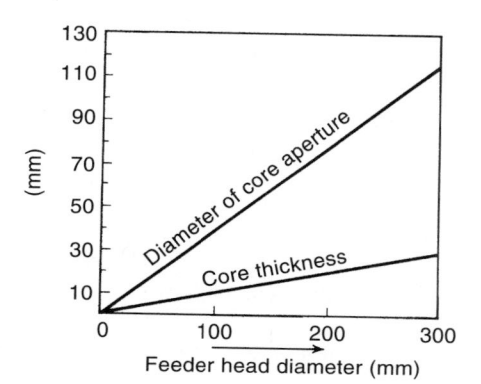

Figure 3.34 Dimensions of Washburn wafer cores (after Wlodawer[34]) (courtesy of Dr R. Wlodawer and Pergamon Press Ltd., reprinted from Directional Solidification of Steel Castings (1966))

The successful use of the Washburn core depends upon maintaining the correct relationship between core thickness, aperture size and metal section; it is evident that the thinner the core the smaller will be the aperture needed for satisfactory feeding. Adequate mechanical strength and resistance to fusion and metal penetration are the other factors determining the thickness.

Dimensional relationships for wafer cores for use with various sizes of steel casting were established by Brinson and Duma[83] and further data was collected by Wlodawer[34]; the latter is summarized in Figure 3.34.

Wafer cores for this application are commonly produced from hard sand mixtures reinforced where necessary with irons; shell moulded cores may also be used. Since metal penetration is prone to occur under these conditions, porosity must be minimized by the use of fine base sands.

Restricted neck feeding greatly reduces the work entailed in feeder head removal and subsequent fettling. With alloys of low as-cast ductility, normal methods of head removal may be discarded in favour of knocking off or hydraulic shearing. These processes are assisted by the provision of a sharply angled edge to the feeding aperture.

Design modifications, padding, chills and insulation

In many castings of complex design, difficulty is encountered in obtaining access to isolated heavy sections for the direct attachment of feeder heads. It then becomes necessary to control the freezing process so as to obtain effective feeding from other heads, separated from the heavy sections by lighter members.

Suitable measures include selective chilling or insulation, the addition of further cast metal in the form of padding, or modification of the basic design of the casting. Similar measures may be needed for parallel walled sections such as bars or plates, where it has already been seen that feeder heads are only effective over limited distances.

Design modifications and padding

In some cases the feeding problem can be solved by means of a fundamental design modification whereby thin sections are thickened or taper is introduced to provide channels for feeding. Heuvers[36] used his technique to verify the adequacy of feeding channels from a head through the body of a casting; by application of the 'inscribed circle' method an indication was obtained of the amount of additional metal needed to ensure a continuously increasing cross-section towards the feeder head. This two dimensional method may only be treated as a rough approximation: an example of its use to determine padding requirements is illustrated in Figure 3.35.

If padding cannot be incorporated as a design modification, it may still be added for casting purposes and subsequently removed by power cutting, grinding or machining. This, however, adds greatly to production costs, so that other measures such as the use of exothermic or indirect padding are to be preferred: these methods will be separately discussed.

The extent of the progressive padding required to ensure the soundness of parallel sections was established for steel castings by Brinson and Duma[84]. Centre line shrinkage in such sections is not however always unacceptable and in many cases no attempt is made to eliminate it altogether.

A similar effect to that of normal padding can be achieved by *indirect padding*, a technique in which molten metal is fed to a position adjacent to the critical section, thereby delaying its freezing (see Figure 3.36). This technique was fully described by Daybell[85].

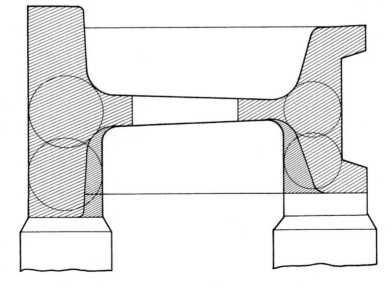

Figure 3.35 Cranewheel casting, illustrating use of padding (from Heuvers[36])

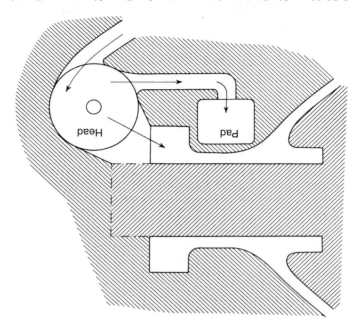

Figure 3.36 Use of indirect pad to promote feeding of heavy sections in valve body casting (after Daybell[85])

Differential cooling: the use of chills

Chills provide a highly effective means of promoting directional solidification: their strong influence in extending the feeding range of feeder heads, demonstrated by Myskowski, Bishop and Pellini in work with castings of parallel cross-section, has been referred to earlier in the present chapter[52]. Since the general influence of surface chilling is to steepen the temperature gradients within the casting, the tendency is for solidification to occur by skin formation rather than in the pasty manner: this function of chills is particularly valuable in promoting the soundness of the long freezing range alloys typified by the bronzes and gunmetals (see Figure 3.3). The close spacing of chills and feeder heads is also used to achieve a high standard of soundness, combined with fine microstructures, in the production of premium quality castings with very high levels of mechanical properties.

The most common use of chills, however, is to encourage selective freezing of thicker members of a casting by accelerated heat extraction. The increased rate of cooling enables such members to draw feed metal through the thin sections whilst these are still liquid. It is thus possible to produce sound sections in situations which would be inaccessible for direct attachment of feeder heads. Equalization of cooling rates can also be an important factor in the suppression of hot tears.

Metal chills may be used either externally or internally to the casting. Internal chills are less commonly used than at one time: the lack of structural homogeneity and the possibility of discontinuities from lack of fusion detract from the metallurgical quality of the casting. Such chills are, however, less objectionable in lightly stressed locations and may be freely used in situations where the chill itself is machined out at a later stage, for example in bosses intended for drilling. External chills are widely employed. They are commonly made of steel or cast iron and are positioned against the pattern during moulding, being provided where necessary with hooks or wires for firm anchorage in the moulding material. Aluminium alloy chills are sometimes employed in light alloy founding.

The effect of a chill in increasing the freezing rate of a casting is a function of its heat diffusivity $(kc\rho)^{1/2}$, where k is the thermal conductivity, c the specific heat and ρ the density of the material[29]. The relative chilling effects of a wide range of materials were determined by Locke *et al*[86] using experiments on the freezing rates of steel spheres. Metal chills were shown to increase the freezing rate by a factor of approximately 4 relative to that obtained in sand moulds, although a much higher factor of about 14 has been indicated in other work with copper alloy castings[87].

The freezing time of a wholly or partially chilled casting is the same as would be associated with a casting of smaller volume in the absence of chilling. Arising from the previously discussed relationship between freezing time and cooling modulus ($V/A = kt^{1/2}$, after Chworinov), it can be seen that a factor of $\frac{1}{4}$ applicable to freezing time is equivalent to a factor of $\frac{1}{2}$ when applied to the modulus. This effect of the chill in reducing the modulus may be alternatively expressed as an apparent increase in the total surface area of the casting: in conjunction with the actual volume, this gives the means of determining an effective modulus value for purposes of feeding calculations:

$$M_{effective} = V/A_{apparent}$$

In a partially chilled casting the value of $A_{apparent}$ is derived by the addition of the area of sand surface to the area of chill multiplied by the appropriate factor, derived from experiments on freezing rate:

$$A_{apparent} = A_{sand} + A_{chill}(t_{sand}/t_{chill})^{1/2}$$
$$= A + A_{chill}[(t_{sand}/t_{chill})^{1/2} - 1] \tag{3.1}$$

In the previously mentioned case of a steel casting, therefore, the apparent doubling of that part of the surface area in contact with a chill would simply require that the area of the latter be added to the total area when calculating the modulus. It has been argued by Wlodawer, however, that the factor of 2 is only applicable in the presence of the air gap commonly formed between a casting and a metal mould: true metallic contact is said to give a factor of 3 for a steel chill.

Briggs, Gezelius and Donaldson[88] demonstrated the use of external chills of various designs for the elimination of shrinkage cavities at L, T, X and Y sections. This technique, together with that of design modification of intersections, is illustrated in Figure 3.37: the latter topic will be further examined in Chapter 7. The differential cooling effect provided by

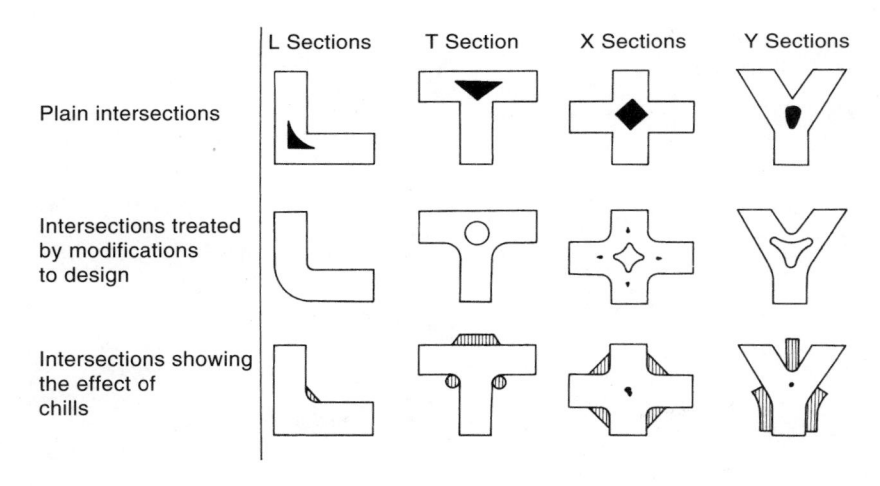

Figure 3.37 Treatments of intersections by design modifications and use of chills (from Briggs et al[88]) (courtesy of American Foundrymen's Society)

chills is particularly suitable in the case of intersections extending for long distances, as exemplified in a plate section reinforced with stiffening ribs: general chilling is in such cases more practicable than direct feeding (see Figure 3.38). Small face chills are also invaluable for local application to isolated bosses and projections where individual feeder heads would be uneconomical.

A further method of using chills is that suggested by Chworinov[89] in which the chill is placed at an angle to the casting surface and exerts its effect through a tapering layer of sand, the degree of cooling varying with the distance from the casting surface. The progressive variation in the cooling influence of an indirect chill can be used to encourage directional solidification in a casting member of uniform section[52].

Whether chills are employed primarily to extend feeding distance or for their local influence in relation to design features, they exert a profound influence on feeding requirements. By creating an earlier demand for feed metal, the freezing time needed in the feeder head is shortened. Thus, the reduction in the effective modulus of a wholly or partially chilled casting enables feeder head dimensions to be reduced towards the level at which volume feed capacity governs the size of head required. Referring to the previously considered series of plate castings fed by a spherical head, the influence of various degrees of chilling on the feeder head volume required to maintain the necessary freezing time is illustrated in Figure 3.39; the considerable shift towards control by volume feed capacity is evident from Figure 3.40. Taking this point in conjunction with earlier discussions, it follows that the extreme case of high yield will be achieved in the case of a gravity die casting or ingot fed by an exothermic head, with its highly efficient utilization of metal (compare Figures 3.39 and 3.22b). Feeder head

Figure 3.38 Use of chills along intersections of ribs. Positions of face chills visible on flat surface

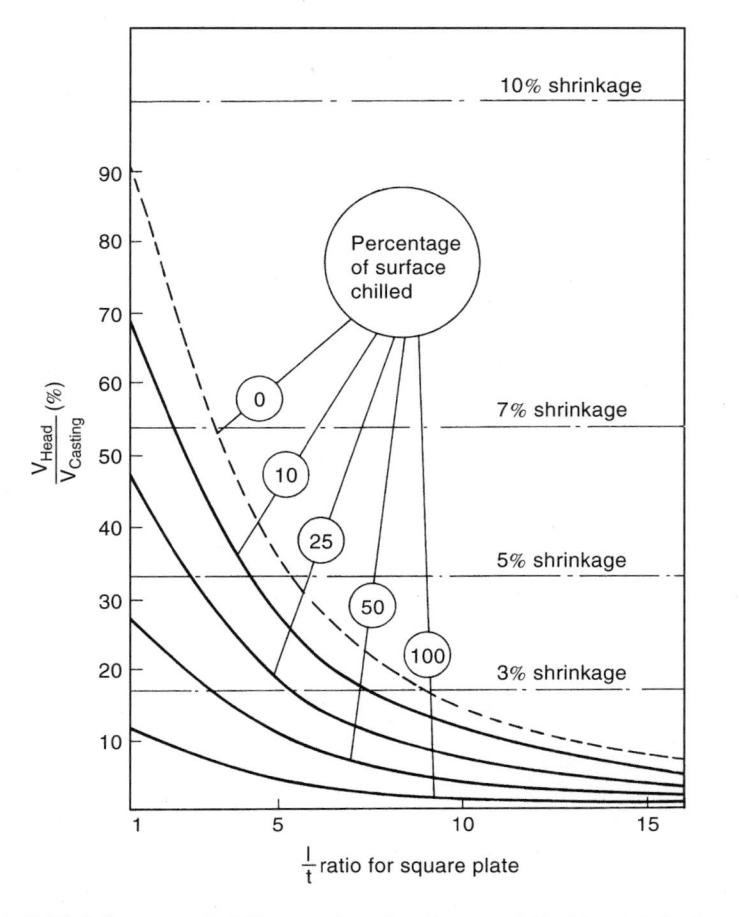

Figure 3.39 Influence of chilling on head volume required to satisfy freezing time criterion (spherical head; plates of various proportions)

volume can in this case approach the ideal of being merely sufficient to compensate for the shrinkage in the system.

Although most chills are of metal, non-metallic substances of high heat diffusivity, such as graphite, carborundum and magnesite, may be used where less pronounced cooling is required. Ruddle[29] presented a large amount of data from many sources on the thermal properties of materials used for chilling and insulation. Based on the experiments of Locke et al[86], Figure 3.41 illustrates the comparative cooling effects of a wide range of substances and gives the factors required to determine the apparent increase in surface area or reduction in effective modulus using the previously mentioned equation 3.1.

For metal chills, steel and cast iron are the most frequently used materials, being both cheap and readily machinable from bar or plate stock. Shaped

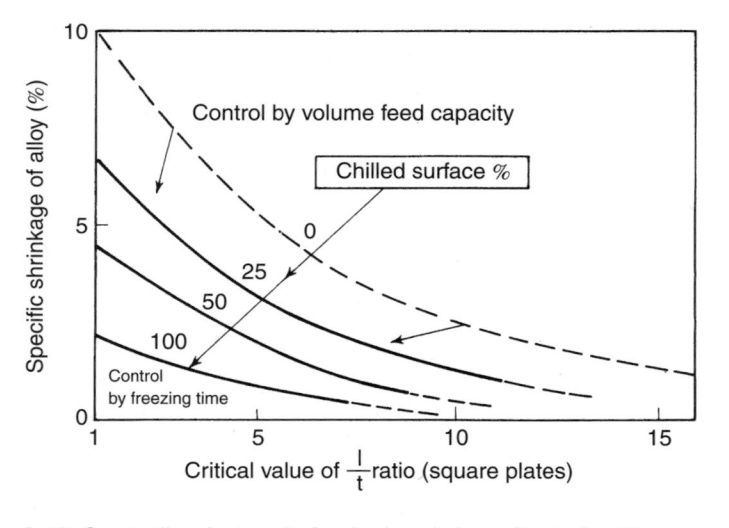

Figure 3.40 Controlling factors in feeder head size: effect of chilling

chills to fit irregular casting contours may be moulded and cast, although it is generally preferable to chill flat rather than contoured surfaces.

A chill must be of adequate mass in relation to the section to which it is to be applied: if the thermal capacity is insufficient the temperature of the chill will rise rapidly to a level at which it will only extract heat at a rate similar to that of the remainder of the mould. Chills must, furthermore, be positioned only where heat can be extracted to useful effect. If a chill is placed in the ingate region, for example, the passage of liquid metal across its surface during mould filling may cause excessive preheating. In either of these cases heat transfer to the chill during the superheat stage may nullify or even reverse its influence by the time the critical stage for feeding is attained. Under such conditions the chill will increase rather than diminish the effective modulus of the casting.

Chills must be dry and free from contamination by oxide scale or grease if blowing defects are to be prevented. Particular care is needed to avoid condensation of steam on the surfaces of chills in greensand moulds, as can occur through the insertion of warm cores. A further potential source of gas defects is porosity or craze cracking of the chill itself: air in the cavities can be expelled into the metal through expansion on casting. Chills should therefore be discarded when they become cracked or heavily oxidised.

Another potential defect is surface tearing near the edges of a chill, due to differential contraction of the casting skin. The danger can be diminished by the use of a number of small rather than a single large chill; this reduces the strain concentration and permits contraction stresses to be absorbed at a larger number of sites between the chills.

Local insulation and exothermic padding

Just as chilling may be used to accelerate the freezing of heavy sections, so may insulation be used to delay solidification of the lighter sections of a casting; the latter may then serve as feeding channels for the heavier sections. This can be accomplished by using mouldable mixtures based upon insulating compounds such as diatomaceous silica. The influence of typical insulating materials upon freezing time and apparent surface area was illustrated in Figure 3.41: it is evident that the effective modulus for feeding can be determined by a similar approach to that adopted for chills. The use of a value below unity for the factor $(t_{sand}/t_{material})^{1/2}$ produces a decrease in the apparent surface area as derived from equation 3.1, giving a corresponding increase in the effective modulus.

The principle of differential insulation has been extended by the introduction of exothermic padding, with mouldable exothermic compounds

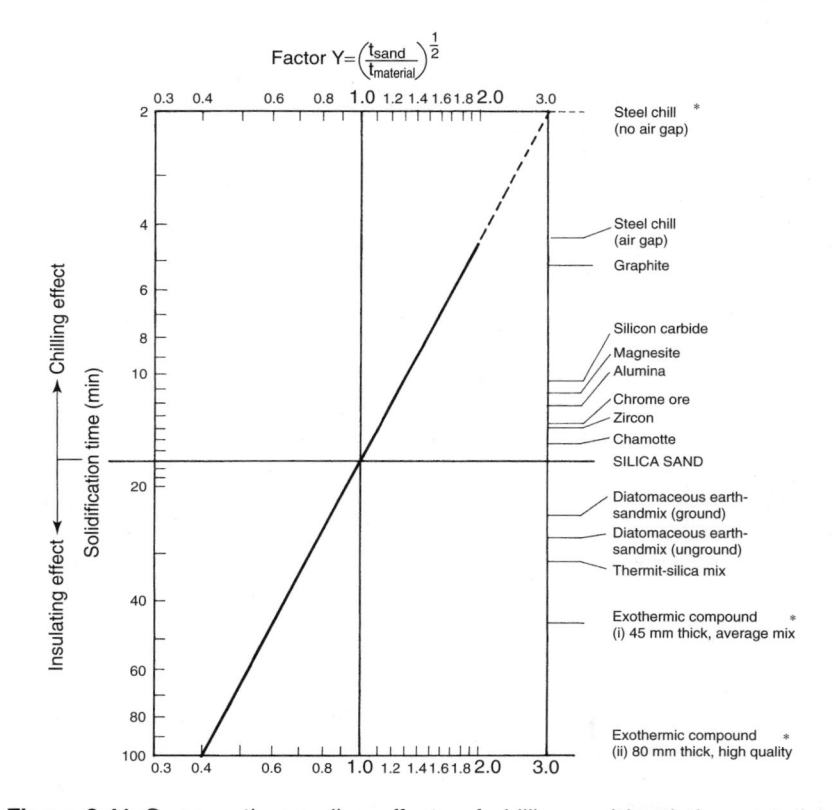

Figure 3.41 Comparative cooling effects of chilling and insulating materials (after Locke *et al*[65] and Wlodawer[34]). Data determined from freezing times for 150 mm steel spheres. Factor Y, for determination of apparent change in surface area = $(t_{sand}/t_{material})^{1/2}$ (see equation 3.1). *Data from Wlodawer, all other data from Locke *et al*

applied directly to the body of the casting[90–92]. It has been established as a rough guide that exothermic materials may be used as a direct substitute for metal padding, whether in the form of local pads beneath the feeder heads or as an alternative to general thickening of a thin member to provide a feeding channel for an isolated heavy section[91]. Similarly, the use of exothermic material has been shown to increase the feeding distance in parallel walled plate sections, in much the same manner as the metallic padding previously discussed[90]. The criteria for exothermic padding and its role in increasing feeding distance in steel castings have been further examined by Wukovich[93].

Exothermic padding is thus suitable in cases where it is not permissible to incorporate metal padding into the design, or where the cost of its removal would be prohibitive.

The feeding of cast irons

The feeding principles reviewed over the preceding sections, although applicable to a wide range of alloys, were developed mainly in the steel castings context, as reflected also in much of the literature. The position with respect to the cast irons was much slower to develop and widely different approaches to the problem were adopted. Despite certain common features, the feeding characteristics of most cast irons differ from those of steel in important respects, as discussed at the beginning of the present chapter and arising from the formation of graphite during solidification. The resulting expansion offsets the demand for feed metal accompanying the contraction of the metallic phases, but the position is also complicated by enlargement of the mould cavity, which varies with the rigidity of the mould; other factors too affect the overall feed requirement, including both composition and pouring temperature.

The feeding of nodular iron was examined by Karsay[94,95] and others, emphasizing the ability, in suitable circumstances, to produce sound castings without substantial provision for feeding. Modulus calculations were, however, also employed with success, although it has to be appreciated that total freezing time is an inappropriate criterion for determination of feeder head size, so that modified modulus values need to be adopted.

This arises from the fundamental difference in the roles of feeder heads for steel and for nodular cast iron, in that the former requires the head to function throughout the freezing of the casting and the modulus approach is designed to achieve that, ensuring directional solidification into the head; the need for nodular iron, by contrast, is to satisfy the feed demand associated with the early liquid contraction and mould cavity enlargement, after which the feeder head should solidify before the onset of the main graphite expansion in the casting. The head, and its modulus value, can thus be considerably smaller relative to the casting modulus.

Further understanding in this field came with systematic investigations undertaken by Gough and Clifford[96], in which a different approach came to be proposed, relying upon the volume feed requirement as the prime determinant for feeder head provision. Various approaches to the feeding of nodular iron were subjected to critical review, and experiments in earlier work[97] had shown wide variations in the ratio of feeder head to casting modulus to be required, depending upon the degree of mould rigidity, metal composition and pouring temperature. Illustrations demonstrated the validity of the observation that the volume feed requirement as a proportion of casting volume remained virtually constant irrespective of the shape and modulus of the casting, provided that other casting conditions, particularly mould rigidity, were standardized. Figure 3.42 shows this linear relationship for nine castings of widely varying shapes, sizes and modulus values, made in resin bonded sand moulds, representing a feeder head to casting volume ratio consistently within the range 0.4–0.45. Further data, however, demonstrated the marked effects of other practical factors which need to be taken into account, just as when using the modulus system.

The modulus approach is included in a modern feeding system for iron castings detailed in Reference 98, which also incorporates the volume feed requirement and thus parallels the dual system already reviewed in detail for steel and other alloys. The crucial difference for the grey and ductile cast irons, however, is the replacement of the normal freezing time criterion with a shrinking time criterion, to determine the effective, as distinct from the geometric, modulus.

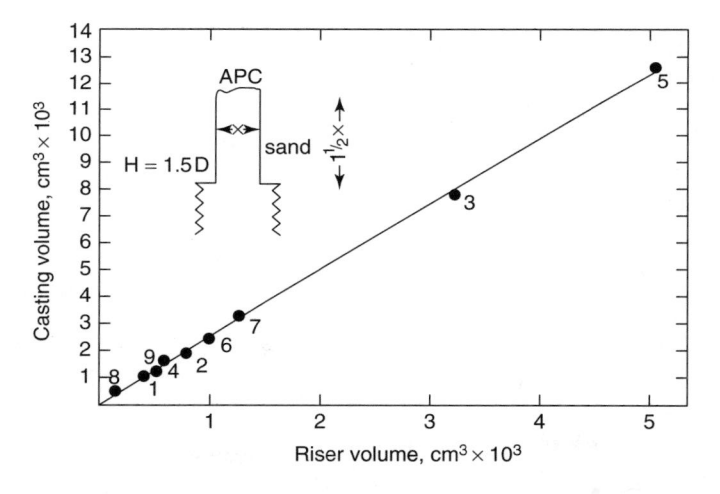

Figure 3.42 Relationship between casting volume and minimum riser volume to achieve sound feed (from Gough and Clifford[96]) (courtesy of Institute of Materials)

The normal modulus value is reduced by applying a factor $\sqrt{ST/100}$, as well as the standard safety factor of 1.2 as customarily applied to steel ie

$$M_F = M_C \times 1.2\sqrt{ST/100}$$

The value of the shrinking time ST, expressed as a percentage of the total freezing time, is derived from charts which interrelate the carbon equivalent, the metal temperature and the magnitude of the geometric modulus itself. As in other practice the modulus of the neck is normally designed to be intermediate between those of the casting and head, achieved by using the factor $1.1\sqrt{ST/100}$.

For the separate volume feed requirement determination, the specific shrinkage value of the alloy to be employed in the calculation is adjusted for the same factors as used for the shrinking time, and also for mould wall movement in the range 0–2% according to the type of moulding material.

As in previous cases the feeder head size needs to be determined from the greater of the two requirements, and will also be governed by the "utilization factor", or proportion of feed metal actually available from the type of head employed. A full range of feeding aids, including insulation and exothermic sleeves, can be applied using similar principles.

Other studies of feeding in nodular iron have paid special attention to the modulus of the neck, to control the flow of liquid between feeder head and casting when high pressure is developed during the graphite expansion stage. This aspect was emphasized by Karsay, the concept being to allow the return of some liquid from the casting to the head, after which freezing of the neck produces a more controlled pressure increase in the casting for the later stages of solidification. Neck dimensions are considered in some detail in References 99 and 100.

Apart from charts or nomograms as referred to above, computer software is available to simplify the time-consuming calculations involved in feeding practice whichever system may be adopted.

Solidification modelling

A significant proportion of modern casting research has been devoted to process modelling and simulation. Aspects of the casting process investigated have included fluid flow and mould filling, solidification and feeding, development of microstructure and the generation of stresses, with their relevance to such problems as distortion and hot tearing. Most progress has been achieved in relation to feeding practice, and has led to extensive commercial use of computer-based systems for the determination of foundry methods for individual castings.

Primary purposes of computer modelling are to increase casting yield, so reducing gross weights and demand for molten metal, to enhance product quality, reducing scrap and rectification costs, and to reduce lead times from design to finished casting. Apart from the energy savings and associated environmental benefits, results also provide a basis for more reliable cost estimates for quotation purposes.

Early developments were restricted to such individual aspects as the determination of accurate casting weights and the calculation of optimum feeder head proportions and feeding distances, but comprehensive simulation of events occurring in the mould on casting has since enabled computer assessments of intended casting methods to replace trial and error in the foundry itself. Potential defects can be predicted and the method modified and reassessed, so that "right first time" castings can be produced, the "scrap" having been confined to the computer. Reduced reliance is placed on sample castings and there is greater opportunity for experiment. Such a facility is increasingly seen as a necessary feature in an overall design and procurement system for cast products.

The common basis of the varied modelling systems that have been developed is the simulation of the solidification process in a proposed casting-feeder head system. The geometry of the casting is first defined in the form of a 3D CAD model. This is then provided with an empirically devised running and feeding system and subjected to the simulation process, commonly presented in pictorial form. The solidification sequence is predicted and shrinkage cavity locations and morphology can be identified. The method can then be modified and re-tested until the required result is achieved. The exercise can precede the manufacture of tooling and can even be integrated into the shape design stage itself.

The systems employed, implemented through dedicated computer software, vary widely in scope, capabilities and cost. A distinction is frequently made between two broad categories, namely knowledge- or rule-based systems, using empirical data derived from experience with actual castings, and more fundamental systems based on mathematical analysis of the physical phenomena occurring in casting, using available data on the relevant properties of the metal and mould.

Particular qualities required in a software system for the modelling of foundry processes were put forward and discussed in a review by Hansen[101]. Special emphasis was given to the need for user-friendliness, but other elements included ease of geometry input and of interfacing with CAD systems, automatic mesh generation, and ease of handling of initial and boundary conditions.

In involving the user, the criteria employed in programs need to be made clear, so that the technical grounds for decisions remain in the hands of the foundry methods engineer. The modeller may also need to be reminded of the economic aims, so that modelling does not become an end in itself, and so that the resources remain focused on the metallurgical and cost benefits of the exercise.

An ideal simulation of the casting process would need to incorporate effects of many interrelated phenomena, including fluid flow in mould

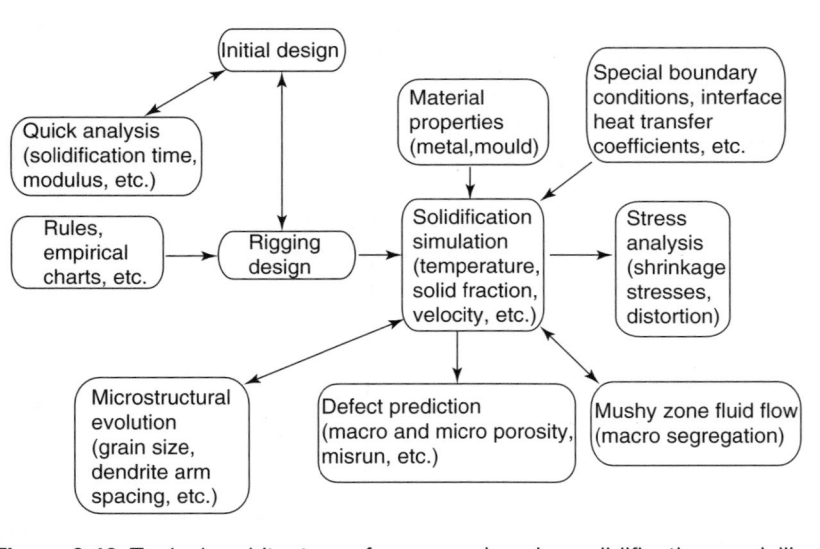

Figure 3.43 Typical architecture of a comprehensive solidification modelling system. (from Upadhya and Paul[103]) (courtesy of American Foundrymen's Society)

filling and during freezing, heat and mass transfer during solidification, microstructure development and volume changes; Jolly listed almost thirty such items in a short review[102]. Comprehensive modelling treatments do indeed incorporate a number of these, as represented in the typical model architecture shown in Figure 3.43, from a major review by Upadhya and Paul[103], in which the principal factors were analysed in detail. The approach used in the advanced systems is to provide specialized modules based on the respective phenomena influencing mould filling, thermal stresses and other contributing factors, using solution methods best suited to each.

Essential to accurate modelling and simulation is the availability of a high quality database, providing property values for use in the heat transfer and other analyses. Examples of such properties are listed in Table 3.8[104], which indicates the magnitude of the data requirement, given the many distinctive cast alloys.

Full physical modelling requires workstation based computer facilities and a substantial range of software. Long times can be involved, using specialist operators and incurring appreciable costs, so that a balance needs to be struck between absolute accuracy and the practical value of the results obtained.

Simpler systems offer rapid analyses at lower cost and can be operated on PC facilities by foundry staff. A typical approach employs the geometric cooling modulus based on the Chworinov rule, further developed by Wlodawer and extensively discussed in the present chapter as the main quantitative technique routinely employed in feeding system design. Its

Table 3.8 Scope of thermophysical data required for the modelling of a casting process (from Quested[104]) (courtesy of Institute of British Foundrymen)

Physics of Casting Process	Thermophysical Data Required	Modelling used for Prediction
Solidification Involving Heat Transfer by Conduction Convection Radiation	Density vs Temperature	Hot spots
	Specific heat vs Temperature	Effectiveness of Riser
	Conductivity vs Temperature	Effectiveness of Chill
	Latent heat of fusion	Effectiveness of Insulation
	Emissivity–metal/mould/furnace wall	Solidification direction
	Liquidus Temperature	Solidification shrinkage
	Solidus Temperature	Microporosity
Metal / Mould/Die/Shell / Chill / Insulation / Furnace	Interface heat transfer coefficient metal/mould metal/chill mould/chill mould/environment	
Micro modelling	Phase diagram	Microstructure morphology
	Chemical species composition	Grain size
	Solid fraction vs Temperature	Grain orientation
	Number of nuclei per volume	
	Growth constant for eutectic	
	Diffusivity of solute in solvent	
	Gibbs Thompson coefficient	
Fluid flow	Viscosity vs Temperature	Cold Shut
	Surface tension	Mould filling time
	Coefficient of friction–metal/mould	Effectiveness of Ingate
		Effectiveness of Runner
		Pouring rate
		Pouring temperature
Stress analysis	Stress/strain vs Temperature	Hot tear
	Thermal expansion coefficient	Casting dimension
		Casting distortion
		Internal stress

adaptation to rapid simulation enables the order of freezing to be determined and feed paths identified. The effects of certain design features such as fins, ribs or cavities involving small cores can be subject to empirical corrections analogous to those originally discussed.

Some of the more rapid methods can themselves incorporate limited heat transfer calculations, to provide better approximation to behaviour under various alloy and process conditions. The added complexities of full modelling need only be introduced in circumstances where the additional resources are warranted by the potential benefits.

Operations in full simulation modelling

The required 3D solid model can be constructed from 2D drawings, using a digitizer facility incorporated in the computer software, or can be input directly in the form of compatible CAD data, so saving computer time. Computation and analysis can then proceed.

The central feature of the physically based systems is numerical heat transfer modelling using finite element, finite difference, boundary element and other methods to solve the relevant partial differential equations. The cast shape is broken down into a 3D mesh or grid of small volume elements, defined by nodal points, for which heat content and transfer calculations can be made and the results integrated into a comprehensive pattern approximating to behaviour in a continuous body. The accuracy depends on the number of elements selected, which can extend to several million. This naturally determines the computer power and time required for the analysis, which in complex cases can be measured in weeks rather than minutes.

Accurate prediction of solidification and feeding, although primarily concerned with heat and mass transfer during freezing, also requires consideration of flow behaviour during filling of the mould cavity. As discussed earlier in the present chapter, the speed and direction of flow during pouring affects the initial temperature distribution in the mould. Modelling of fluid flow using the appropriate equations is therefore incorporated in the more comprehensive systems, the temperature profile at the end of filling providing the initial condition for the subsequent cooling and solidification stages.

Similarly, the modelling of thermal stress development in the solidified parts of the casting enables account to be taken of phenomena which change the heat transfer coefficient at the solid metal–mould interface, including airgap formation. The wider problems of distortion, residual stresses and hot tearing can also be addressed: elastoplastic behaviour and elevated temperature properties of metal and mould are involved in this aspect of modelling, posing a major further requirement for provision of data.

Of the principal analytical systems, the *finite difference method* is very effective for heat transfer and fluid flow calculations and has been widely used for solidification simulation, but is unsuitable for thermal stress and deformation analyses. The *finite element method* is more flexible and is particularly suitable for the calculation of thermally induced stresses, although more costly in computer time. It is, however, possible to reduce this by generating a fine mesh in the most critical regions of the casting and a coarser mesh in other areas.

The interface zone is itself a major consideration in the modelling calculations. Not only can the onset of an airgap cause a discontinuous drop in the heat transfer coefficient, but the occurrence depends on the position and local geometry of the interface; mould and die coatings too affect the coefficient. Airgap formation is a particular feature in die casting processes; modelling aspects have been the subject of numerous researches, including those reported in References 105–107.

Other special aspects include the modelling of microstructure development: some researches in this field, each including reviews and references, are contained in References 108–110.

The broad implications of modelling were reviewed in Reference 111, which examined a number of approaches to particular problems in casting simulation and the types of method employed. Valuable reviews of activities in the field were contained in References 112 and 113, and various commercial systems were referred to, most of which have since been subject to further development. Much outstanding progress has been made in collaborative group projects involving research organizations, industrial companies and academic bodies.

Widely used commercial systems include ProCAST, a product of UES Software Inc, USA and MAGMAsoft, originating at the Aachen University of Technology, Germany. Simpler systems capable of rapid studies include the AFS Solidification System (3D); a widely applied system originating in the UK is SOLSTAR, developed by the foundry supply organization Foundry International Ltd.

The ProCAST system, based on 3D finite element analysis, provides one base module containing the thermal/solidification solver, pre- and post processors and a material database. Seven additional modules are concerned with mesh generation, fluid flow, stress, radiation, microstructure, electromagnetics and inverse modelling, with coupling of the heat transfer, fluid flow and stress analysis: the system is thus capable of modelling a full range of phenomena influencing the behaviour of metal in the mould.

The inverse modelling facility enables the numerical analysis of heat flow to be combined with results of experimental temperature measurements. This permits the direct determination of an interface heat transfer coefficient, or of a property such as thermal conductivity, that may not be available from the database.

The fully automatic 3D solid mesh generator provides for variable mesh densities on edges and surfaces, a finer mesh being used in zones of rapid change on the lines previously mentioned.

The radiation module is particularly important in relation to the modelling of investment casting, in which critical events occur at very high mould and metal temperatures, with heat transfer mainly by radiation. A significant parameter under these conditions is the viewfactor, which is an outcome of complex interactions between radiating surfaces, determined by the casting geometry, in which shadow effects are obtained and radiation is impeded by other surfaces. Some approaches to this and other special problems were reviewed in References 103 and 111.

A further capability for the investment casting application is automatic shell generation, which facilitates the addition of layers of shell material and provides for shells of varying thickness.

The successful use of the ProCAST system for the modelling of super-alloy investment castings was described in Reference 113, which reported work by the ICCA, a US investment casting consortium. In this case the finite element mesh for the mould was created from the surface of the finite element mesh for the metal. The generation of the complete casting model also employed a gating library, comprising a store of solid models for standard systems. The appropriate gating pieces could thus be selected and mathematically integrated with the model for the component.

A complete finite element mesh for the ceramic shell mould for the test casting used is shown in Figure 3.44. The modelling sequence of the casting events begins with the mould at uniform temperature on removal from the preheating furnace, followed by differential heat loss, predominantly by radiation, during transfer to the casting point. This determines the critical

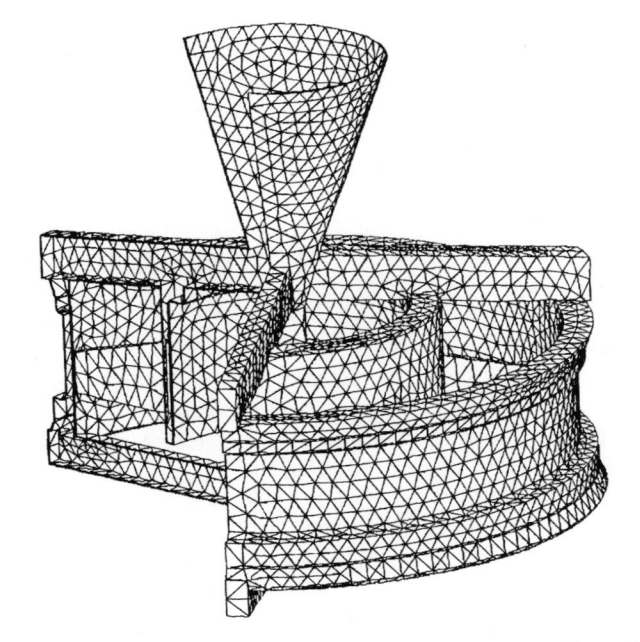

Figure 3.44 Finite element mesh for ceramic shell mould (from Tu *et al*[113]) (courtesy of The Minerals, Metals and Materials Society)

temperature profile before pouring. Mould filling and solidification are then modelled, as affected by the changing heat transfer coefficients between metal and mould, together with the development of stresses in the solidified material and consequent deformation.

Figure 3.45 (see colour plate section) shows intermediate mould filling simulations derived from the thermal and fluid flow analysis. A notable feature was the substantial temperature differences developed within the metal by the completion of mould filling, emphasizing the importance of coupling the fluid flow with the heat transfer analysis.

The ability to model stresses generated within the metal during the later stages of solidification and further cooling is demonstrated in Figure 3.46 (see colour plate section). This also reveals the pattern of distortion produced by the interactions of ceramic mould and casting members due to differential contraction: the actual deformation is exaggerated to assist the observations.

Included in the cited examples of production investment castings was a single crystal gas turbine blade component, produced by the mould withdrawal technique discussed in other chapters of the present work. The nature of the solidification process is confirmed in Figure 3.47 (see colour plate section), where the mushy or pasty zone is seen to be clearly defined. The flat solidus isotherm confirms the sharp delineation of the fully solid material produced by the high temperature gradient in the system.

Further work with much larger industrial gas turbine blades, of lengths up to 75 cm and with equiaxed structures, was described in a later reference[114]. This included attention to the design of stiffening ribs and the geometry of internal cores to eliminate porosity. Simulation of the injection moulding of the wax patterns was also undertaken, together with work on other aerospace components.

Comprehensive modelling capability is similarly represented in the widely adopted MAGMAsoft system, seen in further examples, featuring heavy castings used in the offshore industrial sector and in car body production. Application to structural nodes of the type previously highlighted in Chapter 7 is illustrated in Figure 3.48 (see colour plate section). This shows the complete sequence from design stage to finished product. Design validation using the OSCAL finite element stress analysis system is followed by solidification simulation with the MAGMAsoft program to optimize the casting technique by an iterative process, and on via pattern construction to the finished casting.

Comprehensive modelling has performed a similar central role in work at the Castings Development Centre, described in Reference 115. This entailed the production of one-off SG cast iron dies for the press forming of car roof panels. Finite element analysis using SDRC IDEAS software was first undertaken to determine the stress distribution developed from the loads applied to the working faces of the castings, of which a typical shape can be seen in the illustration of the punch die in Figure 3.49. Casting simulation

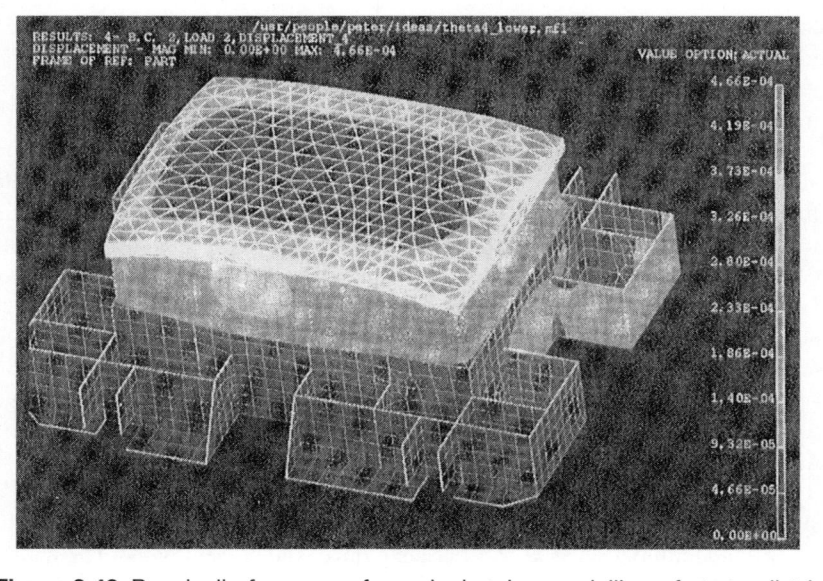

Figure 3.49 Punch die for car roof panel, showing modelling of stress distribution (from Haigh *et al*[115]) (courtesy of CDC)

was carried out using the MAGMAsoft system, with a thermophysical properties database. This was used to predict local solidification rates, shrinkage and, using an associated program, the consequent mechanical property distribution in the workface and supporting rib structure of the die.

Figure 3.50 (see colour plate section) shows the hardness distribution, the higher values of both hardness and tensile strength being obtained, appropriately, in the more highly stressed supporting rib structure. Optimal thickness and dispositions of the ribs were determined to achieve a light, strong design. This was incorporated in the production of foam patterns by NC machining directly from the CAD file: these were used to produce lost foam castings.

The above examples relate to gravity casting in refractory moulds. In pressure die casting the optimization of metal flow to avoid premature solidification, entrapped air and oxide film inclusions is assisted by mapping of the simulated pressure contours during the injection sequence. Gating technique and injection parameters can then be designed to ensure a favourable solidification front. Figure 3.51 (see colour plate section) shows a modelled pressure profile, using ProCAST, for a large transverse car chassis member die cast in magnesium alloy.

Other comprehensive modelling systems include MERLIN, centred on the University of Wales at Swansea. Again finite element meshes are automatically generated and fluid flow, solidification and thermal stress profiles predicted; the thermal stress module is fully coupled with the solidification module to achieve simulation of phenomena at the interface, including air gap formation. Validation on representative long and short freezing range commercial cast alloys was undertaken.

The emphasis on progress with full modelling should not obscure the extensive and successful application of rapid modelling in systematizing foundry methods throughout much of the industry. The well known SOLSTAR system was specifically designed for direct and easy use by foundry staff at modest cost. The package is produced with its own solid modeller to generate the 3D representation of the casting and provisional feeding system. The model is automatically divided into elements of a selected size, and thermal analysis is performed to determine their order of solidification. The simulation incorporates the effects of liquid and solidification shrinkage for the particular alloy, and the predicted final cavities are portrayed in the fully solidified model. Features of the system and procedures were detailed by Corbett[116] and are also included in Reference 117.

The general picture of process modelling in metal founding, also represented in the literature and in numerous examples of the types cited above, shows simulation techniques to have become fully established as an aid to the running and feeding of castings. There is routine industrial use of heat transfer and fluid flow modelling for these purposes, in processes ranging from investment and die casting to the production of individual heavy sand castings, and across the whole alloy range.

There is growing attention to microstructure and mechanical properties prediction, and to stress phenomena including dimensional effects. The latter featured prominently in a critical review by Campbell[118] who also identified internal stress from quenching in the solution treatment of aluminium alloys, and the study of various flow defects, as potentially rewarding subjects for future modelling. Some of the difficulties of realistic simulation of complex phenomena were examined and the need to make the best use of modelling resources was emphasized.

The increasingly ambitious aim of modelling brought much greater emphasis on the need for reliable and accurate thermophysical property databases. Several contributions to the 1997 Sheffield Solidification Processing Conference[119] were devoted to the major task of data collection. The broad modelling situation, including the data problem, was reviewed by Hogg[120], with stress on the need for experimental validation of systems adopted. Several material property measurement techniques developed at the National Physical Laboratory were described by Quested et al[121]: these formed part of an extensive DTI project for database provision. Pain and Jansen[122] pointed to the highly alloy- and process-specific nature of the data required as well as their temperature dependence. The determination of relevant properties by the use of direct thermocouple measurements in conjunction with the simulation process itself, ie the previously mentioned inverse modelling system, was advanced as an effective route to realistic and accurate data; other papers too embodied a similar approach.

References

1 Smithells, C. J., *Metals Reference Book*, 7th Edn. Brandes, E. A. and Brook, G. B., (Eds.), Butterworth-Heinemann, Oxford (1998)
2 Brandt, F. A., Bishop, H. F. and Pellini, W. S., *Trans. Am. Fndrym. Soc.*, **62**, 646 (1954)
3 Dunphy, R. P. and Pellini, W. S., *Trans. Am. Fndrym. Soc.*, **59**, 425 (1951)
4 Bishop, H. F., Brandt, F. H. and Pellini, W. S., *Trans. Am. Fndrym. Soc.*, **59**, 435 (1951)
5 Flinn, R. A. and Kunsman, H., *Trans. Am. Fndrym. Soc.*, **68**, 593 (1960)
6 Campbell, J., *The Solidification of Metals.*, I.S.I. Publication 110, London, *Iron Steel Inst.*, 18 (1968)
7 Piwonka, T. S. and Flemings, M. C., *Trans. Met. Soc.*, AIME **236**, 1, 157 (1966)
8 Campbell, J., *Cast Metals Res. J.*, **5**, 1, 1 (1969)
9 Campbell, J., *Br. Foundrym.*, **62**, 147 (1969)
10 Dahle, A. K., Arnberg, L. and Apelian, D., *Trans. Am. Fndrym. Soc.*, **105**, 963 (1997)
11 Flemings, M. C., *Solidification Processing*, McGraw-Hill, New York (1974)
12 Johnson, W. H., Bishop, H. F. and Pellini, W. S., *Trans. Am. Fndrym. Soc.*, **62**, 243 (1954)
13 Cibula, A., *Fndry Trade J.*, **122**, 337 (1967)
14 Kutumba Rao, G. V. and Panchatharan, V., *Br. Foundrym.*, **66**, 135 (1973)
15 Von Richards, L. and Heine, R. W., *Trans. Am. Fndrym. Soc.*, **81**, 571 (1973)
16 Sciama, G. and Jeancolas, M., *Cast Metals Res. J.*, **9**(1), 1 (1973)
17 Devaux, H. and Jeancolas, M., *Cast Metals Res. J.*, **8**(3), 103 (1972)
18 Chinnathambi, K. and Prabhakar, O., *Trans. Am. Fndrym. Soc.*, **89**, 455 (1981)
19 Reddy, G. P. and Pab, P. K., *Br. Foundrym.*, **69**, 265 (1976)
20 Berry, J. T., Taylor, R. P. and Overfelt, R. A., *Trans. Am. Fndrym. Soc.*, **105**, 465 (1997)
21 Nicholas, K. E. L. and Roberts, W. R., *Br. Foundrym.*, **56**, 122 (1963)
22 Morgan, A. D. and Greenhill, J. M., *Br. Foundrym.*, **54**, 45 (1961)
23 Nicholas, K. E. L., *Br. Foundrym.*, **52**, 487 (1959)
24 Nicholas, K. E. L. and Hughes, I. C. H., *Br. Foundrym.*, **51**, 428 (1958)
25 Hughes, I. C. H., Nicholas, K. E. L. and Szajda, T. J., *Trans. Am. Fndrym. Soc.*, **67**, 149 (1959)
26 Lee, R. S. and Volkmar, A. P., *Trans. Am. Fndrym. Soc.*, **85**, 299 (1977)
27 Tiorello, L. I. and Wallace, J. F., *Trans. Am. Fndrym. Soc.*, **70**, 811 (1962)
28 Tiorello, L. I. and Wallace, J. F., *Trans. Am. Fndrym. Soc.*, **71**, 401 (1963)
29 Ruddle, R. W., *The Solidification of Castings:* Monograph No. 7, Inst. Metals, London (1957)
30 Brandt, F. A., Bishop, H. F. and Pellini, W. S., *Trans. Am. Fndrym. Soc.*, **61**, 451 (1953)
31 Briggs, C. W. and Gezelius, R. A., *Trans. Am. Fndrym. Ass.*, **43**, 274 (1935)
32 Ruddle, R. W. and Skinner, R. A., *J. Inst. Metals*, **79**, 35 (1951)
33 Chworinov, N., *Giesserei*, **27**, pp. 177, 201 and 222 (1940)
34 Wlodawer, R., *Directional Solidification of Steel Castings*. Translation Hewitt L. D. and Riley, R. V., Pergamon Press, Oxford (1966)
35 Batty, G., *Trans. Am. Fndrym. Ass.*, **43**, 75 (1935)

36 Heuvers, A., *Giesserei*, **30**, 201 (1943)
37 Duma, J. A. and Brinson, S. W., *Trans. Am. Fndrym. Ass.*, **48**, 225 (1940)
38 Chursin, V. M. and Kogan, L. B., *Russ. Cast. Prod.*, 136 (1961)
39 Heuvers, A., *Stahl und Eisen*, **49**, 1249 (1929)
40 Caine, J. B., *Trans. Am. Fndrym. Soc.*, **56**, 492 (1948)
41 Caine, J. B., *Trans. Am. Fndrym. Soc.*, **60**, 16 (1952)
42 Adams, C. M. and Taylor, H. F., *Trans. Am. Fndrym. Soc.*, **61**, 686 (1953)
43 Bishop, H. F., Myskowski, E. T. and Pellini, W. S., *Trans. Am. Fndrym. Soc.*, **63**, 271 (1955)
44 Wright, T. C. and Campbell, J., *Trans. Am. Fndrym. Soc.*, **105**, 639 (1997)
45 Briggs, C. W., *Trans. Am. Fndrym. Soc.*, **63**, 287 (1955)
46 Wlodawer, R., *Fndry Trade J.*, **114**, 251 and 283 (1963)
47 Jeancolas, M., *Fndry Trade J.*, **120**, 255 (1966)
48 Johnson, W. H. and Kura, J. G., *Trans. Am. Fndrym. Soc.*, **67**, 535 (1959)
49 Pellini, W. S., *Trans. Am. Fndrym. Soc.*, **61**, 61 (1953)
50 Bishop, H. F. and Pellini, W. S., *Trans. Am. Fndrym. Soc.*, **58**, 185 (1950)
51 Bishop, H. F., Myskowski, E. T. and Pellini, W. S., *Trans. Am. Fndrym. Soc.*, **59**, 171 (1951)
52 Myskowski, E. T., Bishop, H. F. and Pellini, W. S., *Trans. Am. Fndrym. Soc.*, **60**, 389 (1952)
53 Myskowski, E. T., Bishop, H. F. and Pellini, W. S., *Trans. Am. Fndrym. Soc.*, **61**, 302 (1953)
54 Flinn, R. A., *Trans. Am. Fndrym. Soc.*, **64**, 665 (1956)
55 Parsons, W. A. and Twitty, M. D., *Fndry Trade J.*, **118**, 10 (1965)
56 Young, R. G., *Br. Foundrym.*, **74**, 56 (1981)
57 Riley, A. P. and Scott, A. W., *Fndry Trade J.*, **116**, 323 (1964)
58 Adams, C. M. and Taylor, H. F., *Trans. Am. Fndrym. Soc.*, **60**, 617 (1952)
59 Skinner, R. A. and Ruddle, R. W., *Fndry Trade J.*, **93**, 181 (1952)
60 Jackson, R. S., *Proc. Inst. Br. Foundrym.*, **49**, A69 (1946)
61 Taylor, H. F. and Wick, W. C., *Trans. Am. Fndrym. Ass.*, **54**, 262 (1946)
62 Miericke, K. A. and Johnson, E. S., *Trans. Am. Fndrym. Soc.*, **56**, 479 (1948)
63 Rama Prasad, M. S., Srinavaran, M. N. and Seshradi, M. R., *Trans. Am. Fndrym. Soc.*, **86**, 431 (1978)
64 Brown, J., Shah, C. D. and Snelson, D. H., *Br. Foundrym.*, **65**, 254 (1972)
65 Poirier, D. R. and Gandhi, N. V., *Trans. Am. Fndrym. Soc.*, **84**, 577 (1976)
66 Umble, A. E., *Trans. Am. Fndrym. Soc.*, **84**, 49 (1976)
67 Lutts, C. G., Hickey, J. P. and Bock, M., *Trans. Am. Fndrym. Ass.*, **54**, 336 (1946)
68 Finch, S. L., *Proc. Inst. Br. Foundrym.*, **40**, A87 (1947)
69 Atterton, D. V. and Edmonds, R. C., *Proc. Inst. Br. Foundrym.*, **47**, A72 (1954)
70 Gotheridge, J. E., Morgan, J. and Ruddle, R. W., *Trans. Am. Fndrym. Soc.*, **71**, 561 (1963)
71 Snelson, H., *Br. Foundrym.*, **51**, 486 (1958)
72 Strauss, K., *Applied Science in the Casting of Metals*, Pergamon Press, Oxford (1970)
73 *Feedex Manual*, Foseco Ltd., Birmingham
74 Corbett, C. F., *Br. Foundrym.*, **67**, 106 (1974)
75 Ruddle, R. W., *Br. Foundrym.*, **71**, 197 (1978)
76 Janshanow, P. I., *Iron Steel, London*, **32**, 5 (1959)

77 Jazwinski, S. T. and Finch, S. L., *Fndry Trade J.*, **77**, 269 and 293 (1945)
78 Desnizki, W. P., *Iron Steel, London*, **31**, 51 (1958)
79 Middleton, J. M. and Jackson, W. J., *Br. Foundrym.*, **55**, 443 (1962)
80 Briggs, C. W. and Taylor, H. F., *Proc. Inst. Br. Foundrym.*, **46**, A207 (1953)
81 Uram, S. Z., Flemings, M. C. and Taylor, H. F., *Trans. Am. Fndrym. Soc.*, **66**, 120 (1958)
82 Hanson, D. and Slater, I. G., *J. Inst. Metals*, **56**, 103 (1935)
83 Brinson, S. W. and Duma, J. A., *Trans. Am. Fndrym. Soc.*, **56**, 586 (1948)
84 Brinson, S. W. and Duma, J. A., *Trans. Am. Fndrym. Ass.*, **50**, 657 (1942)
85 Daybell, E., *Br. Foundrym.*, **46**, B46 (1953)
86 Locke, C., Briggs, C. W. and Ashbrook, R. L., *Trans. Am. Fndrym. Soc.*, **62**, 589 (1954)
87 Pell-Walpole, W. T., *Proc. Inst. Br. Foundrym.*, **46**, A175 (1953)
88 Briggs, C. W., Gezelius, R. A. and Donaldson, A. D., *Trans. Am. Fndrym. Assoc.*, **46**, 605 (1938)
89 Chworinov, N., *Proc. Inst. Br. Foundrym.*, **32**, 229 (1938–9)
90 Murphy, R. J., Low, R. D. and Linley, A. G., *Trans. Am. Fndrym. Soc.*, **71**, 682 (1963)
91 Bishop, H. F., *Trans. Am. Fndrym. Soc.*, **71**, 771 (1963)
92 Griffiths, M. H., Hall, C. and Neu, M. G., *Br. Foundrym.*, **55**, 98 (1962)
93 Wukovich, N., *Trans. Am. Fndrym. Soc.*, **98**, 261 (1990)
94 Karsay, S. I., Ductile Iron Production Practices, *Am. Fndrym. Soc.*, Des Plaines, Ill. (1975)
95 Karsay, S. I., *Int. Cast Met. J.*, 45 (1980)
96 Gough, M. J. and Clifford, M. J., Solidification Technology in the Foundry and Casthouse, Proceedings of International Conference, Book 273, The Metals Society, London (1983)
97 Gough, M. J. and Morgan, J., *Trans. Am. Fndrym. Soc.*, **84**, 358 (1976)
98 *Foseco Foundryman's Handbook*, 10th Edn. Ed Brown, J. R., Butterworth-Heinemann, Oxford (1994)
99 Hummer, R., *Cast Metals*, **1**, 2, 62 (1988)
100 Louvo, A., Kalavainen, P., Berry, J. T. and Stefanescu, D. M., *Trans. Am. Fndrym. Soc.*, **98**, 273 (1990)
101 Hansen, P. N., *Cast Metals*, **4**(3), 155 (1991)
102 Jolly, M., *Foundry International*, **21**, 3, 16 (1998)
103 Upadhya, G. and Paul, A. J., *Trans. Am. Fndrym. Soc.*, **105**, 69 (1994)
104 Quested, P., *Foundrym.*, **90**, 15 (1997) (after Jain)
105 Tadayon, M. R. and Lewis, R. W., *Cast Metals*, **1**(1), 24 (1988)
106 Abdullah, Z., Salcudean, M. and Davis, K., *Cast Metals*, **3**(1), 7 (1990)
107 Gethin, D. T., Tran, D. V. and Lewis, R. W., *Cast Metals*, **3**(3), 149 (1990)
108 Spittle, J. A., *Cast Metals*, **4**(3), 171 (1991)
109 Sasikumar, R., *Cast Metals*, **2**(4), 214 (1990)
110 Brown, S. G. R. and Spittle, J. A., *Cast Metals*, **3**(1), 18 (1990)
111 Paul, A. J., ASM Handbook, Vol. 20 Materials Selection and Design, 705, ASM International, Materials Park OH (1997)
112 Adams, D., Butlin, G., Higginbotham, G., Katgerman, L., Hills, A. W. D. and Charles, J. A., *Metals and Materials*, **8**, 9, 497 (1992)
113 Tu, J. S., Foran, R. K., Hines, A. M. and Aimone, P. R., JOM **47**(10), 64 (1995)

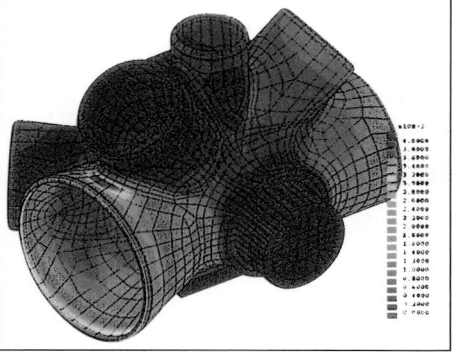

(a)

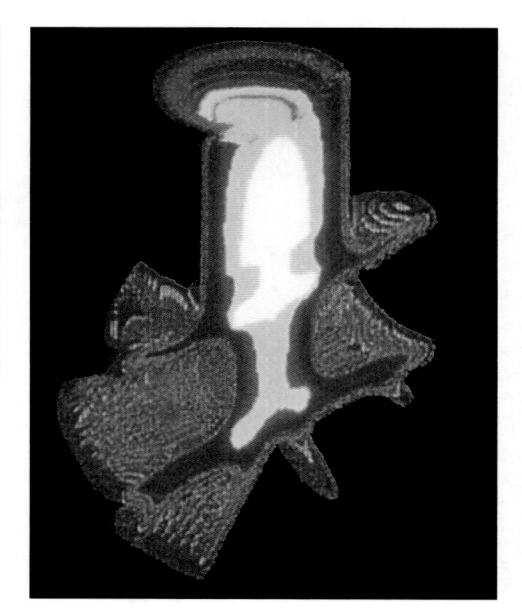

(b)

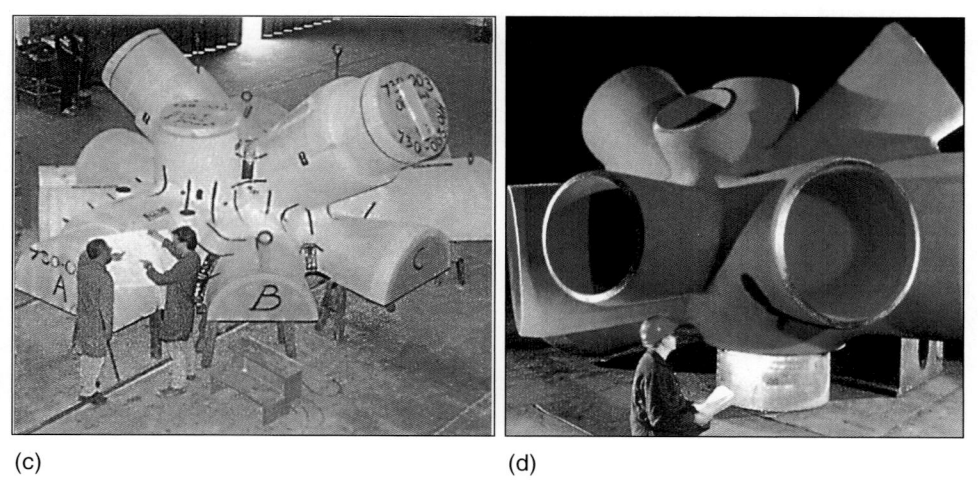

(c)

(d)

Figure 3.48 Process optimization sequence for production of cast steel node: (a) Design stage: 3D finite element stress analysis; (b) Solidification modelling: temperature profile for hot spot observation and feeding path confirmation; (c) Pattern making; (d) Finished casting; (courtesy of River Don Castings, now part of Sheffield Forgemasters Engineering Ltd)

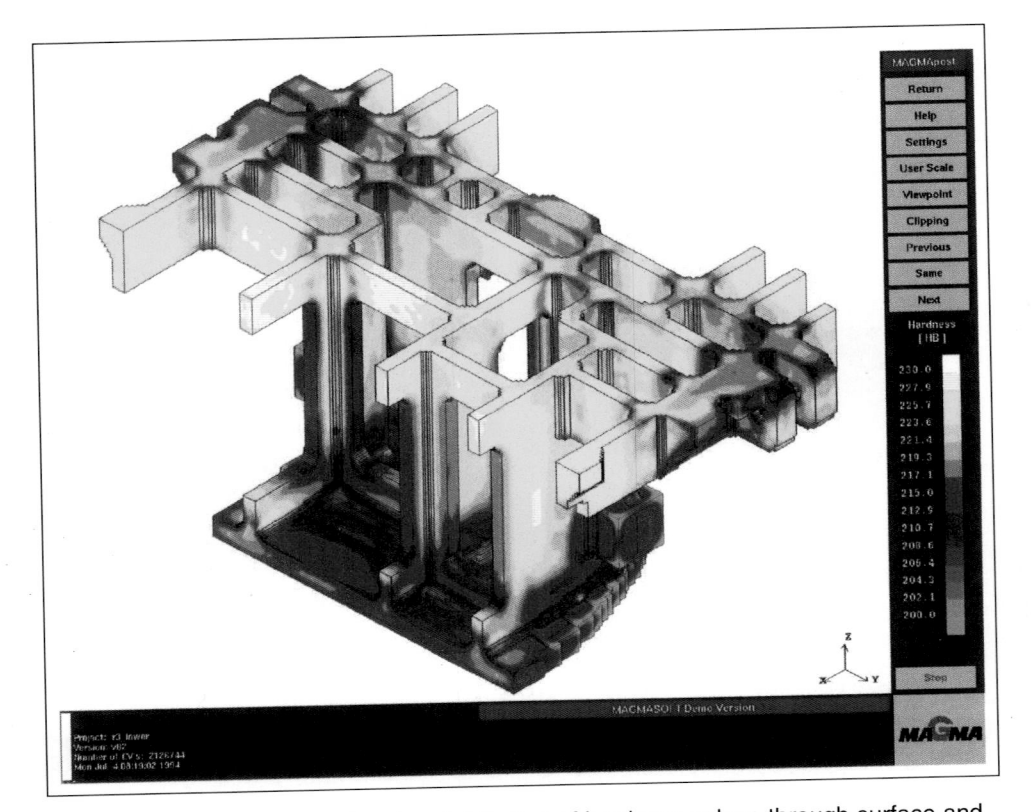

Figure 3.50 Section of die showing distribution of hardness values through surface and web structure, based on solidification modelling (after Haigh et al [115]) (courtesy of CDC)

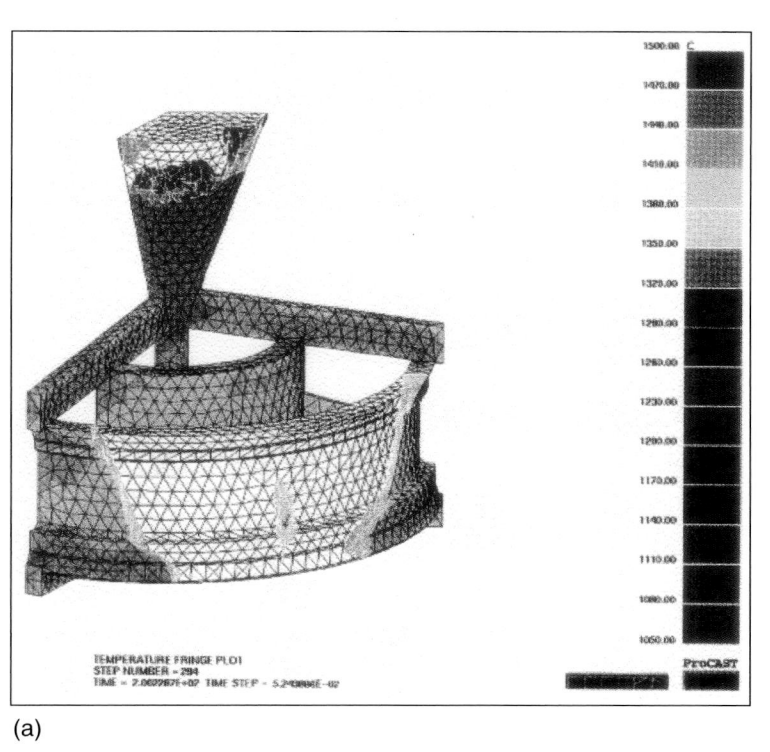

(a)

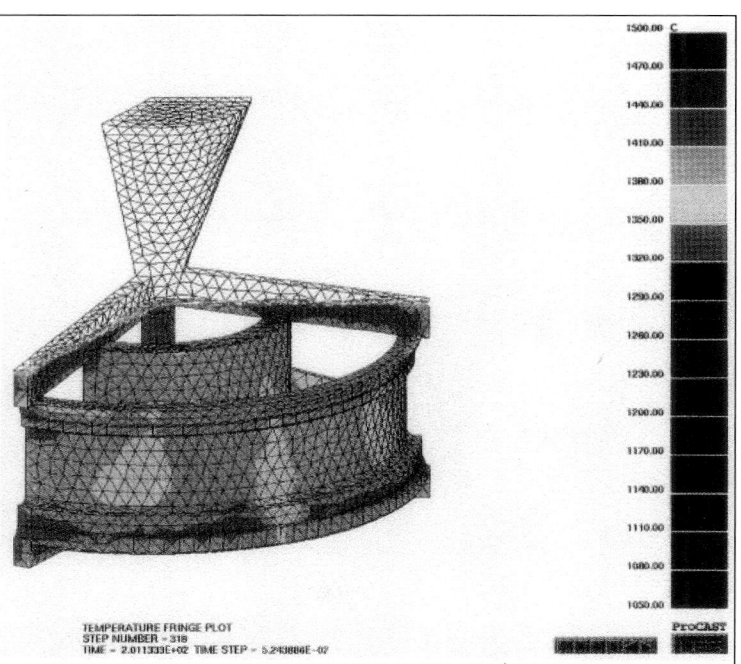

(b)

Figure 3.45 Typical temperature profiles during the mould filling sequence, from the fluid flow simulation: (a) 5.23 seconds, (b) 6.13 seconds, showing the order of filling (from Tu et al [113]) (courtesy of The Minerals, Metals and Materials Society)

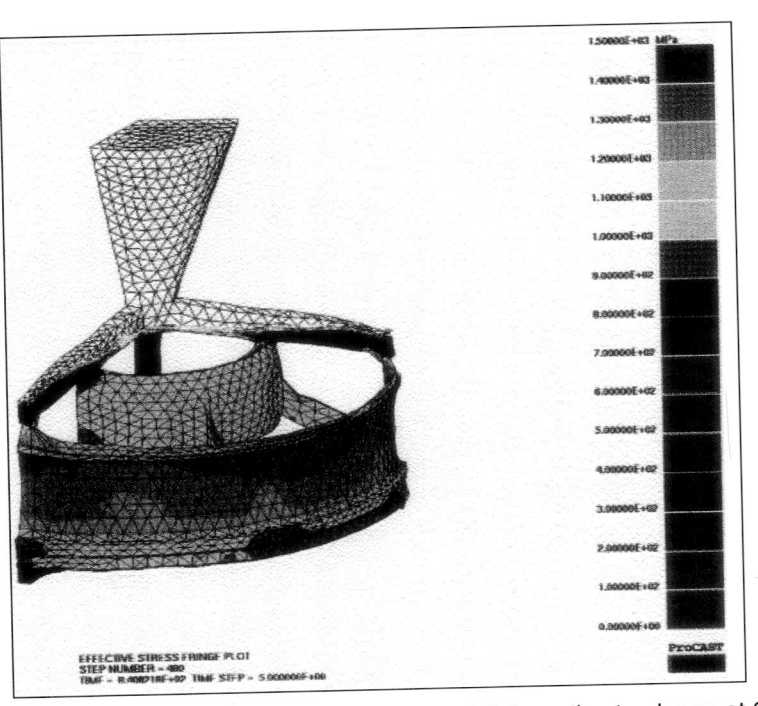

Figure 3.46 Predicted effective stress, and deformation tendency, at 645 seconds after pouring (after Tu et al [113]) (courtesy of The Minerals, Metals and Materials Society)

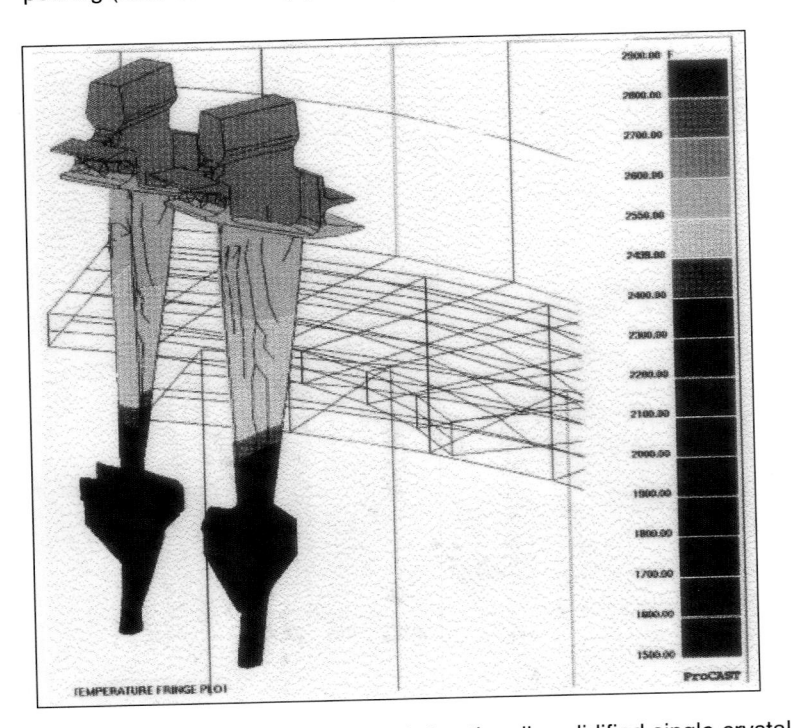

Figure 3.47 Temperature profile of directionally solidified single crystal turbine blade casting, defining mushy zone (after Tu et al [113]) (courtesy of The Minerals, Metals and Materials Society)

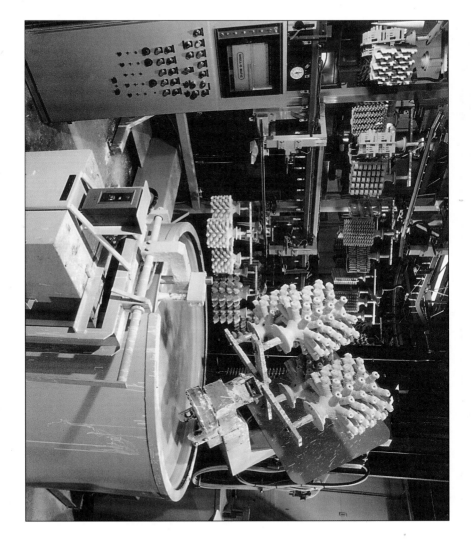

Figure 9.17 Shell-O-Matic dedicated robot coating plant for the production of ceramic shell moulds (courtesy of P.I. Castings Ltd.)

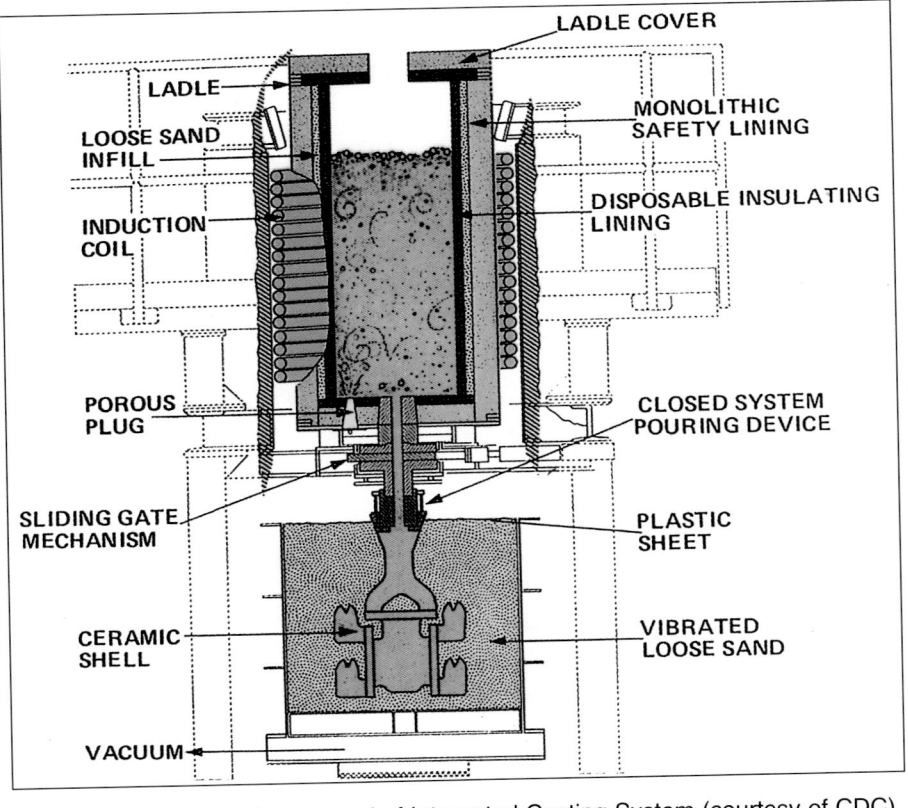

Figure 10.29 Schematic portrayal of Integrated Casting System (courtesy of CDC)

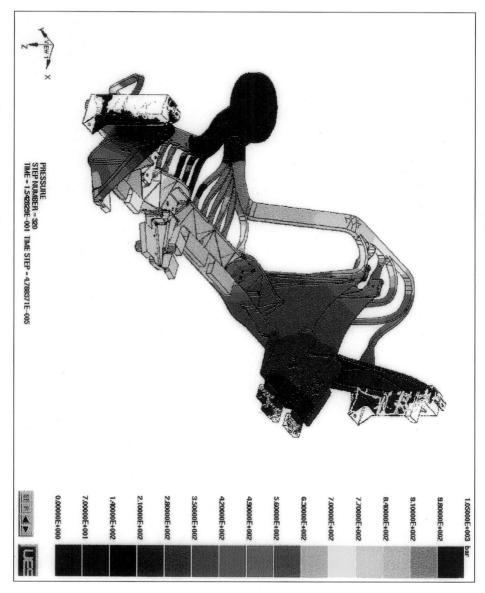

Figure 3.51 Pressure contours during injection of a magnesium alloy car chassis transverse component: length 1.3m approx (courtesy of Alusuisse-BDW)

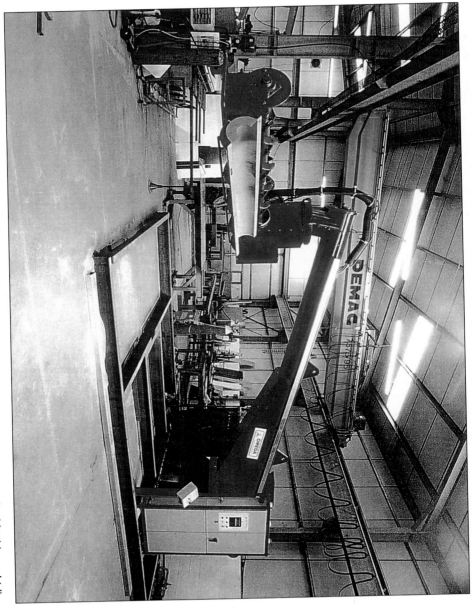

Figure 4.26 Large articulated mixer, showing mixing trough with cover removed (courtesy of Omega Foundry Machinery Ltd)

114 Mueller, B. A., Foran, R. K., Hines, A., Hirvo, D., Simon, T. and Tu, J. S. Solidification Processing 1997: Proc. 4th Decennial Int. Conf. on Solidification Processing, Ed. Beech, J. and Jones, H., Dept. of Eng. Materials, University of Sheffield, 170 (1997)
115 Haigh, P., Oxley, S. and Bond, P. *Ibid.*, 155
116 Corbett, C., *Foundrym.*, **85**, 11 (1992)
117 *Foseco Foundryman's Handbook*, 10th Edn. Brown, J. R., Butterworth-Heinemann (1994)
118 Campbell, J. Modelling of Casting, Welding and Advanced Solidification Processes VII. Ed. Cross, M. and Campbell, J., *Minerals, Metals and Materials Soc.*, 907, 1995
119 Solidification Processing 1997: Proc. 4th Decennial Internat. Conference on Solidification Processing, Ed. Beech, J. and Jones, H., Dept. of Engineering Materials, University of Sheffield (1997)
120 Hogg, J. C., Ref. 119, 116
121 Quested, P. N., Mills, K. C., Brooks, R. F., Day, A. P., Taylor, R. and Szelagowski, H., Ref. 119, 143
122 Pain, A. P. and Jansen, D. J., Ref. 119, 160

4

The moulding material: properties, preparation and testing

Castings can be produced in either permanent metal or expendable refractory moulds. The use of metal moulds or dies for shaped components is subject to certain major limitations. The exacting conditions imposed by repeated contact with molten metal restricts useful mould life to a level which has so far largely confined the technique to metals of low melting point. In addition the cost of die manufacture can only be sustained by a large production requirement. Thus the greater part of the output of the foundry industry consists of castings made in refractory moulds, chiefly sand castings. Since the properties of the moulding material are crucial to the production of sound, dimensionally accurate castings, its selection and testing constitutes one of the vital steps in founding. The present chapter will be devoted to materials of this type, whilst the moulding processes themselves will be further examined in Chapter 8. Die casting will be considered with other special techniques in Chapter 9.

The properties of moulding materials can be measured by a wide range of techniques, which provide the basis both for routine foundry control and for the development of new materials; the significance of various tests will be discussed in the present chapter. Initially, however, consideration will be given to the property requirements of moulding and coremaking materials, in relation to the conditions encountered during the production of a casting.

Functional requirements of moulding materials

A foundry moulding mixture passes through four main production stages, namely preparation and distribution, mould and core production, casting, and cleaning and reclamation. The property requirements of the materials are determined by moulding and casting conditions; the preparation and reclamation stages will, however, also be considered, with particular reference to integrated sand systems.

The principal properties required at the moulding stage are *flowability* and *green strength*: the former is a measure of the ability of the material to be compacted to a uniform density. The ideal balance of these properties depends largely upon the intended method of compaction, which may vary from hand ramming with tools to jolt, squeeze and impact ramming on moulding machines and high velocity delivery on sandslingers and coreblowers. High flowability is particularly necessary in the case of the non-selective ramming action of the moulding machine, where the energy for compaction must be transmitted throughout the sand mass.

The need for green strength arises when the pattern is withdrawn and the mould must retain shape independently without distortion or collapse. The stress to which the moulding material is subjected at this stage depends upon the degree of support from box bars, lifters and core irons and upon the shape and dimensions of the compact: less green strength is needed for a shallow core supported on a coreplate or carrier than for a cod of sand forming a deep mould projection. In many cases, however, dimensional stability and high accuracy may be achieved without the need for appreciable green strength, as when the mould or core is hardened in contact with the pattern surface, a common circumstance when modern bonding systems are employed.

Moving to the pouring stage, many moulds are cast in the green state, but others, including most of those for heavy castings, are hardened to generate greater rigidity under the pressure and erosive forces of the liquid metal. This state was formerly achieved by the high temperature drying of clay bonded sands or the baking of traditional coresands, but this has been largely superseded by the chemical hardening of sands containing reactive binders of the modern organic and silicate types. At this stage, therefore, *dry strength* i.e. strength in the hardened or dried condition is significant; even in greensand practice dry strength is required, to avoid friability should the mould partially dry out during standing before casting.

The other main requirement at the casting stage is for *refractoriness*, or the ability of the mould material to withstand high temperatures without fusion or other physical change. This property is primarily important in the manufacture of high melting point alloys, especially steel; for alloys of lower melting point, refractoriness can be subordinated to other requirements.

In the production of very heavy castings, a considerable layer of moulding material rises to a temperature at which normal mechanical properties are no longer the main criterion governing dimensional stability and resistance to contraction. Depending upon the mass of the casting, the sand may require an appropriate combination of high temperature properties, including *hot strength* to withstand distortion and the capacity for *deformation* to yield to the contraction of the casting. *Collapsibility* determines the readiness with which the moulding material will break down in knockout and cleaning operations.

A further feature of the casting stage is the gas evolved and displaced from the mould. Much of this can be exhausted through open feeder heads and vents, but a large volume must also be dissipated through the pore spaces of the sand. This problem is greatest for greensands and coresands. The evaporation of each 1% of moisture from green moulding sand can be shown to generate over 30 times its own volume of steam; this is paralleled in other types of sand by gases from volatilization and decomposition of organic compounds. To provide a path for the escape of gas, *permeability* is an essential property, giving protection against surface blows and similar defects.

Fineness is required for the prevention of metal penetration and the production of smooth casting surfaces. Since both permeability and fineness are functions of grain size and distribution, the two properties are in conflict and a compromise is usually necessary. Fineness may be achieved by using fine grained sands, by continuous grading or by the incorporation of filler materials, but all these measures also reduce permeability. An alternative approach is to use a highly permeable moulding material and to obtain surface fineness by the use of mould coatings.

Moulding materials need certain further qualities which are not necessarily measurable by standard tests. Examples are *bench life*, the ability to retain moulding properties on standing or storage, and *durability*, the capacity to withstand repeated cycles of heating and cooling in integrated sand systems.

It is thus evident that the qualities required in a moulding material cannot readily be defined in terms of simple physical properties. For complex aggregates bulk properties are of greater significance and some of these can be measured directly by simple tests upon sand compacts. Other qualities are represented in specially developed empirical tests designed to reproduce conditions encountered in the foundry. These tests, in conjunction with the direct measurement of more fundamental characteristics such as mechanical grading and chemical composition, provide the basis for the control and development of moulding material properties.

Moulding practice and the special requirements of coresands

Many castings, including most of those made by machine moulding, are cast in greensand moulds, and the introduction of high pressure moulding machines enabled even castings in the tonnage weight range to be produced to acceptable quality standards. There are strong economic incentives to use this low cost system, but hardened moulds are preferred in many cases, particularly for heavier castings. The advantages of these alternatives can be briefly summarized.

Greensand practice

1. Clay binders involve low material costs and avoid the additional costs of mould hardening, whether by chemical or thermal means.

2. The rapid turnround of moulding boxes and the smooth moulding and casting cycle are advantageous in mechanized systems.
3. The sand is readily reconditioned, since there is little dehydration of the clay bond.
4. Greensands, having lower compression strength, offer less resistance to contraction than hardened moulds, so that the risk of hot tearing is reduced.
5. Moulds joint closely, leaving little flash for removal by fettling.
6. The process is environmentally friendly.

Hardened mould or drysand practice

1. Hardened moulds offer maximum resistance to distortion under metallostatic pressure, and to mould erosion during prolonged pouring. They are therefore suitable for castings of the largest dimensions and provide high standards of accuracy. In the production of cast irons in particular, mould rigidity contributes to internal soundness as well as to dimensional accuracy.
2. The venting problem is reduced in the absence of steam generated from moisture.
3. Impervious mould surfaces are readily attainable, since sands of lower permeability can be used, with coatings where required.
4. Surface chilling is greatly reduced, facilitating metal flow in thin sections.
5. Problems of drying-out during delays in casting, leading to surface friability, are avoided.

Mould hardening

Chemical hardening is accomplished either by liquid or gaseous reagents. Liquid reagents are commonly employed in the cold-set mode and are blended into the sand at the mixing stage, in some cases at the point of entry to the mould, so that no separate hardening operation is required. Gas or vapour hardening is performed after compaction and is mainly applied to cores and smaller mould parts. The materials used in these systems will be reviewed later in the chapter.

Full mould drying has been largely superseded by the above practices, but stoves are still used in some circumstances. These are usually operated in the temperature range 200–400°C. Surface drying can also be carried out, using gas torches or hot air drying hoods. Surface hardening can be enhanced by prior treatment with sprays of water or dilute binder solutions, and inflammable mould coatings can also assist in the process.

The requirements of coresands

Although coremaking techniques are substantially similar to those used in moulding, core and mould conditions diverge during closing and casting.

Irrespective of whether greensand or drysand practice is being employed for the mould, cores are normally dried or otherwise hardened to develop the high strengths demanded by their special situation.

A core, often of delicate and complex construction, must be handled as a separate entity without the support of a moulding box. High dry strength is also required in casting, to sustain the stresses on the core, often largely surrounded by liquid metal and supported at few points. Greensand cores are required in special cases, however, although the handling of these in closing is particularly difficult.

The need for green strength depends on the coremaking operation. Modern processes which eliminate separate baking by effecting the hardening reaction within the corebox render green strength unimportant, but modest green strength is required for cores which are to be stripped and separately stoved on plates or carriers.

The other basic requirement is collapsibility. This facilitates extraction of the core material from interior cavities during cleaning, hence the selection of bonding materials designed to break down after exposure to high temperature. A core must also offer minimum resistance to contraction. This is not necessarily associated with ultimate collapsibility, but rather with short term high temperature deformation; methods for measuring this characteristic will be discussed.

Sand testing techniques

The need for systematic evaluation of the working qualities of moulding materials under foundry conditions has led to the development of a wide range of tests, the work of many committees and individuals. Some of the tests are concerned with basic chemical and physical characteristics, but the majority are designed to measure bulk properties. The relevance of some of these properties has already been stressed, and an account of the associated testing techniques will now be given. Many of the tests are long established and reflect the pioneering work of H. W. Dietert, the American Foundrymen's Society, BCIRA and other individuals and organizations who undertook the design of tests, the manufacture of the necessary equipment and the publication of reports, handbooks and papers, many of which still remain relevant. Further developments include the availability of closely similar metric and imperial versions of much of the equipment, and increased attention to the key properties of chemically bonded sands. Sources containing details of numerous tests include References 1–6; other references on some of the tests and related points will be included in the following sections.

The bulk properties of an aggregate are sensitive to small variations in mixing conditions and specimen preparation, so that rigid standardization is needed at all stages. Even under these conditions results usually need to be derived from two or more determinations. Original studies showed reproducibility to be best in permeability and least in strength tests on some of the more friable sands[7,8].

In the routine testing of materials taken directly from production lines great care is also needed to minimize errors by the use of systematic techniques for collection and storage of samples[2].

Specimens for bulk testing

Mechanical properties and certain other characteristics are determined on specimens compacted to a bulk density similar to that encountered in a well rammed mould. Many tests utilize the 2 in × 2 in cylindrical AFS specimen or its 50 mm × 50 mm DIN equivalent, prepared by subjecting a weighed quantity of sand to a selected number of blows from a compatible standard rammer, transmitted to a close fitting piston in a tubular mould (Figure 4.1).

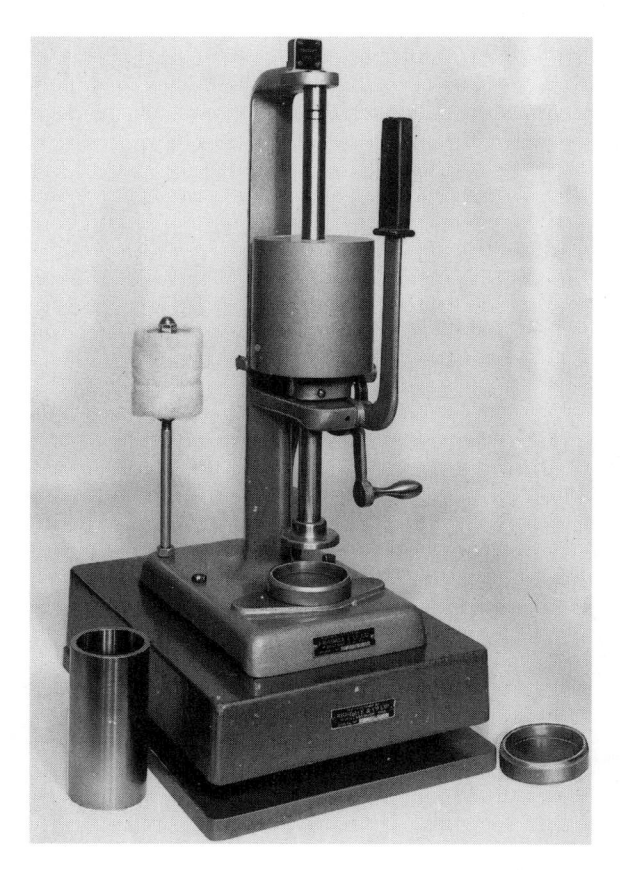

Figure 4.1 Standard rammer for specimen preparation (courtesy of Ridsdale & Co. Ltd.)

The weight of sand is adjusted to produce a close tolerance specimen, which can be expelled from the tube on a stripping post.

This specimen may be used for permeability testing and for a number of green and dry strength measurements; behaviour during ramming can be used as a criterion of compactability or flowability. The standard rammer is also used, in conjunction with suitable moulds and rammer head attachments, for the production of specially shaped specimens for the tensile and transverse testing of high strength hardened sands.

Green and dry strength tests

The principal strength tests measure stress to failure under a constant rate of loading. In the low green strength range, compression testing can be carried out on a simple, manually operated, spring loaded machine, but most strength testing is carried out on universal machines such as that shown in Figure 4.2. In the machine illustrated, the load is applied by means of a pivotted weight, progressively brought to bear on the specimen as the motorized pusher arm climbs the calibrated rack. The machine is fitted with accessories enabling the lever ratio to be changed for selection of the appropriate strength range; alternative types of specimen holder enable any of the individual strength tests to be performed. For very high strengths machines such as that illustrated in Figure 4.3 are available. In this example motorized loading up to 20 kN (4500 lb) can be applied to the selected test

Figure 4.2 Universal sand strength testing machine (courtesy of Ridsdale & Co.)

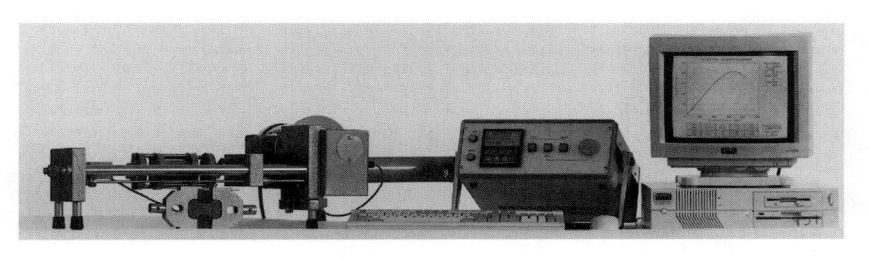

Figure 4.3 Testing machine for higher strength materials (courtesy of Ridsdale & Co.)

specimen, the force being measured through a strain gauge load cell. The unit embodies computer display and recording of test results.

Green strength tests are carried out on newly made specimens and dry or cured strength tests on specimens hardened in a standard manner appropriate to the practice for the particular bonding material.

Compression tests

The cylindrical specimen is axially loaded through flat faced holders as illustrated in Figure 4.4a. Standardization of the rate of loading is particularly important in the green compression test, since creep of greensand occurs readily under constant load.

Shear tests

The compressive loading system is modified to provide offset loading of the specimen as shown in Figure 4.4b. Under most conditions the results of shear tests have been shown to be closely related to those of compression tests, although the latter property increases proportionately more at high ramming densities[9].

The tensile test

A special waisted specimen is loaded in tension through a pair of grips (Figure 4.4c).

The transverse test

A plain rectangular specimen is supported on knife edges at the ends and centrally loaded to fracture (Figure 4.4d).

Tensile and transverse tests are commonly applied to high strength sands, the conditions being especially relevant to the stresses incurred in cores during handling and casting. Both tests, unlike the compression test, provide values well within the working ranges of normal types of universal sand

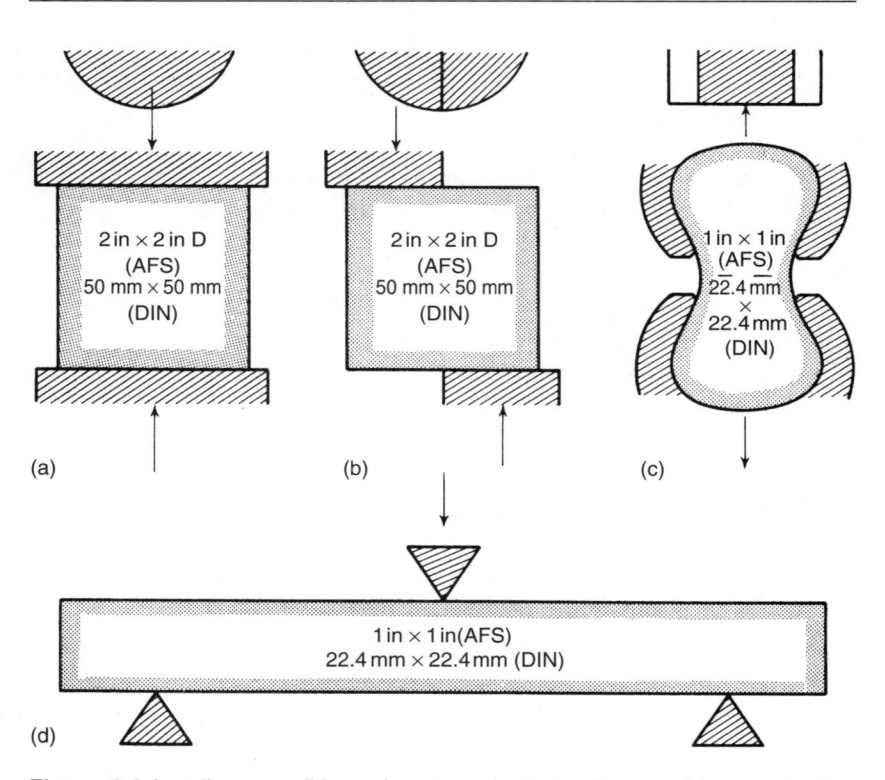

Figure 4.4 Loading conditions in strength tests for moulding materials. (a) Compression, (b) shear, (c) tension, (d) transverse

testing machine. Versions of these two tests are also used for the evaluation of shell moulding mixtures, being carried out on cured specimens approximating to the thickness of a production shell.

Little is to be gained from routine use of the whole range of strength tests, although each may be useful in individual circumstances. Studies of correlations between various strength properties are particularly valuable and can greatly reduce the volume of testing[9,10].

Other green tests relating to moulding quality

Compactability and flowability

The compactability test[11] is widely accepted as both simple to perform and directly related to the behaviour of sand in machine moulding, particularly when involving squeeze compaction. A fixed volume of loose sand is compacted under standard conditions and the percentage reduction in volume represents the compactability. Knowledge of this property also

provides a basis for estimating the amount of sand required for the satisfactory packing of a production moulding box.

The test employs the previously illustrated standard rammer: the cylindrical specimen tube is filled with loose riddled sand, struck off level and
subjected to three rams. The compactability is expressed as the percentage
reduction in length of the specimen and is influenced primarily by the clay
and water contents. Values in the range 35–55% are normally achieved,
the lower levels being considered suitable for high pressure moulding and
the higher for hand ramming. An alternative technique employs a standard
squeeze pressure to compact the specimen.

Numerous tests have been employed for the determination of flowability,
which seeks to represent the capacity of a sand to be compacted to uniform
density under varied moulding conditions, including the need to overcome
frictional forces in lateral flow and in the filling of confined pockets and
pattern features of uneven depth. Since the same frictional forces assist
bonding, the property tends to bear an inverse relation to strength properties.

The most convenient flowability test[12] again employs the standard
rammer, but is based on the movement occurring between the fourth and
fifth blows in the production of a compression test specimen. An arbitrary
scale of values is assigned to this movement, with zero travel corresponding
to maximum flowability, the full compaction of a highly flowable sand
having been accomplished by the earlier blows. The measurement is
influenced by the grain shape and mechanical grading of the base sand
as well as by the clay and water contents.

Other methods include the measurement of surface hardness differentials
across the sloping base of a cylindrical specimen rammed in the modified
tubular mould[13], and observation of the incidence of 'supervoids' in the
surfaces of compacted specimens[14].

Differences in the flow properties of sands can also be observed by
measurements of bulk density, which ranges from $0.60–0.75 \times 10^3$ kg/m^3
for unrammed sand to $1.6–1.7$ kg/m^3 after prolonged ramming. Bulk density
distribution under varied modes of compaction can be determined using
segmented specimen moulds which can be progressively dismantled for the
removal and weighing of successive layers of compacted sand. Examples
of bulk density gradient plots are seen in Figure 4.13, and in others in
Chapter 8.

The shatter index

The long established shatter test[3] is used as an indicator of sand toughness, especially the ability to deform rather than fracture under shock
loading. The standard compression test cylinder is ejected from its mould
and allowed to fall from a fixed height of 1.83 m (6 ft) on to a flat steel
anvil. Fragments retained on a concentric 13.2 mm aperture BS410 sieve
are weighed, together with the residual core from the anvil, and the shatter

index is this weight expressed as a percentage of the total weight of the specimen. A low value is an indication of poor lift, or friability in pattern withdrawal and subsequent handling, whilst too high a value is associated with unsatisfactory moulding qualities resulting from excessive clay or water content. Values of 50–85 represent the mouldable range[5].

Surface hardness

The hardness of a compacted sand surface can be determined using portable spring loaded indentation testers[2]. For the measurement of green hardness a spherical or conical indenter is used depending on the expected hardness level: the depth of penetration from the flat reference surface of the instrument corresponds to an empirical scale of hardness in the overall range 0–100. Instruments with different springs, indenters and scales enable sensitivity to be maintained over the full range normally encountered under the widely varied conditions encountered in greensand moulding practice. The hardness can be measured on actual mould surfaces to check the degrees of uniformity and the ramming efficiency. Other surface tests are required for the evaluation of chemically hardened moulding and core sands, and will be further referred to in that context.

Deformation

The deformation behaviour of moulding sands has received less consideration than the properties reflected in the above tests. Deformation can be determined during the green compression test by measurement of the decrease in length of the cylindrical specimen to the point of failure, whilst more comprehensive data can be derived from full stress–strain curves, an issue examined by Morgan[15] in a broad review of the potential for greensand system controls.

Permeability

Permeability is determined by measuring the rate of flow of air through a compacted specimen under standard conditions. The test for green permeability is carried out on the A.F.S. standard cylindrical specimen, retained in its ramming tube. The permeability meter, illustrated in Figure 4.5 incorporates a graduated bell of 2 litres capacity containing a volume of air over water. A tube from the air enclosure communicates directly with the specimen tube, placed over an 'O'-ring seal, so that the air can escape through the specimen as the bell descends. The time for exhaustion of 2 litres of air is determined.

The permeability number P is defined as the volume of air in cm^3/min passing through a specimen of length 1 cm and cross-sectional area $1\,cm^2$, under a pressure difference of 1 cm water gauge:

$$P = Vh/atp$$

Figure 4.5 Permeability meter (courtesy of Ridsdale & Co. Ltd.)

where V = volume of air, cm^3
h = height of specimen, cm
a = c.s.a. of specimen, cm^2
t = time, min
p = pressure difference, cm water.

Using the standard apparatus and technique, the permeability may be derived directly from the formula

$$P = 3007.2/t$$

where the time t is expressed in seconds.

A more rapid although less accurate measurement of permeability may be made using a modified technique. A standard orifice, of small cross-section in comparison with the porosity of the specimen, is placed between the pressure chamber and the specimen. The pressure between orifice and specimen now lies at a value intermediate between the chamber pressure and atmospheric pressure, depending on the permeability of the specimen. The value is derived directly from this pressure, measured by calibrated water manometer. A similar principle is embodied in the portable quick-reading instrument illustrated in Figure 4.6. The required air pressure is in this case generated by a high speed electric fan and the pressure drop is indicated on a sensitive gauge calibrated directly to read the permeability number.

Either type of permeability meter can be used in conjunction with a flexible tube and contact pad to provide a direct indication of the level of permeability at any flat mould surface.

Permeability determinations on dried or hardened sands require a modified specimen holder enabling the cylindrical walls of the test piece to be sealed against the bore surface of an enlarged tube, using molten wax or an inflatable rubber sleeve.

Additional tests for chemically bonded sands

The foregoing tests are in the main applicable both to clay bonded and chemically bonded sands. In the latter case, however, the emphasis in

Figure 4.6 Compact direct reading permeability meter (Ridsdale–Dietert electric Permmeter, courtesy of Ridsdale & Co. Ltd.)

mechanical tests is on those versions most suited to the very high strength levels of the cured materials as compared with clay bonded sands: tensile and transverse tests are widely employed, although use is still made of compression testing on machines with the necessary load capacities. Laboratory specimen preparation is designed to parallel the conditions used for the curing of production moulds and cores, with standardization of temperature, mixing conditions and specimen preparation, including the curing procedure, whether at room or elevated temperature or by gassing. For the testing of production mixtures samples are taken directly from the mixing plant.

The principal special feature in the testing of these sands is the significance of the time factor in the development of properties, and thus of the point in the interval from mixing to full hardening at which a test is undertaken. For this reason some of the tests involve sequential testing to establish the time dependence itself.

Most of the individual tests, including detailed procedures, were included in major publications of the American Foundrymen's Society[2] and the Institute of British Foundrymen[4]; essential features of the main tests will be summarized.

Bench life

Compression tests are made on specimens rammed at successive intervals of five minutes from mixing. The bench life is defined as the time to reach the value of $10 \, kN/m^2$, beyond which the strength of subsequently compacted material will be prejudiced by interruption of the bonding process already under way in the loose sand. Other relevant time intervals can be applied, depending on the sensitivity of the mixture under test.

Strip time

Numbers of compression test specimens are moulded in multi-gang boxes immediately after mixing. These are tested at intervals to detect the time at which the strength reaches $350 \, kN/m^2$.

A full strength–time plot of the type shown in Figure 4.7 can be developed from this form of test to characterize the setting behaviour in full, including the maximum potential strength of the material.

Scratch hardness

A four point spring loaded penetrator is contained within a concentric reference surface which is pressed against the hardened sand surface. The penetrator is manually rotated though a fixed number of revolutions, normally two, and the depth of penetration is shown on a dial gauge as an indication of relative surface density and scratch resistance.

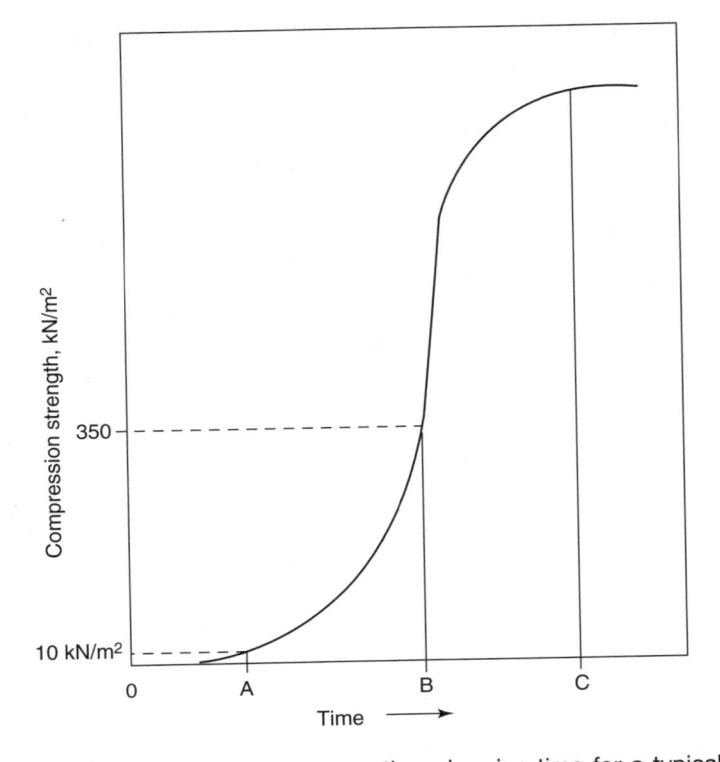

Figure 4.7 Relationship between strength and curing time for a typical cold setting resin bonded sand. Point A defines bench life, B setting time and C time to develop maximum strength (from Reference 4) (courtesy of Institute of British Foundrymen)

Impact penetration

A sharp spring actuated probe (see Figure 4.8) graduated in 1 cm divisions, is repeatedly triggered by manual pressure against the sand surface, and the number of blows required to reach a given depth represents the relative resistance of the sand body to penetration. The test can be carried out on production moulds or on sample 10 cm cubes, moulded and tested whilst in the corebox. A valuable feature of this test is its ability to explore the rate and depth of hardening and to detect any lack of through cure[16]; plots of numbers of impacts against penetration depth enable the degree of uniformity to be investigated.

Gas evolution

The rate and volume of gas evolved from a dried sample of hardened sand can be determined using a sealed, temperature controlled, silica tube furnace with provision for a nitrogen atmosphere. The weighed sample is propelled

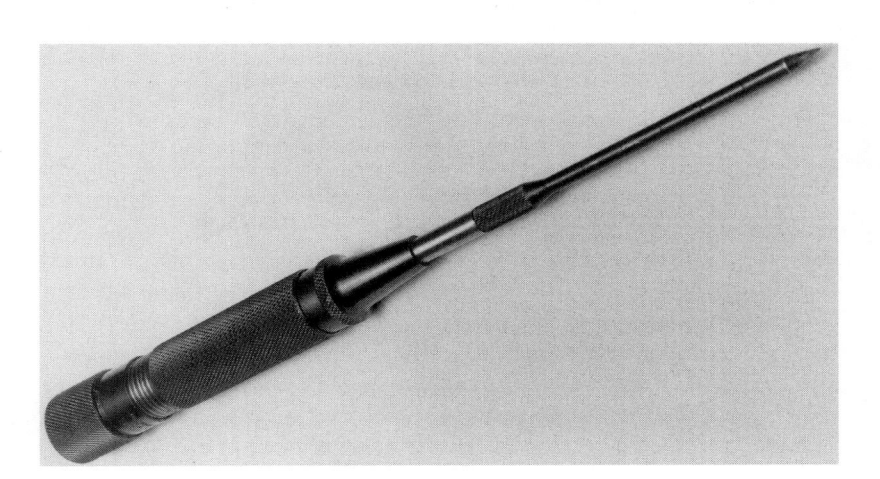

Figure 4.8 Impact penetration tester with internal spring loaded hammer (courtesy of Ridsdale & Co. Ltd.)

into the hot zone of the inert gas-filled tube, normally at a temperature of 850°C, and the pressure rise is continuously recorded until a maximum is attained. The pressure readings can be converted to volumes, using a calibration chart derived from the known total evolution from a standard substance. Both rate and total volume are significant in relation to potential gas defects in castings.

Hot distortion

The hot distortion test[2,17] is carried out on a flat strip sand specimen which is fixed at one end and loaded in cantilever mode at the other (see Figure 4.9) The underside is heated and the deflection at the loaded end continuously measured. Upward movement due to initial expansion of the lower layers of the strip is followed by sagging with the progressive thermal softening and eventual breakdown and collapse of the material. Strain–time curves provide a guide to the behaviour of different binders and curing conditions. Typical curves representative of hot box, cold set and silicate bonded sands, including the influence of iron oxide additives, were produced in a comprehensive study by Morgan and Fasham[17]: an example is shown in Figure 4.10.

High temperature properties

High temperature tests are particularly relevant to the study of mould and core materials, since they simulate in some degree conditions met in the casting process. They are, however, generally more suited to longer term laboratory assessments than to routine testing. Gas evolution and hot distortion tests have been included previously, given their particular value in

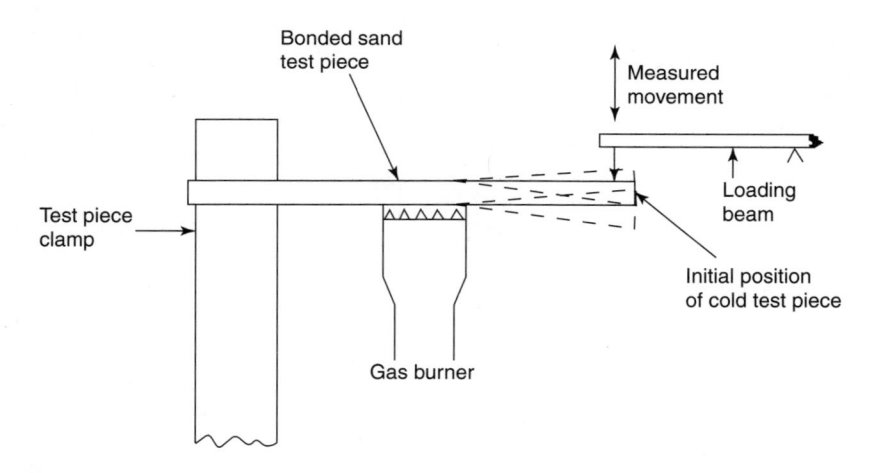

Figure 4.9 Schematic arrangement of hot distortion test (from Morgan and Fasham[17]) (courtesy of America Foundrymen's Society)

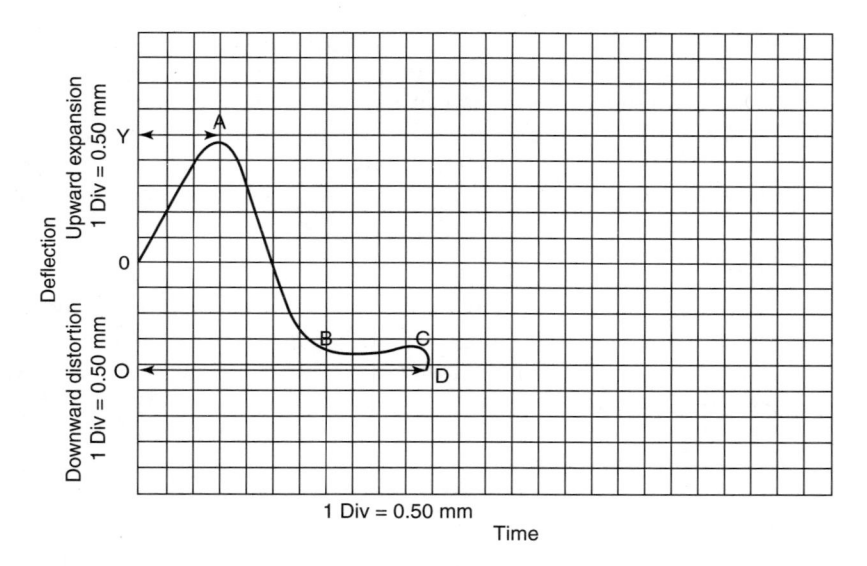

Figure 4.10 Typical hot distortion test strain–time curve. Stages: O–A: upward expansion against load; A–B: thermoplastic softening; B–C: thermosetting; C–D: breakdown and collapse (after Morgan and Fasham[17]) (courtesy of American Foundrymen's Society)

relation to the chemically hardened sands. Similarly, special equipment is available for the hot tensile testing of shell moulding sands, embodying integral curing.

Various forms of hot strength test include the determination of compressive strength to failure at specified temperatures, involving prior soaking periods for the specimen to reach the test temperature, whilst measurements of total deformation can also be obtained from the same test. To assess the retained strength, a property relevant to knockout behaviour, the specimen is put though the thermal cycle, but cooled to room temperature before testing to failure: typical high and retained temperature test results for a sodium silicate bonded sand are shown in Figure 4.18 (see page 212).

Collapsibility can be assessed through determination of the time to failure under constant load: this test is more suitable than isothermal hot strength tests for organically bonded sands, in which the bond can be completely decomposed during the soaking period.

Free expansion in the absence of loading can also be measured using a dilatometer: this type of measurement can be relevant to certain types of sand defect in castings, a topic considered further in Chapter 5.

These and numerous other individual high temperature tests will not be further reviewed here: full details are to be found in Reference 2.

Composition, grading and other tests

In addition to measurements of bulk properties of moulding sands, analysis may be carried out for determination of the basic chemical and physical constitution of the materials. Moisture content is regularly determined as the most sensitive and useful production variable for clay bonded sands. Mechanical and chemical analyses are less frequently required since they are not susceptible to short term fluctuation: techniques for the chemical analysis of refractories will not be considered here. Other frequently used control tests include those for active clay, pH and acid demand, and loss on ignition.

Moisture testing

Moisture content is readily determined by loss in weight on drying at 110°C. Various rapid aids are employed, including balances directly calibrated to read the moisture content of a standard sample after drying; samples can be dried by the passage of hot air through a filter cloth tray.

For shop control the *Speedy* moisture tester can be employed. This makes use of the reaction

$$CaC_2 + 2H_2O = Ca(OH)_2 + C_2H_2$$

The sample is placed in the cap of a metal flask fitted with a pressure gauge and an excess of the carbide reagent placed in the flask body. The flask is closed, clamped and shaken and the acetylene pressure provides a direct reading of moisture content on the gauge.

Electrode probe devices have been used to determine the moisture content of loosely heaped sand, but are not employed for the accurate assessment of samples. However, an electrical method employing measurements of microwave absorption in compacted samples is employed with success to control moisture in mechanized system sands.

Active clay

The live bentonite clay present in a sample of reconditioned sand can be determined by the methylene blue test, which discriminates between active clay and other particles of similar size which are included in the normal clay grade determination. Methylene blue dye, when added to an acidified slurry of clay and water, is adsorbed by the clay to a point at which the appearance of excess dye can be observed by spotting drops of the liquid on to a filter paper. A weighed sand sample is agitated and heated in the selected initial solution, using stirring or ultrasonic means. The standard methylene blue solution is then progressively added from a burette until the spot tests reveal the end point, which is represented by a halo around the spot. The volume of the standard solution required can be directly related to the active clay content using an appropriate calibration curve.

pH and acid demand

The pH value of a solution, a reciprocal function of the hydrogen ion concentration, is the standard representation of the degree of acidity or alkalinity on a scale from 1 to 14. Values from 1–7 represent acid, and 8–14 alkaline, conditions. The pH value can influence the behaviour of clay binders and is readily measured using a meter depending on electrochemical potentials.

In chemically bonded sands employing acid catalysts, account needs to be taken of the presence of alkalis already present in the base sand before the addition of the binder. Since pH measurements are only influenced by substances in solution, the purpose of the acid demand test is to assess the full effect of insoluble alkalis as well.

The test involves the introduction of a standard volume of hydrochloric acid of known concentration to a sample of the sand suspended in water. The acid reacts with the whole of the alkali content, leaving an excess of HCl. This can be quantified by titration with a standard solution of sodium hydroxide, so enabling the true acid demand value of the original sample to be determined by difference.

Loss on ignition

Loss on ignition is employed to determine the presence of organic and other gas forming materials present in the sand mixture or its individual constituents, including new and reclaimed sands. A weighed sample of pre-dried material is fired in a silica crucible held in a muffle furnace at 925°C for 2 hours. The percentage loss in weight is determined and arises from the volatilization, oxidation and decomposition of substances forming gaseous products. These include additions or residues of carbonaceous additives such as coal dust, chemical binders, cereals, and carbonates in sea sands. Although the test does not discriminate between such sources, it does in practice provide a check on the consistency of binder contents and the condition of new and reclaimed sands. Actual volatiles can if necessary be separately determined by similar tests conducted in inert atmospheres.

Mechanical grading

Grain size distribution is the principal factor controlling permeability and surface fineness, and has an important influence upon strength and other properties. Grading is based principally on the use of standard sieves[18] to separate selected fractions for weighing.

The first step is the separation of adhering grains and removal of clay by boiling in dilute ammonia, cooling and siphoning off the resulting clay suspension in repeated rinses. The standard A.F.S. clay grade determination is designed to separate particles less than 0.22 mm diameter by washing in dilute caustic soda in a mechanical agitator, followed by siphoning.

The sand residue is dried, weighed, transferred to a nest of sieves and mechanically shaken until fractions of constant weight are achieved. Further particle size analysis of the separated fines may be undertaken if required by elutriation or sedimentation tests. The latter are based on determination of the rate of settling of the solids in water using hydrometer or pipette techniques[1,19,20]. Particle sizes are deduced from Stokes' Law (q.v. Chapter 5).

Mechanical grading may be represented by direct plots of frequency distribution or by cumulative grading curves based on the total weight of material coarser than each individual sieve. In the former case a regular series of sieves is required since the shape of the plot may otherwise be misleading. The cumulative curve is relatively independent of the sieve apertures selected, so that individual sieves can be omitted without invalidating the result. Additional data from sub-sieve size analysis can be incorporated in the same plots if required. Typical cumulative grading curves for coarse and fine silica sands are illustrated in Figure 4.11.

Some use is made of measures of average grain size derived by calculation from the results of sieve analyses, although this gives no indication of the underlying distribution provided by the full analysis. The long established A.F.S. Grain Fineness Number[2] is based on the number of openings per inch of a sieve which would just pass the average size calculated from the analysis using ASTM Sieves; a parallel system employed the former British Standard series for the same purpose.

Since the introduction of metric sieves in the British Standard[18] and its international equivalents, a similar method is adopted but the fineness is expressed as the Average Grain Size in micrometres μm.

An account has been given of many of the tests in common use, but others have been developed to evaluate behaviour under actual or closely simulated casting conditions. They include tests for resistance to mould erosion, metal penetration and scabbing.

Apart from the control of production mixtures, sand testing is also used to assess base sands and bonding materials, using both direct chemical and mechanical analysis and bulk tests on mixtures with standard binders and sands respectively. Other tests are applied to binders alone, including gelling index, swelling capacity, liquid limit and the cone fusion test for clays, and viscosity for liquid and semi-solid chemical binders. Methods, and the values for typical materials, can be found in the respective specialized literature.

Two valuable critical reviews of modern applications of sand testing, focused on the green sand and chemically bonded sand areas respectively, are given in References 21 and 22, whilst Reference 23 presents a further useful survey of many of the individual properties and tests in the quality control context.

Foundry sands and binders

The combination of properties required in the moulding material depends upon the weight and composition of the casting and upon the moulding practice to be adopted, itself partly governed by quantity and quality requirements. Properties are developed by blending sands or other refractories with bonding materials, water and special additions.

Moulding sands fall into two broad categories according to the type of base sand employed.

Naturally bonded sands are those in which the refractory grain is associated in its deposits with the clay needed for moulding. Such sands often develop good moulding properties with the addition of water alone but their relatively high clay contents reduce refractoriness and permeability.

Synthetic sands are based principally upon silica sands containing little or no binder in the natural state, the strength properties being developed by separate additions. This gives greater freedom in the control of properties and the proportion of binder required is usually much less than that present in naturally bonded materials.

The refractory base

The critical characteristics of the base sand or refractory are its chemical composition, mechanical grading and grain shape.

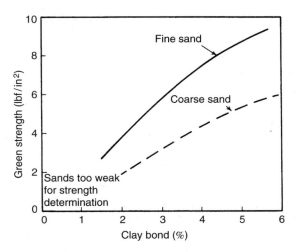

Figure 4.12 Effect of grain size on strength of clay bonded sand (from Davies[5]) (courtesy of British Steel Corporation)

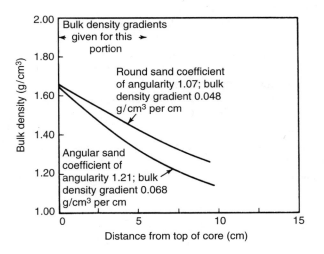

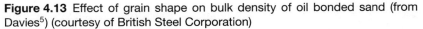

Figure 4.13 Effect of grain shape on bulk density of oil bonded sand (from Davies[5]) (courtesy of British Steel Corporation)

that the strength disparity will be narrowed and ultimately reversed as a greater proportion of flat contacts is generated by prolonged ramming.

The grain size and shape of a moulding material together determine its *specific surface*, defined as the total surface area of grains contained in unit mass. This gives a rough guide to the amount of binder required to coat the grains. Specific surface can be determined from an empirical relationship based on measurements of permeability and bulk density[5,24]. The resulting value can be used in conjunction with mechanical grading data to determine the coefficient of angularity.

Bonding materials

The function of the binder is to produce cohesion between the refractory grains in the green or hardened state. Since bonding materials are not highly refractory, the required strength must be obtained with the minimum possible addition.

Many substances possess bonding qualities, including clays, starch compounds, silicates and numerous organic resins and oils, both synthetic and natural: they may be used singly or in combination. Clay bonded sands are distinguished by the fact that they can be recirculated in closed systems and the bond regenerated by the addition of water; the action of most other binders is irreversible and the moulding material has to be discarded after a single production cycle, although it is normally reclaimed at least in part for further use after suitable treatment.

Binder content. The binder is normally either in liquid form or in a much finer state of division than the sand grains. Ideal distribution is as a thin film around each grain; the primary object of sand preparation is to achieve a uniform coating. On compaction, the binder coatings form lens shaped masses at the points of contact of the grains. The strength of the bond is a function of the number and area of these contacts, although the actual bonding mechanism is not the same in every case. As the binder layers become continuous and then progressively thicker, proportionately less advantage is to be gained from further additions and a graph of strength against binder content shows the general form illustrated in Figure 4.14. The fall in permeability with increasing binder content is a further reason for restricting the addition to the minimum consistent with adequate strength.

Clays

Moulding sands for green and traditional dry sand practice are commonly bonded with clay. In the natural moulding sands the clay occurs in association with the sand grains, whilst the synthetic sands are bonded with selected clays from separate deposits.

Clays take the general structural form of minute plates in the approximate particle size range 0.01–1 µm in breadth, and plasticity and bond are developed by the addition of water. Net attractive forces are generated between charged hydrated clay particles or micelles and between these and the quartz surfaces. The strength of the ionic bond is a function of the total surface area of the particles and is strongly influenced by adsorption of

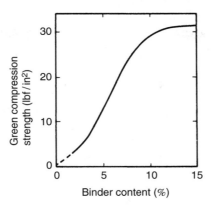

Figure 4.14 Influence of binder content on strength of moulding sand (silica sand bonded with Wyoming bentonite. Properties for optimum moisture contents; clay–water ratios ranging from 2.0 to 4.0) (after Grim and Cuthbert[25])

exchangeable cations at the free surfaces: these modify the balance of local forces between the particles.

On drying, loss of adsorbed water produces shrinkage of the lattice and further strengthening of the bond, so that clay binders are effective in both green and dried condition. Hydration is reversible to temperatures well above the drying range: thus moulds may be dried or cast and the bond regenerated by addition of water after each cycle. Heating to progressively higher temperatures, however, removes chemically combined water and causes permanent loss of bonding capacity. The temperature at which this occurs varies with the particular clay but the loss begins at approximately 400°C and is in all cases complete at 700°C. At still higher temperatures the clays undergo drastic mineralogical changes involving crystallization of alumina and cristobalite and the formation of mullite.

The principal mineral constituents of clays are kaolinite, montmorillonite and illite. Kaolinite corresponds to the general formula $Al_2O_3 \cdot 2SiO_2 \cdot 2H_2O$ and is relatively plentiful as the major constituent of china and ball clays and fireclays. Their relatively high alumina content makes these clays reasonably refractory, but irreversible dehydration occurs in the temperature range 400–650°C. The bonding properties are not so high as those of some other clays so that a higher binder content is required, often in the range 10–20%.

Montmorillonite has the basic formula $Al_2O_3 \cdot 4SiO_2 \cdot H_2O$ but a proportion of the Al^{+++} ions are replaced by Mg^{++} ions in isomorphic substitution. This gives capacity for absorption of exchangeable cations, for example Na^+ and Ca^{++} to which the properties of the clay are particularly sensitive: the typical formula can be represented in the form $M_{0.33}{}^+ \cdot (Al_{1.67}Mg_{0.33})^- Si_4O_{10}(OH)_2$. Montmorillonite is the principal mineral constituent of the bentonites. These clays have a high capacity for water absorption and exceptionally favourable bonding characteristics: strength properties can therefore be derived from additions as low as 3–5%. Maximum swelling power is obtained in the sodium or sodium treated bentonites: these are notable for the high level of dry strength which they confer, whilst the calcium bentonites give maximum green strength. Many bentonites retain their capacity for water absorption up to 550–700°C and can thus be regarded as thermally more stable than alternative clays, with a potentially longer life in closed systems.

Illite clays are produced by the weathering of micas and form the principal source of bond in the natural moulding sands. They do not swell in the same manner as the bentonites but give reasonable strength properties. Irreversible dehydration occurs in the temperature range 500–550°C.

Since the development of bond strength depends upon hydration of the clay, the green strength of a moulding mixture increases with water content up to an optimum value determined by the proportion of clay. Above this value, additional free water causes the green strength to diminish again as illustrated in Figure 4.15. Dry strength, however, continues to increase to much higher original moisture contents, probably due to improved distribution of the binder and the higher bulk densities attainable. Thus by determination of the optimum water content the required strength properties can be obtained with minimum use of clay. A typical combined relationship between clay and water contents and bond strength is illustrated in

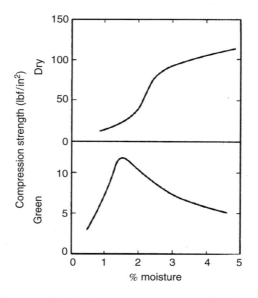

Figure 4.15 Influence of moisture content on green and dry strengths of moulding sands (bentonite-bonded silica sand)

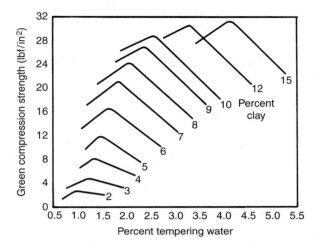

Figure 4.16 Influence of clay and moisture contents on green strength of moulding sands (from Grim and Cuthbert[25]) (courtesy of American Ceramic Society)

Figure 4.16: the peak values can be plotted in the manner of Figure 4.14. Similar relationships for various types of clay are shown in the literature[25].

The constitution and properties of foundry bonding clays were comprehensively examined by Davidson and White[26,27], with particular attention to the comparative effects of heating through the successive stages of dehydration, decomposition and fusion. Thermal analysis was used as a basis for comparison of decomposition temperatures. The properties of important bonding clays were detailed in References 25 and 28.

Cereals

Cereal binders, mainly starch and dextrin, develop a gelatinous bond with water. They are normally employed in conjunction with other binders, for example in clay bonded moulding sands and in traditional oil and resin bonded coresands. Their behaviour was systematically evaluated in controlled investigations[29,30].

In clay bonded sands the cereals confer increased dry strength and contribute to green toughness as measured by the shatter index; an associated advantage is increased resistance of greensand moulds to friability on air drying[29]. In addition, the strong moisture retention of cereals delays the air drying of sand mixtures and contributes to improved bench life. A cereal also acts as a buffer against scabbing and other expansion defects, facilitating deformation of the sand in the critical temperature range.

In coresands the principal function of cereal binders is to improve green strength; they also contribute to collapsibility and ease of knockout, especially in sodium silicate bonded sands. They do, however, increase gas evolution and diminish the resistance of dried cores to the absorption of moisture.

Chemical bonding systems

The modern chemical bonding systems had their roots in the field of core-making. Traditional corebinders based on natural drying oils, most notably linseed oil, provide an effective combination of the main properties required by cores, particularly high dry strength and collapsibility; blends of oil with cereals, water and other additives provide green strength as required. Hardening of the drying oils is accomplished by polymerization or cross-linking, the formation of larger molecules being dependent upon absorption of atmospheric oxygen. Optimum strength is developed in the temperature range 200–240°C. The rate of oxygen supply through the sand controls the rate of hardening[31], emphasizing the importance of permeability, core venting and corestove ventilation. Early thermally cured synthetic resins of the urea and phenol formaldehyde types came to be similarly employed, hardening in this case by condensation polymerization involving loss of water, whilst alkyd oils too can be used in the baking mode, with an oxygen requirement analogous to that of linseed oil for hardening.

Not only coremaking but the entire picture of foundry moulding practice became transformed following the evolution of the new generation of chemical binders, in which setting and hardening could be achieved is relatively short times through controlled reactions. This resulted in the extensive replacement of two of the main traditional practices, namely the drying of clay bonded sand moulds and the baking of cores, by alternatives which dispensed with the stoving requirement. Some drying and thermal curing applications remained, but greensand and chemically bonded sand practices were left as the predominant production systems in the industry.

In the chemical bonding field the distinction between moulding and coremaking sands was itself diminished, many of the modern systems being suitable for both applications. A further advantage as compared with traditional practice is the partial hardening of the sand whilst in contact with the pattern or corebox, which makes an important contribution to increased casting accuracy, as compared with moulds or cores stripped in the green state.

Early impact in the latter direction had been made using sodium silicate binders, rapidly hardened by the passage of carbon dioxide gas though the compacted sand. In a more specialized field the shell moulding process too had shown what was possible using thermal curing of organic resins in direct contact with the tooling. In the conventional sand foundry, liquid organic resins began to be hardened, not only by thermal curing but by the addition of hardeners, or 'catalysts', introduced during mixing.

The next major breakthrough was the development of the first vapour curing organic resin system; with the additional ability to harden silicate binders using liquid reagents as an alternative to gas, a framework existed for the wide and versatile choice of bonding practices since available to the industry.

The use of the modern binders in these different categories was also facilitated by the development of versatile new types of sand mixer, capable of dealing with materials undergoing rapid and varied setting reactions. High speed mixing, self cleaning, and simultaneous mixing and dispensing are some of the important features which will be further reviewed at a later stage. Advanced automatic coreblowing and coreshooting plant was similarly developed to exploit the available properties and characteristics of the materials.

The use of chemical binders, particularly those based on organic compounds, has significant environmental and health implications, with the need for close control of reagents and of evolved vapours, gases and fume generated during mould production and in casting. Safety risks to operatives can arise from skin or eye contact and by ingestion as well as inhalation, calling for gloves, goggles and other means of personal protection as required. Effective ventilation, extraction, treatment and disposal systems are essential, to degrees depending on the particular materials. Silicate binders have fewer problems in this respect and have in some cases been preferred for this reason; the sands are, however, less readily reclaimed than some of those using organic binders, which again has an environmental impact although of a different kind. These and other health and environmental issues will be further considered in Chapter 11.

The chemical bonding systems are normally classified in three groups, defined by the respective approaches to hardening. These are (1) hot curing, (2) cold setting and (3) gas or vapour hardening. Major process applications will be referred to in Chapter 8, but the main operating characteristics of the most important types will now be reviewed. The three classes of binders were the subject of authoritative surveys which contain many valuable practical details relevant to their uses[32,33,34]; charts embodying details of the binder types within these groups and summarizing their most significant properties were given in Reference 35, which also contains proprietary designations and particular areas of application. Other aspects are reported in the literature, including more detailed treatments of some of the bonding reactions[36-39].

Heat curing binders

These may be regarded as derivatives of the previously mentioned traditional stove hardening corebinders, some of which are still used, although on a greatly reduced scale. The main modern additions to this group are the fast-curing, catalyst hardening, organic resins developed for use with the hot and warm box processes, employed for the mass production of blown cores. The liquid resins used in the hot box process employ various combinations of urea formaldehyde (UF), phenol formaldehyde (PF) and furfuryl alcohol (FA), with acid or acid salt catalysts, forming stable sand mixtures which harden immediately on contact with the heated corebox, held in the temperature range 180°–260°C. This permits extraction of the core within a fraction of a minute, after which curing continues under the residual heat in the core. A short period of further heating can be added if required to complete the curing of larger cores.

The subsequently developed warm box system employs high furfuryl alcohol PF or PF/UF resins with metallic salt or acid catalysts, forming mixtures of greater heat sensitivity, which readily harden in the temperature range 150°–190°C.

The hot and warm box systems are particularly well suited to the production of strong and complex cores embodying very thin sections, which cure rapidly in contact with the heated box.

Hot cured bonding materials, although of a different type, are also employed in the shell moulding or Croning process. This distinctive casting system is included with other more specialized techniques in Chapter 9. Shell cores, however, which share some features with hot and warm box cores, are also widely used in mainstream sand casting production.

Cold setting binders

The cold-set or no-bake bonding systems can be of either organic or silicate type. The sand mixture is prepared with additions of liquid binder and an appropriate hardening reagent or catalyst, the setting time being controlled by the nature and amount of the latter addition. The reaction rate can also be sensitive to the temperature, and in some cases to minor constituents, of the base sand. Rapid setting rates can be achieved, but the rate can also be controlled to allow time for the production of large moulds. Suitable designs of mixer facilitate continuous metering and blending of the constituents and rapid delivery to the production point, although batch mixing is also feasible for some applications.

A particular feature of the organic cold-set systems is the very low binder contents required to achieve the appropriate properties: additions of well below 2% and in some cases less than 1% can suffice, with additional reagent or catalyst usually representing a fraction of the resin content. Silicate binders are normally used at levels of 3–4%.

The individual cold setting binder systems were comprehensively reviewed in Reference 33, which examined nine basic types. Some of the systems have undergone modification and, given the active research role of the chemical suppliers, progressive improvements and further new materials can be expected. The full range of available binders will not be covered here but features of some of the more important systems will be summarized. The complex chemistry of the materials and many individual properties and comparative characteristics can be found in the quoted references and in the wider literature.

Furan binders.   With furfuryl alcohol contents in the range 30–85%, these widely used binders are available as three main types based on UF/FA, PF/FA and UF/PF/FA combinations, hardened by acid catalysts: phosphoric acid can be used for the urea furan resins and mixtures based on sulphonic acids for the other two types. Problems of nitrogen porosity can be encountered in ferrous castings, particularly with the UF/FA system, whilst the presence of urea can also cause nitrogen build-up in reclaimed sands. This problem diminishes, although at higher cost, with increasing proportions of furfuryl alcohol in the resins, and disappears with the use of PF/FA resins as an alternative. Newer types of furan binder have also emerged, in which nitrogen, phenol and formaldehyde are all absent: these are also partly aimed at reduction in emissions, including sulphur dioxide, which can be associated with the conventional furan systems.

The sands have good breakdown properties and can be readily reclaimed: they have found wide use over a range of jobbing and mechanized production, including many applications formerly employing drysand practice.

Ester hardened phenolic resins.   Alkaline phenolic resole resin binders can be hardened by the action of various organic esters, for example methyl formate or butylene glycol diacetate; the choice of ester determines the setting time, ranging from a few minutes to several hours as required. This alkali system is less sensitive to the sand composition than the acid hardening resins, which require the use of sands with low acid demand. The sands are reclaimable, with use in proportions up to 85%, and the system is characterized by low emissions and little odour. The system was widely adopted for steel castings, combining good erosion resistance with excellent breakdown properties; a typical report of one such application, to castings of average weight 45–50 tonnes, involved substitution for the ester silicate system, with considerable advantage[40].

Phenolic urethane binders.   In this cold-set system a three-part binder formulation is employed. A phenol novolak resin is mixed with di-isocyanate in the presence of a tertiary amine liquid catalyst, which induces cross-linking at a rate determined by the amount of the addition: this combines rapid curing with the ability to select adequate working time to suit varied production circumstances. Full reclamation is feasible, subject to the avoidance of excessive nitrogen build up. The system is widely used, especially in the USA, but less so in the UK.

Ester hardened silicates.   This system employs sodium silicates with $SiO_2 : Na_2O$ molecular ratios in the range 2.5–2.8:1, the bond being developed by reaction with a liquid organic ester, for example a glyceryl acetate. This lowers the pH value in the hydrolysed solution, forming a silica gel, which then becomes partially dehydrated; additives are normally incorporated in the sand mixture to enhance the modest breakdown properties of this class of binder. The system became widely employed in the manufacture of steel castings and has favourable environmental characteristics, with little fume evolution on casting. Reclamation is, however, limited to 50–75% of useful recovery in normal circumstances. The use of the system has declined with the emergence of the ester hardened phenolic alternative.

Factors influencing the uses of these and other cold-set systems have been critically examined in References 41 and 42 and the effects of foundry variables on some of the types were reviewed in Reference 43.

Cement.   The use of cement as a binder warrants separate mention as the basis of a distinctive sand moulding practice, the Randupson process. Silica sand containing approximately 11% Portland cement and 6% water can be employed in a boxless block moulding system using mounted patterns in temporary wood frames, which are left in place until the initial set of the cement. Support for handling of large moulds is provided by internal grids and irons. Following hydration of the cement, time is allowed for evaporation of excess water and the final mould is strong, rigid and refractory. The system has proved highly suitable for the production of heavy steel castings as well as other alloys. There are no major environmental problems. The hydration reaction is irreversible but the used moulding material is crushed, screened for the removal of spent fines, and reclaimed for use in backing mixtures.

Fluid or castable sands.   The fluid sand system was developed mainly for the production of large moulds and cores, originally employing sodium silicate as the binder, with dicalcium silicate as hardener and vigorous mixing with a foaming agent to produce a pouring consistency requiring no further compaction. A similar principle has been employed using a furan resin based binder system.

The fluid sand system has been successfully used in Russia[44,45], but has not been widely adopted elsewhere.

Gas and vapour hardening binders

These binder systems employ a separate cold hardening operation after the core or mould part has been formed. Setting is accomplished by passage of a gas or vapour through the compacted sand, so that its timing is determined by the operator. The systems are extensively used for the rapid volume production of cores, especially by blowing, but also for larger cores and jobbing work and in varying degrees for moulds.

The CO_2 silicate system.   This versatile development was the first to embody cold hardening of sand, often in contact with the corebox or pattern, although not necessarily so. Sodium silicates, with molecular ratios in the range 1.5 to 3:1 are employed, in amounts mainly within the range 2.5–4%, and usually with breakdown additives, since silicate binders have poor intrinsic qualities in this respect. Sugars, wood flour, cereals and coal dust have been employed for this purpose, but proprietary one-shot high ratio silicates containing starch hydrolysate were later developed to combine high bond strength with stability and excellent breakdown properties.

Passage of CO_2 gas generates carbonic acid in the aqueous solution and the water content itself is reduced, and the viscosity increased, by the flow of the dry gas. The reduction in pH causes a rise in the $SiO_2 : Na_2O$ ratio and the formation of a colloidal silica gel, which hardens and forms the bond. The complex reactions are fully described in the literature but the overall effect can be summarized in the equation

$$Na_2SiO_3 + CO_2 \rightarrow Na_2CO_3 + SiO_2$$

The influence of gassing time on bond strength was illustrated by Haley and Leach[46] as in Figure 4.17, from which it is seen that the bond becomes impaired by overgassing. A given quantity of gas produces higher strength when applied at low pressure and for a longer time than under reverse conditions, whilst the strength continues to increase with ageing.

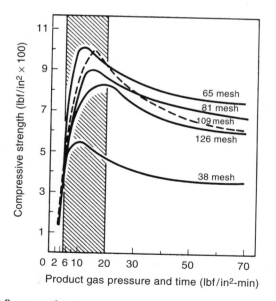

Figure 4.17 Influence of carbon dioxide flow on compression strength, illustrating reduction of strength on overgassing. Sands of varying mesh with constant binder content of 4% (from Haley and Leach[46]) (courtesy of American Foundrymen's Society)

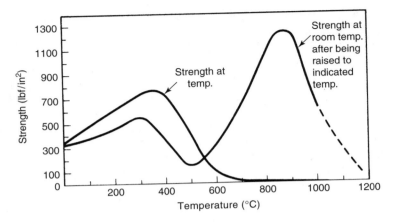

Figure 4.18 High temperature and retained strengths of sodium silicate bonded sand (from Taylor[47]) (courtesy of American Foundrymen's Society)

The effect of casting on the properties of the bonded sand is summarized in Figure 4.18[47], where the very high retained strength explains the need for breakdown agents in silicate bonded sands. Reclamation is limited to around 50%, which can be increased by wet methods: this remains as the most serious limitation to a process otherwise noted for its relatively slight environmental impact. Many aspects of the process have been reviewed in the literature[48–52].

Despite the growth in the application of the organic bonding systems, the CO_2 process retains a significant role in the production of moulds for non-ferrous castings, and in general coremaking, especially for larger castings.

Phenolic urethane cold box system. This system was originally introduced as the Ashland process and was the first organic cold box development using a similar sequence to the CO_2 silicate process. As in the case of the cold-set phenolic urethane binder system, a three-part formulation is employed. A liquid phenol novolak resin is mixed into the dry sand with an isocyanate reagent, but the tertiary amine catalyst is in this case transmitted through the mix as a vapour after compaction, producing rapid hardening within the corebox or on the pattern. The system offers good strength properties with low gas evolution, good breakdown characteristics and suitability for reclamation. It acquired great importance in the cold box coremaking field, being widely preferred for high volume production of blown cores as well as for general coremaking and moulding. Enclosed systems are required for the amine vapour curing stage, residual vapour being purged and exhausted before the hardened core is transferred out of the system. Small traces of amine and solvents produce odours but various modifications have been introduced to diminish this problem. These and other aspects of the vapour cured resin binders are discussed in References 53 and 54.

Ester cured phenolic cold box system. As in the case of the analogous cold-set binder, alkali phenolic resole resin is hardened by the action of an ester, but in this case methyl formate, applied as a vapour. Setting times of less than one minute are achieved, although the strengths are lower than those attained with phenolic urethane. The system is relatively free from the odour problem and fume generation is low. The ester cured cold box system attained wide acceptance for its adaptability to jobbing production of both cores and moulds. Good breakdown qualities facilitate reclamation of 75% and the materials are nitrogen and sulphur free.

Several other gas hardening systems are available, although little used in the UK. Most notable are those employing sulphur dioxide as the hardener, in conjunction with furan or modified epoxy resins containing organic peroxides. The systems give good foundry performance and have been extensively employed in the USA and elsewhere[54]. Close attention is required to the hardening and environmental aspects of the SO_2 gas, as in the case of the organic vapours used in other systems.

Two further organic systems employ carbon dioxide as the hardening gas, with phenolic or acrylic resin binders; although environmentally friendly they have not displaced the highly effective systems previously described.

Selection between the main types of chemical bonding system is influenced by both technical and production characteristics, but the final choice may well be determined by one predominant factor. As an example, the organic binders as a group offer superior breakdown qualities under the influence of heat from the casting, which produces charring followed by complete oxidation and burn-out. This ready collapsibility is particularly advantageous for the removal of confined cores, a prime consideration in the production of castings embodying deep pockets and internal cavities. The particular organic system will then be determined with reference to other considerations.

Unbonded sand systems

Apart from the major progress made in the foregoing field of bonding materials, other developments have pursued the achievement of compacted moulds without the use of conventional binder additions. One system which attracted much interest introduced the use of frozen moulds, in which the main bonding action is derived from the water content of the sand, enabling block moulds of high strength to be produced with the aid of liquid nitrogen or carbon dioxide sprays. A small binder content is, in fact, still required to provide sufficient initial strength for green stripping, necessary because of the expansion of water on freezing. Although high quality aluminium alloy, cast iron and steel castings were produced the process was not adopted on a major scale.

Unbonded moulding material also featured in the magnetic moulding process, in which a magnetic granular material is held in form by a strong magnetic field: again, only limited development followed. The most effective and widely adopted systems have been those in which unbonded sand is held in shape by the application of vacuum suction to the sand mass within the mould container. Two specialized and successful developments based on this principle, the lost foam and V processes, will be examined in Chapter 10.

Special additions to moulding and coresands

The principal constituents of a moulding mixture are the base refractory and the bonding material, but other substances are added to confer special qualities, for example improved resistance to specific defects. Such additions are too numerous for comprehensive review, but Table 4.2 lists examples.

Moulding mixtures

Moulding and coremaking practice is mainly based upon mixtures of the sands, bonding materials and other additions referred to in the foregoing

Table 4.2 Special additions to moulding mixtures

Purpose of addition	Substance
Enhancement of bench life and resistance to drying out	Molasses Sulphite lye Cereals Ethylene glycol
Hot strength development	Iron oxide Silica flour
Surface finish and resistance to metal penetration	Silica flour Coal dust and substitutes
Inhibition of metal–mould reaction	Sulphur Boric acid Ammonium bifluoride
Collapsibility and resistance to expansion defects	Cereals Sawdust Wood flour

sections. The great range of mixes cannot be detailed here but some general points can be noted.

The greatest tonnage of castings is accounted for by greensand production, which includes most of the output of the mechanized and automatic moulding plants, using clay bonded moulding sand and a variety of compaction techniques, which will be characterized in Chapter 8. Synthetic sand–clay mixtures have almost entirely displaced the naturally bonded sands, which remain in use only to a limited extent in non-ferrous production and a few ironfoundries. The synthetic sands are preferred, not only for their refractoriness, mainly relevant in ferrous casting, but also for their consistency and the readiness with which their properties can be controlled in recirculating systems.

Additions of 2% and upwards of coal dust are used in the ironfoundry sands to enhance stripping quality and improve surface finish. A coal with high volatile content is required, to create a reducing atmosphere at the metal–mould interface; purpose designed coal dust substitutes are also available, some of which produce less fume. Special reaction inhibitors are similarly used in the casting of magnesium alloys.

Cores used in greensand production are based predominantly on chemically bonded sands, traditional baked oilsand practice having been largely superseded by systems based on the materials and alternative hardening techniques discussed earlier; production techniques and machines will be further examined in Chapter 8.

For larger castings moulds are commonly produced in self-setting chemically bonded sands, which can be of either organic or silicate type, offering controlled and relatively long setting times: these sands have largely replaced the drysand system, with its reliance on the stoving of clay bonded moulds. They have also come to be frequently preferred in place of greensand for lighter jobbing work, so that greensand practice has gradually become more closely identified with rapid production machine moulding.

In the areas of large castings and jobbing quantities of smaller castings, therefore, the division between moulding and coremaking has become less distinct in terms of the sands employed, which can simplify the problems of reclamation. A further development is the wider use of the core assembly block mould system for small castings. Again, however, the greensand block mould plays the major role where high volume automatic production of such castings is required.

Some further moulding materials

Apart from sand based mixtures, numerous other moulding materials are employed in special circumstances.

Compo and chamotte

In the manufacture of very heavy steel castings conditions call for a material with exceptional refractoriness, volume stability and mechanical strength.

A material traditionally employed for this purpose was steelmoulders composition, a bonded aggregate made from crushed used refractory products with additions of graphite and fireclay. A later development was chamotte, a calcined aluminous clay crushed, graded and bonded with raw clay. Moulds in these types of practice are finished with thick slurries of fine refractory and dried at high temperatures. These materials once dominated the field of heavy steel castings production but they gave way almost entirely to conventional and chemically bonded sands, some of which are based on other types of refractory such as zircon and chromite.

Non-siliceous refractories

Although silica is the most universally available material suitable for the foundry, other refractory minerals provide better thermal stability under extreme conditions. The mineral compositions of three of these materials are shown in Table 4.3.

Zircon is widely employed: highly refractory, it is also free from sudden volume changes such as accompany phase transformations during the heating of quartz. As a sand, it can be bonded with chemical binders or bentonite and used as a general facing material, or more locally in confined pockets or as separate cores, in positions susceptible to metal penetration. Reclamation is seen as essential given the cost, an order of magnitude greater than that of silica sand, a difference accentuated by its higher density.

Chromite sand, slightly lower in cost, is employed for similar purposes, particularly as a facing material in the production of heavy steel castings, for which it provides strong resistance to metal penetration. Olivine from crushed olivine rock has been used to a lesser extent, and is particularly suitable for resistance to slag reaction in the casting of high manganese steels.

Although these sands have been employed primarily for their refractoriness and low expansion (Figure 4.19), they offer the further advantage of a moderate chilling action (Chapter 3, Figure 3.41). This property is potentially valuable for the production of sound surface regions in non-ferrous

Table 4.3 Non-siliceous refractory sands

Sand	Source	Principal mineral constituents
Zircon	Australia; USA	Zirconium silicate, $ZrO_2 \cdot SiO_2$
Chromite	Southern Africa	Chromite, $FeO \cdot Cr_2O_3$
Olivine	Norway	Forsterite $2MgO \cdot SiO_2$
		Fayalite $2FeO \cdot SiO_2$

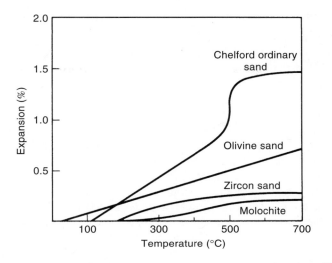

Figure 4.19 Thermal expansion of mould refractories (from Middleton[55]) (courtesy of Institute of British Foundrymen)

alloys of long freezing range. The low expansion characteristic largely determined the choice of zircon sand in the specialized Cosworth process (q.v. Chapter 10).

The properties of the materials can be sensitive to the effects of impurities, requiring the use of formal specifications, particularly those defining pH and acid demand, which affect the performance of binders such as the acid catalysed furan resins. References for several relevant specifications were included in a review by Middleton[56]. Mechanical grading too is important, affecting both binder requirement and surface finish; in chromite the adverse impurities tend to be concentrated in the fines. Loss on ignition is used to detect combined water and oil contamination and the associated risks of pinholing.

Zircon and several other refractories such as chromite, magnesite, and alumina are also used in more finely divided state, both as ramming mixes and as constituents of mould paints (Table 4.4).

Loam moulding mixtures

Loam practice utilizes material of a highly plastic consistency in conjunction, with strickle moulding. The constituents of loams can vary widely and include fine sands, finely ground refractories, clays, graphite and fibrous reinforcements. These materials are mixed with up to 20% water to form thick slurries capable of adhering to an underlying structure of bricks, plates and irons. The finished moulds are completely dried out for casting. Little use is now made of this once widely established practice.

Mould surface coatings

The contradiction inherent in the need to provide reasonable permeability combined with a dense, low porosity mould surface is resolved in many cases by using refractory mould dressings. These are designed to provide smooth, impervious surfaces which prevent metal penetration into intergranular voids in the mould face.

Coatings are usually based on finely divided refractory oxides and silicates or carbonaceous materials, although less refractory substances can be used for lower melting point alloys. Ideally, a coating should be mechanically stable, resistant to thermal shock, inert to the liquid metal, and harmless as a contaminant in the moulding sand.

A coating may be applied either as a wash or as a dry solid. In the former case the refractory particles are suspended in a liquid medium, to which other substances are added to assist suspension and bonding. Some refractories used as dressings are listed in Table 4.4.

Silica flour, once widely used, has been largely discarded in favour of alternative refractories, particularly zircon, for the higher melting point alloys, whilst talc and the alumino-silicates are applied principally to nonferrous alloys of low melting point. Carbonaceous dressings are widely favoured in ironfounding for their effect in producing a reducing atmosphere at the metal–mould interface: wetting of the mould material is inhibited and metal penetration suppressed.

The liquid medium or carrier employed in coatings can be either water or spirit: various organic solvents such as ethanol, isopropyl alcohol, acetone and trichlorethylene are used in the latter category. Water based coatings need to be dried before casting, using gas torches, infra-red heaters or warm air systems in stoves or drying tunnels. These operations are time consuming and involve energy costs. Spirit based coatings can be much more readily dried off by ignition and evaporation but incur additional material costs, fire and health hazards and the need for environmental measures and secure storage.

To delay settling of the solids on standing, all constituents need to be in a fine state of division: refractory powders are typically in the range 200–300 mesh. Settling is further restricted by suspending agents such as clays or cellulose compounds: swelling bentonite is especially effective.

Table 4.4 Mould coating refractories

Silica	SiO_2	Coke	
Alumina	Al_2O_3	Graphite	Carbonaceous
Magnesia	MgO	Plumbago	dressings
Chromite	$FeO \cdot Cr_2O_3$	Blacking	
Zircon	$ZrO_2 \cdot SiO_2$	Sillimanite	
Talc	$3MgO \cdot 4SiO_2 \cdot H_2O$	Chamotte	Aluminosilicates
		Molochite	

If however the bentonite content is too high or the refractory particles excessively fine, an increased tendency to cracking on drying is encountered. In controlled investigations 3–4% bentonite and 0.2–0.3% sodium alginate solutions proved most satisfactory as suspending agents[57]. Bonding too is assisted by the clay additions, but other substances including oils, cereals, sulphite lye and resins can also be used: combinations of such constituents were found to avoid the drying cracks arising from excessive bentonite or cereal alone.

A uniform paint consistency can be maintained by continuous agitation in the container, either by mechanical stirring or by slow passage of compressed air through the suspension. Although coatings can be satisfactorily mixed on site given suitable equipment and controls, the use of separate constituents has been widely superseded by the supply of proprietary pastes and slurries for simple dilution, or of ready-to-use mixtures, obviating potential problems of consistency.

Various aspects of the use of mould coatings were examined in References 58–61 and the subject was included in a report on ancillary operations in mould production[62].

Coatings can be applied by brush or swab, by dip or flow coating or by spraying, and the consistency needs to be controlled to suit the particular mode of application. The Baumé hydrometer is commonly used to monitor the mixtures, although specific gravity alone is of limited value in representing the qualities required. Thixotropic properties are suitable for brushing, allowing free flow to avoid dragging of the sand surface and enabling brush marks to disappear before the viscosity increases again and strengthens the film. Spray coating requires a thinner consistency, more suited to atomization for the formation of fine droplets.

Dip coating requires pseudoplastic behaviour, which means more immediate response to shear forces encountered as the core enters the bath. This is the most effective technique for rapid production, being widely employed for cores, including complex units and assemblies. Flow coating or "overpour" is essentially similar, but more suitable for larger cores and moulds, the wash being pumped from a tank and discharged over the required surfaces through a fan-shaped nozzle, the surplus being collected and returned to the tank. Dip and flow techniques can involve the coating of coreprints as well as metal contact surfaces unless special measures are taken to shield these areas during the operation.

Although water based coatings are unsuitable for use with greensand and silicate bonded sands, in which the bond is water-soluble, there would be a stronger incentive for them to replace spirit based coatings in other cases but for the more onerous drying requirement. Davies[63] examined the problem of drying water based coatings in some detail. Difficulty partly arises through water from the coating wicking into the surface layers of sand. The clay often employed as a suspending agent causes the coating itself to form an impervious layer, inhibiting evaporation of the absorbed water beneath.

New types of proprietary coating were therefore designed to reduce wicking and to omit the clay. These provide a more open coating, and drying times were roughly halved. More efficient drying systems, using higher temperatures and recirculation of exhaust air, gave further reductions, sufficient to encourage change from alcohol to water based systems. Microwave drying has also been successfully employed for the same purpose.

Although liquid coatings based on spirits do find use in greensand practice, dry dressings have traditionally been used in this field as powders dusted and sleeked on to the mould surface. A notable modern development of this principle has been the electrostatic dry powder system. Zircon flour particles, lightly coated with very thin films of thermosetting resin, are blown through a gun in which they acquire a positive electrostatic charge through contact with a plastic rod. The emerging particles are attracted to the moist sand surface, where they adhere and form a continuous layer, which is consolidated with the approach of the liquid metal on casting. The system has been adopted with success in rapid production greensand moulding.

Sand preparation and systems

The quality of a moulding sand depends partly on the manner of its preparation. In addition, economic utilization requires the recovery and processing of the large quantities of materials employed. The sand system is therefore of great importance, its central feature being the mixing plant.

Mixing and aeration

The mixing plant is normally provided with facilities for storage of constituents including new and reclaimed sands, bonding materials and special additions. In many plants provision exists for feeding bulk materials directly into the mixers in measured quantities without intermediate handling, and for the direct metering of water additions. The water content of new sand is normally low enough for immediate use but sands intended for use with chemical binders often require preliminary drying.

The purpose of mixing is to secure distribution of constituents and a smooth, lump-free consistency. To minimize cost the binder should be finally distributed as a thin film round each sand grain. In the case of clay binders distribution of clay and water is commonly accomplished by the gentle squeezing action of rollers, which progressively expose more clay flakes to the adsorption of water. The Simpson batch mixer, illustrated in Figure 4.20, typifies this action, being provided with alternate rollers and ploughs for continuous agitation. Such mixers can be provided with automatic loading skips and discharge directly into the sand distribution system. Many variations of roller action mills are manufactured, some

Figure 4.20 Simpson type batch sand mixer (courtesy of August's Ltd.)

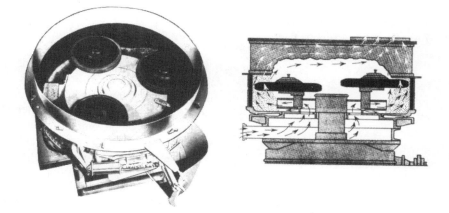

Figure 4.21 Speedmullor high speed sand preparation unit, with provision for cooling hot sand (courtesy of Herbert Morris Ltd.)

using the principle of a rotating pan and fixed rollers: mixing times are normally in the range 3–6 min.

The demand for more rapid batch mixing in mechanized systems can be met by using a high speed mill. The Speedmullor, illustrated in Figure 4.21, uses a rubber lined mixing chamber with rollers rotating in the horizontal plane. Circulation of the sand is assisted by ploughs, supplemented in some cases by an air blast designed to provide forced cooling when processing hot sand. The unit is powered for operation at high speeds; the intensive milling action enables sand preparation to be completed in 60–90 s.

Further designs of high speed mixer dispense with rollers altogether. The general principle of intensive mixing and some characteristic operating features, for example counter-rotation, were examined by Townsend[64], who demonstrated that high mixing efficiency could be achieved with sufficiently vigorous agitation. One modern design of intensive mixer is illustrated in Figure 4.22(a). A rotating mixing pan with a stationary scraper directs the sand into the path of the two high speed rotary mixing tools, and homogeneous distribution of clay and additives is achieved within one minute, providing a throughput of approximately 150 tonnes/hr. In some variants of this machine an inclined mixing pan is also used to enhance the flow pattern, which can in many cases be optimized using a single rotor. An example of this design too is shown in Figure 4.22(b).

A further feature of these machines is the availability of simultaneous vacuum cooling and mixing, in which accelerated activation of the bentonite binder is achieved within the sealed mixing chamber through steam generation and the virtual absence of air: enhanced temperature control with both energy and binder savings are claimed for this system.

In the batch mixing of sand, properties increase progressively as the ideal distribution of the binder is approached, the relation of strength to milling time following the general form shown in Figure 4.23. Excessive milling time, however, leads to frictional heating and subsequent loss of moisture on standing.

In integrated sand plants with high tonnage throughput, continuity of sand flow through the preparation and reclamation systems is aided by continuous sand mills. In this type of mill continuous sand feed is balanced by continuous discharge, water and binder additions being introduced either separately or as a slurry. Although this system is convenient for the processing of large tonnages of reclaimed sand, control is less exact and the bond less efficiently distributed than in batch milling, necessitating higher additions than would otherwise be necessary. However, such mills have been developed to a high pitch of efficiency: one design claimed to give results equivalent to those from batch mills is illustrated in Figure 4.24. The fast batch mills now available can themselves give a close approach to continuous sand supply: fully continuous delivery to the distribution system can be achieved with an intermediate hopper and rotary feed table in the circuit. The intensive mixers as referred to above are themselves available in continuous as well as batch form.

Mixers employed for clay bonded sands, especially those employing rollers, may be supplemented by aerators designed to deliver the final mixture in light condition, free from lumps, so encouraging uniform packing density in the moulds. Simple aeration can be effected by power operated riddles or by passing the sand stream over toothed belts, but in mechanized lines power driven rotor agitators can be incorporated in the conveyor sequence between mixer and moulding stations.

The mixing of sand with liquid or semi-liquid binder systems can be readily achieved by the action of ploughs, blades, paddles and rubbing plates. Many batch units employed variations of the traditional coresand

(a)

(b)

Figure 4.22 (a) Interior of Eirich large capacity intensive mixer. (b) Eirich Type R mixer, featuring inclined rotating pan and stationary wall–bottom scraper to prevent accumulation of residues and accelerate discharge (courtesy of Orthos Projects)

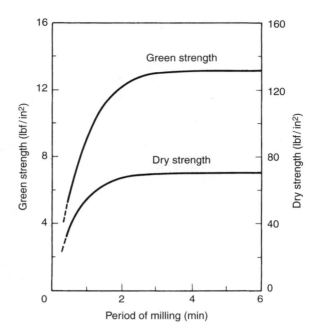

Figure 4.23 Relationship of bond strength to milling time (from Davies[5]) (courtesy of British Steel Corporation)

Figure 4.24 Continuous sand mixer, illustrating charge–overlap–discharge mixing action (courtesy of August's Ltd.)

mixer design as illustrated in Figure 4.25. These are only suitable for chemically bonded sands with long setting times, but later designs of high speed bowl mixer combine mixing times of less than one minute with self cleaning actions.

The principal type of unit employed for chemically bonded sands, however, is the continuous trough or tube mixer which simultaneously

Figure 4.25 Coresand mixer with rotor and rubbing plates (courtesy of Fordath Ltd.)

mixes and delivers the sand through an enclosed cylindrical body. Mixing and propulsion are achieved by rotating screw or multiple blade elements. Metering pumps are provided for binder and hardener additions and the units can be operated on either a continuous or batch basis.

Units of this type can consist of a single mixing chamber, but articulated versions are available to deliver the mixture over a radius of several metres, either to large moulds or to several moulding or coremaking stations.

The large articulated mixer illustrated in Figure 4.26 (see colour plate section) has a reach of 8.5 metres and a delivery capacity of 60 tonnes per hour of mixed sand. The primary arm is essentially a conveyor feeding the secondary arm or mixing trough through a small superimposed hopper. In the hopper is a level probe and an accurate metering gate. The probe regulates the sand flow from the primary arm to maintain a constant feed through the gate throughout mixing. Liquid catalyst is pumped in at the rear end of the trough, quickly followed by the resin a few centimetres further along; accurate timing of this sequence governs the quality of the first sand to emerge.

The mixing trough is designed for zero retention and runs at approximately 300 rpm, with a dwell time of around 12 seconds, but smaller units run at speeds up to 900 rpm, with dwell times as little as 3 seconds. Although self emptying, the mixing chamber is periodically cleaned using water, or dilute ammonia for acid cured systems; the primary arm remains full of dry sand ready for re-starting.

Machines of this type can be operated with binder systems giving setting times of 10 minutes or less, as required in fast loop automatic layouts requiring rapid stripping, but much longer times are normal for larger moulds. Sand mixtures can be readily changed given the self-emptying mixing chamber, but if a different base sand is to be introduced the primary arm too needs to be emptied, in which case a single chamber mixer is faster and more convenient.

Many different designs of mixer are employed, of which only some major examples have been mentioned. Further mechanical details and operating features of these and other mixer types have been reviewed in comprehensive working group surveys, particularly those in References 65–67. Valuable aspects of calibration techniques for the maintenance of accurate pump deliveries of bonding additions are provided in Reference 68.

Sand systems

The mixing plant is frequently incorporated in a comprehensive reconditioning system through which the main bulk of the foundry sand circulates in a continuous cycle. The most primitive system is based upon simple batch reconditioning without the refinement of automatic handling and reclamation plant. The moulding sand is prepared by milling or even by hand. After casting, the moulds are knocked out, the floor sand being watered and re-used after riddling, aeration, or remilling. At intervals the sand heap is reconditioned with additions of new sand and binder. Natural moulding sands are best adapted to this practice since the binder is already well distributed and the moisture content less critical. The practice still finds some application in small jobbing foundries.

In contrast with such simple systems are the fully integrated sand processing plants used in modern mechanized foundries: a typical sequence of elements in such a system is shown in Figure 4.27. Closely controlled mixtures are produced from blends of new and reclaimed sand, binders and other additions, the mixture being delivered by skip or conveyor directly to the moulding stations; moulding machines and sandslingers are served through feed hoppers. After casting, sand from the knock-out is fed into the reconditioning system. Sand systems of this type can be highly automated, with centralized push button control of the entire handling, mixing and distribution cycle.

The operations embodied in the system vary greatly with the type of production and with the sand practice, but the aims are to remove metal spillage, nails and core irons, break down lumps and achieve maximum cooling. Whilst it is desirable to minimize the amounts of coresand and spent binder recirculating through the system, recovery of unchanged binder in the moulding sand should be one of the main objectives.

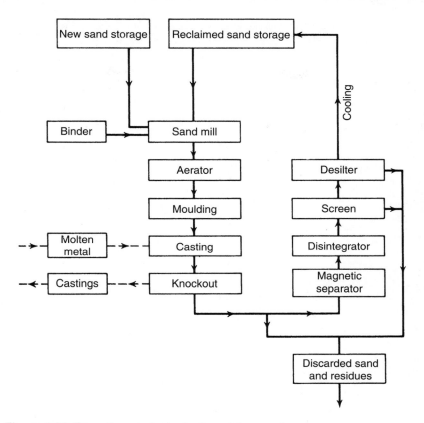

Figure 4.27 Flow diagram for typical moulding sand system

Plant employed in reclamation includes magnetic separators, disintegrators, vibratory screens and de-silters, the sand being transported between the stages by belt conveyor and bucket elevator. The processed sand is finally fed into a storage hopper from which it can be discharged directly into the mixing plant for the next cycle. The normal movement of the sand through the conveyor system provides some opportunity for air cooling, but water sprays may be required for very hot sand, and a specifically designed cooling unit is frequently incorporated in the plant. The general principle of a tower system using counterflow air currents in a cascade is illustrated in Figure 4.28, but the controlled use of water additions is much more effective than simple air movement: it has been emphasized that the evaporation of 1% of water will reduce the temperature of sand by about 26°C. This and other aspects of sand cooling were fully reviewed in Reference 67, with examples of several different types of plant. These include rotary screens, fluidized beds and the use of agitation under reduced pressure to enhance evaporation: the dual mixing and cooling role of the

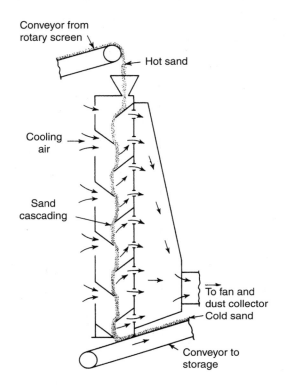

Conveyor from rotary screen

Hot sand

Cooling air

Sand cascading

To fan and dust collector

Cold sand

Conveyor to storage

Figure 4.28 The general principle of one type of sand cooling and desilting plant (from Gardner[69]) (courtesy of Institute of British Foundrymen)

high speed intensive mixer when operated under vacuum has previously been mentioned. Rotary drums for the separation of sand and castings also perform a cooling function.

The strength properties of the reconditioned sand in systems can be maintained at an equilibrium level with very small additions of binder on each passage through the mixer. Since adsorption of water by the clay bond is reversible up to temperatures in the range 400–700°C, the only permanent loss of bond occurs in the zone immediately adjacent to the casting; the extent of this zone depends on the mass and temperature of the casting and the interval before knockout. Rapid knockout, therefore, is an important means both of conserving binder and of minimizing the serious problem of hot sand in systems. Such losses of bond as do occur, apart from that in the layer against the casting, are caused by physical loss of sand from the system, by dilution with coresand which contributes no renewable bond, and by separation of unchanged binder during screening and de-silting: this loss increases with drying and is at a minimum in high production light greensand systems.

The properties of the reconditioned sand are also influenced by changes in the sand grains and by contamination with metal oxides, slag reaction products, spent binder and coresand which have not been eliminated in reclamation. Breakdown of quartz grains and accumulation of fusible fines reduce both permeability and refractoriness. These changes can be kept under control by systematic additions of new sand. This may be introduced either as a regular constituent of the reconditioned mixtures or in reclaimed facing sand containing a high proportion both of new sand and of binder. Less systematic is the sporadic renewal of larger masses of sand, the system being allowed to continue for long periods with binder additions alone. Although feasible where refractoriness is relatively unimportant, the periodic fluctuation of sand quality is an undesirable feature of the latter practice.

Systematic introduction of new sand into the plant requires regular dumping of used sand in quantities equivalent to those of the incoming material, maintaining an equilibrium volume in circulation.

In fully integrated sand systems, characterized by close control, synthetic mixtures based entirely on silica sands are most suitable, bentonite clays being frequently employed as the primary binder. Table 4.5 gives the composition and properties of two synthetic system sands used for the production of machine moulded steel castings. The first is a unit mixture based on regular additions of new sand as well as of binder, whilst the second mixture is primarily used as a backing sand and derives its properties from regular influx of the facing mix also shown in the table. The synthetic unit sands employed in cast iron and non-ferrous practice differ principally in their use of finer base sands; coal dust is also usually present in the former case.

Table 4.5 Reconditioned system sands

1 System unit sand	2A System backing sand	2B New facing mixture
90% Reclaimed sand	Reclaimed sand	50% Chelford silica
5% Chelford silica	0.35% Bentonite	50% Arnolds 52 silica
5% Arnolds 52 silica		4.0% Bentonite
0.65% Bentonite		0.5% Starch
0.15% Starch		
Moisture content, % 4.3	3.5	4.0
Permeability 130	160	150
Green compression strength 45 kPa(6.5 lbf/in^2)	52 kPa(7.5 lbf/in^2)	48 kPa(7.0 lbf/in^2)
Dry compression strength 600 kPa(100 lbf/in^2)	760 kPa(110 lbf/in^2)	600 kPa(100 lbf/in^2)

Davies[5] applied a theoretical treatment to the maintenance of bond in sand systems, demonstrating the eventual stabilization of effective binder content at a level determined purely by the cyclical addition and independent of initial content, assuming stable loss rates. The loss of bond in each cycle, due to irreversible dehydration and physical separation, is stated to fall normally in the range 5–20%, representing a requirement for additions of 0.2–0.8% of binder to maintain the properties of a 4% mixture.

Comparing the unit mixture in Table 4.5 with the equivalent 'starter' facing mix, 0.4% of the 0.65% clay addition could be regarded as being required to balance the 10% addition of new unbonded sand, the remaining 0.25% being sufficient to maintain equilibrium in respect of bond losses in the 90% of reclaimed sand. This, the equivalent of an addition of 0.28% when expressed as a percentage of the reclaimed sand fraction, represents a renewal of 7% of the original clay binder content in each cycle, or a 'life factor' of approximately 14 cycles. In this case, therefore, the major part of the overall 16% clay renewal addition arises from dumping loss necessitated by the new sand addition rather than from deterioration of the binder in casting. As might be expected from the lower decomposition temperature of the cereal binder, its consumption rate is approximately three times that of the clay binder, proportionately greater cyclical additions being required.

The above conditions, encountered in the production of machine moulded steel castings in the weight range 20–50 kg, would be modified in the direction of increased binder life in systems with lighter castings and more rapid knockout. Reduction in the new sand addition, with corresponding conservation of binder, can be made in cases where refractoriness of the mixture is of less consequence.

Controlled experiments on rate of bentonite loss were carried out by Zrimsek and Vingas[30] and the results presented in graphical form: the shape and size of casting were shown to be critical in determining the rate of new addition required.

Unit and duplex sand practice

Integrated sand systems can most conveniently be operated using a single or unit sand mixture which passes round the cycle of mixing, handling, moulding and reconditioning in the manner already discussed. Such a system is ideally adapted to mechanized production; sand hoppers can discharge directly into the moulding boxes and both mixing plant and moulders are only concerned with a single material.

Unit sand thus offers the customary features of standardization and where refractoriness is not an overriding consideration the advantages are great. Its use, however, does represent a compromise. Apart from reduced refractoriness compared with uncontaminated sand containing a high proportion of new quartz grain, the material may contain metal shot too small for separation by the system: this can cause surface blemishes on the castings. The system sand also tends to be warm, producing drying out at the mould face.

The binder addition to a unit sand must be maintained at a level determined by its strength requirements as a facing material, when lower properties might suffice in the main bulk of the mould. There is, moreover, no opportunity to tailor the properties of the standard mixture to the varying requirements of individual types of casting. Most of these disadvantages can be overcome by separate facing and backing mixtures; the choice of practice must be determined by the relative importance accorded to standardization.

Facing sands may consist wholly of new sand or may contain proportions of reclaimed sand smaller than those in unit sands. The backing sand usually consists of reclaimed sand with the sole addition of moisture. In duplex practice, therefore, the new sand and binder additions are deployed to maximum advantage by being introduced into the system at the point where refractoriness and high strength are most required. The chief disadvantage lies in the need for separate handling of the small quantities of facing sand.

The economics of facing-backing sand practice and the balance of properties in such systems depend heavily upon the relative consumptions of the two materials. Facing sand consumption partly depends on the type of casting being produced and tends to form a greater proportion of the whole in small moulding box production. Usage also depends on the discretion of the operator, so that cost control requires a close watch to be kept on the relative consumption.

Further aspects of sand reclamation

The foregoing discussion has been concerned with the reclamation and processing of used clay bonded sands to form reconditioned mixtures, in which a substantial proportion of the original binder remains unchanged and so makes a major contribution to the properties of what is commonly a unit or backing sand. This concept has come to be referred to as primary reclamation and is standard practice in greensand foundries. The necessary introduction of a proportion of new sand and binder has been seen to entail the discard of enough used sand to maintain the mass balance of moulding material circulating in the system, as well as its quality.

More radical reclamation of used sand can be undertaken, to achieve full recovery of the base sand for use as a direct substitute for new sand. This entails a more exacting process cycle, aimed at the removal of coatings of spent binder adhering to the sand grains, and the separation of the resulting fines. This is referred to as secondary reclamation and is less widely used.

Two factors have, however, brought growing attention to this practice. Firstly the costs of dumping waste sand in landfill sites are increasing, with an additional contribution from the landfill tax: taken together with new sand and double transport costs the break-even point for plant investment is brought much closer. Secondly there is much greater use of chemically bonded sands, for moulds as well as cores, with irreversable reactions affecting the whole bulk of the mould. This is in marked contrast to the limited changes in a thin layer against the casting surface such as occur in a greensand system.

The chemically bonded sands need more intensive treatments to achieve breakup of strong networks of spent binder, whether of organic or silicate type: mechanical attrition, thermal reclamation or wet scrubbing may be used, singly or in combination.

The complete removal of binders and additives, and restoration to a condition equivalent to that of new sand, will normally be required only in respect of the fraction of the used sand that would otherwise be discarded from the system, but in special circumstances the whole of the used sand may be processed in this way, for example when using a high cost material such as zircon in the manufacture of more specialized cast products.

Mechanical attrition. Reclamation techniques employed to break down lumps and achieve particulation of the used sand most commonly rely on vibratory screens, but rotating cylinders fitted with internal baffles and blades to tumble and erode lumps are also used: in this case the sand passes along the cylinder to be screened at the exit end. A further variant is the use of shotblast equipment for simultaneous casting cleaning and sand reclamation: the sand and shot are subsequently separated by magnetic means.

The materials emerging from these low-energy primary treatments retain binder coatings on the individual sand grains. In the case of greensand much of this is live clay and provides a significant proportion of the useful bond in the subsequent mix. In resin bonded sands the residual coatings can have a minor but positive effect on subsequent strength, subject to avoidance of excessive build-up on the grains, and of harmful accumulation of contaminants such as nitrogen.

These simpler mechanical treatments are not very effective in removing silicate binder residues. Soda build-up occurs, so that the proportion of used sand that can be employed in new mixes is much lower than with clay or resin bonded sands, although some specially developed ester-silicate systems have achieved greater success in this respect.

More intensive mechanical systems remove increasing proportions of spent binders from the grain surfaces, and thus allow higher percentages of reclaimed sand to be used in subsequent mixtures. These systems, with thermal treatments and wet scrubbing, also provide bases for full secondary reclamation, used to generate clean material virtually equivalent to new sand.

Many types of intensive mechanical attrition plant are available. Pneumatic and impeller driven sand streams are thrown at high speed against target surfaces and each other, so that the sand grains impact and abrade the residual coatings in a process of dry scrubbing, whilst the resulting fines are separated and collected. High velocity pneumatic scrubbing plants are sometimes employed for the reclamation of clay bonded greensands, to achieve separation and return of clay to the system, whilst producing base sands suitable for use in cold box coremaking with resin or silicate binders. Limited air preheating may be used: this and a wide range of other systems and their applications were reviewed by McCombe[70] and by Stevenson[71].

Thermal reclamation. The most radical process, especially effective for the full reclamation of organically bonded sands, involves high temperature treatment in an oxidizing atmosphere. This achieves complete combustion of the main binder residues and a close approach to the condition of new sand. Mechanically particulated used sand is fed into the plant, most commonly a fluidized bed system with integral gas, oil or electrical heaters. Air flow from a blower or compressed air supply provides fluidization and supports combustion; operating temperatures in the range 700°–800°C are usual although not universal. Emissions carrying dust and fines are processed though bag filter systems and the cleaned sand proceeds though a cooling stage to the final storage hopper. One such system[72] embodies screw feed to two successive fluidized bed stages, the first heated by a horizontal gas burner and the second cooled by a bank of submerged water pipes, giving exit temperatures of around 35°C. Pneumatic conveyors are employed for sand transfers, so such systems have few moving parts.

Combinations of calcining and dry scrubbing are featured in rotary kilns with internal baffles and end burners. Combustion of binder coatings is not the only effect of high temperature treatment: embrittlement and disintegration of coatings can also occur, so eliminating clay and inorganic residues as well as organic compounds.

The particular suitability of thermal reclamation for resin bonded sands is enhanced by the major contribution of combustion of the organic materials to the total heat requirement: this aspect was closely examined, and the conditions for successful thermal reclamation reviewed, by Reier and Andrews[73]. Thermal reclamation is, however, unsuitable for silicate bonded sands, as the binders do not degrade at high temperature.

Capital and operating costs of thermal reclamation are relatively high, but correspondingly high quality reclaimed sands are obtained and overall cost benefits have been demonstrated, with modest payback times for the plant investment.

Wet reclamation. The remaining secondary reclamation technique is wet scrubbing, in which suspended sand grains are subjected to vigorous motion and mutual rubbing by water currents, jets and agitation. Binder films and additives are dislodged and soluble constituents dissolved. The sand must then be separated and dried, and the sludges and water effluent disposed of. Wet reclamation can also be integrated with hydraulic cleaning and decoring of castings (qv. Chapter 8). The system is unsuitable for resin bonded sands and, whilst clay bonded greensands can be processed, costs are much too high for practical application. The main use of wet reclamation has been for sodium silicate bonded sands, where the whole of the binder residues can be effectively removed. High costs, however, again greatly restrict the use of this practice.

Practical issues concerning the suitability of individual reclamation techniques for use with the various bonding systems are examined in References 74–77 in the light of relative costs and other factors.

Increasing environmental stringency, and exacting economics of raw materials consumption, seem certain to impose full and efficient sand reclamation as an integral feature of all castings production. The case has already been advanced for communal reclamation facilities using large scale plant, but installations tailored to the individual foundry requirement appear likely to remain as the norm for the foreseeable future. The same pressures will continue as a powerful factor in the design and selection of future binder systems.

Despite these pressures, one further factor seems likely to influence the future economics of sand reclamation, and especially of those variants involving high capital costs. This concerns the serious efforts now being undertaken to find alterative markets for discarded foundry sands. Examinations of other uses for industrial sands, for example in concrete and asphalt building and paving products, indicate that a substantial proportion of new sand in such applications could be replaced by foundry sand subjected to minimal processing. This would clearly shift the balance from the situation in which landfill dumping is the sole alternative to reclamation.

Sand utilization in the foundry

It is clear from previous discussions that sand consumption is increasingly important from both economic and environmental viewpoints. Although much can be achieved by maximizing reclamation, there is scope for more direct savings by reducing the volumes of sand employed in the casting process itself. The measure of performance in this respect is the ratio of sand mixed to molten metal poured. Since this is determined by the volume of the mould container relative to that of the cast form, the crucial factor is that of moulding box selection, and the packing density of patterns obtainable in the available space.

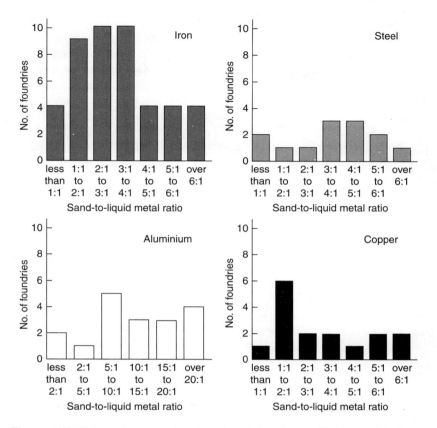

Figure 4.29 Total mixed sand-to-liquid metal ratios in UK foundries (from Reference 75) (courtesy of Department of Trade and Industry)

Results of a major industrial survey undertaken to determine sand to metal weight ratios for the main moulding systems and metal sectors were reported in References 78 and 75, for greensand and chemically bonded sands respectively. This revealed very large differences within as well as between the cast alloy sectors. It is clear that the ratio is influenced by such factors as alloy density, product shape and the degree of box standardization. In the greensand survey, for example, the average ratio for the ironfoundries was 9:1, as against 4:1 for the copper alloy sector, no doubt reflecting the greater proportion of machine moulding in standard boxes for iron, and the predominance of low volume work in more closely tailored boxes for the copper alloy.

The ratios were generally lower in the chemically bonded sand field, with the iron ratio peaking below 4:1. In this case the values obtained for aluminum were largely in the range between 5 and 15:1, reflecting the low metal density.

Despite these explainable differences, however, the striking feature of the analysis was the wide scatter in the ratios reported for each sector, typified in the group of distribution charts shown in Figure 4.29. This emphasizes the potential scope for savings in binder, new sand and other costs, by close analysis of mould size and pattern layouts to produce castings with minimum sand thickness around the mould cavity. The use of loose blocks to fill waste space in the mould container is a further way of reducing consumption.

The target sand-to-liquid metal ratios suggested as realistic in Reference 75 were as follows:

Alloy sector	Sand to liquid metal ratio
Iron	up to 4:1
Steel	up to 5:1
Copper	up to 4:1
Aluminium	up to 12:1

The attainable ratio will be subject to the special factors previously discussed. The survey also provided comparative statistics on amounts and costs of sand purchased per tonne of finished castings produced.

References

1 Foundry Sand Handbook, 7th Ed., *Am. Fndrym. Soc. Des Plaines, Ill.* (1963)
2 Mold and Core Test Handbook, *Am. Fndrym. Soc., Des. Plaines*, Ill (1978); 2nd Ed. (1989)
3 Third Report Joint Comm. on Sand Testing, *Inst. Br. Foundrym.* (1966)
4 Methods of testing cold setting chemically bonded sands: first report of Joint Committee on Sand Testing (1973), *Br. Foundrym.*, **68**, 213 (1975)
5 Davies, W., *Foundry Sand Control*, Sheffield, United Steel Companies (1950)
6 B.S. 1902., *Methods of testing refractory materials*, Br. Stand. Instn., London
7 Booth, B. H., Rosenthal, P. C. and Dietert, H. W., *Trans. Am. Fndrym. Soc.*, **58**, 611 (1950)
8 Morey, R. E. and Ackerlind, C. G., *Trans. Am. Fndrym. Ass.*, **65**, 288 (1947)
9 Heine, R. W., King, E. H. and Schumacher, J. S., *Trans. Am. Fndrym. Soc.*, **67**, 229 (1959)
10 Heine, R. W., King, E. H. and Schumacher, J. S., *Trans. Am. Fndrym. Soc.*, **66**, 59 (1958)
11 Hofmann F., Dietert H. W. and Graham A. L., *Trans. Am. Fndrym. Soc.*, **77**, 134, (1969)
12 Dietert, H. W. and Valtier, F., *Trans. Am. Fndrym. Ass.*, **42**, 199 (1934)
13 Tipper, A., *Br. Foundrym.*, **50**, 62 (1957)
14 Report of A. F. S., Flowability Comm. *Trans. Am. Fndrym. Soc.*, **61**, 156 (1953)
15 Morgan, A. D., Foundry Practice for the 80's. BCIRA conference paper (1979)
16 Phillips, D. R., *Trans. Am. Fndrym. Soc.*, **83**, 75 (1975)

17 Morgan, A. D. and Fasham, E. W., *Trans. Am. Fndrym. Soc.*, **83**, 75 (1975)
18 B.S. 410. Test sieves, Br. Stand. Instn., London
19 Waterworth, A. N., *Br. Foundrym.*, **54**, 215 (1961)
20 Price, M. and Krynitsky, A. I., *Trans. Am. Fndrym.*, **54**, 402 (1946)
21 Pedicini, L. J. and Mizzi-Krysiak, M. B., *Trans. Am. Fndrym. Soc.*, **97**, 373 (1989)
22 Busby, A. D. and Stancliffe, M. R., *Foundryman*, **90**, 37 (1997)
23 Bailey R. *Modern Casting*, **73**, 6, 19 (1983)
24 Hoffman, F., *Trans. Am. Fndrym. Soc.*, **67**, 125 (1959)
25 Grim, R. E. and Cuthbert, F. L., Illinois State Geol. Survey Report 103: *J. Am. Ceram. Soc.*, **28** No. 3 (1945)
26 Davidson, S. and White, J., *Proc. Inst. Br. Foundrym.*, **46**, A79 (1953)
27 Davidson, S. and White, J., *Fndry Trade J.*, **95**, pp. 165 and 235 (1953)
28 Data Sheets on Moulding Materials, 3rd Ed., *B.S.C.R.A.*, Sheffield (1967)
29 Middleton, J. M. and Bownes, F. F., *Br. Foundrym.*, **56**, 473 (1963)
30 Zrimsek, A. H. and Vingas, G. J., *Trans. Am. Fndrym. Soc.*, **71**, 36 (1963)
31 Marek, C. T. and Wimmert, R. J., *Trans. Am. Fndrym. Soc.*, **61**, 279 (1953)
32 First Report of Working Group T30. *Foundrym.* **86**, 94 (1993)
33 Third Report of Working Group T30. *Foundrym.*, **87**, 50 (1994)
34 Fourth Report of Working Group T30. *Foundrym.*, **88**, 269 (1995)
35 Chemical Sand Binders. *Foundry International*, **21**, 2, 17 (1998)
36 Morley, J. G., *Br. Foundrym.*, **76**, 183 (1983)
37 Bushnell, R. S. and Parkes, E., *Br. Foundrym.*, **55**, 325 (1962)
38 Metals Handbook 9th Ed. Vol. 15, Casting, Am. Soc. Metals, Metals Park, Ohio (1988)
39 Chemically Bonded Cores and Moulds. Am. Fndrym. Soc., Des Plaines Ill. (1987)
40 Gorby, W. A., Markow L. W. and Winters D. L., *Trans Am. Fndrym. Soc.*, **103**, 87 (1995)
41 Busby, A. D., *Foundrym.*, **86**, 319 (1993)
42 Stevenson, N., *Foundry International*, **17**, 2, 100 (1994)
43 Robins, J., Tiorello, L. I. and Schafer, R. J., *Trans. Am. Fndrym. Soc.*, **99**, 623 (1991)
44 *Fndry Trade J*, **119** (14 and 17) (1965)
45 Chernogorov, P. V., Vasin Yu P. and Nikiforov, A. P., *Russ. Cast. Prod* 413 (1963)
46 Haley, G. D. and Leach, J. L., *Trans. Am. Fndrym. Soc.*, **69**, 189 (1961)
47 Taylor, D. A., *Trans. Am. Fndrym. Soc.*, **69**, 272 (1961)
48 Atterton, D. V. and Stevenson, J. V., *Trans. Am. Fndrym. Soc.*, **89**, 55 (1981)
49 Martin, G. J. and Ennis, C. S., *Foundrym.*, **79**, 376 (1986)
50 Middleton, J. M. and Bownes, F. F., *Br. Foundrym.*, **57**, 153 (1964)
51 Thompson, R. N. and Hugill, R. W., *Fndry Trade J.*, **119**, 529 (1965)
52 Atterton, D. V., *Proc. Inst. Br. Foundrym.*, **48**, B45 (1955)
53 Busby, A. D., *Foundrym.*, **86**, 324 (1993)
54 Hoffman, M. C. and Archibald, J. J., *Trans. Am. Fndrym. Soc.*, **105**, 289 (1997)
55 Middleton, J. M., *Br. Foundrym.*, **57**, 1 (1964)
56 Middleton, J. M., *Br. Foundrym.*, **66**, 258 (1973)
57 Middleton, J. M. and McIlroy, P. G., *Br. Foundrym.*, **53**, 429 (1960)
58 First report of working Group P8. *Br. Foundrym.*, **69**, 25 (1976)

59 Broome, A. J., *Br. Foundrym.*, **73**, 96 (1980)
60 Pursall, F. W., *Br. Foundrym.*, **75**, xii (1982)
61 George, R. D., Harris, J. A. and Sandford, E., *Br. Foundrym.*, **75**, XVI (1982)
62 Fifth Report of Working Group T30. *Foundrym.*, **88**, 417 (1995)
63 Davies, R. W., *Foundrym.*, **89**, 287 (1996)
64 Townsend, R. D., *Trans. Am. Fndrym. Soc.*, **82**, 369 (1974).
65 First Report of Working Group P12. *Br. Foundrym.*, **73**, 174 (1980)
66 Second Report of Working Group P12. *Br. Foundrym.*, **77**, 84 (1984)
67 Sixth Report of Working Group T30. *Foundrym.*, **89**, 3 (1996)
68 Cost-effective management of chemical binders in foundries. Guide 66 10 4, Environmental Technology Best Practice Programme, Harwell (1998)
69 Gardner, G. H. D., *Br. Foundrym.*, **56**, 145 (1963)
70 McCombe, C., *Foundrym.*, **88**, 39 and 77 (1995)
71 Stevenson, M., *Foundry International*, **15**, 3, 96 (1992)
72 *Foundrym.*, **88**, 206 (1995)
73 Reier, C. J. and Andrews, R. S. L., *Trans Am. Fndrym. Soc.*, **97**, 783 (1989)
74 Optimizing sand use in foundries. Guide GG119, Environmental Technology Best Practice Programme, Harwell (1998)
75 Chemically bonded sand: use and reclamation. Guide EG4, Environmental Technology Best Practice Programme, Harwell (1995)
76 Stancliffe, M. R., *Foundrym.*, **90**, 261 (1997)
77 Leidel, D. S., *Trans. Amer. Fndrym. Soc.*, **102**, 443 (1994)
78 Foundry greensand: use and reclamation guide EG5, Environmental Technology Best Practice Programme, Harwell (1995)

5

Defects in castings

Under practical conditions castings, like all metallurgical products, contain voids, inclusions and other imperfections which contribute to a normal quality variation. Such imperfections begin to be regarded as true defects or flaws only when the satisfactory function or appearance of the product is in question: consideration must then be given to the possibility of salvage or, in more serious cases, to rejection and replacement.

This type of decision is dependent not only upon the defect itself but upon its significance in relation to the service function of the casting and, in turn, to the quality and inspection standards being applied. The question of acceptance standards and the methods used for the detection of defects will be further reviewed in Chapter 6; the present chapter will be devoted to the causes and prevention of defects, to their significance, and finally to an appraisal of the techniques of rectification.

The general origins of defects lie in three sectors:

1. the casting design,
2. the technique of manufacture–the method,
3. the application of the technique–'workmanship'.

A defect may arise from a single clearly defined cause which enables the remedy to be sought in one of these sectors. It may, however, result from a combination of factors, so that the necessary preventive measures are more obscure. All foundrymen are familiar with the persistent defect which defies explanation and finally disappears without clarification of its original cause. Close control and standardization of all aspects of production technique offers the best defence against such troubles. More specific precautions can be taken in those cases where there is a known susceptibility to a particular defect, whilst the radical approach of design modification may need to be considered in the extreme cases which do not respond to changes in foundry technique.

It should be clear that any approach to the elimination of casting defects must be on an economic basis. Costs of preventive measures and control procedures must be balanced against the alternative costs of salvage operations and replacement castings, whilst the confidence factor and production planning repercussions must not be overlooked. Defects can, above all, be minimized by a clear understanding of their fundamental causes.

The logical classification of casting defects presents great difficulties because of the wide range of contributing causes, but a rough classification may be made by grouping the defects under certain broad types of origin: this arrangement has been adopted for the present chapter and seven categories of defect are considered, as follows:

1. Shaping faults arising in pouring.
2. Inclusions and sand defects.
3. Gas defects.
4. Shrinkage defects due to volume contraction in the liquid state and during solidification.
5. Contraction defects occurring mainly or wholly after solidification.
6. Dimensional errors.
7. Compositional errors and segregation.

Although these groups are not mutually exclusive, they afford reasonably clear lines of division in most cases.

It will be noted that the defects considered in the present chapter are mainly those of a general character, common to a wide range of types and compositions of casting. Defects associated with particular cast alloys or with special casting processes have mostly been omitted, although some of the latter will be mentioned in Chapters 9 and 10. Examples of common defects will be illustrated in context; more comprehensive collections have been illustrated in atlas and other forms in the literature, in some cases using classifications different from that selected here (see References 1–8). Major defects and their origins also provide a principal theme in Campbell's work Castings[9].

Shaping faults arising in pouring

When the liquid metal enters the mould, the first requirement is that it should satisfactorily fill the mould cavity and develop a smooth skin through intimate contact with the mould surface. Gross failure to meet these conditions produces the most serious defect in this group, the *misrun* or *short run* casting, in which the metal solidifies prematurely and some limb or section of the casting is omitted. *Cold laps* (see Figure 5.1) are a less severe manifestation of the same fault. These arise when the metal fails to flow freely over the mould surface; the intermittent flow pattern is retained on solidification due to lack of coalescence of liquid streams. *Cold shuts* are more serious, the discontinuity extending completely through a casting member in which streams of metal have converged from different directions.

The first sign of conditions giving rise to such defects is the occurrence of rounded corners and edges and a general lack of definition of sharp features and fine mould detail. The defects are most generally associated with metal temperature, cold metal being the usual cause in castings for

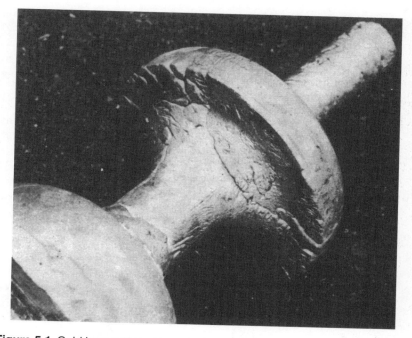

Figure 5.1 Cold laps and shut in a steel casting (courtesy of Institute of British Foundrymen)

which the production method is normally satisfactory. A further cause can be excessive chill from the mould face; this may arise from heavy chilling or from too high a moisture content in greensand. Laps may be encountered, for example, when a method developed for drysand practice is used in conjunction with greensand instead.

A contributing cause to these defects can be an inadequate rate of mould filling relative to the freezing rate of the casting: especially susceptible, therefore, are extended castings of high surface area to volume ratio. Slow mould filling may result from low pouring speed, from an inadequate gating system, or as a result of back pressure of gases in a badly vented mould cavity: several aspects of the casting method are therefore involved in prevention.

As previously discussed in Chapter 1, alloys showing poor fluidity, especially those carrying strong oxide films, are particularly prone to defects resulting from improper flow. In these cases special techniques of gating and high superheat are the principal aids to sharp outlines and completely filled moulds.

Inclusions and sand defects

Non-metallic inclusions in castings may be considered in two main groups. The first are the *indigenous*, or *endogenous*, inclusions, the product of reactions within the melt. These are relatively small particles which remain suspended in the alloy at the time of pouring, or which may be precipitated due to changes in solubility on cooling. They can be regarded not as defects in the macroscopic sense but as to some degree inherent and characteristic of the alloy and the melting practice. They are normally dispersed throughout the casting.

The second group are the *exogenous* inclusions, which result from entrainment of non-metallics during pouring. These vary widely in size and type and include dross, slag and flux residues, formed and separated in the melting furnace but carried over with the metal stream; other sources are refractory fragments from furnace and ladle linings. A further group of exogenous inclusions originates in the mould itself, consisting of moulding material dislodged during closing or pouring. Exogenous inclusions can be regarded as specific defects and tend to be concentrated in certain regions of the casting, for example at the upper surfaces and adjacent to the ingates.

Indigenous inclusions

Reactions forming indigenous inclusions involve common impurities such as oxygen, nitrogen and sulphur, together with the more reactive metallic constituents of the alloy. The most universal reaction is oxidation, whether incidental to air melting or deliberately sought for refining. In the latter case the final oxygen content is normally stabilized as insoluble oxides by addition of deoxidants before pouring. High melting point oxide inclusions may retain their simple composition, but complex slag melts may be subsequently formed by heterogeneous reactions at the inclusion–metal interfaces. Fluxes and furnace refractories can be involved in similar reactions. Other inclusions result from shifts in melt equilibria with falling temperature. In the casting itself precipitation can follow segregation of impurity elements during freezing.

Indigenous inclusions can be minimized by using clean charge materials and melting conditions. Very low inclusion contents are also attainable by melting in vacuum or inert atmospheres. It is possible in special cases to deoxidise with carbon or hydrogen to form gaseous products in place of the normal precipitates.

As soon as indigenous inclusions are formed they begin to be eliminated by gravity separation. This tendency is expressed in Stokes' Law for small spherical particles suspended in a liquid medium:

$$V = \frac{2}{9} \frac{r^2 g(\rho_1 - \rho_2)}{\mu}$$

where V = velocity of separation, m/s
r = radius of particle, m
ρ_1 = density of liquid, kg/m^3

ρ_2 = density of particle, kg/m^3

μ = viscosity of liquid, Ns/m^2

The accuracy of predictions based on this expression must be limited by other influences, notably that of convection in the liquid, but the radius of the particle is seen to be the predominant factor in separation.

Since large particles separate relatively rapidly, to allow molten metal to stand in furnace or ladle for an interval before casting is a viable method for reducing total inclusion content, although elimination of the smallest inclusions is not to be expected in the short times available. The importance of particle size has been demonstrated in deoxidation practice for steel, where a particular combination of manganese and silicon, in the ratio 4:1, is found to produce particles which readily coalesce into large globules for rapid separation.

Gravity separation is not very effective in the case of light alloys because of the relatively small differences in metal and inclusion densities (e.g. Table 1.1 of Chapter 1): in this case it is important to disturb the metal surface as little as possible during melting. However, melts can be treated with fluxes which absorb and dissolve suspended non-metallics: detailed fluxing practice for magnesium alloys has been described by Emley[10].

Inclusions precipitated within the casting have no opportunity for separation: they must therefore be suppressed by maintaining soluble impurity contents at low levels by charge selection, melting and refining techniques.

Thus indigenous inclusions are symptomatic of the general condition of the cast metal rather than of a particular casting, and their control lies largely in the field of melting practice. The nature of the inclusions occurring in various types of alloy and the influence of their form and distribution on mechanical properties is the subject of an extensive but scattered literature: major works in the field are listed as references and these contain extensive bibliographies[12-16].

Exogenous inclusions

Individual grains of foreign matter are inevitably dislodged from furnace ladle and mould, but many exogenous inclusions are complex aggregates of considerable size. Wide variations in form and composition are encountered according to origin.

Slag; dross; refractories. Due to their liquid origin, slag inclusions are of smooth rounded form whilst dross and refractory fragments are of more irregular shape. Such inclusions can best be prevented by retention in the furnace and by careful skimming at the pouring stage. Clean and well maintained teapot or bottom pouring ladles give a high degree of protection; lip pouring ladles require removal of non-metallic matter by skimming, or the thickening of slag to a consistency at which it can be held back during pouring.

Moulding material. Sand inclusions can originate as loose material in the mould cavity, so careful closing is essential. Moulding material can also be eroded during pouring, leading either to massive inclusions or to a widespread distribution of separate grains. In an extensive investigation of sources of exogenous inclusions in steel castings, Middleton and Cauwood[17] found sands to be the predominant constituents, although erosion and slagging of refractories and reactions between moulding material and metal oxides contributed to many inclusions. Metal–refractory interaction in the formation of 'ceroxides' has also been stressed by Flinn *et al*[18].

Mould erosion may result from inherent friability or from low ramming density and soft patches in the mould face. Friability has been shown to be associated primarily with inadequate dry strength[19]. However, it has also been shown that grains can become detached even from strong and well rammed surfaces due to small regions of low density in the mould face[20]: these can occur in high strength moulding materials because of inadequate flowability, emphasizing the need for a suitable balance between the two properties.

Controlled experiments have shown that fine sand, fillers, hard ramming and mould paints can all contribute to the prevention of mould erosion[21]. A useful measure against surface friability is the spraying of mould surfaces with water or dilute solutions of molasses or sulphite lye. Greensand moulds should be cast without delay, although the material also requires reasonable dry strength in case of drying out.

Of particular importance is the design of the gating system, which should avoid impingement of metal against mould and core surfaces, and embody smooth curves rather than sharp corners and abrupt changes in direction. The passage surfaces are of paramount importance: once these are destroyed erosion is progressive and inclusions are distributed throughout the casting. To withstand the passage of so large a volume of metal, gates and runners need to be especially well rammed and can be reinforced where necessary with preformed sleeves, tiles and cores.

Oxide films. This aspect of inclusion formation has received much attention and has been a central theme in the extensive work of Campbell, especially in relation to aluminum alloys[22,23]. The importance of minimal disturbance of the metal surface during melting operations has already been stressed, since gravity separation is relatively ineffective, but oxide films are also produced by various flow patterns in the mould itself, and can form serious discontinuities in the solid casting. Tubular forms of defect result from downhill flow, whilst other crack-like defects are produced by the meeting of confluent steams, each liquid surface carrying a film. Other film defects can occur in the wake of air bubbles passing through the liquid during mould filling.

The avoidance of surface turbulence and air entrainment are thus seen as the overriding need, with particular attention to gating practice to ensure uphill flow; low metal velocity is also required. The more radical solutions involving counter-gravity casting systems will be reviewed later.

Lustrous carbon.   A further type of surface film is responsible for the lustrous carbon defect encountered in the production of ferrous castings, especially cast irons. A lapped and wrinkled graphite skin is formed on the casting surface, particularly the upper surface, and can in some cases become enfolded within the metal itself, producing internal discontinuities and the danger of failure. Occurrence of this defect followed the widespread adoption of modern chemical binders, particularly of the urethane type, but it can also arise in the special circumstances of the lost foam or EPC process, where high concentrations of hydrocarbon vapours are formed in the mould cavity on entry of the molten metal. The vapour decomposes on reaching higher temperatures, to form the thick graphite films which are retained on or below the solidified metal surface.

Lustrous carbon defects can be countered by increased sand permeability and enhanced provision for mould cavity venting, whilst iron oxide (Fe_2O_3) additions have proved effective in reducing the problem. As with other surface film and lap defects it is important to minimize turbulence and to maintain adequate superheat during mould filling. Lower binder contents and modified compositions have also proved successful, whilst alternative foam constituents to the normal polystyrene type have been used to reduce lustrous carbon formation in the lost foam process[24]

The role of filtration

Filtration makes a major contribution to the suppression of inclusions, including oxide films, formed at stages before the molten metal enters the mould cavity. Early application to light alloys employed porous filter beds of granular material located at the exit of the melting unit and aimed primarily at the suppression of indigenous inclusions[11,25,26]; one such system is illustrated in Figure 5.2. Alumina, fluorspar, magnesite, quartz and graphite were all used as constituents. The filters use the impingement principle and can separate particles much below the pore size of the bed. Very low inclusion contents can be achieved, and units can also incorporate inert gas counterflow for degassing.

More general adoption of filters placed in the mould has already been discussed in Chapter 1 with other aspects of gating practice. Such filters exercise a dual function: inclusions, including oxide films formed during metal transfer, are intercepted before the metal reaches the mould cavity, whilst quiescent filling reduces erosion and turbulence as further potential sources within the cavity itself.

Inclusions are frequently concentrated at or immediately below the upper surfaces of castings, where they are revealed on machining or after scaling

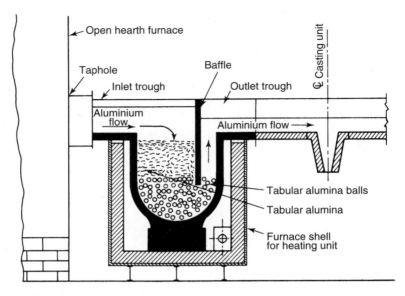

Figure 5.2 Diagram of filtration system for molten aluminium (from Brondyke and Hess[11]) (courtesy of Metallurgical Society of AIME, Journal of Metals)

in heat treatment. It is therefore advisable for the most important faces to be placed downwards; increased machining allowances are usually made on the top surfaces of castings destined for machining.

Erosion scabs and mould crush

These defects are often associated with major inclusions. An erosion scab is a rough projection formed on the casting where sand has been washed from the mould wall. Although the scab itself can be removed by fettling, its presence implies widespread inclusions within the casting.

Mould crush occurs when a mould is closed over an ill fitting core or badly finished joint surface. When the boxes are assembled and clamped, the pressure shears a section from the edge of the mould cavity, producing a large inclusion and a corresponding projection on the casting. The defect can be avoided by careful maintenance of pattern equipment to eliminate distortion and the need for manual patching.

Expansion defects

Expansion scabs result from partial or complete spalling of a section of mould face and penetration of liquid metal behind the surface layer of sand. The defect is seen as an irregular metal scab (Figure 5.3), which can be removed to expose the underlying sand trapped within the casting. In some cases the expansion is insufficient to bring about spalling and metal

Figure 5.3 Expansion scab (courtesy of Editions Technique des Industries de la Fonderie)

penetration, but bulging and shear cracking of the mould face produce a surface fissure or line defect known as a *rat tail* (Figure 5.4). In this case the casting surface shows either a step or a shallow indentation along the path of the incipient mould failure, often with a short metal fin representing the original crack.

The defects often occur on upper surfaces, particularly where moulds embody large flat areas without intermediate points of support. Other sites are associated with intense local heating produced by flow of metal over the surface during pouring. Important factors include the constituents, grading, moisture content and ramming density of the sand and the pouring conditions in casting.

Expansion defects are associated mainly with siliceous moulding materials, which show high expansion on heating to relatively low temperatures: identically graded sands based on more volume–stable minerals are not affected. The effect of temperature on the volume of quartz is illustrated in Figure 5.5, which shows the sudden expansion accompanying the $\alpha \rightarrow \beta$ phase change occurring at 575°C; although quartz undergoes further transformations to tridymite and cristobalite at higher temperatures, it is the mechanical shock associated with the sharp change in volume at 575°C which produces the tendency to cracking and spalling.

The nature of the scabbing process is illustrated diagrammatically in Figure 5.6. During pouring, the upper surfaces of the mould are subjected to intense radiation from the rising liquid metal, so that the surface layers of sand can reach the transformation temperature before being submerged. Expansion of the mould face sets up compressive stresses causing the surface successively to bulge, crack, peel and spall, enabling metal to penetrate behind the original surface. Alternatively, the sand expands and shears along a plane close to the edge of a zone of intense heating as shown in Figure 5.6b.

Figure 5.4 Example of rat tail defect in grey iron casting (courtesy of Institute of British Foundrymen)

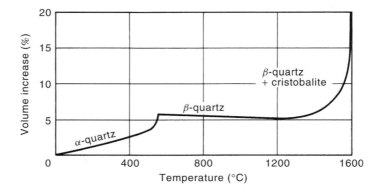

Figure 5.5 Effect of temperature on volume stability of quartz (from McDowell[34])

A contributing factor to scabbing is the occurrence of a zone of weakness in the material behind the mould face. Expansion scabs in greensand moulds have been found to be associated with a wet layer produced by evaporation of moisture from the interface and its condensation on meeting cooler conditions further back[27,28]. The importance of 'wet strength' in countering

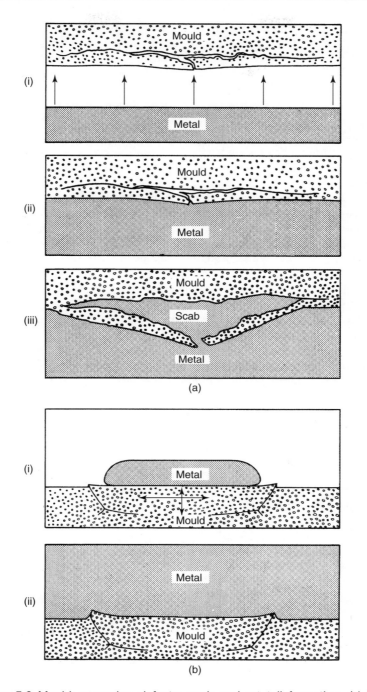

Figure 5.6 Mould expansion defects: scab and rat tail formation. (a) After Levelink and Van den Bergh[35], (b) after Kilshaw, from Reference 36

this effect has been emphasized[29]. However, moisture migration is not the primary cause of scabbing since the defect is not confined to greensand practice.

It is accepted that expansion defects are governed by the extent to which the initial expansion of the quartz can be offset by deformation of the sand mass under compressive stresses. Thus the relationship between hot deformation and confined expansion test results has been found to give a measure of scabbing tendency[30,31]: further relevant tests were detailed in a valuable review by Morgan[32]. The influence of mould variables was demonstrated in comprehensive investigations by Vingas[33]. The deformation capacity of the sand is affected by the amount and type of bonding material and can be increased by special additives. High clay contents are beneficial since they increase high temperature deformation, firstly by their drying shrinkage and subsequently by sintering (Figure 5.7a). A high clay content also gives improved wet strength. Although moisture content should be kept to the minimum consistent with adequate moulding properties, the best type of clay is one retaining reasonable green tensile strength at moisture contents above the optimum. Sodium bentonites with their high swelling capacity combine this quality with superior hot deformation characteristics.

Additives to counter expansion include wood flour, coal dust and other organic materials which soften or decompose to cushion the sand grains. Cereal binders such as starch and dextrin exercise a similar effect, additions of approximately 0.5% being frequently made to clay bonded sands for the purpose. The value of cereals in diminishing scabbing tendency is illustrated in Figure 5.7b, which summarizes the results of investigations over a range of cereal contents[33]. Fibrous reinforcements have also been used but involve practical difficulties in mixing.

The incidence of expansion defects is found to be greatest in the case of uniformly graded sands. Under these conditions all the grains in the surface zone tend to undergo transformation at approximately the same time, producing maximum thermal shock within the sand mass. If, on the other hand, the sand is of widely distributed grain size, the smaller grains will attain the transformation temperature at an earlier stage and the sand will undergo its major expansion over a longer time interval.

High ramming density is a further adverse factor since it reduces the capacity of the bulk to absorb the expansion of the grains: high flowability can thus contribute to scabbing. This problem is accentuated in high pressure moulding, for which additives and high clay contents are essential. The influence of ramming density is demonstrated in Figure 5.7.

The scab susceptibility of castings with broad horizontal surfaces in the cope arises from radiation during pouring, producing simultaneous expansion of large unsupported spans of sand. The radiation interval can be diminished by fast pouring which is found to reduce the scabbing tendency. The interval of exposure to radiation can also be shortened by casting on an incline: the mould is then heated progressively across its upper surface.

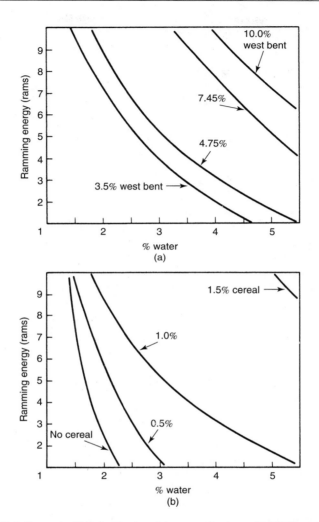

Figure 5.7 Influence of binder and moisture contents and ramming density on scabbing tendency of moulding sand (from Vingas and Zrimsek[33]). (a) Sand mixtures bonded with 1% cereal and various amounts of bentonite, (b) sand mixtures bonded with 4.75% bentonite and various amounts of cereal. Scab-free area to left and below, scabbed area to right and above, lines (courtesy of American Foundrymen's Society)

Reinforcement of susceptible surfaces with flat headed clout nails is widely practiced.

A radical solution to persistent scabbing can be sought by a change to other types of moulding material: reference to Figure 4.19, Chapter 4 shows that non-siliceous materials are inherently safer because of their lower rates of expansion.

Surface roughness and sand adherence

The ideal mould would possess a smooth, impervious surface capable of being faithfully reproduced on the casting. This situation is closely approached with alloys of low melting point but at higher temperatures surface roughness and sand adhesion can be encountered, with loss of appearance and increased dressing costs. In the worst cases the moulding material becomes impregnated with metal to form a solid mass. This can be difficult to remove and if the condition occurs in a confined pocket the casting itself may be scrapped.

Despite earlier conjectures it has been firmly established that metal penetration is primarily responsible for sand adhesion. The controlling factors have been extensively investigated and reviewed[37-39]; both metal and mould variables are involved. Penetration into voids in the moulding material depends primarily upon the pressure head of liquid and thus upon the design, orientation and feeding technique associated with a particular casting. Penetration has been shown to occur at a critical pressure directly proportional to the surface tension of the metal[40] and is therefore also affected by alloy composition.

Since metal flow depends upon superheat, a low temperature minimizes surface roughness and sand adherence. In castings of very heavy section and high thermal capacity, deep penetration may result from the attainment of liquid metal temperatures to a considerable depth within the sand (Figure 5.8). A further factor in penetration can be the occurrence of a pressure wave resulting from the explosive evaporation, and in some cases ignition, of volatiles generated on entry of the molten metal. This phenomenon was examined in References 41 and 42 and attributed to excessive mould filling rates, coupled with inadequate permeability and venting capacity in the mould. Volatiles, especially moisture content, need to be minimized, as does the accumulation of fines in the moulding material.

Surface quality is directly influenced by the mould material and finish, the predominant characteristic being the initial porosity. Whilst low porosity can be achieved by fine grain size, continuous grading or the use of fillers, mould coatings or facings enable surface porosity to be reduced without sacrificing the permeability which is essential in the bulk material. The functions and requirements of mould coatings were reviewed in the previous chapter.

The formation of liquid oxide and slag has been thought in certain cases to reduce the pressure needed for metal penetration by causing wetting of the sand grains: this has been held to account for the improved surfaces of iron castings under the reducing atmospheres created by coal dust additions. This is not a universal effect, however, since oxide films have been shown in some cases to inhibit penetration, no doubt by increasing the effective surface tension of the liquid metal[40]. Much must, therefore, depend upon the characteristics of the oxide film as established by the elements present in

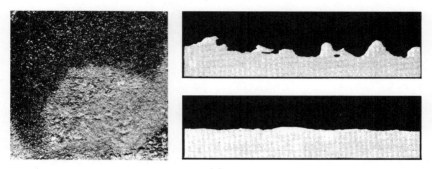

(a)

(b)

Figure 5.8 Metal penetration. (a) Direct view and profiles of casting surface showing local penetration, (b) deep penetration behind mould face

the alloy. Carbonaceous dressings containing volatile matter are also held to improve casting surfaces by creating a back pressure opposing penetration.

The nature of slag reactions at the metal–mould interface was extensively studied in the case of ferrous castings[43–46]. In the absence of penetration this slag is found to assist stripping. The metal–mould reaction occurring in the case of certain non-ferrous alloys is in a different category in that a gaseous product is generated: this will be considered separately in the following section. The reactions occurring in both ferrous and non-ferrous alloys have been the subject of a detailed and comprehensive review by Flinn et al[47].

Gas defects

Gases may be present in castings in solution, as chemical compounds or as included cavities: the latter are the true gas defects. The gas may result from entrapment of air during pouring, from evolution on contact between liquid metal and moulding material, or may be precipitated during solidification as a result either of chemical reaction or of a change in solubility with temperature.

Defects take the form of internal blowholes, surface blows, airlocks, surface or subcutaneous pinholes or intergranular cavities, depending upon the immediate cause. The gaseous origin is frequently evident from rounded contours but in some cases the shape of the cavity is governed by other factors: in the case of intergranular porosity, for example, concave walled cavities can result from constraint by solid–liquid interfaces existing at the time of precipitation.

Other defective conditions include embrittlement and cracking following retention of gases in solution in the solid state; the occurrence of gas–metal compounds as solid inclusions has been referred to in an earlier section. Gases also affect the distribution of shrinkage cavities and segregates in castings. The present discussion will, however, be concerned mainly with specific defects rather than with more general influences of gas content. Although the causes of the defects vary widely they may be conveniently considered in two main groups roughly analogous to those already used for non-metallic inclusions: those caused by physical entrapment on pouring and those resulting from precipitation by the metal on cooling.

Entrapment defects: retention of mould gases

Air may be trapped within a casting when excessive turbulence or aspiration in pouring is combined with metal with little superheat. Bubbles may then be unable to escape before partial freezing of the casting. Defects may also result from failure of the mould to exhaust air displaced by the liquid metal, as when a confined mould pocket is filled at a faster rate than the venting system and mould permeability can sustain. The cushioning effect of the trapped air delays flow long enough for solidification to prevent further flow into the recess: this produces an 'airlock' defect in upward facing projections. Confined air in mould pockets and cores is also capable of expanding sufficiently to blow back through the metal and produce internal cavities; in extreme cases metal can be ejected from the mould.

Prevention depends primarily on effective venting of the mould cavity. Vents for mould pockets and cores need to be of the whistler type, communicating directly with the outside atmosphere, and must be large enough to sustain high rates of gas flow.

Other sources of pressurized mould gas are moisture and volatile organic compounds in moulding mixtures. Excessive moisture in greensand and underbaked or damp cores are common causes of gas cavities, especially when associated with low permeability and inadequate venting. Surface blows and pinholes can also result from pellets of unmilled clay, from oxidized metal shot in reclaimed sand, from improperly dried mould coatings from porous chills and from condensation or rust on chills and chaplets. Precautions are principally designed to reduce gas forming substances present in the mould, whilst ensuring adequate permeability and venting.

Precipitation defects: gases evolved by the metal

The second group of defects is caused by precipitation of gases from the metal on cooling. The gases may have been dissolved during melting or as a result of interaction between liquid metal and mould surface, or may result from reactions involving elements already in solution.

The gas most frequently precipitated as a single element is hydrogen, which shows appreciable solubility in most casting alloys. Nitrogen can behave similarly in iron and steel but is virtually insoluble in the principal nonferrous casting alloys, for which it can be employed as a scavenging gas (q.v.). Oxygen is not usually precipitated directly, tending to form stable oxides with many metals. Where appreciable solubility exists, however, the element can play a major role in reaction porosity through the precipitation of compound gases, mainly carbon monoxide and steam, formed in combination with other solutes.

Origins of gas solution

The solubility of a gas in a liquid metal is a function of the external pressure. For simple diatomic gases such as hydrogen and nitrogen, solution is in accordance with Sievert's Law, which states that the solubility of a gas in a metal at constant temperature is proportional to the square root of its external partial pressure.

i.e. for the gas–metal equilibrium

$$H_2 \rightleftharpoons 2[H]$$

$$K = \frac{[H]}{(P_{H_2})^{1/2}}$$

Molten metals will thus absorb gaseous elements from atmospheres with which they are in contact up to a level determined by the gas–metal equilibrium, defined by the value of the constant K for the particular system.

By the same principle, gas is evolved from a solution in contact with a gas-free atmosphere. Surface adsorption or desorption is an essential step in such a transfer: the rate of approach to equilibrium thus depends upon conditions at the surface as well as within the liquid and gaseous phases. The rate depends upon the degree of mixing in the liquid and the presence or absence of stable surface films of oxide, which reduce the number of sites for adsorption and so diminish the rates of transfer of other gases.

The other major variable determining gas solubility is temperature. The metals providing the basis of most casting alloys form endothermic solutions of gases and so exhibit increasing solubility with rising temperature. Figure 5.9, showing the influence of temperature upon the solubility of hydrogen in aluminium, typifies this behaviour and illustrates the abrupt fall in solubility accompanying solidification. Data for other combinations of gas and metal are available[48–52].

The value of the constant K is thus temperature dependent and varies in accordance with the Van't Hoff equation

$$\ln K = -\frac{\Delta H}{2RT} + c$$

where ΔH is the heat of solution and R and c are constants. This produces a straight line plot of $\ln K$ against $1/T$.

Gaseous elements may also be absorbed by dissociation of compound gases in contact with molten alloy. The most common example is water vapour, present in normal atmospheres and formed by fuel combustion and

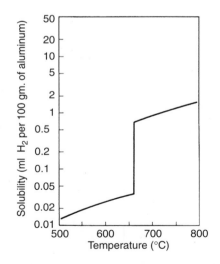

Figure 5.9 Influence of temperature on solubility of hydrogen in aluminium (from Eastwood[52], after Ransley and Neufeld[53]) (courtesy of American Society for Metals)

by evaporation from furnace additions and moulds. Hydrogen absorption occurs more readily through this type of reaction than from an atmosphere of molecular hydrogen. Depending on the composition of the melt, the oxygen either dissolves in the alloy or combines with its reactive components to form a stable oxide phase. Such reactions will shortly be further examined.

The most significant factor in relation to precipitation defects is the relative solubility of gas in the liquid and solid alloy. Very low solid solubility can lead to porosity even with low initial gas contents. This is again illustrated by the case of hydrogen in aluminium and explains the special susceptibility of this metal and its alloys to gas porosity. A solubility of $0.6 \, cm^3$ per $100 \, g$ just above the melting point falls by a factor of 20 to approximately $0.03 \, cm^3$ per $100 \, g$ during freezing: gas evolution from saturated liquid would thus represent 1.54% by volume of the metal.

Precipitation of compound gases

Apart from the precipitation of simple gases, defects may occur through the formation of compound gases by reactions between elements dissolved in the liquid. Such elements, principally carbon, hydrogen and oxygen, may be present at the outset or may be absorbed by reaction between the molten metal and its surroundings. (Absorption of gases in melting and by metal–mould reaction will be subsequently discussed.)

A gas forming reaction may occur within the liquid when an existing relationship between solute elements is disturbed either by a change in temperature or by concentration due to differential freezing or surface absorption. Such reactions include those between carbon and oxygen in iron and nickel alloys and between sulphur and oxygen in copper. To these may be added reactions between hydrogen and oxygen where either or both elements have been absorbed by the liquid.

The well known reaction between carbon and oxygen in iron typifies the conditions for compound gas precipitation. Where both elements are present in solution on casting, the pseudo-equilibrium existing in the liquid is overthrown by differential freezing: the resulting concentration of solutes in the residual liquid causes a resumption of the reaction

$$[C] + [O] \longrightarrow CO_g$$

The resulting porosity, a normal feature of rimming and semi-killed steel ingots, must be excluded from castings by full deoxidation of the melt before pouring; this is normally accomplished by additions of elements forming stable oxides. The presence of excess deoxidant also provides protection against surface pinholing which can otherwise result from reaction between carbon and local oxygen concentrations produced by surface oxidation[54].

In a reaction of the above type, involving dilute solutions of the elements, the equilibrium can be represented by the expression

$$K = \frac{P_{CO}}{[C] \cdot [O]}$$

Since the external pressure of CO is maintained at a low and virtually constant level characteristic of an open system, $K' = [C] \times [O]$ and the relationship between the two elements can be represented by a hyperbolic graph as shown in Figure 5.10. A similar relationship occurs in nickel-base alloys.

The same type of relationship exists between hydrogen and dissolved oxygen in iron, copper and nickel alloys, accounting for the term 'steam porosity'; sulphur and oxygen react similarly in copper. A detailed account of these and other gas forming reactions was given by Phillips[55]. The main precaution against compound gas porosity is to minimize the final oxygen content of the melt, as in the specific case cited above. Whilst strongly oxidizing conditions can be used to eliminate hydrogen, carbon and other impurities during melting, residual oxygen must be eliminated at the final stage.

The form and distribution of gas cavities. The nature of precipitation defects depends not only upon the amount of gas present, but upon the mechanism of solidification of the casting and the conditions for gas nucleation in the liquid. The pressure within a suspended gas bubble depends

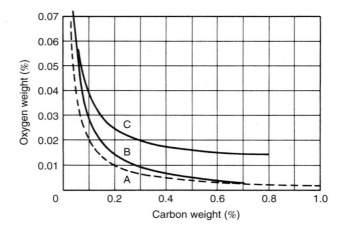

Figure 5.10 Relationship between carbon and oxygen in solution in molten iron (from Reference 56). A. Equilibrium 1600°C. B. electric furnace (after Marsh), C. open hearth furnace (after Fetters and Chipman) (courtesy of AIMMPE)

upon the metallostatic pressure and the surface tension of the metal:

$$P_g = P_L + \frac{2\gamma}{r}$$

where P_L is the total pressure due to the liquid head and atmosphere, r the radius of the bubble and γ the surface tension.

Bubble nucleation is thus opposed by both fluid pressure and surface tension and cannot readily occur in a simple system. In practice, however, heterogeneous nucleation can occur on solid surfaces, for example non-metallic inclusions and metal–mould interfaces, aided by the existence of minute crevices containing occluded molecular gas. The role of inclusions in nucleating pores in aluminium alloy castings has been investigated by Mohanty et al[57], with particular reference to the interfacial energies between inclusion, solid and liquid phases. The solid–liquid interface of a growing crystal is itself a frequent site for bubble nucleation, partly as a result of the high local concentration of gas due to differential freezing. Void nucleation can also be initiated by turbulence or vibration, the associated reduced pressure waves facilitating the creation of new interfaces as discussed in Chapter 2. The interaction of gas pressure and solidification shrinkage in creating pores in closed systems was referred to in Chapter 3.

Many factors thus affect the form and distribution of gas porosity. In some cases gas can concentrate by diffusion into the central region, the final defect appearing as major blowholes, whilst in other cases dispersed porosity may be encountered throughout the casting. Other defects may be localized in surface regions due to high concentrations of gas resulting from metal–mould interactions.

In alloys of short freezing range, dissolved gas rejected at the solid–liquid interface has maximum opportunity to diffuse into the residual liquid, reaching high concentrations which favour the eventual formation of large bubbles from few nuclei. Alternatively, nucleation in the surface region may be followed by progressive inward growth of the gas bubble concurrently with the crystallization front, producing an elongated pinhole or wormhole running parallel with the columnar grains. In alloys of long freezing range which solidify largely by independent nucleation, the gas tends to accumulate in interdendritic pockets of liquid until the last stages of freezing. Local precipitation then produces discrete porosity which may only be revealed on microscopic examination: such pores can in some cases be distributed in layers. Solidification shrinkage and gas pressure contribute jointly to this type of unsoundness: the individual cavities often exhibit an angular form due to the pre-existing solid grains. (See Figure 5.19.)

Gas absorption in melting

The difficulty of securing low gas contents at meltout is evident from consideration of the potential sources of gas contamination:

Metallic constituents of the furnace charge. Gas may be present in solution either from previous melting cycles or from electrolytic sources: many alloy additions are liable to contain hydrogen resulting from the reducing conditions employed in smelting. Hygroscopic corrosion products and lubricants are prolific sources of hydrogen when using metal scrap, a problem accentuated in the case of finely divided materials such as swarf and turnings.

Furnace refractories, fluxes and slagmaking additions. Refractories used in certain regions, for example furnace spouts, tundishes and ladles, are not always sufficiently preheated before contact with liquid metal. They form sources of gas absorption through dissociation of steam at the contact surface.

Atmosphere. Furnace atmospheres contain water vapour and products of combustion as well as nitrogen and oxygen: correlations have frequently been established between atmospheric humidity and hydrogen absorption. Gas absorption from the atmosphere may be aggravated by conditions peculiar to the melting furnace, for example by dissociation of gas molecules in electric arc melting.

Metal–mould reaction

Reference has already been made to gas blows resulting from sudden volatilization of mould constituents. Other forms of gas porosity occur through a less direct mechanism involving reaction between metal and mould and absorption of elements into the surface liquid. Gas is subsequently precipitated, either in the surface region or more generally, to form zones of porosity (Figure 5.11). The site of the porosity will depend on the rate of surface absorption compared with the rate of diffusion into the interior of the casting. Such reactions are prevalent in light alloys containing magnesium and in phosphor–bronzes, the critical reaction occurring between the alloy constituent with the highest oxygen affinity and the free or combined water in the mould. Reactions are of the type

$$[Mg] \ + \ H_2O \ \longrightarrow \ 2[H] \ + MgO$$
$$\text{(in Al)} \quad \text{(mould)} \qquad \text{(in Al)}$$

The problem of metal–mould reaction was the subject of much research, resulting in the development of a number of countermeasures; the work was largely summarized in accounts by Whitaker[58], Ruddle[59] and Lees[60]. The factors influencing the reactions were postulated by Lees. It was observed that such reactions can proceed only when there is direct contact between metal and gas, and will be impeded by oxide films on the liquid metal: continued contact will be maintained when such a film is discontinuous.

Figure 5.11 Example of severe metal–mould reaction in aluminium–magnesium alloy casting (courtesy of Institute of British Foundrymen)

Mould reaction tendency will thus be high in cases where the alloy constituent with the highest oxygen affinity has an oxide with a Pilling–Bedworth ratio* of less than unity. Magnesium is such a constituent, which accounts for the special susceptibility of the aluminium–magnesium alloys to this type of unsoundness. The influence of magnesium is not confined to alloys containing major amounts of the element and has, for example, been associated with reaction pinholing in magnesium-treated spheroidal graphite cast irons[61].

It follows that metal–mould reaction can be suppressed by the substitution of a surface film having the required continuity. This can be accomplished by the introduction of an alloying element with a still higher oxygen affinity than the offending element, but having an oxide with the required properties. An addition of 0.004% beryllium was found, for instance, to eliminate metal–mould reaction in light section Al–10% Mg alloy castings[58]; small additions of aluminium exercise a similar effect in phosphor–bronzes. In some cases, however, complications arise from other elements which themselves influence the film: such an effect is produced by silicon in leaded bronzes due to the formation of a liquid slag[62].

---

* The Pilling–Bedworth ratio is the ratio of the volume of a metal oxide to that of the metal from which it has been formed.

Metal–mould reaction in light alloys is most commonly suppressed by addition of inhibitors to the moulding sand, the aim being to replace the oxide with an alternative compound film. Boric acid and ammonium bifluoride are suitable for this purpose[60]. Paint coatings can also be used as a vehicle for inhibitors. In magnesium-base castings, the strong reaction tendency is suppressed by the use of sands containing approximately 1% sulphur in conjunction with boric acid.

In the case of magnesium treated iron castings it was shown[61] that reaction pinholing could be prevented by maintaining inert conditions at the metal surface with the aid of clay-free moulding dressings, by the elimination of free moisture and by the use of coal dust to provide a reducing atmosphere.

Pinholing in steel castings (Figure 5.12) may result from a number of causes but metal–mould reaction can again be a major factor[63]. The iron reacts with free moisture in greensand to produce hydrogen which diffuses into the metal. Following concentration by differential freezing, the hydrogen combines with dissolved oxygen, pinholes being produced by water vapour as represented by the equation

$$2[H] + [O] \rightleftharpoons H_2O$$

Deoxidation with aluminium is an effective counter to pinholing from this and other causes.

The susceptibility of a casting to gas porosity from metal–mould reaction will depend partly upon the initial gas content of the melt. Defects

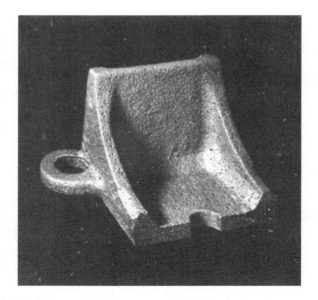

Figure 5.12 Pinhole porosity (courtesy of Dr W. J. Jackson)

may be produced, even with normally acceptable levels of moisture in the moulding material, if the general hydrogen content is already close to the critical level; slight surface absorption may then be enough to produce local supersaturation on freezing.

This combined effect of metal and mould, with liquid metal predisposed to gas defects, can be a source of perplexity in the foundry. The relative responsibilities of moulder and melter for blowholes formed a subject for argument as long ago as 1910[64].

A further form of metal–mould reaction is that resulting in the lustrous carbon defect, involving a solid rather than a gaseous product, referred to previously when discussing inclusions.

Preventive measures

The most important safeguard against gas defects is a low gas content in the metal when poured. In some cases this can be achieved by carefully designed melting techniques and precautions; in other cases the molten metal may need to be degassed in a separate operation.

Melting precautions

The production of gas-free molten metal can be simplified by the use of a gas-free charge. This requires the use of selected constituents from known sources, whilst contaminated materials are either processed separately for the preliminary removal of gas or are used in melts which will themselves be degassed.

Preheating of charge materials to evaporate surface moisture can be extended to higher temperatures to volatilize oils, paints or other organic contaminants and to remove water of crystallization from hydrated corrosion products. More prolonged heating may be used for the elimination, by diffusion, of hydrogen in solution in alloying additions. In the batch melting of cold charges, the benefits of preheating are derived during normal heating to the melting temperature, eliminating the need for separate treatment for evaporation. Many charges are however too bulky for more than a proportion to be charged initially; some furnaces are operated on a semi-continuous basis with a heel of liquid always present. Serious danger of gas absorption accompanies the addition of cold materials to an already liquid bath. A bulky charge often contains a high proportion of swarf and turnings, contaminated with cutting fluid over a large surface area; such materials must be pretreated if gas absorption is to be avoided.

Reclamation of large quantities of turnings from machine shops, vital in the case of expensive alloys, may justify special degreasing plant embodying a centrifuge, followed by treatment with organic solvents. Alternatively the material can be remelted for the production of pigs. This provides opportunity for rejection of gas from the open surface on solidification, whilst the heats can be analysed for use in subsequent melts. Against these advantages must be set the cost of remelting and additional melting losses. In many cases the swarf is disposed of for remelting by specialist concerns, so that foundry melting can be confined to purchased ingot, virgin alloys and heavy scrap returns.

Since gas–metal contact in melting is a function of time and of the surface area of metal exposed, precautions can include fast melting and the use of compact charges: these measures also minimize melting losses. Protective fluxes can be employed, whilst in many cases protection against hydrogen is sought through strongly oxidizing conditions using active fluxes, furnace atmosphere control, and injection of gaseous oxygen. Dissolved hydrogen is thus eliminated by a combination of chemical and physical action before final deoxidation. Oxidizing conditions are beneficial only in alloys with relatively low oxygen affinity but with appreciable oxygen solubility: they are thus applicable to many copper and nickel alloys and to steels, but cannot be used to remove hydrogen from aluminium and magnesium or from alloys containing appreciable amounts of these or similarly reactive elements.

During the last stages of melting great care is needed to avoid the introduction of further gases: final alloy additions must be completely dry and gas-free. Metal temperature should be maintained at the lowest level consistent with adequate fluidity and holding time reduced to a minimum. Furnace spouts, ladles, shanks and furnace tools should be completely dried and preheated to avoid the danger of gas absorption at the final stage.

The production of molten metal of low gas content is facilitated by provision of covered storage for metal stocks and furnace materials. Underfired hotplates and rotary driers are suitable for the treatment of non-metallic materials, which should always be stored in dry, warm atmospheres.

Various melt quality tests are available for estimation of gas content. These range from full analysis, involving vacuum fusion and hot extraction, carrier gas techniques and chemical methods for determination of individual gases[65,66], to semi-quantitative estimations of total gas content, for example by density measurement or by visual comparison of samples solidified in air or under reduced pressure. The latter techniques are discussed in Chapter 8.

Vacuum melting

The most radical technique for production of gas-free metal is vacuum melting. Four specific effects of vacuum conditions are relevant:

1. atmospheric contamination is excluded,
2. dissolved gases are extracted through the influence of reduced pressure on gas–metal equilibria,

3. pressure dependent reactions assist the elimination of elements which precipitate compound gases during freezing,
4. elements with high vapour pressures are preferentially evaporated.

These phenomena are invaluable for refining but can render compositional control more difficult through their effects upon alloying elements.

The technique most generally applicable to the production of castings is vacuum induction melting, in which the molten metal may be removed from the furnace and cast in air, or in which the entire casting cycle can be conducted under vacuum. In the latter case the moulds must be free of volatile constituents and the process is largely confined to the production of precision investment castings. Vacuum arc melting also has limited application in this field: in conjunction with skull melting, shaped castings can be produced in highly reactive metals such as titanium, q.v. Chapter 10.

Although vacuum melting offers clear technical advantages, its application in founding is limited by considerations of cost and scale. Capital and operating costs being extremely high, the process can only be adopted where exceptional quality requirements justify a substantial premium on the selling price. This has so far been principally for high duty precision castings in special alloys. It seems probable that vacuum techniques for general casting production will be confined to the more economical degassing treatments.

The degassing of molten metal

Although gas absorption can be reduced by precautions in material selection and melting, it is difficult to avoid the presence of some gas at melt-out under routine foundry conditions. This can be eliminated by various types of degassing treatment.

Most degassing treatments make use of the equilibrium existing between a melt and a gas atmosphere. Contact with an atmosphere having a low partial pressure of the gaseous impurity brings about transfer of the gas in accordance with the equilibrium. Slow degassing thus occurs from the surface of any bath in contact with a clean atmosphere.

Gas scavenging. One of the most effective methods for gas extraction uses a flushing or scavenging gas which, bubbled through the melt, provides a large gas–metal interface and general agitation. Flushing treatment is most extensively applied in aluminium alloy founding, where there is particular susceptibility to hydrogen absorption and where the high oxygen affinity of the alloy precludes protection by oxidizing conditions. Argon, nitrogen or chlorine can be employed. Chlorine is especially effective and can be generated within the melt through the decomposition of unstable compounds such as hexachloroethane. Tablets of compound are plunged and held within the melt and controlled amounts of chlorine are liberated: this process has, however, declined in favour of inert gas treatment on environmental and health grounds. A further example of gas scavenging from a solid compound entails the use of calcium carbonate to generate carbon dioxide in copper base alloy melts, a much less toxic operation.

Flushing treatments with inert gas can be applied through a simple immersed tube or lance, but measures to generate smaller bubbles greatly increase the efficiency of the process. Booth and Clegg demonstrated marked improvements obtained by attachment of porous ceramic plug diffusers for this purpose[67]. An important further development in the treatment of aluminium alloys came with the introduction of *rotary degassing*, in which inert gas is injected into the bath through a lance provided with a rapidly rotating impeller. This shears the emerging bubbles to generate a fine dispersion, greatly reducing the diffusion distance and increasing the surface area through which the dissolved impurity gas can be transferred. The available contact time is also extended, since the smaller bubbles rise more slowly through the melt. Inclusion separation is also assisted as in a froth flotation system.

Rotary degassing of aluminium alloys has been subject to detailed comparsion with simple lance degassing[68] and shown to be much more effective, subject to selection of the optimum process parameters. Figure 5.13 illustrates the importance of adequate rotational speed, whilst the avoidance of excessive treatment temperatures is another major requirement, as in other degassing systems.

A further variation of flushing treatment is the jet degassing process described by Hoyle[69]: hydrogen contents in steel were successfully reduced to below $2 \, cm^3/100 \, g$ using argon jets on the surface of the metal bath. Inert gas purging can also be used in conjunction with filtration in installations designed for the simultaneous reduction of gas and inclusion contents.

The function of a scavenging gas is also performed by the oxygen injected for removal of carbon and other elements in steelmaking: hydrogen and nitrogen contents are reduced during the process. A carbon monoxide boil promoted by slag oxidation is similarly valued for its physical side effect in eliminating dissolved gases.

Oxidation, deoxidation and the precipitation of compounds. The exposure of liquid metal to oxidizing conditions is a frequent feature of melting practice for iron, copper and nickel alloys. Oxidation is derived from slags and fluxes containing unstable oxides, from oxidizing furnace atmospheres and from injection of gaseous oxygen. Such treatments, by increasing the concentration of dissolved oxygen, remove hydrogen by direct chemical reaction arising from relationships similar to that shown in Figure 5.10 and represented in this case by the equation

$$2[H] + [O] \longrightarrow H_2O_g$$

whence $k' = [H]^2 \cdot [O]$

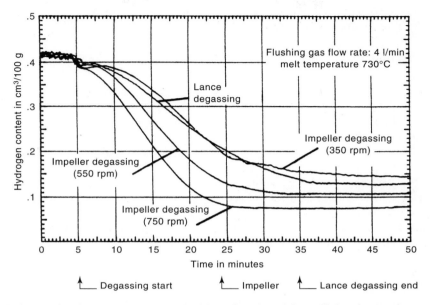

Figure 5.13 Comparative lance and impeller degassing efficiencies at three different rotor speeds (from Chen *et al*[68]) (courtesy of CDC)

The treatments normally leave an excess of dissolved oxygen, which may in other cases result from initial contamination of the charge. Deoxidation is thus required and is normally accomplished by additions designed to precipitate stable oxides.

The principle of stabilization of the impurity by precipitation as a compound can be employed for nitrogen too, particularly in nickel-base alloys: nitrogen pinholing can be suppressed by stable nitride formers such as lithium, titanium and zirconium[70].

Vacuum degassing. Extraction of gases under vacuum occurs comparatively rapidly, so that this particular advantage of vacuum melting can be obtained by brief vacuum treatment of molten metal. Several processes of vacuum degassing have been developed, all depending on overthrow of the gas–metal equilibrium by reduction of the external pressure. The rate of degassing depends partly upon the geometry of the system – the surface area and mass of metal – and partly upon the extent of the agitation and stirring which accelerate the transport of dissolved gas to the nearest surface. Processes are designed to improve these conditions and provide effective treatment in the limited time available during the cooling of superheated metal to the pouring temperature. Since most measures which accelerate gas removal also accelerate cooling, possibilities are restricted by the need for heat conservation.

A number of industrially established processes are schematically illustrated in Figure 5.14:

(a) *Static bath treatment.* The simplest form of vacuum treatment involves the enclosure of a ladle of molten metal within a chamber which can be sealed and evacuated. Given suitable modifications, the ladle itself may be sealed with a cover, the space above the molten metal forming the vacuum chamber. The latter system gives the advantage of a small pump down volume, so reducing the pump capacity requirement.

(b) *Induction degassing.* Molten metal is transferred from the melting unit into a preheated bath surrounded by an induction stirring coil, the whole being located within a chamber which can be sealed and rapidly evacuated. This type of unit is essentially a simplified vacuum induction melting furnace. A still simpler arrangement is to utilize an induction stirring coil in conjunction with an evacuated ladle.

(c) *Fractional degassing.* The contents of a ladle are progressively degassed by treatment of fractions in a separate vacuum vessel. Two notable examples are the method developed by Dortmund Hörder Hüttenunion AG, in which successive fractions of molten metal are raised through a suction nozzle into a vacuum chamber by alternate raising and lowering of the chamber[71], and the continuous process developed by Ruhrstahl, AG, in which metal flows up a tube into the vacuum chamber under the influence of inert gas bubbles bled into the base of the tube; the degassed metal returns to the ladle through a second tube so that continuous circulation is achieved[72]. The circulation can alternatively be effected by electromagnetic pumping. These processes also enable reactive alloy conditions to be made under vacuum.

(d) *Stream droplet degassing.* A ladle is tapped through a sealed annulus into a previously evacuated chamber containing a second ladle, degassing being accomplished during exposure of the falling stream to the vacuum. Gas evolution disintegrates the stream into small droplets so that the geometrical conditions are suited to rapid gas extraction even though the period of exposure is extremely short.

Of the established processes, fractional and droplet degassing have hitherto been confined to the bulk tonnage treatment of liquid steel[73], where the fall in temperature is minimal. Heat loss does however present an obstacle in treating smaller batches of foundry metal, where the problem of maintaining temperature is critical.

Induction degassing can offer the combined advantages of induction heating and circulation, the sharp distinction between brief degassing and full vacuum melting disappearing in favour of a wide choice of treatment

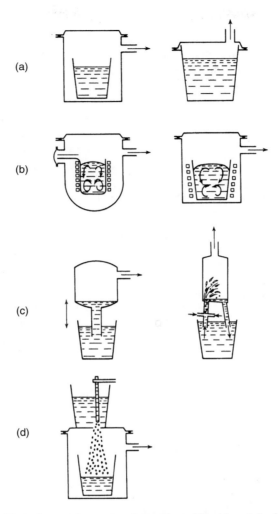

Figure 5.14 Techniques of vacuum degassing of liquid metals. (a) Static bath, (b) induction degassing, (c) fractional degassing, (d) stream droplet degassing

times. The high cost of generating equipment, however, makes the adoption of this process a major undertaking unless existing facilities can be used, as in foundries already operating induction melting plant.

The ideal process for general foundry application would accomplish the treatment in either the melting or the casting vessel, eliminating heat losses involved in transfer to a further vessel. Simple static bath treatment can be effective with small volumes and shallow baths: low cost installations could make feasible the routine treatment of much foundry metal used in small batches. Installations of this kind, involving treatment in the ladle or the crucible, have been described in the literature[74,75].

The efficacy of static bath treatment is limited by lack of circulation other than that provided by natural convection, degassing depending on transport of gas to the upper surface except where the initial content is so high as to induce bubble nucleation. The ideal process should therefore include a system for circulation or agitation, for example by seeding the melt with bubbles of inert gas injected at a low level of the bath, or by the use of an external induction stirring coil of the type previously mentioned. One method of introducing a seeding gas is through a porous plug in the refractory lining of the ladle. This has the advantage of dispersing the gas as a large number of small bubbles.

The pressure dependence of the gas content of liquid metals suggests that the ideal state of freedom from gas porosity in castings might be achieved by the use of a vacuum–compression cycle, in which vacuum treatment of the molten metal for gas extraction would be followed by solidification of the casting under high pressure to suppress precipitation of any remaining gas. Russian work with aluminium and titanium[76] showed that metal density increases steeply with pressure, porosity having been virtually eliminated at pressures exceeding 3 atm. The typical relationship between density and solidification pressure is illustrated in Figure 5.15. Retention of the gas in solution can, however, bring its own problems, as will later be shown.

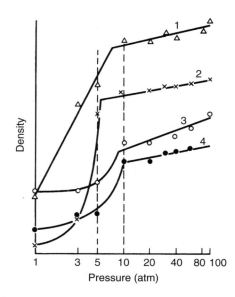

Figure 5.15 Relationship between density and solidification pressure in aluminium alloys. Curve 1 AL2, 20 mm thickness; curve 2 AL9, 20 mm; curve 3 AL2, 80 mm and curve 4 AL9, 80 mm (from O. N. Magnitskii[76]) (courtesy of Plenum Publishing Co. Ltd.)

In considering measures for the elimination of gas porosity from castings, it must be borne in mind that in special cases benefit is derived from limited gas evolution on freezing. The widest use of this effect is for reduction of the primary pipe and cropping discard in ingots for mechanical working. Complete elimination of gas can cause a severe increase in piping. In shaped castings, gas evolution in some alloys of long freezing range can be used to produce a wide dispersion of microporosity rather than a more localized void concentration, with benefit to pressure tightness. Such evolution may in certain copper alloys be obtained from controlled mould reaction[77]. Controlled gas evolution is also employed as a counter to hot shortness in the die casting of aluminium alloys: the slight expansion offsets the linear contraction which produces hot tearing under restraint.

Influences of gas in the solid state

Although the main concern has been with cavities formed during freezing, gas remaining in solid solution or as stable gas–metal compounds can cause reduced mechanical properties or even cracking of the alloy. Reduction in solid solubility on further cooling causes diffusion of dissolved gas into existing voids and imperfections, where molecular atmospheres are created. Supersaturation of the metal thus generates extremely high internal pressures, which lower the level of applied stress required for fracture. The stress to initiate cracking may result from structural change, from resistance to contraction, or in service. The embrittlement of solid metals by gases has been the subject of many investigations and was treated in several specialized works and reviews[48–52].

Residual dissolved gas may be eliminated by high temperature diffusion treatments but radical measures for production of gas-free liquid metal can eliminate these lengthy treatments in the solid state.

An extensive review and bibliography of many aspects of gases in metals, but particularly relating to hydrogen in aluminium, magnesium and copper alloys, was given in Reference 78.

Shrinkage defects

Shrinkage defects arise from failure to compensate for liquid and solidification contraction, so their occurrence is usually a symptom of inadequate gating and risering technique. The actual form of defect depends upon design factors, cooling conditions and the mechanism of freezing of the alloy. Various types of internal cavity or surface depression are encountered: these are illustrated schematically in Figure 5.16 and some examples are shown in Figures 5.17 to 5.21.

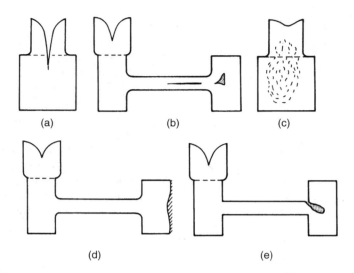

Figure 5.16 Forms of shrinkage defect. (a) Primary pipe, (b) secondary cavities, (c) discrete porosity, (d) sink, (e) puncture

Figure 5.17 Centre line shrinkage in plate section

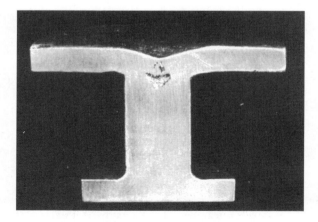

Figure 5.18 Sectioned casting exhibiting internal shrinkage cavity and sink

Major shrinkage cavities

Sharply defined cavities occur primarily in those alloys which solidify by skin formation and result either from premature exhaustion of the supply of feed metal or from failure to maintain directional solidification throughout freezing. The most conspicuous example is the primary shrinkage cavity or pipe resulting from an inadequate feeder head: due to lack of feed metal the final pipe extends into the casting, becoming visible on head removal (Figure 5.16a).

Sporadic occurrence in a casting with a well established method must be attributed to some change in practice, for example omission of a feeding compound or cessation of pouring before the head was completely filled; drastic changes in pouring speed or casting temperature are other possible causes. Since the defect is localized rectification by welding is sometimes feasible.

Unlike primary shrinkage, secondary shrinkage is wholly internal and occurs in positions remote from the feeder head. Depending on the severity of the conditions, the defect may be a massive cavity or a filamentary network. Typical sites include the central zones of extended parallel walled sections and local section increases where no provision has been made either for direct feed or selective chilling (Figure 5.16b). Although this form of cavity is inaccessible for repair, its location near to the neutral axis of stress diminishes its influence on the strength of the casting. Typical examples of internal shrinkage cavities are seen in Figures 5.17 and 5.18.

Discrete porosity

Whilst alloys of short freezing range tend to produce clearly defined cavities, the longer freezing range alloys are subject to scattered porosity. This takes the form of intercrystalline cavities occurring in a large zone and frequently extending to the surface. Surface porosity results from the absence of an intact solidified skin in this type of alloy: intercrystalline liquid remains accessible to atmospheric pressure and to the additional pressure of gases generated at the metal–mould interface. Discrete porosity is aggravated by gas rejected from the metal on freezing, the gas pressure tending to oppose capillary feeding. Susceptible alloys include bronzes and gunmetals, numerous light alloys and phosphorus-containing cast irons.

One form of occurrence is the *layer porosity* encountered in aluminium alloys. Campbell has explained this in terms of the difficulty of pore formation with contraction in the pockets of residual liquid: this is delayed until a critical pressure is reached, as determined by the nature of the inclusion nuclei and the local gas concentration. Once pore nucleation does occur, growth of interconnected pores continues along a plane of constant pressure. The laterally displaced liquid feeds the adjacent zone until further contraction again generates the critical pressure, producing repeated nucleation and the growth of a further layer of pores parallel to the first.

Studies of microporosity in cast aluminium alloys were conducted by Gruzleski *et al*[79], with particular attention to the roles of pouring temperature, with its effect on cooling rate, and of gas concentration. Very high or very low gas concentrations were found to be associated with uniform porosity distribution, as against local occurrences at intermediate levels.

Severe intergranular or interdendritic porosity can be detected radiographically, but in the case of fine microshrinkage neither radiography nor visual examination of machined surfaces will necessarily reveal the defect. The porosity can, however, be detected by density measurements and by attenuation of ultrasonic signals; microscopic examination shows cavities of concave outline reflecting the late freezing of pockets of liquid (Figure 5.19). Unsoundness may come to light only when leakage occurs under hydraulic pressure, indicating the continuous filamentary nature of some forms of microshrinkage. Finely dispersed shrinkage can also exert an adverse influence upon mechanical properties, especially ductility. The significance of discontinuities in determining the behaviour of a casting under service stresses will be considered later in the chapter.

Scattered shrinkage porosity in alloys which freeze in a pasty manner cannot readily be eliminated by simple feeding. As stressed in Chapter 3, chilling is especially effective and moulding materials with a mild chilling action can be used to combat surface porosity in these alloys.

Sinks and surface punctures

Shrinkage cavities in skin forming alloys do not always occur at the thermal centre. As the residual liquid becomes isolated from atmosphere by a continuous envelope of solid, low pressure conditions occur within the casting. The skin may at this stage deform under atmospheric pressure, so that the shrinkage defect appears as a sink in the casting surface. In some cases the defect takes the form seen in Figures 5.18 and 5.20. The depression may

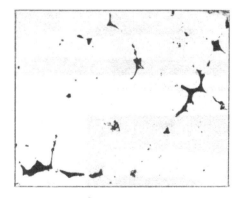

Figure 5.19 Example of microporosity in casting: note concave form of cavities. ×66 approx (courtesy of Dr W. J. Jackson)

Figure 5.20 Surface sink

be so slight as to become evident only on dimensional checking. In other cases a local puncture occurs; access of atmospheric pressure to the liquid then produces an inkwell cavity or 'draw' (Figure 5.21). A frequent site for this type of defect is a re-entrant angle in the casting contour: the hot spot and slow rate of skin formation create the conditions for local failure of the solid layer, an effect similar to that used in atmospheric feeder heads.

Puncture is most likely to occur with high pouring temperatures. Under these conditions a sand projection approaches thermal saturation during cooling to the liquidus temperature, producing a major difference in local rates of skin formation.

These defects can be avoided by ensuring access of atmospheric pressure to the liquid metal in the feeder head, through adequate temperature differentials, feeding compounds, breaker cores, and other measures as discussed in Chapter 3. Local chilling may be used at susceptible points to accelerate the formation of a strong surface layer.

Although the basic causes of shrinkage defects are well understood, the technique of high energy X-ray fluoroscopy offers attractive possibilities for future research into the nucleation and growth of cavities.

Contraction defects

The cooling of cast metal from the solidus to room temperature is accompanied by considerable further contraction. The magnitude of this contraction is indicated in Table 5.1 where the coefficients of thermal expansion of a number of metals provide the basis for a rough estimate of the total linear contraction in the solid state.

Figure 5.21 Two examples of shrinkage puncture or 'draw' (courtesy of Institute of British Foundrymen)

Unlike the liquid and solidification shrinkages, which can be compensated by an influx of liquid, solid contraction affects all linear dimensions of the casting, hence the need for standard pattern allowances in accordance with the expected contraction behaviour of the alloy.

Dimensional change due to solid contraction would be expected to begin as soon as a coherent solid mass is formed, whether a surface layer or a more

Table 5.1 Solid contraction of some metals

Metal	Coefficient of linear* expansion deg $C^{-1} \times 10^6$	Melting point °C	Approximate total linear contraction to 20°C %
Aluminium	29.2	660	1.9
Copper	20.6	1083	2.2
Iron	17.3	1536	2.6
Lead	31.6	327	1.0
Magnesium	31.4	649	2.0
Nickel	18.4	1453	2.6
Tin	24.4	232	0.5
Zinc	35.9	420	1.4

* Mean value for range 20°C–T_M, estimated by extrapolation from data in Reference 80. No allowance is made for phase changes in the solid state.

extensive network of crystals. Under practical conditions, however, castings never contract completely freely and the metal must develop sufficient cohesive strength to overcome significant resistance. Hindrance to contraction may be offered by the mould, by hydrostatic pressure of residual liquid and by other parts of the casting itself due to differential cooling. Stresses can thus arise either from external restraint or from thermal conditions alone. (Figure 5.22.)

The effect on the casting depends upon the severity of restraint relative to the mechanical properties of the cast metal at successive stages during cooling. If the restraint is readily overcome, the casting will contract in a predictable manner and no defect will occur. More rigid hindrance produces tensile strain and a defect may ensue. The form of the defect will depend

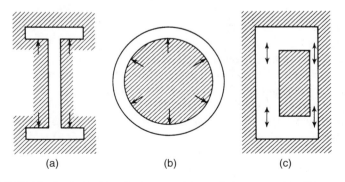

(a) (b) (c)

Figure 5.22 Typical design features giving rise to contraction stresses. (a) Mould restraint, (b) core restraint, (c) differential contraction of casting members

upon the mechanical properties at the point in the cooling sequence when the restraint becomes critical.

During the cooling of a casting from the molten state down to shop temperature the mechanical properties can be related to four stages of behaviour:

1. That period of solidification during which there is no long range cohesion and when sufficient liquid is present for mass and interdendritic feeding to compensate for contraction. This may be termed the liquid–solid stage and the metal has negligible strength but infinite ductility.
2. The stage when solidification is well advanced but when small amounts of residual liquid delay the development of full cohesion. Contraction behaviour is essentially that of a solid but the metal is weak and ductility almost though not entirely absent. This may be termed the solid–liquid or coherent brittle stage.
3. The high temperature solid region in which the metal has developed limited strength and a capacity for plastic deformation without strain hardening.
4. The lower temperature solid region associated with relatively high strength and elastic behaviour.

If the resistance to contraction becomes critical at the highest temperatures, when the alloy is in a relatively brittle condition due to liquid films, hot tearing occurs. If the casting survives this vulnerable stage, the development of contraction hindrance at lower temperatures may cause plastic deformation, residual elastic stress or low temperature cracking, depending upon the severity and timing of the restraint. This is governed by the rate of contraction and by the mechanical properties of both metal and mould.

Strengths of moulding materials, although low when compared with those of metals at normal temperatures, are comparable to those in the solidus region. As the temperature falls this source of hindrance becomes less significant relative to the growing strength of the metal and the main resistance occurs within the framework of the casting itself. Differences in cooling and contraction rates between separate parts of the casting may arise from section thickness variations, from initial temperature gradients produced by the gating technique, or from the cooling environment. The cooling situation may, for example, be such that heat loss from certain sections is accelerated by partial exposure or by draughts, whilst that of other sections is retarded by sand insulation or radiation shielding.

The development of stresses through differential cooling was the subject of several notable studies[81–85]. The basic mechanism is best explained with reference to a simple stress grid casting of the type illustrated in Figure 5.23a, in which the thin members A tend to cool more rapidly than the thick members B.

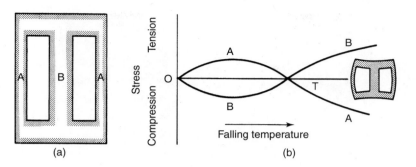

Figure 5.23 Development of stresses through differential contraction. (a) Stress grid casting (b) schematic plot of stresses in thick and thin members, illustrating change in sign during cooling (after Reference 81)

Initially the thinner and more rapidly cooled sections pass through the brittle temperature range, during which stage hot tearing may occur due to mould restraint. As the whole casting cools into the plastic region, high stresses are not generated since differential contraction is readily accommodated by plastic strain. On further cooling, however, the thinner members acquire elastic behaviour and their further contraction places the more slowly cooled and still plastic regions B under compression, producing permanent deformation. As these compressed regions themselves become elastic and complete their cooling, they encounter restraint from the now cold thinner sections and residual elastic stresses are produced in the system. The more rapidly cooled sections A carry a residual compressive stress whilst the last sections to cool B are now in tension. In general, therefore, light sections and the surface zones of castings tend to be left under compressive stress and heavy sections and interior zones in tension.

This development of the stress pattern is schematically illustrated in Figure 5.23b. The initial phase of the diagram represents the period during which any failure will occur in the thin members at a hot spot or stress concentration. To the right of the point of inversion the risk changes to one of residual stress or of cold cracking in the regions now under tension. In alloys which undergo transformations on cooling, the stress pattern may be modified by accompanying temperature and volume changes[83].

The individual types of defect will now be further examined.

Hot tears

Hot tears or 'pulls' are characterised by irregular form, partial or complete fracture following an intergranular path. Tears are often located at changes in section, where stress concentration is associated with locally delayed cooling (see Figure 5.24). In some cases, however, a tear may be wholly internal; the distinction between such a tear and some forms of shrinkage

Figure 5.24 Typical hot tear at change of section (courtesy of Editions Techniques des Industries de la Fonderie)

is indefinite, since restraint of solid contraction and lack of liquid feed may both be involved during the later stages of freezing.

It is well established that the critical temperature range for tear formation lies in the region of the solidus, where the contraction behaviour of a solid is combined with extreme weakness and little capacity for plastic flow (stage 2 in the previously outlined cooling cycle)[86,87]. Thus, even the comparatively low strength of the moulding material may produce fracture.

The occurrence of hot tears is influenced by three factors, namely alloy composition, the design of the individual casting and foundry technique.

Alloy composition

Susceptibility to tearing is closely associated with the mode of freezing and thus with alloy constitution. The tendency is loosely related to the length of the temperature range over which the alloy possesses a certain cohesion yet ductility is minimal. This range is related to the freezing range of the alloy, since cohesion is attained at some critical proportion of solid to residual liquid. The extent of the brittle range determines the total contraction occurring in a cast member of a given length whilst the alloy is in the vulnerable condition. Alloys with very short freezing range, therefore, including those of near eutectic composition, show little tearing tendency. Alloys containing small amounts of eutectic, with their relatively long temperature range of primary dendrite cohesion, are especially prone to tearing. In a comprehensive early investigation of the aluminium–silicon system Singer and Cottrell[86] showed the hot shortness temperature range to reach a maximum in the region of 1.8% Si and thereafter to decrease rapidly with increasing amounts of the eutectic constituent. Rosenberg, Flemings and Taylor[88] compared the hot tearing tendencies of a number of alloys using flanged bar castings of varying length, poor resistance to tearing being indicated by the appearance of the defect in the shortest of the bars. Minimum resistance to tearing in eutectic systems was found to be associated with alloys containing just enough eutectic to surround the primary grains.

Larger amounts of eutectic enable liquid flow to heal incipient tears until the eutectic temperature is attained, accounting for the low incidence of hot tearing in grey cast irons and in the eutectic aluminium–silicon alloy. The comparative susceptibilities of various alloys, derived from contraction rate characteristics in relation to liquid film stage duration, were evaluated by Bishop, Ackerlind and Pellini[89]. Figure 5.25 illustrates this comparison for a number of major types. In this figure the unrestrained contraction of a cast member is plotted against time, the latter being expressed as a multiple of the solidification interval of the particular alloy. The factors of characteristic contraction and brittle range are thus both incorporated in the comparison.

Clyne and Davies[90] similarly emphasized the capacity of alloys to undergo strain without cracking during the liquid and mass feeding stages, and for cracks to appear only in the later stages when there is insufficient

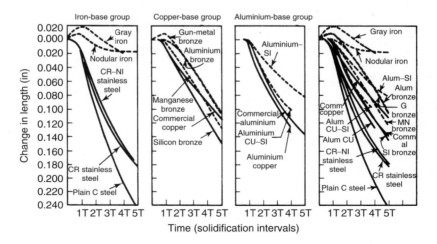

Figure 5.25 Linear contraction as a function of solidification interval of alloy. Alloys represented by solid curves developed hot tears in restrained test bar (from Bishop et al[89]) (courtesy of American Foundrymen's Society)

liquid to allow further deformation in a plastic mode. They introduced the concept of the crack susceptibility coefficient t_V/t_R, where t_V represents the vulnerable period during which cracks can develop between the solid grains or dendrites, and t_R the earlier time available for stress relaxation by liquid and mass flow.

In later experimental work with aluminium–copper alloys, Guven and Hunt[91] confirmed that alloys of low copper content tear in thin liquid films, the coherent state beginning at around 4% residual liquid and cracks appearing around 2%. Campbell and Clyne[92] considered the results of several investigators using various tests with these alloys, finding that they conformed to the t_V/t_R criterion, the relaxation and crack-susceptible periods being associated with liquid fractions in the ranges 0.6–0.1 and 0.1–0.01 respectively. The peak susceptibility occurred at copper contents of around 0.8% rather than at 5.7%, the composition with the maximum freezing range.

Apart from the basic alloy composition, susceptibility to hot tearing is influenced by segregation of alloying elements and impurities. Microsegregation or coring lengthens the true freezing range and accentuates the tendency for prolonged retention of intergranular liquid films, with consequent increase in the accumulated strain during the critical period of weakness. This accounts for the serious lowering of resistance to hot tearing by high sulphur and phosphorus contents in steel[93,94]. It has also been shown[86] that the termination of the vulnerable temperature range in steel accompanies freezing of the heterogeneous carbon regions resulting from differential solidification. The importance of the segregation effect was similarly demonstrated in work with copper alloys[95]. The tearing tendency of an alloy is, on the other hand, diminished by precipitation of gaseous impurities during freezing, since the resulting volume expansion partly offsets the contraction during the critical film stage.

In major reviews of this long-standing problem Sigworth[96] and Campbell[9] have evaluated a large volume of modern research and literature. Mechanical properties within and just below the normal freezing range of alloy systems have been increasingly appreciated as central to the incidence of hot tearing, and the continuing need for further data has been stressed by Hattel et al[97]. and other investigators. Such measurements are clearly essential to successful modelling, and form a part of the general database requirement already referred to in Chapter 3.

The pioneering work of Singer and Cottrell[86] focused on the short time tensile strengths of various aluminium–silicon alloys as summarized in Figure 5.26 whilst later work by Williams and Singer[98], Prokhorov[99] and other investigators emphasized the deformation aspect. Several contributions to the 1997 Sheffield Solidification Processing conference[100] indicated the renewed importance attached to the direct measurement of properties at these temperatures. Suvanchai et al[101] studied both ductility and strength, relating these to the fraction solid. Of particular interest was the work of Chu and Granger[102], in which full stress–strain curves were determined

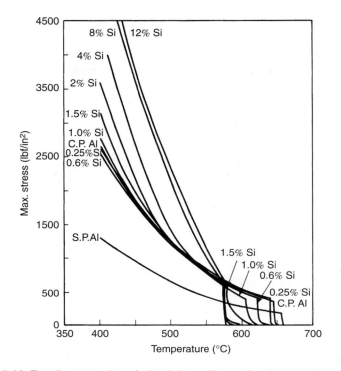

Figure 5.26 Tensile strengths of aluminium silicon alloys at temperatures in the solidus region (from Singer and Cottrell[86]) (courtesy of Institute of Metals)

and related to temperature, fraction solid, strain rate and microstructure for different aluminium alloys; temperatures were expressed as a proportion of the full freezing range for the particular alloy. The alloy known from earlier work to have the most pronounced tearing tendency was that showing the greatest rate of strengthening during freezing, and thus having reduced ability to accommodate the total strain imposed during the longer period of finite and measurable strength. This situation is portrayed schematically in Figure 5.27.

Full determination and classification of the stress–strain–time relationships for the many cast alloys at elevated temperatures, and particularly in the mushy or pasty zone, will greatly assist in optimizing future alloy selection and casting conditions to minimize the risk of hot tearing, as well as of other defects.

Design and production factors in hot tearing

Design and production conditions influence hot tearing mainly through effects upon temperature distribution and resistance to contraction.

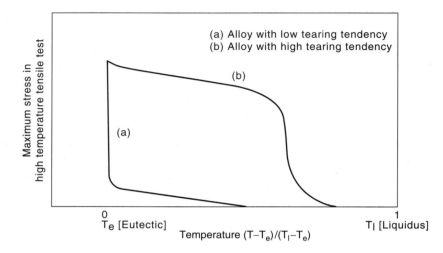

Figure 5.27 Development of strength during solidification in alloys of contrasting behaviour (schematic, after Chu and Granger[102])

Temperature distribution is of crucial importance in respect of all types of contraction defect, since it governs distribution of mechanical properties and at the same time creates the pattern of stress–strain relationships due to differential contraction. As a general rule the probability of tearing is increased by longitudinal temperature differentials, which produce hot spots during cooling. The more confined and intense the hot spot the greater the tearing tendency, since the strain resulting from the hindered contraction of the whole member is then concentrated within a narrow zone of weakness, where it must be accommodated by relatively few films of residual liquid.

Temperature distribution and hot spot intensity are influenced by design, by gating technique, and by selection of pouring speed and temperature, all of which can be used to reduce the strain concentration.

Design. A hot spot occurs at any intersection or local section increase. The basic aim, therefore, should be to equalize cooling rate by designing for uniform section thickness. Where this is impossible cooling can be equalized with the aid of chills. Alternatively a direct source of feed metal may prevent tearing. A highly unorthodox solution to certain hot tearing problems has been demonstrated by Schmid and Crocker[103]. Rather than equalizing the cooling rate, exothermic padding is used to accentuate the hot spot. This then remains well above the solidus whilst most of the contraction is occurring in those members which place the hot spot under strain.

Gating technique. Hot spots may also exist as relics of the pour, the ingate regions forming frequent sites for tears. Gating practice can be designed to reduce this risk by the use of multiple ingates positioned to achieve wide distribution of the molten metal, so avoiding intensive heating at a single ingate. Care is required, however, to avoid creation of a stress framework embodying the gating system itself.

Middleton[104] drew attention to the contradictory nature of the requirements for uniform cooling rate and for the temperature gradients preferred for feeding: a compromise may be needed to produce a defect-free casting.

Pouring conditions. A high casting temperature is associated with increased incidence of tearing. This can be explained partly by its effect upon transverse temperature gradients in the casting member. Pouring with excessive superheat, by preheating the mould, produces shallow transverse gradients during solidification, so that an increased proportion of the cross section passes simultaneously through the brittle range. Steep transverse gradients, on the other hand, favour progressive inward growth of a solid layer, of which the weaker inmost region has constant access to the residual liquid whilst the skin develops full solid cohesion. These conditions are favoured by moderate pouring temperatures and by greensand moulds. The influence of a high pouring temperature can also be partly attributed to a grain size effect, since a coarse structure produces concentration of longitudinal strain at fewer intercrystalline liquid film sites[89].

The influence of pouring speed on hot tearing arises primarily from its effect upon long range longitudinal temperature gradients: rapid pouring promotes uniformity and reduces hot spot intensity, enabling the strain to be accommodated at a maximum number of sites.

The mould. The resistance to contraction of a mass of moulding material depends partly upon its bulk properties at normal temperatures and partly upon properties in the heated layers adjacent to the casting. In a greensand mould bulk properties predominate and there is little tendency to hot tearing when compared with drysands and coresands: contraction is accommodated by deformation of the whole mass.

The traditional assumption that organically bonded cores suppress hot tearing is only valid if sufficient time elapses for breakdown of the binder to an adequate depth before the casting cools through the susceptible temperature ranges. Before this stage heating of the sand actually aggravates the problem through expansion. Owing to the low temperature diffusivity of moulding materials, the conditions for collapse are only achieved with castings of extremely heavy section. For light section castings, which cool rapidly to the tearing temperature, it is the high dry strength of the unchanged coresand that is significant, producing a strong tearing tendency[82]: the contrasting situations arising from the time dependence of the sand properties are schematically illustrated in Figure 5.28a; elevated temperature properties of actual materials are compared in Figure 5.28b.

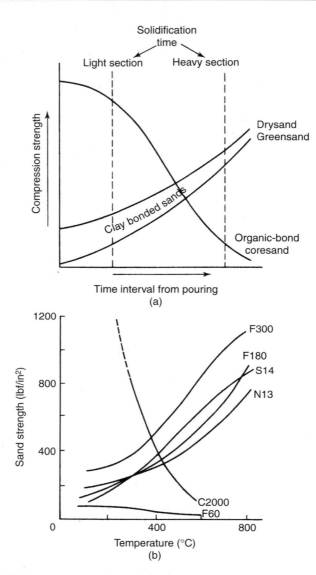

Figure 5.28 Strength changes in moulding materials after casting. (a) Schematic comparison of clay bonded and organically bonded sands, (b) compression strength–temperature relationships for some moulding materials (from Parkins and Cowan[82]). F300, F180, S14, N13: clay bonded moulding sands: C2000: oil and cereal bonded coresand; F60: sand mixture containing 5% sawdust for collapsibility (courtesy of Institute of British Foundrymen)

Although the strengths give a rough indication of tearing tendency, capacity for deformation is the crucial factor. For a sand of high initial dry strength, the capacity for high temperature deformation in the surface layers alone must be sufficient to accommodate the whole of the contraction. The extent of this deformation is a function of metal temperature, binder composition, ramming density and other variables and can be determined by deformation–time testing. One such test maintained a sand specimen under constant load against a heated metal surface, producing conditions analogous to the contraction of a hot casting[105]. The introduction of the time parameter is a major advantage of this type of test, since the occurrence of tearing is governed by the rate of sand deformation relative to the unhindered contraction rate of the casting.

Deformation can be enhanced by special additions such as sawdust, wood flour, or chopped straw; this is also one of the functions of cereal binders. The contraction resistance of larger moulds and cores may also be reduced by centre fillings of loose materials such as coke and unbonded sand. Certain mould and core sections may be completely hollowed out by the use of dummy pattern blocks in moulding.

The mechanical construction of the mould must be designed to avoid hindrance from box bars, reinforcing grids and irons. Downrunners and feeder heads emerging at the cope surface must be kept well clear of these obstructions, a considerable amount of free movement being required for long dimensions. This movement can be helped by removal of moulding material from between heads and runners immediately after casting, a practice aided by relief cores which can be extracted after a suitable interval.

Specific precautions against hot tears can be embodied in the casting design. Apart from the overall aim of section uniformity, section changes should be gradual and all corners provided with adequate radii to reduce stress concentration. Reinforcement can be introduced in the form of thin webs and brackets across critical regions; these may be a temporary expedient to be removed in fettling or may be retained as a permanent feature of the design. Their function is to accelerate cooling and so to strengthen vulnerable zones during their cooling through the critical temperature.

Corrugated surfaces have been suggested as one means of sharing the strain concentration between a number of sites in members prone to tearing.

Cracking, distortion and residual stress

Hot tearing has been seen as the outcome of contraction resistance encountered at an early stage during cooling, when the presence of liquid films renders the metal incapable of plastic strain. When the casting has cooled below the tearing temperature but lies wholly above the temperature for elastic behaviour, resistance can, for most alloys and rates of cooling, be accommodated by continuous plastic deformation. Fracture will only occur at this stage if cooling is so rapid as to cause the ultimate stress of the alloy to be exceeded; hot cracking in the plastic range is comparatively unusual. Plastic strain may however cause dimensional errors through failure to contract to rule, although such errors can be offset by individual contraction allowances based on prior experiments with similar designs. Machining allowances may suffice to correct minor errors on critical dimensions.

The influence of moulding material hindrance in plastic deformation was demonstrated by Parkins and Cowan[82]; uneven mould resistance also contributes appreciably in some instances to the level of residual stress. As previously emphasized, however, the main resistance at lower temperatures arises through the interaction of members cooling at different rates. The magnitude of the stress developed can be expressed as follows:

The strain $\Delta l/l$ due to differential contraction is determined by the temperature difference ΔT and by the coefficient of thermal expansion of the alloy α.

$$\varepsilon = \frac{\Delta l}{l} = \alpha \Delta T$$

Since
$$\frac{\sigma}{\varepsilon} = E,$$

$$\sigma = E\alpha\Delta T$$

where E is the Young's Modulus of the alloy.

Residual stress is thus proportional to Young's Modulus, to the coefficient of expansion and to the temperature difference between the members.

Very high stresses can be developed, especially where sections are of widely varying thickness. The stresses can exceed the elastic limit of the material in extreme cases. If the alloy has poor ductility, fracture can then occur during the late stages of cooling: the typical cold crack or 'clink' (see Figure 5.29), often of catastrophic length, is straight and clean in contrast to the hot tear.

If the casting cools without fracture but retains a high level of residual stress, it remains in a vulnerable condition with respect both to strength and dimensional stability. This condition is illustrated in an exaggerated fashion in the small diagram in Figure 5.23. If the internal stress is close to the ultimate for the material, a small superimposed stress of the same sign will suffice to cause fracture. This may occur during further processing, for example through rough handling or through local heating occurring in head removal, grinding or welding. Similarly, failure may result from a change in temperature in a direction accentuating the original stresses. Thus a casting may crack, paradoxically, at the beginning of a heat treatment intended for stress relief because the position of the casting is such that the first effect of heating is to expand a section already in compression, increasing the tensile stress elsewhere.

If, finally, working stresses are superimposed on an already high level of internal stress, premature failure may occur in service.

Figure 5.29 Typical cold crack or 'clink' in a casting (from Reference 2) (courtesy of Editions Techniques des Industries de la Fonderie)

A casting in a state of residual stress exhibits elastic distortion and is dimensionally unstable: its dimensions may change either spontaneously on ageing or if heating occurs in service. Such change may also be initiated by removal of any part of the cast surface in machining, due to a redistribution of stresses. This behaviour forms the basis for methods of investigating the magnitude of residual stresses by measuring the movement occurring when a section is removed from a closed shape such as a ring casting[84,106]. It is thus inadvisable to attempt to correct elastic distortion by machining: internal stress should be avoided or eliminated for full dimensional stability.

Measures against residual stresses and low temperature cracking again seek the equalization of cooling rates throughout the casting. As in the case of hot tearing, design for uniform section is a fundamental requirement. Gating and pouring techniques also play some part, although the relative importance of initial temperature gradients diminishes at lower temperatures.

Castings known to be particularly susceptible to internal stresses can be protected by extraction from the mould whilst still in the plastic region. This alone may in some cases reduce residual stress[85,107]. More favourable conditions can be achieved by charging the casting immediately into a hot furnace, where cooling can be controlled and equalized throughout the critical period; care must be taken to avoid plastic deformation from lifting stresses or due to support at too few points on the furnace hearth.

Another approach to the equalization of cooling rate is that described by Longden[108] in which ducted air was used for selective cooling within the mould.

Distortion or cracking in flame cutting or welding can be suppressed by preheating to minimize temperature differences.

Stress relief by heat treatment

Residual stresses, whether generated in casting or subsequently, can be removed by heat treatment. This involves heating to a temperature at which relaxation of the elastic stress is brought about by plastic deformation corresponding to the elastic strain. The process being essentially one of creep, the temperature required for rapid stress relief is determined by metal composition and generally lies in the region 0.3–0.4 T_m where T_m is the melting temperature expressed in Kelvin.

A comprehensive study of stress relief in steel castings was carried out by Jelm and Herres[109], using data from many sources. This showed temperature to be the principal factor governing the rate of relief, little advantage being derived from prolonged treatment times (Figure 5.30). Stress relief should thus be carried out at the highest temperature compatible with maintaining the desired structure and properties. It will be evident that slow and uniform

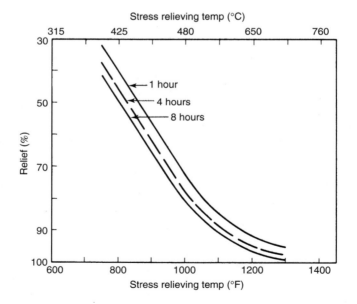

Figure 5.30 Effects of temperature and time on relief of residual stresses in steel castings (from Jelm and Herres[109]) (courtesy of American Foundrymen's Society)

cooling from the stress relief temperature is required if renewed stresses are to be avoided. A further major contribution to the understanding of stress relief was that of Rominski and Taylor[110].

Stress relief is incidental to heat treatments applied for separate metallurgical reasons. Full annealing, carried out above the recrystallization temperature and followed by slow and uniform cooling, is automatically accompanied by full stress relief so that no further treatment is required. Treatments using rapid cooling, on the other hand, such as normalizing and quenching, introduce new stresses although partial relief may accompany subsequent ageing or tempering treatments. Campbell[9] has reviewed the problem of quenching stresses in some detail, emphasizing the beneficial role of polymer solutions in preference to water or oil whenever the somewhat reduced cooling rates are acceptable.

The need for stress relief may follow casting, head removal, weld repair, heat treatment or machining operations and the treatment must therefore be coordinated with the overall production sequence. Machining may initiate dimensional movement by changing the balance of stresses already present, or may itself introduce stresses by deformation of the surface layers: machining should accordingly be so integrated into the heat treatment sequence that cuts are made under reasonably stress free conditions. A typical simple sequence would be:

1. casting,
2. annealing or stress relief,
3. rough machining,
4. stress relief,
5. finish machining.

For ultimate precision, especially of thin walled components, several intermediate stress relief treatments may be required at successive stages of machining, the depths of the cuts being progressively reduced between treatments. The later treatments need to be carried out at a temperature low enough to avoid oxide discolouration of the product.

Apart from heat treatment, stress relief of castings has been sought by ageing or weathering for prolonged periods before use, and by vibration. There is little agreement as to the effectiveness of ageing at ambient temperature: whilst dimensional changes may be detected, it is probable that they represent the relief of only a small proportion of the residual stress. Experiments have indicated[111] that up to 25% of residual stresses in iron castings can be relieved by vibration treatment, although the conditions are evidently critical since other investigations have shown smaller or negligible amounts of relief[112]. Vibration has been claimed to accelerate the attainment of long term dimensional stability in heavy castings for precision engineering[113]; a further review of the nature and viability of vibratory stress relief for shape stabilization was undertaken in Reference 114.

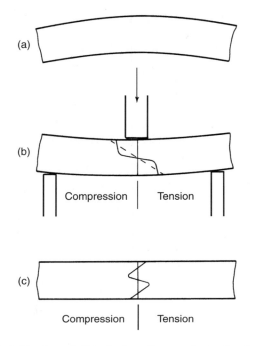

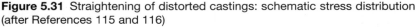

Figure 5.31 Straightening of distorted castings: schematic stress distribution (after References 115 and 116)

Straightening

Distorted castings can sometimes be rectified by straightening operations involving 'setting' in a press or hammering. Since the procedure involves plastic deformation the alloy must possess limited ductility and may be manipulated either hot or cold: the latter is preferred where possible. After cold straightening, the casting is left in a state of residual stress, so that further stress relief is required for permanent dimensional stability. Successful straightening of distorted castings requires considerable patience and skill, largely because of the elastic recovery of part of the press deflection: this circumstance is illustrated in Figure 5.31. A background of experience with particular types of casting assists the choice of press fixtures and packing to achieve the correct final dimensions.

Dimensional errors

The dimensions of a casting are subject to variation from minor changes in production conditions within the limits of normal working practice. Apart from this variation, to be considered further in Chapter 7, individual errors may result from specific faults in equipment and practice; such errors can occur in patternmaking, moulding and casting, or fettling. Pattern-making and fettling errors are relatively uncommon and most dimensional faults originate during mould production or in casting. Principal causes are misalignment of mould parts and cores, mould distortion, anomalous contraction and distortion in cooling: the latter phenomenon was examined in the preceding section.

Alignment faults for example cross-jointing and misplaced cores, are amongst the commonest yet the least necessary of all casting defects, arising mainly from deterioration of pattern equipment and moulding tackle. Although cross-jointing can originate from errors in the plate mounting of patterns and from dowel slackness on split patterns used in floor moulding, the principal cause lies in location systems, including box pins and coreprints: worn pins and pinholes give opportunity for lateral shift along the joint line. Many moulding boxes are now fitted with hardened steel bushes to reduce pinhole wear and loose pins are preferred to fixed, the pins themselves being regularly checked and renewed. Loose pins must be retained in position until the moulding boxes are clamped together to prevent movement when clamping.

A high standard of box maintenance is a major factor in eliminating dimensional errors; boxes should be jig-checked at regular intervals for pin-centre dimensions and for pin and pinhole clearances. These precautions, normally practised to a high standard in mechanized foundries, tend to be neglected when boxes are used at irregular intervals.

Defects from misplaced or ill fitting cores can be avoided by atten-tion to coreprint design and clearances. These should obviate the need for rubbing print surfaces to achieve fitting, a frequent cause of eccen-tricity. Print dimensions should be such that a core can only be inserted in the correct position, eliminating the need for reference to drawings in closing.

Mould distortion a common source of inaccuracy, can have many causes. At the moulding stage the mould cavity can be enlarged by excessive rapping in pattern withdrawal, whilst local discrepancies can result from manual patching. Low green strength and soft ramming can produce sagging. Lack of rigidity in the assembled mould, whether as insecure cores, an inadequately reinforced cope, or badly rammed mould parts, allows movement through pressure and buoyancy forces on casting. Core shift, swelling, and growth across the joint line result from these faults. Measures such as high pressure moulding and the hardening of moulds and cores in contact with the pattern make such forms of distortion much less likely.

Most gross dimensional errors, except those originating in contraction anomalies, can be eliminated with adequate equipment and workmanship.

Compositional errors and segregation

Most compositional errors arise from simple causes, for example melting losses of reactive elements or the use of incorrect furnace charges. Such errors can be avoided by careful melting practice but segregation presents different problems. The term is used to describe compositional differences arising during solidification and persisting in some cases as a permanent feature of the cast structure.

The process of differential freezing was discussed in Chapter 2, where it was shown that in alloys forming solid solutions the solute concentration in the solid at any stage during freezing differs from that in the residual liquid. Since equilibrium and homogeneity require prolonged time at temperatures high enough for diffusion, differences arising in freezing persist in the solidified casting. Segregation may occur with respect either to alloying or impurity elements, the final pattern being seen either as a compositional gradient in a single phase or as a local concentration of a second phase.

The inherent segregation tendency in an alloy system can be represented by the equilibrium distribution coefficient K_o. This is defined as the ratio of the concentration of solute in the solid to that in the liquid phase with which it is in equilibrium. As will be clear from Figure 5.32, values of K_o are less than unity for solutes which lower the freezing point and greater than unity where the freezing point is raised. Major deviations from unity tend to produce high local concentrations of certain impurities even though these may be present only in small amounts in the parent solution, and may cause the precipitation of second phases which would find no place in the equilibrium structure.

Constitutional factors which produce a strong segregation tendency are long freezing range, gentle liquidus slope and low solid solubility. The wide variation in the possible magnitude of the distribution coefficient is illustrated in Table 5.2, where values are given for various elements in

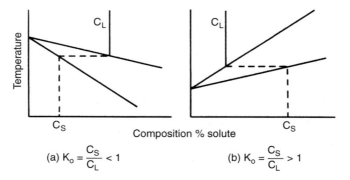

(a) $K_o = \dfrac{C_S}{C_L} < 1$ (b) $K_o = \dfrac{C_S}{C_L} > 1$

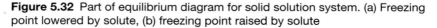

Figure 5.32 Part of equilibrium diagram for solid solution system. (a) Freezing point lowered by solute, (b) freezing point raised by solute

Table 5.2 Distribution co-efficients for some alloying elements and impurities in iron (From Reference 117. Revised values reported in Reference 56 are given in parentheses)

Solute	K
C	0.13 (0.20)
Mn	0.84 (0.90)
Si	0.66 (0.83)
S	0.05 (0.02)
P	0.13
O	0.10 (0.02)

solution in iron. The table explains the strong segregation tendency of sulphur and phosphorus in steels.

Although the inherent segregation tendency of a solute is determined by the distribution coefficient K_o, the degree and pattern of segregation depend upon solidification conditions in the casting, including freezing rate, the mode of development of the grain structure, and motion of crystals and residual liquid under various forces.

Segregation occurs in two types of distribution. Compositional differences may be on a microscopic scale, extending over dimensions of the order of a single grain or less, or there may be major zonal segregation between one part of a casting and another; these are referred to as micro- and macrosegregation respectively. Apart from metallographic examination, macrosegregation can be studied by autoradiographic techniques[118] and by various methods of microanalysis, some of which were reviewed in Reference 119.

Microsegregation

Intergranular or dendritic segregation results from accumulation of rejected solute between the growing crystals and its failure, due either to inadequate time or to a physical barrier of solid, to diffuse or mix into the main body of residual liquid. The final segregation pattern thus follows the form of the grain or sub-structure, the growth centres being depleted and the terminal zones enriched in solute. This is the basis of the typical cored microstructure revealed by etching contrast, and explains the frequent presence of non-equilibrium phases in interdendritic regions. Microsegregation is also seen as an inherent condition in multiphase alloys freezing under equilibrium conditions.

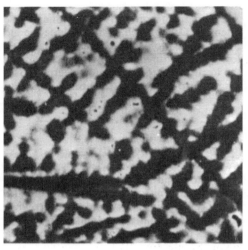

(a)

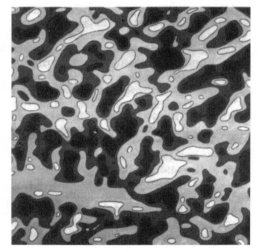

(b)

Figure 5.33 Cored microstructure in carbon–chromium steel, showing correlation between dendrite morphology and composition. (a) Structure, (b) contour map of chromium content as established by microprobe analysis (from Johnson *et al*[120])

The pitch of the compositional fluctuations in microsegregation may be the same as the grain size but is frequently on a finer scale determined by the spacing of subgrain features such as dendrite arms or cell walls. Figure 5.33 shows the relationship between composition and dendritic microstructure in a carbon chromium steel, in which the dendrites are depleted and the interdendritic spaces enriched in chromium. In this case the segregation ratio

$$\frac{C_{Cr\,max}}{C_{Cr\,min}} \approx 2.$$

The mechanical properties of a cast alloy are naturally sensitive to microsegregation, since strength, tensile ductility, impact properties and fatigue resistance are all affected by intercrystalline conditions which differ from the matrix. Apart from functional properties, the microsegregation of alloying elements and impurities can affect strength and ductility in the solidus region and can thus govern susceptibility to hot tearing.

Macrosegregation

Segregation on a macroscopic scale is produced by various mechanisms depending upon the mode of freezing. However, the basic factor is the accumulation of rejected solute by transport over relatively long distances through the casting. In the simplest case, usually termed *normal segregation*, the final parts of the casting to freeze contain high concentrations of solute elements, whilst the term *inverse segregation* is used to describe the opposite condition.

Because diffusion is slow relative to the distances involved, the extent of macrosegregation is determined partly by the degree of mixing arising from pouring turbulence and from convective flow due to local differences in temperature and density. Other mechanisms include the movement of growing crystals from their nucleation sites by gravity or turbulence and the long range capillary flow of liquid during the final stages of freezing. These phenomena influence the location of zones of macrosegregation and can radically change their basic pattern. Typical distributions of segregates can now be considered.

Factors governing segregation patterns

The balance between short and long range segregation is determined essentially by the degree of opportunity for transport of solute from the solidification interface. Normal segregation most readily occurs when there is no solid barrier between the interface and the main mass of residual liquid, i.e. when a plane interface advances progressively in a single direction as in certain types of columnar growth. Solute transport by simple mixing becomes more restricted as the depth of the solidification zone increases, as in columnar dendritic growth. Mixing is also impeded when freezing occurs in a pasty manner by the growth of independently nucleated equiaxed grains. These modes of freezing might be expected to produce maximum microsegregation and minimum macrosegregation. However, this takes no account of capillary flow of residual liquid under various influences during the later stages of freezing. The latter aspect will be further discussed in relation to inverse segregation, but consideration of the simpler case of normal segregation will first be completed.

Since time is required for solute segregation the rate of cooling, coupled with the degree of physical disturbance, exerts a major influence on the segregation pattern. The macroscopic distribution of solute under various conditions of freezing was considered in detail by Pfann[121] and other investigators. Except in the case of extremely slow cooling, diffusion in the solid can be disregarded. The distribution of solute in a progressively solidified section for the assumed cases (a) of continuous and complete mixing into the residual liquid and (b) of transport by diffusion alone, may then be represented by curves of the types shown in Figure 5.34.

With rapid cooling there is little opportunity for solute transport and the profile approximates to curve b: a steady state condition during freezing results in a uniform distribution of solute except for transient zones at the beginning and end of freezing, and the effective value of the distribution coefficient $K = 1$. With slow cooling, on the other hand, time is available for rejected solute to become mixed into the residual liquid, with increasing effect upon its average composition as its volume diminishes. The distribution now corresponds to curve a: during much of the freezing process the deposited solid remains relatively deficient in solute, which attains a high concentration only in the last liquid to solidify. (The solute profile within the spaces between the dendrite arms in microsegregation is similarly affected by the local growth conditions.)

The theoretical distribution of solute under conditions of complete mixing in the liquid fraction is given by the Scheil equation:

$$C_s = K_o C_o (1 - g)^{K_o - 1}$$

where C_s = local concentration in solid formed,

C_o = solute content of the alloy,

g = fraction solidified,

K_o = equilibrium distribution coefficient.

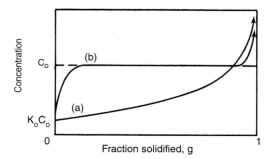

Figure 5.34 Distributions of solute in a progressively solidified section. (a) Complete mixing in liquid fraction, (b) transport in liquid by diffusion alone (from Pfann[121]) (courtesy of John Wiley and Co. Ltd.)

Complete mixing is the condition known to exist within the small bodies of interdendritic liquid during solidification.

Where only partial mixing occurs, the effective value of the distribution coefficient is no longer K_o but K_E, where $K_o < K_E < 1$. However, although the above expression gives a broad guide to segregation behaviour, the quantitative prediction of macrosegregation presents a difficult problem under the complex conditions in the freezing of actual castings.

The slow cooling associated with the strong tendency to macrosegregation is encountered in very heavy sections; a further contributing factor is a high pouring temperature. The tendency is most pronounced when these conditions are combined with active stirring, accounting for the marked segregation associated with gas evolution in rimming quality steel ingots.

Several further mechanisms were mentioned as contributing to macrosegregation, which occurs in many patterns besides the 'normal' type resulting from progressive concentration of segregates towards the centre of the casting. Two patterns occur frequently under practical conditions: these are gravity and inverse segregation.

Gravity segregation. The positions of zones of macrosegregation are frequently influenced by mass movement of precipitated phases due to differences in density compared with the parent liquid. Metal crystals growing independently in the melt tend to sink and produce a corresponding upward displacement of solute enriched liquid, whilst non-metallic inclusions rise or fall to form local concentrations. Gravity segregation is mainly encountered in heavy sections, where solid phases can be suspended in the liquid for prolonged times.

Inverse segregation. Alloys inherently prone to segregation frequently exhibit a form of macrosegregation in which the high concentration of solute which might have been expected in the central region of the casting appears in the outer zone or even at the surface.

It is well established that inverse segregation occurs by flow of solute rich liquid through interdendritic channels from regions deeper within the casting, where it has been formed by the normal processes of differential freezing. The movement results from capillary or dendritic feeding during the continued contraction of the outer layers of grains and is accentuated by precipitation of dissolved gases and by a high metallostatic head, both of which increase the feeding pressure. Despite its long recognized role in inverse segregation, however, the wider significance of this form of interdendritic liquid flow was only later appreciated.

Regions of coarse columnar dendritic structure have been found to be especially susceptible to this type of macrosegregation because of their relatively straight capillary feed passages; major concentrations of segregate and patches of surface exudation are sometimes associated with incipient hot tears in the outermost skin. Since the outside zone also contains the purest solid in the first crystals to freeze, very large local differences in composition can be encountered.

Despite a large volume of early work on macrosegregation in steel ingots, much of it relating to the widely recognized A and V segregate patterns, the major significance of the movement of liquid through the semi-solid mesh forming the solidification zone only became evident following later renewal of research on macrosegregation. This was particularly relevant to certain problems encountered in shaped castings.

Forms of what became known as channel segregation were identified in both ingots and castings, and were attributed to the preferential flow of low melting point liquid through 'pipes' developed in the semi-solid zone. The movement is driven by density differences in the impurity or alloy segregated interdendritic liquid, and as this flows into a hotter region local remelting occurs. The direction of the liquid flow is determined by both thermal and compositional effects on the local density, whilst the preferential channels arise from the unstable flow conditions.

This thermosolutal convection effect has since been used to explain the A segregation in steel ingots, but has also been associated with channel segregation in numerous other alloys and was reproduced experimentally by Hunt and his colleagues[122–124] and other groups. A particular example of channel segregation met in production castings is the 'freckling' phenomenon in high temperature alloy products, a problem investigated by Giamei and Kear[125]

In this case the residual liquid becomes enriched in the low-density aluminum and titanium alloy components, and the resulting convection can cause the nucleation of unwanted equiaxed grains in the directionally solidified structures. The problem can be minimized by the use of steeper temperature gradients to reduce the extent of the semi-solid zone, and by reducing the dendrite arm spacing, but a radical solution involved the introductions of a heavy element, tantalum, which partitions preferentially to offset the effect of the lighter segregates.

Numerous further references to work in this field were included in a short review in Reference 126 and investigations and quantitative analyses were summarized by Flemings[127] and by Mehrabian[128].

Homogenization

The attainment of an equilibrium structure requires opportunity for solute diffusion in the solid state, but the rate of cooling from the casting temperature is normally too high for this to be completed: some degree of coring is therefore the rule rather than the exception.

Homogenization treatment consists of reheating the casting and maintaining it for a prolonged period at high temperature. Concentration gradients are gradually eliminated by diffusion, provided that the distances concerned are short as in microsegregation or coring. The time required for homogenization is proportional to d^2 where d is the spacing of the compositional variations. The coarseness of the original cast structure thus greatly affects the time: a structure with a dendrite arm spacing of 1 mm, for example, will require one hundred times the homogenization time needed for a spacing of 0.1 mm. In general, therefore, the greater the mass and the slower the freezing rate of the casting the longer the time required. The highest possible temperature is necessary consistent with other metallurgical requirements, since the rate of diffusion increases exponentially with temperature in accordance with the relationship.

$$D = D_0 e^{-Q/RT}$$

where D is the diffusion coefficient for the system, T the absolute temperature and D_0, Q and R are independent of temperature.

There is no practical possibility of eliminating macrosegregation by heat treatment because of the large distances involved.

The significance of defects

Casting defects, whatever their cause, cannot be considered in isolation; their significance can only be established in relation to the function of the casting. Behaviour under service stresses and environment is in most cases the overriding consideration but appearance can also be important. Defects can influence these characteristics through mechanical properties, hydraulic soundness and surface condition.

Mechanical properties. The significance of a local defect depends not only upon size and shape but upon position relative to the stress pattern in the casting. Knowledge of the distribution of working stresses is thus desirable at the inspection stage, where much is to be gained from the provision of drawings indicating highly stressed zones.

Generally the skin of the casting contains the regions of highest stress, hence the importance of freedom from surface defects and of adequate surface finish, especially under fatigue conditions. Amongst surface defects, a flaw on an extensive flat surface is potentially less serious than one situated at a change of section where there may already be appreciable concentration of the applied stress. Internal defects, particularly centre line porosity or segregation, are often less serious because of location close to a neutral axis of stress. They are however more difficult to detect and, unlike surface defects, are less accessible for radical repair.

Any defect produces an increase in the mean stress by reducing the cross-sectional area carrying the load, but this effect, except in the case of very large defects, is of secondary importance to that of the redistribution of stresses. Stress concentration at surface and internal discontinuities is illustrated schematically in Figure 5.35, which shows the typical pattern of stress distribution across a section loaded axially in tension. The stress concentration factor is the ratio of maximum to nominal stress and its magnitude is greatly influenced by the shape of the discontinuity.

The theoretical stress concentration factor at the edge of a discontinuity was considered by Peterson[129] for the two dimensional case of an elliptical hole in a plate and the three dimensional case of an ellipsoidal cavity. The factor increases with the relative elongation of the defect in the direction normal to the applied stress or, expressed differently, with reduction in the minor radius of the ellipse. Referring to Figure 5.36,

$$K_t = 1 + 2\frac{b}{a} \text{ for the two dimensional case}$$

$$K_t = 1 + 1.25\frac{b}{a} \text{ for the three dimensional case.}$$

These considerations presume the discontinuity to be remote from any edge or surface. The maximum stress at the edge of a defect, however, increases as the defect approaches the surface (Figure 5.37), a further reason for the sensitivity of surface regions to discontinuities. If the cavity is filled with a foreign substance, for example a non-metallic inclusion, a further factor is introduced: depending upon the ratio of the elastic moduli of inclusion and matrix, the stress concentration is less than that for an empty cavity. A fuller treatment of geometric effects, with many further examples of specific

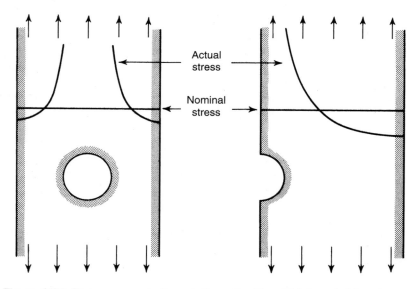

Figure 5.35 Stress concentration at discontinuities. (a) Internal, (b) surface

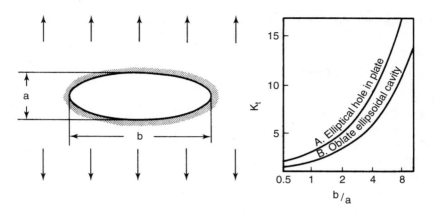

Figure 5.36 Stress concentration factor K_t for ellipsoidal cavity (after Peterson[129])

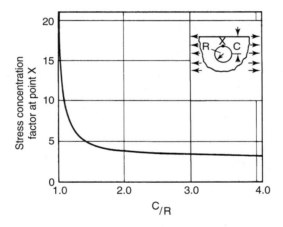

Figure 5.37 Stress concentration at hole near edge of plate (after Mindlin[130])

features and components, is given in Reference 131. These include multiple holes, stepped bars with fillets, gear teeth, turbine blades, crankshafts and cylindrical pressure vessels.

The theoretical treatments show that although an approximately spherical defect such as a blowhole or slag inclusion introduces a stress concentration, the factor increases steeply with angularity. This is characteristic of shrinkage cavities and of some major sand inclusions, the extreme case being that of a fine crack, where the danger of failure is high.

The use of fracture mechanics and the availability of fracture toughness data for alloys facilitate calculations to evaluate the critical defect size in relation to the design of a casting in a particular alloy, an aspect to be discussed further in Chapter 7. Taken in conjunction with non-destructive testing (q.v. Chapter 6), this can nowadays assist in finding whether a particular defect can be tolerated without impairing the function of the component.

Where large numbers of small defects are present their influence can be regarded as a general one affecting material properties in a manner analogous to that of alloy phases. This applies to microporosity, to non-metallic inclusions and to brittle segregates. The influence of microporosity can be quantified by density measurements and its drastic effect in lowering the mechanical properties of cast alloys has been repeatedly demonstrated[132–136]; typical results for ductility and tensile strength are shown in Figure 5.38. Discrete defects mainly influence the phenomena of plastic deformation and fracture, becoming successively more serious in relation to tensile strength, tensile ductility, impact properties and fatigue resistance[52,136]. Yield stress, on the other hand, is comparatively unaffected, being governed by the inherent properties of the matrix.

The influence of such defects on properties is not constant since the dispersion is determined by the spacing of the particular cast structure. Similarly, the loss of mechanical properties due to the segregation of alloying elements or impurities depends upon the metallurgy of the individual alloy. One example of the adverse influence of segregation was reported in Reference 137, where the fracture paths in tensile tests on a cast steel were shown to follow interdendritic segregates, accounting for the lower tensile strength and ductility obtained in the central regions of large masses; impact properties did not however follow the same path. Grain boundary embrittlement by segregated impurities is well documented for many alloys[138].

Other attributes affected by the presence of defects include hydraulic soundness and suitability for surface treatment.

Hydraulic soundness. Capacity for retention of fluids under pressure is a frequent requirement for hydraulic cylinders, valves and pipework, proof pressure testing being usually undertaken for this class of component. Intergranular porosity is the commonest cause of leakage and several copper and nickel based alloys noted for their corrosion resistance are prone to this condition because of their long freezing ranges. Even very fine porosity may produce weeping under high pressure due to its filamentary character. Measures for the control of porosity in such alloys have been previously discussed: they largely concern feeding and control of gas content.

Surface finish. Cast or machined finish is always relevant to appearance, but freedom from surface flaws becomes a prime requirement for castings designed to operate under corrosive conditions or to undergo certain finishing treatments.

Surface pits or inclusions form nuclei for corrosive attack. They also become conspicuous after operations such as honing and polishing, where

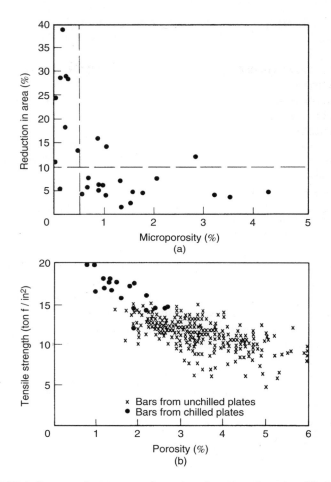

Figure 5.38 Influence of microporosity on tensile properties of castings. (a) On ductility of high strength steel (from Uram *et al*[133]) (courtesy of American Foundrymen's Society), (b) on tensile strength of gunmetal (from Glick and Hebben[136]) (courtesy of Institute of British Foundrymen)

they may prevent the attainment of the lustre and finish required. The problem is accentuated when special finishes are applied, since surface defects are emphasized by enamelling, plating and anodizing; for these purposes a sound surface is a prerequisite to successful treatment.

Salvage and rectification

Many types of casting can be regularly produced without significant defects, save in a small proportion of cases. In such conditions it is possible to institute an inspection standard under which the few defective castings are scrapped and replaced. This procedure is often preferred, for example, in the mass production of light castings, where the low marginal cost of replacements compares favourably with the cost and inconvenience of salvage. In jobbing foundries, producing few castings of a particular type, or in the case of heavy castings, the complete avoidance of defects is much more difficult, whilst the cost of replacement is extremely high. This provides a strong incentive to salvage castings which can be made serviceable by repair.

Rectification can be undertaken for either of two purposes: the restoration of properties and service performance to a standard equivalent to that if no defect were present, or the improvement of appearance in cases where performance is in any case unimpaired. It is important that the distinction between these aims should be clearly drawn, since different measures might be considered reasonable in the two cases.

Rectification should only be carried out in accordance with clearly understood principles and a systematic policy, since indiscriminate repairs might foster a relaxation of standards in the earlier production stages as well as undermining the confidence of the user. With the availability of modern welding techniques, on the other hand, it would be illogical to object in principle to salvage, particularly when fabrication by welding is well established and when many castings are themselves built into fabricated assemblies. Repairs are widely accepted as a matter for consultation between manufacturer and user, provision for this being recognized in many specifications and conditions of sale. The question of rectification was ventilated by the early work of committees of the Institute of British Foundrymen, whose reports reviewed the practices and techniques then available[139]; further welding techniques have since become well established.

Salvage techniques of widely varying types are available, each with its own field of application. The fusion welding processes give the closest approach to complete homogeneity but less radical repair processes can be useful in less critical situations.

Welding

The fundamental difference between welding and all other methods of rectification is the achievement of true metallurgical union between filler metal and parent casting. This union is not superficial as in soldered and brazed joints but features continuity of crystal structure across the original interface. Given satisfactory welding technique, inspection and appropriate post heat treatment, a weld repair can possess properties fully equivalent to those of the original cast metal.

Ideally, the weld should match the parent casting in respect of corrosion resistance and appearance as well as mechanical properties. This is best achieved by use of a filler metal approximating to the composition of the casting, although in many cases the best mechanical properties can be obtained by using a different composition. It must, however, be borne in mind that differential composition at the site of a weld repair may accelerate corrosion, whilst slight hardness or colour differences may become evident on machined surfaces.

The welding operation itself is only one stage in the execution of a successful repair. Irrespective of the process to be used, the defect must first be removed by grinding or machining or by gouging with arc, gas flame or chisel. Close attention is required to ensure complete removal of defective material, since attempts to weld over porous metal lead to gas evolution and porosity in the weld deposit. Incompletely removed cracks are liable to extend under thermal stresses in the path of the weld; the elimination of fine cracks before welding may be confirmed by penetrant crack detection at the root of the prepared cavity.

Fusion welding processes vary principally in the method of heating, application of filler metal, and shielding of the weld from oxidation. Gas and arc welding are most commonly used for casting repairs, although the old established technique of burning-on may also be regarded as a true welding process since it too embodies fusion of the parent metal surface.

In *gas welding* the flame, usually oxyacetylene or oxypropane, can be controlled to provide an atmosphere suited to the particular alloy. Control of the relative proportions of fuel and oxygen fed to the torch enables oxidizing, neutral or carburising conditions to be maintained. Neutral conditions are preferred for most alloys, although oxidizing conditions can be used for the suppression of zinc volatilization and porosity in the welding of brasses. Fluxes are employed in gas welding to provide a sink for the removal of impurities and to give additional protection against contamination. The filler alloy is applied as a separate rod fed into the hot zone.

Gas welding is extensively applied to non-ferrous metals, the softer and more diffuse heat source being especially suitable for alloys of relatively low melting point. The comparatively cool outer zone of the gas flame can provide a preheating and post-heating effect when required. The heat affected zone in the final structure is larger than that produced by the more intense and localized electric arc and the tendency to distortion is therefore greater.

Of the many variations of *arc welding*, the metal arc process is most widely used for casting repairs, a consumable electrode providing the source of filler metal. Gas and slag shielding are derived from mineral coatings on the filler rods. The essential features of metal arc welding with coated electrodes are illustrated in Figure 5.39. For reactive alloys there is considerable use of inert gas shielding, as in the tungsten and consumable electrode variations of the argon arc process. The relatively small heat affected zone in arc welding minimizes distortion. A further variant, giving deeper penetration, is the submerged arc process, whilst electroslag welding too has found limited use in joining operations involving heavy castings, although not for rectification.

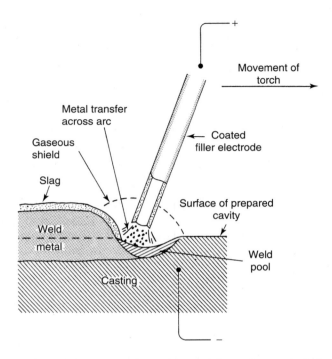

Figure 5.39 Essential features of metal arc welding using coated electrodes

Arc welding is applied to most types of alloy, although filler alloys containing zinc are an exception because of problems of volatilization and porosity.

Details of the many techniques used in the welding of particular groups of alloys are contained in the wide body of general literature devoted to welding and welding metallurgy. The fusion welding of steel castings in particular is the subject of a special standard, BS4570[140], which covers process requirements for both rectification and fabrication operations (q.v. Chapter 7) Consumables or filler materials, preparation, preheating and post-weld heat treatment are included. Separate standards cover such aspects as inspection, testing and terminology.

An authoritative work and review of the welding of steel castings, again including their role in fabricated structures, is that by Ridal[141]: this details the alternative processes and the effects of the main variables, including joint design and preparation, deposition techniques and potential defects. Numerous examples are given, with a substantial bibliography.

Although not devoted to the welding of castings, a further review by Davies and Garland[142], again with references, was concerned primarily with metallurgical effects in fusion welding, including solidification of the weldpool and effects on microstructure and properties.

The final properties of a welded casting depend not only upon the structure and soundness of the weld deposit but upon the structure of the heat affected zone in the parent metal. Residual stress is also an important factor.

The structure of the heat affected zone, i.e. that region attaining a sufficient temperature for evident metallurgical change to occur, depends upon parent metal composition and initial structure, temperature attained, and cooling rate. Within the zone, all temperatures up to the melting point will have been encountered, the temperature distribution taking the form shown in Figure 5.40. It is common to find coarse grained structures immediately adjacent to the fusion zone. The cooling rate depends upon the relative rates of heat dissipation and heat input, being governed partly by the relative masses of weld and parent metal: a small weld undergoes the most rapid cooling. Other factors are the thermal properties of the alloy and the welding technique, particularly power input and rate of deposition. Cooling rate can also be modified by pre- and post-weld heating.

Weldability.   The weldability of an alloy is its capacity to undergo welding without cracking or serious embrittlement. Hot and cold cracking in the weld metal or heat affected zone result from differential cooling stresses or transformation stresses, coinciding with inadequate ductility. In the case of hot cracking the mechanical properties of the alloy in the region of the solidus are critical, cracking being associated with the brittle condition resulting from films of liquid constituents or impurities. As in the case of hot tearing, crack susceptibility is a function of the length of the brittle temperature range, itself roughly proportional to the freezing range of the alloy.

Cold cracking, for example underbead and toe cracking, can occur in intrinsically brittle alloys or as a result of transformations leading to brittle

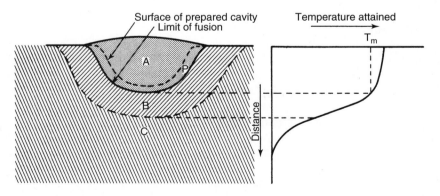

Figure 5.40 Thermal conditions and structural zones in a weld repair. A Fusion zone, B Heat affected zone, C Original zone and P Penetration

microstructures, when restraint is imposed through differential cooling. Stresses can be minimized and transformations suppressed by pre- and post-heating to retard cooling, but in some cases full subsequent heat treatment may be needed for development of the required structure and properties. Embrittlement in the heat affected zone can be aggravated by hydrogen or nitrogen diffusing from the weld deposit: special precautions are therefore required to reduce gaseous impurities in alloys which develop susceptible microstructures.

As might be expected, weldability is associated primarily with alloys of high ductility. In carbon steels the carbon content itself is the principal factor. Steels containing less than 0.35% carbon, with their predominantly ferritic structures, can be welded without difficulty but above this level preheating is necessary to avoid cracking. High strength low alloy steels are more difficult to weld due to the increased probability of martensite formation, the most adverse conditions being associated with the internally twinned plate structure of higher carbon martensites: a contributing factor to cracking is hydrogen absorbed from filler coatings and contaminants. Cracking thus ensues from the combination of a susceptible microstructure, hydrogen and stress. In certain cases this combination is achieved after slow diffusion of hydrogen from the weld deposit to the heat affected zone, producing delayed fracture on attaining a critical hydrogen content.

The major precaution in the welding of low alloy steels is the reduction of cooling rate by preheating and post-heating; preheat temperatures in the range 150–400°C are usual according to composition. Hydrogen absorption is minimized by the use of low hydrogen electrodes. Where the weld repair lies in a relatively lightly stressed region conditions may be alleviated by the use of a ductile low strength filler metal, in this case frequently an austenitic stainless steel.

Austenitic stainless and heat resisting steels are themselves readily welded, the best compositions from this viewpoint being those giving structures containing small amounts of ferrite, which reduces susceptibility to cracking.

Cast iron has relatively poor weldability, although repairs can be carried out with suitable precautions, of which preheating is the most important. Heating to the range 400–480°C reduces differential contraction and, by lowering the cooling rate, inhibits the more brittle chilled and martensitic structures. Post-heat treatment at 600–630°C is followed by slow cooling. In gas welding, chilled carbide structures are discouraged and machinability maintained by the use of special high silicon (3.0–3.5% Si) filler rods. Arc welding is also employed, using electrodes of steel, nickel, and nickel alloy of the Monel type.

Spheroidal graphite iron, being ductile, is more readily welded than grey iron: it is normally preheated to approximately 300°C and arc welded with 60% nickel iron electrodes.

Most non-ferrous alloys are amenable to weld repair using suitable techniques and precautions. Copper alloys are readily weldable with the exception of the leaded brasses and gunmetals, which exhibit hot shortness. The remaining groups of alloys can be welded by gas, metal arc or inert gas processes using electrodes of similar composition to the parent casting; a notable exception is the brasses, which require gas welding using an oxidizing flame.

Aluminium alloy castings are most satisfactorily welded using inert gas tungsten arc or inert gas metal arc processes, no difficulty being encountered with the low and medium strength alloys containing silicon or magnesium or both these elements. In the heat treatable alloys softening occurs in the heat affected zone but the properties can be largely recovered by subsequent solution and precipitation treatment. Preheating to 300°C is carried out for complicated castings. Welding is avoided wherever possible for the high strength alloys since they are more susceptible to cracking and the general level of properties is not attained in the weld.

Of the other important groups of cast alloys, the magnesium alloys are weldable using inert gas shielding, preheating being normally employed. Nickel alloys can mostly be arc welded with coated electrodes or inert gas shielding, but the corrosion resistant nickel–silicon alloys are brittle and difficult: castings in this material are normally gas welded with preheating.

Although weldability is related primarily to mechanical properties, corrosion resistance is another sensitive characteristic. Local corrosion can result from dissimilarity in the compositions of parent and filler metals or from structural changes in the heat affected zone. A notable example is the carbide precipitation encountered in certain austenitic stainless steels: this produces severe intergranular corrosion and embrittlement. The effect can be countered by solution heat treatment or by modification of the composition to include weld stabilizers such as titanium or niobium. Stress corrosion is a further danger in some alloys and can be inhibited by stress relief heat treatment.

Apart from metallographic structure the most critical factor in weld repairs is the generation of long range stresses which can cause cracking, distortion and reduction in strength. The stresses originate in differential expansion and contraction produced by intensive local heating. Residual stress can be minimized by preheating, either of the entire casting or of the zone surrounding the weld; these measures are most important for alloys with inherently low ductility. Cooling rate can also be reduced by post-heating or furnace cooling.

Other measures to reduce differential contraction include selective pre-heating of individual members of a cast framework, whilst in multiple repairs a symmetrical sequence of weld deposition can be used to balance opposing stresses. Prestressing in fixtures and peening of the weld metal to introduce compressive stress can also be employed, although such measures require a high degree of special knowledge and skill.

Where full heat treatment is carried out after weld repair, relief of welding stresses is achieved incidentally. In cases where no such treatment is required, separate low temperature stress relief heat treatment can be employed to eliminate residual stress without significant change of properties.

For maximum properties it is vital that rectification by welding be kept under close metallurgical supervision. A useful measure of quality control in relation to weld repairs can be maintained through records of filler metal consumption relative to the tonnage of castings produced.

Burning-on

Burning-on is essentially a welding process in which a continuous flow of superheated molten metal from an external source is used to produce fusion of the surface of the prepared cavity in the parent casting. At the appropriate stage flow is stopped and the pool of metal allowed to solidify and form a permanent filling. The filling is of identical or similar composition to that of the parent metal.

The casting is embedded in sand, which is built up to surround the prepared cavity and provide a reservoir of liquid. This small mould is provided with an overflow lip enabling flow to be maintained until surface fusion is achieved. The volume of retained liquid must be somewhat greater than that of the original cavity to allow for solidification shrinkage, so that the finally dressed repair will be fully sound. As with other types of weld repair, successful burning-on requires considerable skill and experience. Preheating is needed in most cases and the casting requires subsequent treatment for stress relief.

Brazing and soldering

Brazing and soldering differ from welding in that there is no fusion of the parent metal, the entire process being carried out below the melting point of the casting using a lower melting point alloy as filler: this alloy is often of a completely different composition from that of the casting, so that hardness and colour differences may be introduced. There is little penetration into the parent metal, the original joint interface remaining as a distinct plane structural transition between the two alloys. Good adhesion can nevertheless be achieved through continuous metallic contact and high strength repairs are feasible.

Structural repairs are confined essentially to the higher temperature processes, variously termed brazing, bronze or braze welding and hard soldering. For these processes copper and nickel based alloys and the silver solders can be used as fillers; their melting points range from 650°C upwards. Soft soldering with lower melting point alloys based on tin, lead, cadmium and zinc is confined to the filling of minor surface cavities where strength is unimportant.

The key to successful brazing and soldering is the achievement of perfect contact between filler and casting through wetting of the parent surface. Apart from preliminary cleaning the principal aid to wetting comes from the flux used in the process. Due to its low surface tension the molten flux flows readily over the surface and is then displaced by the filler alloy. The flux dissolves or floats off residual grease and oxide contamination and assists in preventing oxidation whilst the joint is at high temperature.

Brazing fluxes are commonly based on borax and on borates and fluorides of the alkali metals; the selected flux must be sufficiently active to remove tenacious oxide films. Since these fluxes are highly corrosive, care is required to remove traces of residue before the casting is put into service.

Some further techniques

Patches and plugs

In some circumstances it is possible to remove a casting defect by machining or drilling out the defective zone and inserting a patch or plug. The insert may be made as a separate casting or may be machined from plate or bar stock of suitable composition. A small insert can most conveniently be made as a circular plug and fitted by tapping and threading, being finished by peening over or tack welding. Patch inserts for larger apertures can be chamfered and fitted by V welding. These techniques are not only applied for rectification but may also be used for the final sealing of temporary coreprint apertures in the original casting.

Caulking and impregnation

A casting designed to withstand hydraulic pressure may possess a high standard of general soundness yet show minor local seepage when pressure tested. In some cases sealing can be achieved by caulking. The metal is hammered using a hand or pneumatic tool, the leak being closed by plastic deformation. High ductility alloys are most readily rectified by this technique, which should however be confined to minor leaks in low pressure equipment.

More widespread porosity can be sealed by impregnation. The black or machined casting is given a general treatment with a sealing compound in liquid form, penetration being assisted by pressure alone or by a combination of pressure and vacuum treatment. In the latter case the casting is placed in a tank which is evacuated, the liquid medium being run in to cover the casting. The vacuum is then released and air pressure applied. The compound is finally cured, normally by thermal treatment, to produce an inert, volume-stable filling. Synthetic resins have been extensively used for this purpose; other materials include linseed oil and sodium silicate. The ideal medium should possess low viscosity to facilitate flow into fine passages, coupled with minimum shrinkage and the formation of a tough, inert solid on curing.

Impregnation is most widely accepted for alloys especially prone to microporosity; the process is operated as a commercial service by specialist firms.

Filling compounds

The use of non-metallic compounds, or compounds containing metal powders blended with a non-metallic bond, has a long established but chequered history in casting manufacture. Hardening putties have limited justification for smoothing insignificant surface blemishes, whether to improve appearance or as a preparation for painting; in this respect their position is analogous to that of the soft solders. With the development of high strength non-metallic materials capable of adhering strongly to metals, for example the epoxy and acrylic resins, repairs of appreciable strength can be made. The properties are, however, so fundamentally different from those of cast metals, and the conditions for adhesion so critical, that such materials cannot be regarded as suitable for structural repairs under most conditions.

Hot isostatic pressing

Although employed for the closure of internal pores, it would be inappropriate to see HIP primarily in terms of rectification of defective castings. Used as a means of upgrading already high mechanical properties in special purpose castings, it can form an integral stage in the production process and will be reviewed later in that context.

References

1 Manual of Defects in Castings (Iron Edition). Institute of British Foundrymen, Birmingham (1994)
2 Album de Défauts de Fonderie. 2 vols. Editions Techniques des Industries de la Fonderie, Paris (1952)
3 Recherche de la Qualité des Pièces de Fonderie. Comité International des Associations Techniques de Fonderie. Trans. from VDG Text, by Hénon G., Mascré, C. and Blanc, G., Editions Techniques des Industries de la Fonderie, Paris (1988)
4 Analysis of Casting Defects, 4th Edn. American Foundrymen's Society, Des Plaines, IL (1997)
5 International Atlas of Casting Defects. Ed. Rowley M. T., American Foundrymen's Society, Des Plaines, IL (1993)
6 Casting Defects Handbook. American Foundrymen's Society, Des Plaines, IL (1984)
7 Metals Handbook, 9th Edn. Vol. 15, Castings. ASM International Metals Park Ohio (1988)
8 Metals Handbook, 9th Edn. Vol. 11. Failure Analysis and Prevention. ASM International Metals Park Ohio (1986)

9 Campbell, J., *Castings*, Butterworth-Heinemann, Oxford (1991)

10 Emley, E. F., *J. Inst. Metals*, **75**, 431 (1949)

11 Brondyke, K. J. and Hess, P. D., *J. Metals*, **17**(2), 146 (1965)

12 Benedicks, C. and Löfquist, H., *Non-metallic Inclusions in Iron and Steel*, Chapman and Hall, London (1930)

13 Sims, C. E., *Trans. metall. Soc. A.I.M.E.*, **215**, 367 (1959)

14 Kiessling, R. and Lange, N., *Non-metallic Inclusions in Steel, Part I*. I.S.I. Publication 90, London Iron Steel Inst. (1964)

15 Kiessling, R. and Lange, N., *Non-metallic Inclusions in Steel, Part II*. I.S.I. Publication 100, London, Iron Steel Inst. (1966)

16 Kiessling, R., *Non-metallic Inclusions in Steel, Part III*. I.S.I. Publication 115, London, Iron Steel Inst. (1968)

17 Middleton, J. M. and Cauwood, B., *Br. Foundrym.*, **60**, 320 (1967)

18 Flinn, R. A., Van Vlack, L. H. and Colligan, G. A., *Trans. Am. Fndrym. Soc.*, **74**, 485 (1966)

19 Savage, J. and James, D. B., *Br. Foundrym.*, **55**, 45 (1962)

20 Caine, J. B., King, E. H. and Schumacher, J. S., *Trans. Am. Fndrym. Soc.*, **70**, 120 (1962)

21 Middleton, J. M. and Savage, J., *Br. Foundrym.*, **56**, 337 (1963)

22 Campbell, J., *Foundrym.*, **86**, 329 (1993)

23 Campbell, J., *Foundrym.*, **92**, 313 (1999)

24 Moll, N. and Johnson, D., *Foundrym.*, **79**, 458 (1986)

25 Korotkov, V. G., *Gases in Cast Metals*, Ed. B. B. Gulyaev, Moscow. Translation Consultants Bureau, New York, 140 (1965)

26 Kurdyumov, A. V. and Alekseev, L. A., *Russ. Cast. Prod.*, 210 (1967)

27 Petterson, H., *Br. Foundrym.*, **48**, A200 (1955)

28 Marek, C. T., *Trans. Am. Fndrym. Soc.*, **71**, 185 (1963)

29 Hoffman, F., *Br. Foundrym.*, **52**, 161 (1959)

30 Report of A.F.S. Committee 8J., *Trans. Am. Fndrym. Soc.*, **63**, 123 (1955)

31 Report of A.F.S. Committee 8J., *Trans. Am. Fndrym. Soc.*, **65**, 128 (1957)

32 Morgan, A. D., *Br. Foundrym.*, **59**, 186 (1966)

33 Vingas, G. J. and Zrimsek, A. H., *Trans. Am. Fndrym. Soc.*, **71**, 50 (1963)

34 McDowell, J. S., *Trans. Am. Inst. Min. metall. Engrs.*, **57**, 46 (1917)

35 Levelink, H. G. and Van den Berg, H., *Trans. Am. Fndrym. Soc.*, **70**, 152 (1962)

36 Morgan, A. D., *Br. Foundrym.*, **59**, 186 (1966)

37 First Report of Sub-committee, T.S. 48., *Br. Foundrym.*, **51**, 17 (1958)

38 Murton, A. E. and Gertsman, S. L., *Trans. Am. Fndrym. Soc.*, **66**, 1 (1958)

39 Vingas, G. J., *Trans. Am. Fndrym. Soc.*, **67**, 671 (1959)

40 Hoar, T. P. and Atterton, D. V., *J. Iron Steel Inst.*, **166**, 1 (1950)

41 Levelink, H. G. and Van den Berg, H., *Modern Casting* **54**, 69 (1968)

42 Levelink, H. G., *Trans. Am. Fndrym. Soc.*, **80**, 359 (1972)

43 Savage, R. E. and Taylor, H. F., *Trans. Am. Fndrym. Soc.*, **58**, 564 (1950)

44 Beeley, P. R. and Protheroe, H. T., *J. Iron Steel Inst.*, **167**, 141 (1951)

45 Flinn, R. A., *Trans. Am. Fndrym. Soc.*, **66**, 452 (1958)

46 Caine, J. B., *Trans. Am. Fndrym. Soc.*, **56**, 26 (1948)

47 Flinn, R. A., Van Vlack, L. H. and Colligan, G. A., *Trans. Am. Fndrym. Soc.*, **94**, 29 (1986)

48 Cupp, C. R., *Prog. Metal Phys.*, **4**, 105 (1953)

49 Cotterill, P., *Prog. Mater. Sci.*, **9**, 201 (1961)
50 Smithells, C. J., *Gases and Metals*, Chapman and Hall, London (1937)
51 Queneau, B. R. and Ratz, G. A., The Embrittlement of Metals, *Am. Soc. Metals*, Cleveland (1956)
52 Smith, D. P., Carney, D. J., Eastwood, L. W. and Sims, C. E., Gases in Metals, *Am. Soc. Metals*, Cleveland (1953)
53 Ransley, C. E. and Neufeld, H., *J. Inst. Metals*, **74**, 599 (1948)
54 Savage, R. E. and Taylor, H. F., *Trans. Am. Fndrym. Soc.*, **58**, 393 (1950)
55 Phillips, A. J., *Trans. Am. Inst. Min. metall. Engrs.*, **171**, 17 (1947)
56 Derge, G. Basic Open Hearth Steelmaking, 3rd Ed *Am. Inst. Metall. Petrol Engrs*, New York (1964)
57 Mohanti, P. S., Samuel, F. H. and Gruzleski, J. E., *Trans. Am. Fndrym. Soc.*, **103**, 555 (1993)
58 Whitaker, M., *Proc. Inst. Br. Foundrym.*, **46**, A236 (1953)
59 Ruddle, R. W., *Proc. Inst. Br. Foundrym.*, **46**, B112 (1953)
60 Murphy, A. J. (Ed.), *Non-Ferrous Foundry Metallurgy*, Pergamon Press, London (1954)
61 Murray, W. G., *Br. Foundrym.*, **55**, 85 (1962)
62 Hudson, D. A., *Br. Foundrym.*, **56**, 50 (1963)
63 Sims, C. E. and Zapffe, C. A., *Trans. Am. Fndrym. Ass.*, **49**, 255 (1941)
64 Roxburgh, W., *General Foundry Practice*, Constable, London (1910)
65 *The Determination of Gases in Metals*, Special Report 68, Iron Steel Inst., London (1960)
66 James, J. A., *Metall. Rev.*, **9**, No. 86, 93 (1964)
67 Booth, C. W. and Clegg, A. J., *Br. Foundrym.*, **77**, 96 (1984)
68 Chen, X.-G., Klinkenberg, F.-J. and Engler, S., *Cast Metals* **8**, 1, 27 (1995)
69 Hoyle, G. and Dewsnap, P., *Fndry Trade J.*, **114**, 71 (1963)
70 Ames, B. N. and Kahn, N. A., *Trans. Am. Fndrym. Ass.*, **55**, 558 (1947)
71 Harders, F., *J. Iron Steel Inst.*, **190**, 306 (1958)
72 Maas, H., *J. Iron Steel Inst.*, **192**, 63 (1959)
73 *Vacuum Degassing of Steel*, Special Report 92, *Iron Steel Inst.*, London (1965)
74 Philbrick, H. S., *Trans. Am. Fndrym. Soc.*, **69**, 767 (1961)
75 Johnson, W. H., Bishop, H. F. and Pellini, W. S., *Trans. Am. Fndrym. Soc.*, **63**, 345 (1955)
76 Magnitskii, O. N., *Gases in Cast Metals*, Ed. B. B. Gulyaev, Moscow. Translation Consultants Bureau, New York, 149 (1965)
77 Baker, W. A., *Proc. Inst. Br. Foundrym.*, **40**, A130 (1946–7)
78 Talbot, D. E. J., Review 201, *Int. Metall. Rev.*, **20**, 166 (1975)
79 Gruzleski, J. E., Thomas, P. M. and Entwistle, R. A., *Br. Foundrym.*, **71**, 69 (1978)
80 *Smithells Metals Reference Book*, 7th Edn, Brandes, E. A. and Brook, G. B., (Eds.), Butterworth-Heinemann, Oxford (1998)
81 Second Report of Sub-committee T.S. 32., *Proc. Inst. Br. Foundrym.*, **49**, A56 (1956)
82 Parkins, R. N. and Cowan, A., *Proc. Inst. Br. Foundrym.*, **46**, A101 (1953)
83 Parkins, R. N. and Cowan, A., *J. Inst. Metals*, **82**, 1 (1953–4)
84 Portevin, A. and Pomey, J., *Proc. Inst. Br. Foundrym.*, **48**, A60 (1955)
85 Angus, H. T. and Tonks, W. G., *Br. Foundrym.*, **50**, 14 (1957)
86 Singer, A. R. E. and Cottrell, S. A., *J. Inst. Metals*, **73**, 33 (1947)

87 Christopher, C. F., *Trans. Am. Fndrym. Soc.*, **64**, 293 (1956)
88 Rosenberg, R. A., Flemings, M. C. and Taylor, H. F., *Trans. Am. Fndrym. Soc.*, **68**, 518 (1960)
89 Bishop, H. F., Ackerlind, C. G. and Pellini, W. S., *Trans. Am. Fndrym. Soc.*, **65**, 247 (1957)
90 Clyne, T. W. and Davies, G. J., *Br. Foundrym.*, **74**, 65 (1981)
91 Guven, Y. F. and Hunt, J. D., *Cast Metals* 1, 2,104 (1988)
92 Campbell, J. and Clyne, T. W., *Cast Metals* **3**, 4,224 (1991)
93 Middleton, J. M. and Protheroe, H. T., *J. Iron Steel Inst.*, **168**, 384 (1951)
94 Bhattacharya, U. K., Adams, C. M. and Taylor, H. F., *Trans. Am. Fndrym. Soc.*, **60**, 675 (1952)
95 Couture, A. and Edwards, J. O., *Cast Metals Res. J.*, **3**, (2), 57 (1967)
96 Sigworth, G. K., *Trans. Am. Fndym. Soc.*, **104**, 1053 (1996)
97 Hattel, H., Andersen, S. and Svoboda, J. M., *Trans. Am. Fndrym. Soc.*, **105**, 31 (1997)
98 Williams, J. A. and Singer, A. R. E., *J. Inst. Metals*, **96**, 5 (1968)
99 Prokhorov, N. N., *Russ. Cast. Prod.*, 176 (1962)
100 Solidification Processing 1997, Ed., Beech, J. and Jones, H. Proceedings 4th Decennial International Conference, Department of Engineering Materials, University of Sheffield (1997)
101 Suvanchai, P., Okane, T. and Uneda, T., Ref. 100, 190
102 Chu, M. G. and Granger, D.A., Ref 100, 198
103 Schmid, R. F. and Crocker, B. W., *Mod. Cast.*, **46**, 462 (1964)
104 Middleton, J. M., Report of T.S. 32. Discussion. *Proc. Inst. Br. Foundrym.*, **49**, A56 (1956)
105 El Mahallawi, S. and Beeley, P. R., *Br. Foundrym.*, **58**, 241 (1965)
106 Report of Sub-committee T.S. 57., *Br. Foundrym.*, **57**, 49 (1964)
107 Report of Sub-committee T.S. 18., *Proc. Inst. Br. Foundrym.*, **42**, A61 (1949)
108 Longden, E., *Trans. Am. Fndrym. Soc.*, **56**, 36 (1948)
109 Jelm, C. R. and Herres, S. A., *Trans. Am. Fndrym. Ass.*, **54**, 241 (1946)
110 Rominski, E. A. and Taylor, H. F., *Trans. Am. Fndrym. Ass.*, **51**, 709 (1943)
111 Kotsyubinskii, A. Yu. *et al.*, *Russ. Cast. Prod.*, 365 (1961)
112 Third Report of Sub-committee T.S. 32., *Br. Foundrym.*, **53,** 10 (1960)
113 Skazhennik, V. A. *et al.*, *Russ. Cast. Prod.*, 304 (1967)
114 Claxton, R. A. and Saunders, G. G., *Br. Foundrym.*, **69**, 651 (1976)
115 Sachs, G. and Van Horn, K. R., *Practical Metallurgy*, Cleveland, *Am. Soc. Metals* (1940)
116 Forrest, G., *Symposium on Internal Stresses*, London, Inst. Metals, 153 (1948)
117 Hayes, A. and Chipman, J., *Trans. Am. Inst. Min. metall. Engrs.*, **135**, 85 (1939)
118 Kohn, A., The Solidification of Metals. Proc. 1967 Joint Conference. Sp. Publication 110, Iron Steel Inst. London, 356 (1968)
119 Martin, P. M. and Poole, D. M., *Metall. Rev.*, **5**, No. 150, 19 (1971)
120 Johnson, M. P., Beeley, P. R. and Nutting, J., *J. Iron Steel Inst.*, **205**, 32 (1967)
121 Pfann, W. G., *Zone Melting*, 2nd Edn, Wiley, New York (1966)
122 McDonald, R. J. and Hunt, J. D., *Metall. Trans.*, **1**, 1 787 (1970)
123 Hebditch, D. J. and Hunt, J. D., *Metall. Trans.*, **4**, 2008 (1973)
124 Fisher, K. M. and Hunt, J. D., Solidification and Casting of Metals, Proc. International Conf. Sheffield. Book 192, 325. The Metals Society. London (1979)

125 Giamei, A. F. and Kear, B. H., *Metall. Trans.*, **1**, 2,185 (1970)
126 Beeley, P. R., Solidification and Casting of Metals, Proc. International Conf. Sheffield. Book 192, 319. The Metals Society, London (1979)
127 Flemings, M. C., Solidification Processing. McGraw-Hill New York (1974)
128 Mehrabian *et al.*, *Metall. Trans.*, **1**, 1209 (1970)
129 Peterson, R. E., *Internat. Conf. on Fatigue of Metals*. Inst. mech. Engrs., and Am. Soc. mech. Engrs, 110 (1956)
130 Mindlin, R. D., *Proc. Soc. Exptl. Stress Anal.*, **5** (2), 56 (1948)
131 Peterson, R. E., Stress Concentration Factors, John Wiley & Sons Inc. N.Y. (1974)
132 Mahadevan, N. S., Seshadri, M. R. and Ramachandran, A., *Cast Met. Res. J.*, **7**, No. 2, 71 (1971)
133 Uram, S. Z. M., Flemings, M. C. and Taylor, H. F., *Trans. Am. Fndrym. Soc.*, **68**, 347 (1960)
134 Larson, H. P., Lloyd, H. W. and Hertchy, F. B., *Trans. Am. Fndrym. Soc.*, **67**, 676 (1959)
135 Jackson, W. J., *Br. Foundrym.*, **50**, 211 (1957)
136 Glick, W. W. and Hebben, P. J., *Br. Foundrym.*, **55**, 54 (1962)
137 Wallace, J. F., Savage, J. H. and Taylor, H. F., *Trans. Am. Fndrym. Soc.*, **59**, 223 (1951)
138 Parker, E. R. *et al.*, *Effect of Residual Elements on the Properties of Metals*, Cleveland, *Am. Soc. Metals* (1957)
139 Reports of sub-commitees TS 23 and 26. *Proc. Inst. Br. Foundrym.*, **43** (1950) and **44** (1951)
140 BS 4570. Specification for fusion welding of steel castings. *Br. Stand. Instn.*, London
141 Ridal, E. J., In *Steel Castings Design, Properties and Applications*, Ed. Jackson, W. J., Steel Castings Research and Trade Ass., Sheffield (1983)
142 Davies, G. J. and Garland, J. G., *Int. Metal. Rev.*, **20**, 83 (1975)

6

Quality assessment and control

Quality systems and management

The approach to casting quality has undergone a long period of change, involving a broadening of the subject to enhance many more aspects of the production process and product than were traditionally regarded as parts of the quality control function. These changes have had major repercussions in the field of internal management organization, as well as in the commercial relationships involved in the procurement of castings. A review of the nature of these developments will provide a background to the more detailed consideration of quality in castings, and of the various measures used in its assessment.

The modern philosophy of quality assurance was already becoming evident around 1970, when a significant paper by the late George Quilter[1] described the revised approach in terms of the contrasting organizational charts shown in Figures 6.1 and 6.2. Here quality management is seen to supersede inspection as the cardinal function in the system, as recognized in the titles of the functions, and implicitly of the staff involved.

It was thus emphasized that a modern quality control organization must be concerned with a great deal more than the routine sorting of properties and attributes in relation to specified limits. Much data is obtained by the use of highly sophisticated equipment and techniques, and more detailed treatment than simple go or no-go tests is required in interpreting the results. Adverse trends may then be detected, and corrective action taken, before quality deterioration leads to a high proportion of rejects. This aspect of quality control, involving feedback as a crucial feature, will be further considered later in the chapter. The role of inspection still remains as a vital element in the modernized system and it too will be further examined.

Despite this relatively early appreciation of the need for change, the most significant development came with the introduction of more formal schemes for quality assurance systems, applicable to a wide range of industrial activity and necessitated by the increasingly stringent commercial, legal and technical backgrounds against which casting quality must be judged. Issues of competition, product liability, short delivery times, safety, and the environment itself all contribute to the need for more rigorous systems

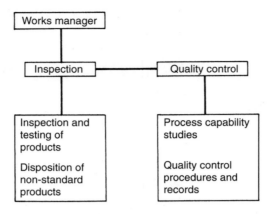

Figure 6.1 Functional organization of inspection and quality control activities

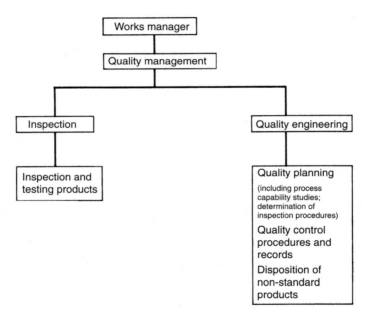

Figure 6.2 Revised concept of organization for inspection and product quality (after Quilter[1])

and procedures. This need was addressed by the introduction of specific published standards which could be adopted by manufacturers, both as guides to systems and as evidence of producer capability.

The principle of certification of producers is not new, having long precedents in such industries as defence and nuclear power, but the advent of BS 5750[2] and subsequently of the closely related BS EN ISO 9000[3], brought the concept to the forefront as a feature of supply to major organizations, with consequent implications along the entire supply chain from raw material to finished product. Some of the features and implications were examined in References 4 and 5.

The central feature of ISO 9000 and its BS equivalent is the requirement for a closely defined and comprehensive quality system. This entails the formulation of written procedures to constitute a Quality Manual, and a system of records to ensure that the procedures have been followed, and the appropriate action taken, to ensure that the product complies with specification. The system must, moreover, be subject to an external audit and certification by an independent accredited outside authority.

The quality system standards do not embody any single set of rules applicable to all circumstances, but essential features include clear statements of policy, objectives and monitoring systems, as well as of the specific procedures for inspection and testing. The stress is on prevention rather than detection of defects, with full two-way flow of information rather than the mere screening out of rejects. Such vital aspects as the need for casting identification and explicit documentation will be stressed later in this Chapter when dealing with typical inspection and testing techniques.

The successful introduction and operation of such a system has numerous implications, foremost of which is that the whole of a company becomes involved in the quality assurance role, at all levels and in every phase of the production process. This in turn requires clear allocation of staff responsibilities and the provision of training, the whole being integral to the central management structure.

A detailed breakdown of the series of quality management standards embodied in BS 5750 and BS EN ISO 9000 is beyond the scope of the present work, although some individual items are listed with other standards in the Appendix. Their contents are largely self-explanatory. Aspects of assessment for registration with accredited bodies and other practical details are included in References 6 and 7, whilst a general commentary on ISO 9000 was produced by Rothery[8]. Further standards relevant to quality management and assurance issues include BS 4778[9] and BS ISO 8402[10], which deal with definitions and vocabulary.

Scrimshire[11] published a detailed account of differences of emphasis between BS 5750 and the evolving ISO 9000. It seems clear that the continuing development of formal standards in this field will place increasing demand for fully documented controls, not just in processing and inspection but on incoming goods and raw materials, design and other aspects in line with the general concept of Total Quality Management[12]. Although such systems remain voluntary, it is frequently pointed out that marketing advantage and other pressures will tend to require their use in all aspects of casting manufacture. Analogous standards for environmental management systems (q.v. Chapter 11) seem likely to form a part of the same framework of management controls.

A comprehensive guide to implementation of the ISO 9000 standard, including the provision of suitable systems and procedures and produced by Hoyle, is detailed in Reference 13.

Definition and assessment of quality

Quality in the broad sense is usually defined in terms of fitness for purpose, or the ability to perform the functions as specified in the design process. For the practical purposes of verification and assessment, however, it is useful to think of the quality of castings in terms of two attributes. Firstly, the material confers bulk characteristics that are largely reproducible between one piece of metal and another, determined by the microstructure associated with the selected composition, casting conditions and heat treatment. These "normal" properties may, however, also be affected by some types of disseminated defect, arising from impurities introduced in the raw materials or during the foundry process, as examined previously in Chapter 5. Indigenous inclusions and dispersed porosity are in this category and can produce common property trends in all the castings in a batch. This "metallurgical condition" can be measured from samples and test specimens intended to be representative of the cast material, an issue to be more fully examined later.

The second quality attribute concerns the casting as an individual shape and relates more specifically to the success of the foundry technique. The mould influences the dimensional accuracy and surface finish, and the gating and feeding practice determine the degree of freedom from local defects in the component. This aspect of casting quality is assessed by direct inspection, using non-destructive techniques from dimensional measurement to full surface and internal examination.

Inspection

The inspection process has the dual purpose of ensuring that the product conforms to design requirements and of providing the information needed for quality control. Although foundry production and technical staff are concerned in many aspects of quality control, the systematic collection and integration of quality data is frequently channelled through an inspection department, which has a special responsibility for the administration of standards embodied in contracts and commercial relationships.

In the latter function the inspectorate is frequently envisaged as being independent of production management. Patently true independence cannot be realized within a single manufacturing organization. Such a principle may in any case be thought to be somewhat at odds with the philosophy of ISO 9000, which would see all from directors to operatives as being concerned with quality as much as with production. However, an inspector does have special and defined responsibilities with direct accountability to senior management.

In some cases the inspector may act in an independent capacity, in cases where an outside body formally delegates authority in implementing its standards: again, however, such delegation is in effect made to a supplier company as a whole if the approved quality management system is in place, and forms the basis of the contract. In yet other cases the function can be exercised directly by the purchaser, or by his delegated authority in the shape of an independent inspection organization, a procedure sometimes followed by public authorities or insurance underwriters concerned with safety standards. The internal inspector is then involved in the delicate relationships between all the bodies concerned.

Standards and their commercial implications

The initial requirement for inspection is the establishment of standards reflecting the intentions of the designer. Design in the broadest sense requires definition of the product in terms of physical shape and dimensions, composition and properties, and general quality. Inspection is based on permissible standards of deviation from the ideal. Some factors, for example dimensions, composition and mechanical properties, are capable of precise quantitative definition so that numerical tolerance limits can be prescribed: in other cases subjective judgements are needed.

Many explicit design requirements can be conveyed in drawings, computerized data or CAD files and specifications. Castings may, for example, be manufactured to a widely recognized specification such as a British Standard, in conjunction with drawings or patterns provided by the purchaser. Many formal specifications are, however, confined to the definition of chemical composition and/or mechanical properties, whereas castings require further qualities implicit in the capabilities of the process: the maintenance of these qualities, for example soundness and surface finish, is frequently dependent on internal standards instituted by the manufacturer. These standards are closely associated with commercial policy and may be set at a higher level than strictly necessary to meet the formal specification. Standards must not, on the other hand, be set at an unrealistic level in relation to the basic capability of the particular casting process; the economic implications of high standards must also be appreciated. Unnecessarily tight compositional tolerances, for example, may entail additional raw material costs or higher rejection rates, whilst the cost of more rigorous inspection procedures must itself be reflected in the selling price of the casting.

Radiographic and other special techniques are particularly expensive and should not be formally specified unless absolutely necessary. It is important that the significance and cost of these techniques be fully appreciated by the purchaser. Whilst of great value to the founder for methods development and for sample checks on quality during production, they should, where stipulated for routine inspection, be separately quoted and invoiced wherever possible.

Numerous standard specifications will be mentioned in the present and following chapters in relation to the design and inspection of castings. Major sources of such standards include the British Standards Institution[14], the American Society for Testing and Materials[15] the International Organization for Standardization (ISO), and the European Standards body CEN. Some relevant specifications from the former source are listed in the Appendix. Such specifications are, of course, subject to periodic revision and amendment.

Material and product testing procedures

Routine casting inspection normally involves geometric checking for shape and dimensions, coupled with visual examination for external defects and surface quality. Representative chemical analyses and mechanical properties are also often available. These measures are increasingly supplemented by various forms of non-destructive testing, hydraulic testing and proof loading. Since all such measures add to the cost of the product, the first consideration must be to determine the amount of inspection needed to maintain adequate control over quality. In some cases this may require full examination of each individual casting but in other cases sampling procedures may suffice.

Whilst it is customary to perform at least a limited visual inspection on all castings, detailed inspection may be confined to critical points at which variations are likely to occur: it is in this respect that a full understanding of the foundry method on the part of the inspector can eliminate unnecessary routine. Certain cast dimensions are susceptible to variation from mould and core movement, whilst others, dependent primarily on a single element of the pattern, are liable to vary only within narrow limits. Similarly, detailed understanding of the service function of a casting clarifies the relative importance of its dimensions and the significance of defects in different zones.

The inspection procedure must therefore be adjusted to suit the particular circumstances surrounding each type of casting. If sampling is used it must be appreciated that there can be no absolute assurance of freedom from defects in an individual case, so that complete coverage is needed in critical applications where no further sorting will occur. In those cases where inspection is to be limited to certain features or to a proportion of total production, full inspection may be carried out on preliminary samples and during the early stages of production, and may be resumed if trouble is encountered during later productions runs.

An important element in inspection and quality control concerns the identity of the castings examined. Whilst many castings are manufactured with no attempt at prolonged identification, a distinguishing mark of origin on each casting is a valuable aid to the maintenance of quality records.

Abbreviated cast and serial numbers can be stamped at an early stage on all save the smallest castings: the marking can often be retained throughout the life of the casting to provide permanent reference to its production and inspection history. Metal stamping must be applied with discretion in view of the frequent incidence of fatigue initiated at such stamps: it is essential to position them away from highly stressed regions. This traceability of the origins and production history of a casting forms an important part of the certification requirements embodied in modern standards for cast products of higher quality grades.

Shape and dimensions

Brief visual examination of each casting ensures that none of its features has been omitted or malformed by moulding errors, short running or mistakes in fettling. Serious surface defects and roughness can be observed at this stage.

Dimensional examination is then carried out against the drawings, aided by gauges, jigs and templates: tolerances on the cast dimensions are normally included on the drawing. In some cases this operation can be combined with marking out for subsequent machining. Where castings are destined for machining without individual marking out, particular care is needed in respect of dimensions incorporating datum surfaces. Errors involving these surfaces can produce consequential errors or inadequate machining cuts elsewhere on the casting. In cases where the weight of the component is critical a specific weight check may be included: this may also be required for costing purposes.

Where dimensional errors are detected in relation to general drawing tolerances, their true significance must be determined. A particular dimension may be of vital importance, but may on the other hand have been fortuitously included in blanket tolerances needed primarily elsewhere. It is strongly recommended that functional dimensions be stated on drawings so that tolerances are not restricted unnecessarily.

In cases where castings have been produced from patterns supplied by the purchaser, these may, subject to the recognized contraction allowances, form the direct basis for the dimensional examination. The owner of the pattern is then implicitly responsible for the cast dimensions, apart from errors due to distortion or fettling. It is nevertheless usual for most foundries to verify the patterns against drawings when these are also available.

Dimensional variation and tolerances are further discussed in Chapter 7.

Surface quality and finish

Quantitative standards are seldom adopted for the surface finish of castings, which are normally judged from visual examination alone. Numerical standards may, however, be specified if required. The simplest evaluation is by visual comparison with a series of standard surfaces: these may be arbitrarily graded, or may themselves have been quantitatively assessed by other means[16]. Alternatively, a stylus measuring instrument may be used to obtain representations of surface profiles. Such instruments are usually only suitable, however, for distinguishing between relatively smooth surfaces such as those obtained in precision casting processes: they provide graphical representation of the profile or an integrated numerical value derived from the total vertical movement of the stylus relative to a datum during a fixed length of traverse. This value is related to the most widely recognized index of surface texture, namely the CLA or centre line average value; this is the arithmetical average value of profile departure both above and below the centre line[17]. Figure 6.3 illustrates the derivation of the CLA value from the profile.

It should be emphasized that surface characteristics cannot be uniquely defined by the CLA or any other single numerical index, which only allow comparison between surfaces of similar origin. The operating conditions of a stylus instrument must be devised to measure texture alone and to omit the effect of more widely spaced irregularities. These factors are discussed fully in the literature including a notable review of 21 methods for surface roughness measurement carried out by Swing[18].

Composition and mechanical properties

The composition and mechanical properties of castings are normally determined by tests carried out on specimens cast from the same heat of metal. The questions of identity and representation are thus critical, the first essential being to ensure valid correlation between test specimens and castings. Many casting specifications stipulate independent values for both composition and mechanical properties but in some cases the partial

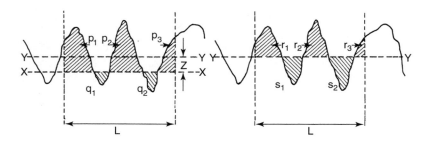

Figure 6.3 Determination of centre line average height from a trace. (a) Centre line Y established by drawing line X and determining areas p and q by planimeter or measurement of ordinates. Displacement $Z =$ (sum of areas p–sum of areas q)/L, (b) Centre line average $h =$ (sum of areas $r +$ sum of areas s)/L (Reproduced from a past issue of BS 1134. The complete current standard can be obtained by post from BSI Customer Services, 389 Chiswick High Road, London W4 4AL)

interdependence of these characteristics is recognized by requiring one or the other alone.

Composition. Chemical composition is usually determined on separately cast analytical samples identified by cast-in tags or by stamping. To be truly representative the sample must be drawn at the same time as the castings are poured, since losses of volatile or readily oxidized elements may otherwise cause the compositions to diverge. Where casting operations are prolonged, or where successive ladles need to be drawn from a single heat or from a holding furnace, frequent sampling is needed at the casting point. Analysis may also be carried out on discard metal from heads and runners.

Composition is normally assessed in relation to specified limits for the major alloying elements and expected impurities, but watch also needs to be kept on occasional tramp elements and on gas content. The need for a particular balance of composition, as required, for example, to maintain a carbon or zinc equivalent, may be superimposed on the limits for individual elements.

The use of spectrographic, X-ray fluorescence and other physical techniques enables a comprehensive range of elements to be simultaneously and rapidly determined. Apart from the advantage for routine analysis, this facilitates bath sample analysis and thereby improves control in melting. The composition of each cast can also be verified before the castings are further processed, minimizing loss of production time. Wet analysis remains available for standardization or for use where advanced physical equipment is not available.

Mechanical properties. Specimens for mechanical testing are normally machined from specially designed cast test bars. These may be poured separately from the same heat of metal as the casting or may, especially in the case of heavy castings, be directly attached to the casting itself.

The status of the test bar should be free from ambiguity: it is intended, with certain exceptions, to be representative of the particular heat of metal and not of any individual casting. The measured properties will reflect influences of dissolved gas, inclusions and melt treatments but will not necessarily be those which would be obtained from test pieces cut from actual castings. since there may be differences in section, structure and degree of feeding. Standardization of test bar conditions permits the use of measures to ensure perfect soundness of the test specimen, so that yield is largely disregarded and most test bars are heavily overfed. Keel block and other typical test bar designs showing this feature are illustrated in Figure 6.4. In some cases, however, more benefit is to be derived from the retention of cast surfaces in close proximity to the final test piece, hence the use in certain test bar designs of a cast shape conforming closely to the contour of the machined test specimen. Bars of this type are illustrated in

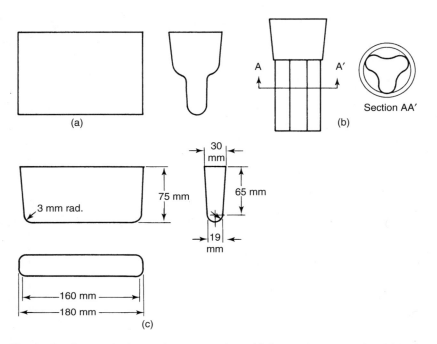

Figure 6.4 Some designs of cast test bar. (a) General purpose keel block, (b) clover leaf test bar, (c) chill cast test bar for light alloys

Figure 6.5. It should be added that, to avoid major differences in cooling conditions, all test bars are normally cast in moulds of similar material to that used for the groups of castings they represent; casting temperature should also be standardized as far as possible.

The more recently issued cast alloy specifications are less prescriptive of the actual designs to be employed for cast test bars but do stress underlying principles. BS EN 1982, for example, recommends that bars for long freezing range copper alloys should be fed at one or both ends, with preference for feeding all along the length for other types. These conditions correspond to Figure 6.5*e* and *a* respectively, but these and the other designs in the illustrations are no longer included in the specifications. The general principle now stressed is the need for the test bar design to be agreed between foundry and purchaser. The specifications also provide for test specimens from actual castings.

Whilst mechanical tests are primarily indicative of metal rather than individual casting quality, special conditions exist in the case of cast iron owing to the marked dependence of structure and properties upon section thickness and cooling rate. Tensile test pieces of various diameters were formerly specified to seek results representative of particular casting thicknesses, even though it was emphasized that the metal rather than the casting quality was being assessed. This practice was later revised in favour of a single

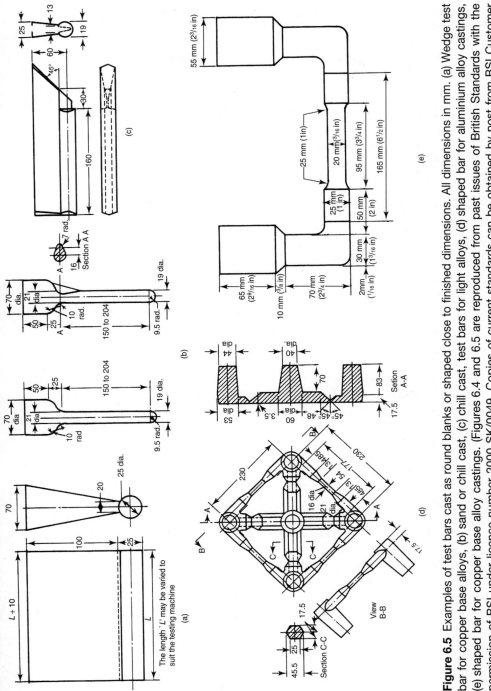

Figure 6.5 Examples of test bars cast as round blanks or shaped close to finished dimensions. All dimensions in mm. (a) Wedge test bar for copper base alloys, (b) sand or chill cast, (c) chill cast, test bars for light alloys, (d) shaped bar for aluminium alloy castings, (e) shaped bar for copper base alloy castings. (Figures 6.4 and 6.5 are reproduced from past issues of British Standards with the permission of BSI under licence number 2000 SK/0049. Copies of current standards can be obtained by post from BSI Customer Services, 389, Chiswick High Road, London W4 4AL)

cast test bar of diameter 30 mm, a guidance chart being appended to enable estimates of properties in different sections to be derived from the test results (see Figure 7.8)

Section dependence of mechanical properties is also a feature of some copper alloys, where a marked effect can be introduced by the dispersed porosity associated with the pasty mode of freezing. The whole position of the relationship between test bars and castings in specifications for various groups of alloys was the subject of a valuable review by Everest[19].

From the foregoing considerations it should be clear that integrally cast test pieces are no more representative of the mechanical properties of a particular casting than are separately cast test pieces from the same piece of metal. They may still be heavily fed, either by direct attachment beneath a heavy section of the casting or by the use of an individual feeder head: in the latter case the test bar is virtually treated as a separate casting, being connected to the parent casting only through a runner bar. Attached bars may be prescribed as a guarantee of authenticity, since an inspection stamp imprinted at the cast stage can be retained on the specimen until the test can be witnessed. This precaution is, however, equally applicable to separately cast bars stamped at the time of casting so that attachment is merely a handicap to the standardization of test bar casting conditions.

Test bars, whether integrally or separately cast, should accompany the appropriate castings through any production heat treatment, ensuring similarity of structure in heat treated alloys.

Premium quality castings

Although mechanical test results derived from test bars cannot be attributed to the casting without reservation, a modern development is the concept of the premium quality casting, in which the founder guarantees the level of properties in certain specified zones whilst the additional production costs and more rigorous inspection standards are reflected in a higher selling price. The principle itself is not new, but its introduction on a formal basis in light alloy founding suggests a general trend towards consolidation of the higher quality made feasible by modern methods of founding, non-destructive testing and metallurgical examination. The value of destructive examination in constructing a basis for control and determination of actual casting properties will be discussed in a subsequent section.

The general philosophy underlying premium castings has been reviewed by Vogel[20] and several of its aspects are referred to in other chapters, particularly those dealing with solidification and casting design. The key features lie in close controls throughout the casting process itself, in reproducibility of mechanical properties, particularly ductility and fatigue resistance, in dimensional tolerances and surface finish, and in the quality assurance principles and techniques applied.

The need for a partnership between designer and foundry producer has been stressed, not least in the present book, and the general approach to procurement and to quality pioneered in the premium castings area has already been reflected more widely, including such aspects as the involvement of outside suppliers envisaged in modern quality standards. A further example is the perception of inspection and quality control functions as an integral part of manufacture, and thus of the duties of all supervisors and operatives involved.

More specific, however, is the application of those distinctive foundry techniques needed to develop, not just reproducible, but high, mechanical properties in the product. In this respect cost becomes secondary to the achievement of the exceptional soundness and structural refinement required: some typical measures have been referred to in Chapters 2 and 3. Flemings[21] instanced dendrite arm spacings below 50 mm as appropriate for an aluminium alloy in this category, with sufficient metal purity to permit solution treatment within 10–20°C of the eutectic temperature, producing complete solution of the second phase in less than 3 hr, a dramatic reduction compared with that normally expected from sand castings.

Such aspects as "hard sand" moulds and the use of melt treatments, filtration and special metal transfer techniques are also likely to be involved. Vogel pointedly invoked the further attribute of 'an attitude of mind'.

Castings may be evaluated on the basis of a wide range of mechanical tests, for which full test procedures are detailed in numerous standards for individual tests and alloys (e.g. References 14, 15 and Appendix). The tensile test is most commonly specified; this is frequently supplemented by hardness determinations, carried out where practicable on the castings themselves.

The tensile test[22] is normally carried out on the universal type of testing machine, results for proof or yield stress, tensile strength, elongation and reduction of area being obtained. Requirements in casting specifications normally include one or more of these properties. The test, together with other mechanical tests carried out on cast alloys, should preferably be carried out on full sized standard test pieces in view of the relatively large grain size of the materials. Anisotropy and specimen size effects become increasingly important as the number of grains across a specimen diminishes[23]. The larger the test specimen in relation to grain size and other metallographic features, the more representative will be the results from a single test. The use of miniature test pieces, on the other hand, produces a high degree of scatter due to structural heterogeneity, so that the results of a single test may give a misleading picture of the properties of the bulk material.

Hardness testing, apart from its value as a guide to wear resistance and machinability, is frequently used to verify the effectiveness of heat treatment applied to actual castings; its general correlation with tensile strength enables a rough prediction of that property to be made.

Hardness is normally determined by indentation tests, of which the Brinell test[24] is most frequently employed for cast alloys, a combination of a large (5 or 10 mm) diameter ball and heavy load (750 or 3000 kg) being preferred for the most effective representation. The deep impression eliminates any undue influence of the immediate surface layer. The test is, however, unsuitable for use at very high hardness levels, where distortion of the ball indenter affects the geometry of the impression.

For high quality and precision castings on which the large Brinell impression cannot be tolerated, or for alloys of extreme hardness, the Rockwell or Vickers test is required. The Rockwell test[25] is the faster and the more commonly used for control purposes, but the highly accurate Vickers pyramid test[26] finds wide application in laboratory work. Because of the very small impressions produced in these tests, which use loads of 150 kg or less, results must be based on the average of a number of determinations; special care is needed to ensure adequate surface preparation and firm seating of the specimen.

Indentation tests normally require that the specimen be placed on the platform of the machine, precluding direct readings on heavy castings, for which representative test specimens must be employed. Certain hardness tests can, however, be applied directly to large castings. The Scleroscope, measuring the rebound of a free falling pointed indenter, can be used to determine the hardness of any horizontal surface. Other portable instruments give readings capable of reasonable correlation with standard indentation tests. These additional tests are used for control purposes rather than for formal inspection.

Notched bar impact values are included in some casting specifications, even though the results have no direct quantitative design significance. Both Charpy and Izod tests are used[27,28]; although much testing is performed at room temperature, sub-zero testing is increasingly specified for castings for cryogenic and other low temperature applications.

In the case of grey cast iron, notched bar impact testing is of little value in discriminating between the quality of different irons, but an unnotched test piece was developed for this purpose: a plain round specimen is held in stepped grips and the fracture energy determined as in the normal types of test[29,30]. Tests based on the same principle can be used for other types of alloy with very low ductility, whilst special tests using non-standard specimens with small, rounded cast-in notches have proved suitable for the impact testing of wear resistant cast alloys with heterogeneous microstructures, containing substantial amounts of brittle second phases.[31]

Other mechanical tests which may be specified include bend and compression tests, both of which can be carried out on the universal type of testing machine. In the bend test[32] a bar specimen supported at its end is bent round a former through which the load is applied.

Although castings are frequently used under fatigue stresses or for high temperature applications, fatigue tests and stress rupture or creep tests are not normally used for routine inspection and release of production castings.

Other tests on representative samples.   Various special tests may be applied to establish the response of the cast material to subsequent manufacturing and service conditions. These include tests for weld decay as applied to austenitic stainless steels, ballistic tests on plates, wearability tests and other tests based on analogies to the conditions to be encountered by the casting.

Destructive tests on castings

Mechanical test results from cast-on or separately cast bars may be supplemented by tests carried out on specimens removed from actual castings. Variation of properties within a single casting results from inevitable differences in microstructure, density, and composition produced by local differences in freezing conditions. Whereas the normal test block is designed to evaluate the metal under ideal conditions of feeding, tests on sectioned castings can provide a more direct indication of properties within a particular design.

Systematic sectioning also facilitates the direct investigation of structure and internal soundness by metallographic techniques, and aids the accurate interpretation of results of radiographic and other methods of non-destructive testing.

Another form of destructive testing involves the direct loading of representative castings to failure in special fixtures simulating service stresses: the deformation and breaking load and the position and nature of the fracture provide a guide to material quality and to the behaviour of the design under highly stressed conditions. In a simpler form of test the casting can be fractured by compression in a press. Large scale fatique testing was employed in the evaluation of the oilrig node castings as referred to in Chapter 7.

Castings for destructive testing may be selected from rejects from other causes, but it may in some cases be necessary to sample a proportion of castings from normal production. Such techniques can form a vital aid to the interpretation of test results and inspection findings derived by more conventional means.

Soundness: The non-destructive testing of castings

Various methods are available for determining the internal soundness of castings without recourse to sectioning. The ability to detect invisible flaws not only assists in maintaining high quality standards, but provides a valuable aid to the development of foundry methods. These factors have greatly enhanced the position of the casting in engineering. However, since non-destructive testing greatly increases the sensitivity of the inspection operation, the techniques need to be adopted and interpreted with discretion if unnecessary rejections are to be avoided.

It will be the object in the following section to review those methods which are widely employed in the foundry field. These include a number of techniques of surface flaw detection and the two main forms of internal examination, namely radiography and ultrasonic testing. The many terms used in connection with these techniques are defined in full in References 33 and 34. Some of the appended references provide practical details, whilst fuller treatments of the techniques can be found in the more general literature on non-destructive testing: classical and modern works in the latter category are represented in References 35–40.

Tests for defects in the surface zone

Cracks and other flaws communicating directly with the surface of the casting can be detected by a number of techniques including penetrant methods, chemical etching and magnetic flaw detection; the latter method is also capable of revealing flaws situated immediately below the surface. All the methods require clean and reasonably smooth surfaces for effective detection.

The penetrant methods are based on the use of fluids to permeate and highlight defects by increasing their contrast with the surrounding surface. The methods are also particularly useful for establishing the soundness and integrity of thin walled castings in cases where pressure testing is not readily applicable. Penetrant solution can be applied to one face and the opposite face examined for signs of seepage.

Dye penetrant testing[41,42]

The surface of the casting is in this case treated with a dye solution based on a low viscosity solvent. The solution can be applied by brush or aerosol spray or the casting can be entirely immersed in a tank of solution; the casting is then thoroughly rinsed and dried. After a time, seepage of residual fluid occurs from cracks and other narrow surface defects to produce a visible stain. For maximum contrast a light background surface is desirable: this can be provided by applying, after the dye and rinsing treatments, a separate developing compound. A fine suspension of a white powder, for example chalk, readily absorbs and indicates the dye.

Various types of penetrant are available. Using a spirit based penetrant, a solvent rinse is required for removal of the excess from the surface. Emulsified penetrants enable water rinsing to be used: the medium can be applied as a prepared emulsion or the excess can be emulsified on the surface by a separate application of emulsifying agent.

The penetrant method is sensitive to small surface irregularities, which tend to retain dye in spite of the rinse. Smooth casting surfaces or machined surfaces, therefore, give more satisfactory conditions for the test; in any event experience is needed for interpretation of the colour indication.

A widely used variation of the technique utilizes fluorescent dyes, enabling the search for flaws to be carried out with an ultraviolet source in a darkened cubicle.

Oil and chalk testing

This long established technique uses paraffin oil as the penetrant medium. The casting is treated, usually by immersion, with heated paraffin; it is then rinsed in detergent solution and water and dried. Using a low pressure air dispenser the surface is coated with dry powdered chalk, which becomes marked after a time by seepage of paraffin from cracks or porous zones.

Chemical etching

Very fine surface cracks, laps and similar defects may be masked by a thin layer of metal resulting from superficial plastic deformation. This can occur in shotblasting, in fettling or even in the machining of a casting. In other cases the masking layer may consist of oxide scale formed in heat treatment. Normal penetrant methods are not always effective in such cases.

Controlled corrosive attack in a pickling solution offers a method of revealing these hidden surface defects and at the same time of exposing subcutaneous pinholes. The technique employs acid or acid–salt mixtures developed for particular alloys, the casting being immersed in the etching bath, rinsed and dried. Defects are rendered visible by local staining against the clean, matt surface produced by the etch. Care must be taken to neutralise excess acid and rinse after completion.

Magnetic flaw detection[43]

Ferromagnetic materials, notably iron and steel, respond to magnetic flaw detection, a highly effective and sensitive technique for revealing cracks and similar defects on or just beneath the casting surface.

When a piece of metal is placed in a magnetic field and the lines of magnetic flux are intersected by a discontinuity such as a crack, poles are induced on either side of the discontinuity. The resulting local flux disturbance can be detected by its effect upon magnetic particles, which become attracted to the region of the defect. Maximum sensitivity of indication is obtained when a crack runs in a direction normal to the applied field and when the strength of this field is just enough to saturate the section being examined.

Although limited use can be made of permanent magnets for small scale testing, commercial crack detection equipment employs d.c. or a.c. current to generate the necessary magnetic fields. The current can be applied in a variety of ways to control the direction and magnitude of the field. The technique is characterized by high currents and low voltages.

One method of magnetization uses a heavy current passed directly through the casting member by placing this between two contacts in the form of solid copper clamps or flexible gauze pads. The induced magnetic field then runs in the transverse or circumferential direction, producing conditions favourable to the detection of longitudinally disposed cracks (Figure 6.6a). For annular parts a field of this kind can be obtained by using an axially disposed conductor bar threaded through the orifice (Figure 6.6b).

Cracks running in the transverse direction are best detected in a longitudinal field, which can be obtained by energising an external electromagnet with the poles situated at the ends of the section (Figure 6.6c). A permanent magnet can be similarly used.

Alternatively, the required field may be conveniently generated by using a flexible braid conductor, which can be coiled round any metal section to induce an axial field when current is passed (Figure 6.6d). This method is particularly adaptable to components of awkward shape.

Small castings can be examined directly on magnetic crack detectors such as those illustrated in Figure 6.7. Critical regions of larger castings can be inspected by the use of coils or contact probes carried on flexible extension cables connected to the source of current, enabling most sections of castings to be reached.

Since the sensitivity of detection varies according to the direction of the crack in relation to the magnetic field, means must be adopted to examine

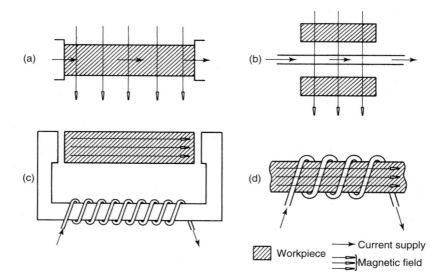

Figure 6.6 Techniques of magnetization for crack detection. (a) Contact method: current flow in workpiece, (b) threading bar: axial conductor, (c) electromagnet, (d) coiled conductor

Figure 6.7 Magnetic crack detector incorporating current flow and magnetic flux testing circuits suitable for casting inspection (courtesy of Fel-Electric Ltd.)

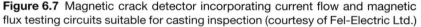

material in more than one direction. This may be done by rotating the specimen, or successive magnetic fields in two directions at right angles may be applied to a fixed specimen, using current and flux magnetism respectively. Examination is also possible under simultaneous flux and current magnetism, using a d.c. energized electromagnet combined with a.c. current flow through the metal section. The sinusoidal variation of the a.c. induced field, superimposed on the permanent field of the electromagnet, produces a resultant field of continuously changing direction; this gives an effectively continuous indication of cracks, irrespective of their direction.

Since distortion of the lines of magnetic flux by a discontinuity extends beyond the boundaries of the actual flaw, a defect just below the surface can still exert a visible influence upon magnetic particles at the surface. This influence becomes weaker with distance, so that sensitivity falls away rapidly with depth. For subsurface defects, the deeper penetration afforded by d.c. magnetism is most suitable, but a.c. magnetism is preferred for maximum surface sensitivity and offers several operating advantages, including straightforward demagnetization after testing.

The magnetic particles for flaw detection may be applied directly to the surface as a dry powder, but are more commonly suspended in a suitable fluid as an 'ink'. The particles in such inks must be large enough, and the flow of fluid over the surface sufficiently slow to facilitate rapid settling under the influence of light magnetic forces. The magnetic powder base may be treated with a dye or fluorescent coating to assist observation against dark surfaces.

The particle test may be applied during the passage of the magnetizing current, but an alternative method uses the residual magnetism following the passage of very high current: this method avoids the danger of oversaturation.

As the most sensitive available method of crack detection the magnetic particle technique predominates in ferrous metal casting, penetrant methods being required for non-ferrous metals and austenitic steels.

The detection of internal defects

Two principal methods have been applied to the detection of internal defects in castings, namely radiography and ultrasonic testing. Of these, radiography is the more highly developed as a technique for detailed examination, being capable of providing pictorial representation of the form and extent of most types of internal defect. Ultrasonic testing, less universally applicable, can give qualitative indications of many defects, its value being enhanced in the case of castings of relatively simple design, for which the signal patterns can be most reliably interpreted.

Both methods are extremely versatile, radiography being applicable to the examination of castings of great thickness and ultrasonic testing to the determination of important structural characteristics in cast alloys. For both techniques, skilled interpretation is the key to full exploitation. Radiography has also proved invaluable in the study of foundry processes: reference has been made in earlier chapters to the use of high energy X-ray fluoroscopy in studies of liquid metal flow and solidification phenomena.

Radiography[44]

Radiographic examination is based upon exposure of the casting to short wavelength radiation in the form of X-rays or gamma rays from a suitable source. The amount of radiation absorbed by a particular alloy is a direct function of its total thickness, so that transmission is influenced by the presence of internal cavities or by inclusions of substances possessing different radiographic density from that of the parent material. The emergent radiation is consequently subject to local variations in intensity, or shadow effects, which can be detected with the aid of photographic films, fluorescent screens or counting equipment. The production of radiographs on film is the technique most commonly applied to castings, although the less sensitive fluorescent screen offers a convenient means for the rapid examination of light alloys in thin section.

On radiographic film, the image of a defect appears in most cases as a dark shadow, representing the local increase in the transmission of radiation

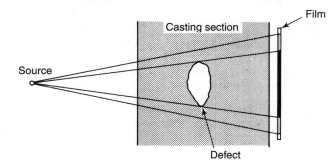

Figure 6.8 Production of radiographic image on film

due to the effective reduction in metal thickness in the path of the beam (Figure 6.8). Non-metallic inclusions in light alloys, on the other hand, reduce transmission and appear in light contrast. To obtain maximum sensitivity of defect detection, therefore, the conditions for the production of a radiograph must be carefully selected to secure the required degree of contrast in the film image. Technical requirements must be combined with a reasonable exposure time.

The search for defect indications in the dense photographic image is assisted by the use of an artificially illuminated viewing screen in a darkened room.

X-rays

X-rays are emitted from a metal anode target bombarded with electrons from a hot filament cathode in an evacuated tube. The penetrating power of the radiation increases as the wavelength diminishes, this being accomplished by increasing the voltage applied to the tube. X-ray sets used for the examination of castings range from approximately 100–2000 kV rating according to the thickness and density of the alloy being examined.

Since radiographic contrast decreases with the wavelength of the radiation, the operating policy is to utilize the minimum tube voltage consistent with penetration and a filament current adequate to deliver the required radiation intensity: the latter determines the exposure time needed to produce a radiograph. These effects are summarized in the following diagram:

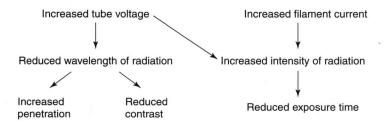

Radiation intensity is itself governed by the inverse square law and thus by the source–object distance. The effect of radiation in producing a photographic image on film, however, may be intensified, and the exposure shortened, by the use of salt or metal screens in direct contact with the film. A salt screen, for example of calcium tungstate, produces visible fluorescence which exerts a direct photographic effect on the film emulsion; a similar but less intense photographic effect is produced by photoelectrons and fluorescent X-rays emitted in the irradiation of a very thin metal screen.

Characteristics of radiographs

An important practical aim in the production of a radiograph is to minimize the exposure time, but the major consideration is to achieve the required characteristics of contrast and definition in the final image.

1. *Contrast* is the visible difference in the darkening of the film produced by local variations in radiation intensity. Apart from the previously mentioned influence of radiation wavelength, it is determined by factors analogous to those in photography. It is influenced by exposure time, film characteristics and development conditions. The optimum exposure time can be derived from a study of the characteristic curve for the particular film under the recommended development conditions. The general form of the typical curve is shown in Figure 6.9a. Maximum contrast is associated with the greatest slope of the density–log exposure curve. For many films, however, contrast and sensitivity continue to increase with exposure time throughout the range of image densities which are practicable for evaluation using an illuminated viewing screen (Curve b); the maximum possible exposure time is therefore used.

Optimum contrast cannot always be achieved, since many radiographs incorporate metal sections of varying thicknesses: exposure must then be a compromise between the requirements of the whole range of section thicknesses covered by the film.

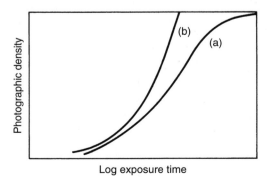

Figure 6.9 Relationship between film density and time of exposure. (a) General curve, (b) common form of curve within practicable range of image densities

General improvement in contrast is secured by measures to diminish scattered as distinct from directly transmitted radiation, thereby reducing the overall level of fog in the photographic image. This can be accomplished by lead filters or screens, positioned either directly over the X-ray tube or between casting and film. Extensive masking with lead sheet can also be used to exclude scatter arising in regions outside the direct radiographic zone.

2. *Definition*, the second quality characteristic, is determined by the size of the radiation source and by the relative positions of source, casting and film. Ideal conditions for definition would be obtained with a parallel beam of radiation, which would project a sharp image of the correct size on to the film or fluorescent screen. In practice, the long source–object distance which would be required for approximation to a parallel beam is excluded because of the low radiation intensity and prohibitively long exposure time: the beam is therefore markedly divergent.

To obtain a perfectly sharp image, moreover, a point source would be needed: the finite size of actual sources results in a penumbral zone at the boundaries of discontinuities and imposes a lower limit to the size of defect which can be resolved. This factor is illustrated in Figure 6.10.

Under practical conditions, therefore, a radiographic image is subject to appreciable distortion and to hazy outlines, the best definition being achieved with a long object–source and a short object–film distance.

The attainable degrees of contrast and definition are influenced to some extent by the photographic emulsions and processing techniques selected. Fine grained film is preferred for maximum contrast and definition, but fast films of coarser grain are utilized where it is important to reduce the exposure time.

Gamma radiography

Gamma radiation, a product of radioactive decay, is extensively used in the non-destructive testing of castings, the general technique being similar to

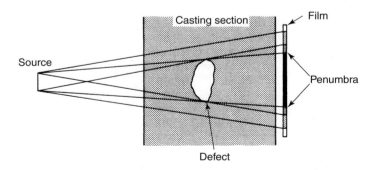

Figure 6.10 Influence of source dimension on radiographic definition

that employed in X-ray work. Radium and radon were originally employed as gamma ray sources, but more convenient sources became available in the shape of artificial isotopes possessing the appropriate characteristics. The small size and portable nature of the source, as compared with the X-ray tube and its ancillary generating equipment, is a valuable feature of gamma radiography.

The important characteristics of a gamma ray source are its energy, strength and half life. The energy of the radiation is constant for the particular isotope selected and determines the penetrating power; the energy of gamma rays emitted by cobalt 60, for example, is equivalent in this respect to X-rays generated by tube voltages well in excess of 1000 kV and the radiation is capable of penetrating heavy sections of steel.

The intensity of radiation, determining the exposure time necessary to produce a radiograph, is in the case of gamma radiation dependent solely upon the strength of the individual source, normally expressed in curies.* The exposure conditions can thus only be changed by variation of the source–film distance.

The strength of a source dwindles continuously due to radioactive decay, the half life, or the time taken to reach half its original strength, giving an indication of its useful life for radiographic purposes (Figure 6.11). The half life may be quite short: the period of only 74 days for iridium 192, for example, imposes the need for renewal at regular intervals.

The characteristics of some important gamma ray sources are given in Table 6.1.

The technique of producing gamma radiographs differs in important respects from that used in X-ray examination. Unlike an X-ray tube, which is usually designed to emit a beam of radiation over a limited field, gamma rays are emitted from the source in all directions. Advantage can be taken

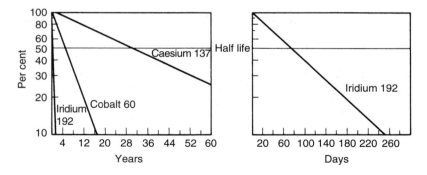

Figure 6.11 Decay of radioactive isotopes used as gamma ray sources (from Reference 45) (courtesy of CDC)

---

* 1 curie represents gamma radiation equivalent to that from the decomposition of 1 g of radium in a sealed container.

Table 6.1 Gamma ray sources

Source	Main Energy kV	Half life (approx.)	Typical application
Cobalt 60	1100, 1300	5.3 yr	40–200 mm steel
Iridium 192	610	74 days	10–100 mm steel
Caesium 137	667	33 yr	20–100 mm steel
Thalium 170	84	129 days	25 mm Al alloy
Radium	600, 1120, 1760	1600 yr	100–150 mm steel
Radon	600, 1120, 1760	3.8 days	100–150 mm steel

of these conditions to produce a number of radiographs simultaneously. These may be radiographs of a number of separate castings disposed in a circle around the source, or the source may be placed in a central position for a comprehensive series of peripheral radiographs to be obtained on a single large casting. By these means, although radiation intensities are low and exposures long compared with those in X-ray practice, a high total output can be obtained.

Radiographic equipment

For X-ray examination, basic requirements are the high voltage transformer and control equipment, together with high tension leads and the X-ray tube itself. Most X-ray sets used in the foundry industry operate at voltages in the range 150–400 kV, giving capacity for the penetration of sections equivalent to 100–150 mm of steel. Sets of 1 MV and even higher later became available, whilst other means exist for generating radiation of yet greater penetrating power, for example the 10 MV betatron or the linear accelerator. These machines extend the potential range of radiography to a point at which there are few castings that cannot be examined. The use of these high energy sources for the study of liquid flow and solidification phenomena was referred to in Chapter 1.

The relationship of radiation energy to metal thickness is illustrated by Table 6.1, which gives recommendations for X-ray energies for the examination of various thicknesses of steel (Table 6.2 includes similar data for gamma radiography). The equivalence to other metals of different radiographic density depends to some extent on the energy level but an approximate indication for typical cases is given in Table 6.3.

Although some low powered X-ray sets are portable, equipment is commonly operated in a laboratory provided with radiation protection, facilities for handling, and film processing installations. Gamma radiography is more flexible, requiring little other than shielded storage facilities and a portable container for the source, together with handling rods for

Table 6.2 Radiation energy for steel

Radiation	Steel thickness range, mm
100 kV	up to 12.5
200 kV	12.5–40
400 kV	40–90
1 MV	50–300
2 MV	60–250
5–31 MV	75–400

Table 6.3 Radiographic equivalence factors

Metal	100 kV X-rays	220 kV X-rays	^{60}Co γrays
Mg	0.6	0.08	
Al	1.0	0.18	0.35
Steel	12.0	1.0	1.0
Cu	18.0	1.4	1.1
Zn		1.3	1.0
Pb		12.0	2.3

From Non-Destructive Testing Handbook, edited by Robert C. McMaster © 1959, The Ronald Press Company, New York.

its safe manipulation. Equipment is also available for remote control of source movement at the beginning and end of the exposures. The operations may be carried out within a protected laboratory or in any convenient factory area which can be isolated during irradiation: advantage can thus be taken of weekend and overnight periods to exploit the continuously emitted radiation.

Processing facilities are identical for X-ray and gamma ray practice, but the overall capital cost for X-ray equipment is high compared with the small outlay for gamma ray sources.

Several further modern developments have enhanced both the capability and speed of operation of radiographic equipment. Image conversion to visible light facilitates viewing whilst the casting is under examination, with digital image processing, intensification and recording, during which the casting can also be moved to obtain further inspection zones or different viewing aspects. An additional technique in this field is X-ray computed tomography: a thin fan of radiation is passed through the casting, which is rotated to obtain successive images of interior volume elements or slices.

These can be computer processed and integrated to achieve full 3D characterization of porosity or other internal defects. The enhanced detection capability of this system as compared with conventional radiography is illustrated in Figure 6.12.

Aspects of these techniques as applied to castings were examined in References 46 and 47.

Factors in defect detection

The sensitivity of radiographic examination depends upon close control of exposure and process conditions to achieve optimum definition and contrast. Apart from the quality of the individual radiographs, however, the overall effectiveness of the technique depends upon discrimination in selecting the number and directions of the various radiographs to achieve a coordinated and systematic examination.

The smallest detectable reduction in metal thickness lies in the range 0.5–2.0% of the total section thickness. It is evident, therefore, that radiography will not detect narrow defects, for example cracks or cold shuts, unless these lie in a plane approximately parallel to the incident beam of

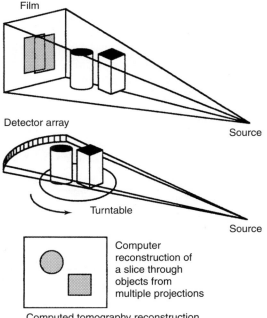

Figure 6.12 Principle of computed tomography as compared with conventional radiography (from Bossi and Georgeson[47]) (Courtesy of American Foundrymen's Society)

radiation. For such defects, ultrasonic testing is a more effective means of detection. Major discontinuities such as hot tears, on the other hand, are readily detected by radiography.

For all types of narrow, elongated defect, the probability of detection is greatly increased if due regard is paid, in selecting the direction of the radiograph, to the likely orientation of such defects in relation to the design of the casting: the typical sites for hot tears, for example, are well known. In some cases, although the main body of a defect may not appear on a radiograph, skilled observation reveals a significant indication at some point where it undergoes a change of direction bringing it into the plane of the radiation; its full extent may then be explored with further radiographs.

Determination of the size and exact position of any defect requires two or more separate radiographs from different angles, their choice and interpretation being a matter of considerable special skill.

The sensitivity obtained under a particular set of conditions can be established with the aid of a penetrameter, which determines the influence of small known thickness variations on radiographic density[48]. A typical penetrameter of step design is illustrated in Figure 6.13. Superimposed on the casting in the path of the radiation, it introduces local thickness increments appropriate to the total thickness of the section being examined; the smallest step that can be distinguished, expressed as a percentage of total section thickness, represents the sensitivity attained.

Radiographic technique and interpretation are closely interrelated. Any radiograph must be examined, not in isolation, but taking account of the whole series of factors in its production, including the direction of examination, nature of the radiation, exposure time, film type and processing procedure. Serious inspection must therefore be regarded as a function for a specialist radiographer. The selection of positions for radiographs and the interpretation of defect indications do, however, require knowledge of the principles of founding and experience of defects based on sectioning of scrapped castings. Interpretation must take account of the possibility of

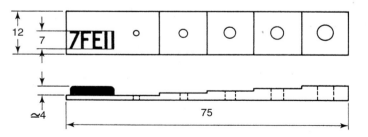

Figure 6.13 Sketch of step/hole type image quality indicator. All dimensions in mm (adapted by permission of BSI from a past British Standard. Copies of current standards can be obtained by post from BSI Customer Services, 389 Chiswick High Road, London W4 4AL)

spurious indications or artefacts which may arise on the film in the form of pressure marks, processing blemishes and diffraction mottling. It should also be noted that surface irregularities as well as internal defects affect the radiographic image.

Various aspects of radiographic examination are included in the general literature on non-destructive testing as previously cited, and in the specialized standard BS EN 444[44], whilst the specific application to the detection and interpretation of defects in castings is treated in Reference 49.

Ultrasonic Testing[50]

Ultrasonic flaw detection is based on the use of acoustic waves with frequencies in the range between 20 kHz and 20 MHz. These waves have the property of being transmitted by solids but reflected by internal surfaces such as those present at defects; the behaviour of an ultrasonic beam thus affords a basis for the detection, location, and size estimation of defects.

Ultrasonic waves are generated and detected by means of crystal transducers. To produce the waves, alternating current of appropriate frequency is applied to a crystal, commonly of quartz, causing it to oscillate through the inverse piezoelectric effect. This source is placed in close contact with the surface of the casting so that the waves are beamed through the metal in the required direction. When the waves meet a discontinuity, for example the metal–air interface at a crack, a reflection is obtained and can be detected by an appropriately positioned receiver: a further echo is generated at the opposite surface of the section. Alternatively, a receiver can be positioned directly in the path of the original beam to detect any local reduction in its intensity.

The technique may therefore involve the employment of separate probes, one to transmit the waves and the other to receive them after passage through the casting; alternatively, since the ultrasonic waves are transmitted as a series of intermittent pulses, the same crystal can serve both as transmitter and receiver (Figure 6.14). Various designs of probe are used, including combined double probes containing separate transmitting and receiving transducers.

The piezoelectric effect at the receiving crystal produces an electrical signal, which can be amplified and fed into a cathode ray oscillograph. On receipt of an echo, a vertical 'blip' is produced on the horizontal trace of the oscillograph. Any defect giving rise to an echo deflection can be accurately located by what is essentially a timing process. The time base of the instrument is adjusted so that the full width of the trace represents the section being examined, the echo from the far surface being used to register the end of the trace. The time for the transmitted pulse to reach the receiver is then represented in relation to the time to traverse the whole section, by the distances of the respective echo deflections along the trace.

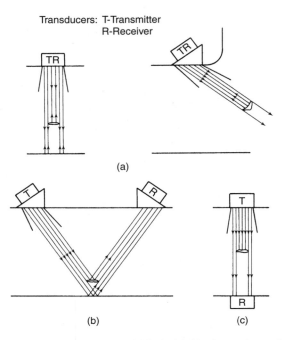

Transducers: T-Transmitter
R-Receiver

(a)

(b)   (c)

Figure 6.14 Ultrasonic testing (schematic). (a) Single probe reflection techniques, (b) two probe reflection technique, (c) two probe transmission technique

A typical trace is illustrated in Figure 6.15. A major defect is represented by a single large deflection on the trace, whilst scattered minor defects give rise to more general fuzziness or 'grass'. A severe defect can cause complete loss of the echo from the far surface.

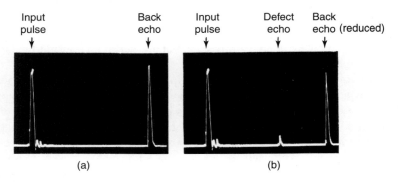

Input pulse   Back echo   Input pulse   Defect echo   Back echo (reduced)

(a)   (b)

Figure 6.15 Typical ultrasonic traces. (a) From sound metal section, (b) from section containing a single defect. (From Sully and Lavender[52]) (courtesy of Institute of British Foundrymen)

Although the amplitude of a particular deflection is directly related to the magnitude of the echo, it cannot be used as an accurate measure of defect size, since it is also influenced by other factors, for example the orientation and texture of the defect. Attenuation of the signal is also produced by structural features such as coarse microconstituents, segregates, grain boundaries and inclusions.

Quantitative assessment of defects must therefore be based on the use of comparison standards containing real or artificial defects such as holes drilled in metal blocks[51]. As in the case of other methods of non-destructive testing, the interpretation of defect signals is assisted by post mortem sectioning of castings containing serious flaws; correlation with radiography is also useful.

The sensitivity of ultrasonic flaw detection is extremely high, being at a maximum when using waves of the highest frequency. The sensitivity must, however, normally be set at such a level as to achieve discrimination between major defects and other less significant discontinuities. The sensitivity of the technique to grain boundary and segregation effects and to other structural features has, however, itself been exploited to predict the properties of cast alloys. Attenuation by such features as graphite flakes and coarse grain boundaries can be correlated with consequent variations in properties: the principle has been variously applied to cast iron and to certain steels[53].

The technique of testing normally involves systematic scanning for the initial detection of significant defects, followed by their more detailed investigation. Two basic types of probe are employed, either singly or in pairs. A longitudinal wave probe is used to transmit waves in a direction normal or at a high angle to the surface: this is unsuitable for examination of a zone within the divergent beam beneath the probe and gives particularly effective detection of defects extending in a plane parallel to the surface. For more inaccessible defects and for cracks and tears running in a direction normal to the surface, a transverse or shear wave probe can be used. This has an angled head with the transducer mounted on a plastic wedge, the waves being refracted into the casting obliquely to the surface. Such probes can also be used to explore the shape and extent of defects detected by longitudinal wave probes.

The technique is particularly sensitive to surface roughness. To avoid the reflection of a large proportion of the waves at the probe-casting interface, it is imperative to obtain good contact and avoid an air gap. This is assisted by smooth casting surfaces and if necessary by local grinding, but coupling is further assured by maintaining an oil film between probe and casting. For curved surfaces, specially shaped probes can be used to achieve good contact. A radical approach to the problem of coupling is provided by immersion testing. The entire casting-probe assembly can be submerged in a water tank, the water readily transmitting most of the waves into the casting.

The principal attraction of ultrasonic testing lies in its low cost and high speed of operation. Rapid and systematic scanning of large areas is possible, the high sensitivity of the method giving a good assurance of defect detection. The principal difficulties are those of interpretation of the oscillograph trace. In castings other than those of simple geometric form, a complex pattern of echoes is generated by reflection from the numerous surfaces, and many other factors combine to require considerable skill and experience on the part of the operator. This form of inspection is, therefore, most readily adapted to castings of relatively simple shape and uniform section and to castings for which long production runs enable the technique to be fully developed.

As with previous techniques, ultrasonic testing is included in the earlier general references, and in a specialized standard BS EN 583[50]. Many observations in relation to castings were contained in contributions by Lavender and various co-authors, particularly References 53 and 54.

Other methods of non-destructive testing

The foregoing sections describe the most generally applicable and widely used techniques for the non-destructive testing of castings. Additional methods suitable for more limited use are treated in the general literature as previously detailed. These include a group of electromagnetic methods in which detector coils are employed to register dissimilarity between a specimen and a standard. This principle is applied in various types of sorting bridge, where any electromagnetic condition departing from standard causes loss of balance in a circuit. The principle can be used for flaw detection or as a means of sensing dimensional errors. In the case of small investment castings an encircling coil can be used, whilst in other cases a probe coil can be placed close to the zone being inspected: a defect is then revealed by a local change in electromagnetic characteristics.

Proof loading and pressure testing

Certain inspection procedures are designed to subject a casting to conditions resembling those which will be encountered in service. Castings may be mechanically loaded in special fixtures to simulate service stresses, but much more common is the pressure testing carried out to verify hydraulic soundness in castings required to contain fluids under pressure.

Equipment for pressure testing is designed to seal the apertures in valves, pipe sections and similar castings for the application of hydraulic or pneumatic pressure. In the latter case the casting is submerged in a water tank or is surface treated with detergent solution to induce visible bubbles at leaks; great care is required in case of explosive failure. Pressure testing of castings destined for machining is of limited value if carried out on the black castings, since further porosity may be exposed in later machining, for example in the drilling of flanges for boltholes. The test is nevertheless useful in proving general wall soundness before machining expenses are incurred.

Quality and process control in the foundry

As emphasized at the beginning of the present chapter, casting inspection can provide the basis of a quality control system in which feedback of information is used to improve quality and to detect significant changes in the manufacturing process at an early stage. Similar objectives are served by systematic observation of process variables from raw material to final product.

The operation of a quality control system simplifies the task of meeting specifications and can reduce the amount of inspection required, whilst the presentation of data in the form of visual control charts reveals trends not readily evident from examination of the data itself. The use of such systems focuses attention on quality and provides a platform for cooperation between inspection and production departments.

Quality control is concerned with variation of a process or product characteristic about some hypothetical standard state. In assessing the significance of deviations from this state, distinction must be drawn between variations due to chance and those resulting from definite changes in production conditions and requiring corrective action. Control charts are provided with limits designed to show when these conditions are reached, the limits being established by statistical methods which take into account the natural variations arising in process and product. These methods also give guidance in the use of sampling techniques to minimize the task of measurement. This statistical process control, or SPC, has become a vital component in modern management systems.

A limited amount of information concerning variation is given by simple classification into categories, for example acceptance or rejection on inspection. Much more information can be derived, however, from measurement of some property which can be represented by a quantitative index, for example a linear dimension, temperature, composition or mechanical property. Collection of such data shows the full nature of the variation and enables trends to be detected before a critical condition is reached.

The spread of measurements obtained when a large number of minor causes combine to produce variations about some peak value is often characterized by the normal or Gaussian frequency distribution curve of the type illustrated in Figure 6.16. Such a distribution can be wholly defined by a measure of magnitude, for example the mean value $(\overline{X})$, and by a measure of the variation or degree of scatter about the mean. The latter measure can be the range between maximum and minimum values (w), the standard deviation (σ) or the coefficient of variation $(v = \frac{\sigma}{\overline{X}} \times 100\%)$.

In a Gaussian distribution, the proportion of values lying within various sections of the scale of measurement bears a known and constant relation to

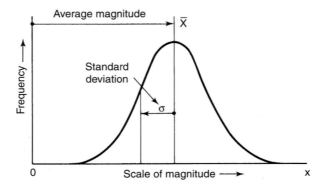

Figure 6.16 Normal frequency distribution of measurements

the standard deviation σ, data for various multiples of σ being ascertainable from published tables. Once the particular Gaussian distribution characteristic of the process is known, therefore, this provides the means of finding whether any further set of readings conforms to the distribution and hence whether the process is operating under control.

The particular Gaussian distribution obtained under stable conditions illustrates the inherent variability of the process, or 'process capability'. Since it represents the normal or chance variation when the process is operating satisfactorily, it should govern the choice of the process in relation to specification requirements: it is necessary to know not only that a casting can be made, but that it can be made with a high probability of success.

Control charts are designed to facilitate continuous observation of the magnitude and spread of measured values and to detect changes in either respect indicating that production is deviating from its characteristic stable state. In a typical control chart, measurements of the selected parameter are plotted on a vertical scale and the horizontal scale represents the successive individuals or samples examined. Horizontal lines are drawn to represent the mean and the upper and lower control limits. To fix the proportion of points which should fall within the limits, these are determined with reference to σ, the standard deviation when the process is operating under control. Separate sets of outer and inner limits may be employed, the former indicating the need for action and the latter to give warning of possible adverse changes. Figure 6.17 illustrates a control chart of this type.

The steps in the construction of such a chart will now be considered in more detail. Individual measurements of the selected property x are first made over an appreciable period whilst the process is considered to be operating under perfectly stable conditions. The overall average $\overline{X}$ and the standard deviation σ can then be determined. The latter value is the root

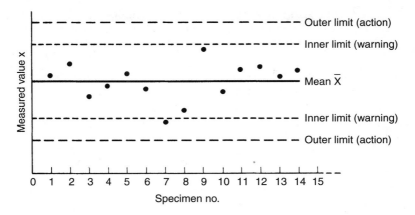

Figure 6.17 Control chart for a process or product characteristic

mean square of the individual deviations from the mean,

$$\sigma = \left[\frac{x_1 - \overline{X})^2 + (x_2 - \overline{X})^2 + (x_n - \overline{X})^2}{n} \right]^{1/2}$$

where n is the number of measurements involved. The values for $\overline{X}$ and σ would be virtually unchanged if measured over further periods of stable operation.

The spacing of the control limits can then be determined as a multiple of σ according to the degree of stringency with which it is intended the control should operate. Table 6.4 gives the proportions of values falling outside limits represented by various multiples of σ, assuming that measurements of every individual (x) continue to be made and entered on the chart. In this case it can be seen that, whilst the process remains under control, some 99.7% of values would be expected to fall within limits represented by $\overline{X} \pm 3\sigma$ and 95.5% within limits represented by $\overline{X} \pm 2\sigma$. These or closely similar proportions are widely used as action and warning limits, respectively.

In practice, however, control is usually based on the examination of selected samples and it is the averages of these samples ($\overline{x}$) and not individual measurements that are to be plotted on the control chart. The control limits for the values of the averages need to be set more closely than in the previous case to allow for the presence of more widely deviating individual values within the samples. To maintain the same degree of control, therefore, the limits are in this case set at an equivalent multiple of $\sigma/\sqrt{n}$ where n is the number of measurements in the sample. Table 6.5 gives control limit data, based on this criterion, for samples of various sizes. Samples of from 4 to 10 are recommended for normal purposes.

As the average for each succeeding sample then becomes available it is plotted on the chart and the occurrence of a point outside the outer limits indicates loss of control. Although one point in twenty is expected to fall

Table 6.4 Gaussian frequency distribution. Proportions of measurements falling outside and within limits determined by various multiples of the standard deviation σ

Multiple of σ (t)	Percentage of measurements below $\overline{X}$ - tσ or above $\overline{X}$ + tσ	Percentage of measurements within limits $\overline{X} \pm$ tσ
0	50	0
0.5	30.9	38.3
1.0	15.9	68.3
1.5	6.7	86.6
2.0	2.3	95.5
3.0	0.13	99.7
1.28	10	80
1.64	5	90
1.96	2.5	95
2.33	1.0	98
3.09	0.1	99.8

Data in second half of table is reproduced from BS 2564:1955 by permission of BSI under licence number 2000 SK/0049. The complete standard can be obtained by post from BSI Customer Services, 389 Chiswick High Road, London W4 4AL.

outside the inner or warning limits, a greater frequency or close grouping of such points suggests the development of a bias in the process which may need correction.

Although the above description referred to the calculation of limits from the standard deviation, it is also possible, using modified conversion factors, to derive these limits directly from the average range ($\overline{w}$) of the samples, thereby eliminating the calculation of σ. These alternative conversion factors are shown in the same table.

In addition to the average magnitude, it is desirable to maintain continuous observation of the spread of results within the samples. This can be accomplished with the aid of a separate control chart for variability, most conveniently expressed as the range (w) between maximum and minimum readings within each sample. Control limits are again established at suitable multiples of the average range (Table 6.6), but only the upper limits are in this case required for practical purposes. A combined average and range chart is illustrated in Figure 6.18.

Control charts for average and range are particularly applicable to the control of metal composition by analysis, to sand testing and to the dimensional inspection of castings where precision is of maximum importance. In some cases, however, a control chart has to be based not on measurement but on classification or counting: such an approach can be used, for example, in the inspection of repetition castings for defects, using visual examination

Table 6.5 Control chart limits for sample average $\bar{x}$. (a) in relation to standard deviation σ, (b) in relation to average range $\bar{w}$

Number in Sample	(a) Values of factor A for limits of $\bar{X} \pm A\sigma$		(b) Values of factor A' for limits of $\bar{X} \pm A'\bar{w}$	
	Outer limit factors (99.8% confidence) $A_{0.001}$	Inner limit factors (95% confidence) $A_{0.025}$	Outer limit factors (99.8% confidence) $A'_{0.001}$	Inner limit factors (95% confidence) $A'_{0.025}$
(1)	(3.09)	(1.96)	(–)	(–)
2	2.19	1.39	1.94	1.23
3	1.78	1.13	1.05	0.67
4	1.55	0.98	0.75	0.48
5	1.38	0.88	0.59	0.38
6	1.26	0.80	0.50	0.32
7	1.17	0.74	0.43	0.27
8	1.09	0.69	0.38	0.24
9	1.03	0.65	0.35	0.22
10	0.98	0.62	0.32	0.20

Values in section (b) are reproduced from BS 2564: 1955 by permission of BSI under licence number 2000 SK/0049. The complete standard can be obtained by post from BSI Customer Services, 389 Chiswick High Road, London W4 4AL.

or non-destructive testing, or in the routine checking of a single dimension using a gauge or template. This type of system necessarily provides less information than that based on measurement, since quality variations within permissible limits are not recorded: there is consequently no opportunity for the observation of trends until the arbitrary quality level used as the inspection standard is crossed.

In a typical control chart using this principle, the numbers of defective castings in successive samples are again plotted against a time scale and observed in relation to suitable limits. Whilst the process is operating under stable conditions, the number (m) of defectives will vary from one sample to another but the average value ($\bar{m}$), determined from numerous samples taken over a long period, will remain constant and representative of the whole bulk of production. The frequency distribution of values of m in the samples is not Gaussian or symmetrical about the average $\bar{m}$. Control limits for a given value of this average can, however, be selected from published data based on the probability of occurrence of different values of m in a series of samples[55] (see Figure 6.19).

The procedure in this case is to select a sample size such that one to two defective castings would be expected in each sample when the process is

Table 6.6 Control chart limits for sample range _w_

| | Values of factor D' for limits at $D'\overline{w}$ | | | |
| | Upper limit factors | | Lower limit factors | |
Number in sample	Outer (99.8% confidence) $D'_{0.999}$	Inner (95% confidence) $D'_{0.975}$	Outer (98.8% confidence) $D'_{0.001}$	Inner (95% confidence) $D'_{0.025}$
2	4.12	2.81	0	0.04
3	2.98	2.17	0.04	0.18
4	2.57	1.93	0.10	0.29
5	2.34	1.81	0.16	0.37
6	2.21	1.72	0.21	0.42
7	2.11	1.66	0.26	0.46
8	2.04	1.62	0.29	0.50
9	1.99	1.58	0.32	0.52
10	1.93	0.56	0.35	0.54

Data for sample sizes up to 6 are reproduced from BS 2564: 1955 by permission of BSI under licence number 2000 SK/0049. The complete standard can be obtained by post from BSI Customer Services, 389 Chiswick High Road, London W4 4AL.

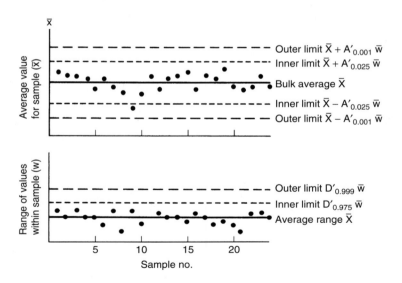

Figure 6.18 Control chart for sample average and range

operating at an acceptable level of control. Some 15–20 samples are then taken at regular intervals and the average number of defectives $\overline{m}$ calculated. Control limits for _m_ are next determined in accordance with some suitable probability level for points occurring outside, for example 5% and 0.2%,

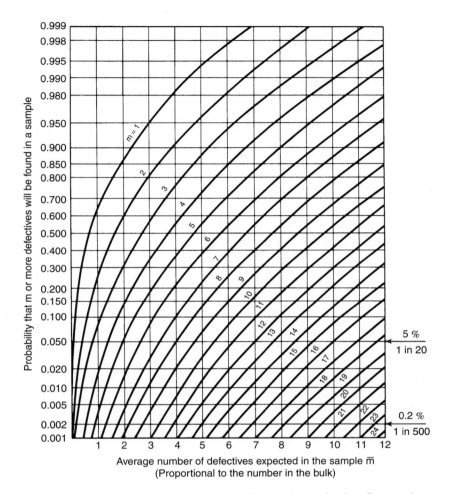

Figure 6.19 Selection of control chart limits for number defective. Curves show probabilities of occurrence of each number of defectives m for various given values of $\overline{m}$ (This extract from BS 2564:1955 is reproduced with the permission of BSI under licence number 2000 SK/0094. The complete standard can be obtained by post from BSI Customer Services, 389 Chiswick High Road, London W4 4AL)

and a similar procedure of plotting and observation of sample counts is subsequently followed.

If a point falls outside the inner limit, a further sample can be taken and the two results compounded and considered against a new limit appropriate to the enlarged sample and the doubled value of $\overline{m}$ which follows. This principle of cumulative sampling is particularly suitable for cases in which the average number of defectives in a practicable sample would be much less than unity, and provides a means of reducing the proportion of castings undergoing examination to the minimum needed to obtain a positive result.

The general form of a control chart based wholly on sequential sampling is shown in Figure 6.20. It is seen that in this case the control limits are no longer represented by horizontal lines, but by gradients which take into account the progressively increasing number of defectives to be expected as the sample accumulates. The sampling sequence is maintained until a point falls either above or below the limits, at which stage the batch of castings can be either accepted or rejected without further testing. The application of simple and sequential sampling schemes to quality control is described more fully in the literature, with consideration of the implications of sampling in terms of the risks to which producer and user are exposed[55-57].

Examples of applications of quality and process control systems to foundry problems, including additional techniques to those described above, were examined in a review by Fuller[56]; classical general treatments of quality control were contained in References 58 and 59, the latter a comprehensive and detailed text. A major review of modern concepts in this field is the work by Oakland[60].

Typical uses of these techniques in the foundry concern the control of incoming materials such as melting stock, base sands and binders, using chemical and physical testing. Coordination with suppliers is an essential adjunct to such controls: modern quality management standards envisage full quality assurance procedures throughout the supply chain. Process control in manufacture embraces such aspects as sand testing, whilst metal temperature and composition, including gas content, are further characteristics suitable for continuous observation and chart control. Pioneering work by Davies in the former field was referred to previously in Chapter 4.

Product inspection has been discussed earlier in the present chapter. Although dimensional examination, mechanical properties and the incidence

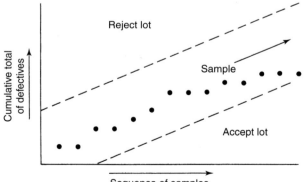

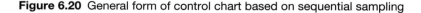

Figure 6.20 General form of control chart based on sequential sampling

of defects are the most obvious points of interest, metallurgical characteristics such as grain size, dendrite arm spacing and non-metallic inclusion
counts are equally relevant targets for such controls.

Apart from the use of the techniques in the pursuit of quality, control
charts based on similar principles may be used to observe management
ratios designed to achieve improved productivity and economy of operations. Examples include facing sand, fuel and power, or refractory consumption per tonne of castings, metal losses, casting yield, grinding wheel
consumption, and production per man-hour. Charts of performance in these
respects can, if interpreted with care, provide a useful guide to production
staff as to the efficiency of plant operations.

References

1 Quilter, G. W., *Br. Foundrym.*, **62,** 319 (1969)
2 BS 5750: Quality systems, Br. Stand. Instn., London
3 ISO 9000: Quality management and quality assurance standards. Int. Stand.
   Org.
4 White J. N., *Metals and Materials*, **4**(3), 165 (1988)
5 Wooton R. and Knight D. F., *Foundrym.*, **82,** 527 (1989)
6 Second Report of Working group T80-Quality Control and Quality Assurance.
   Foundrym., **83,** 210 (1990)
7 Third Report of Working Group T80-Quality Control and Quality Assurance.
   Foundrym., **87,** 94 (1994)
8 Rothery B., *ISO 9000*, 2nd Ed. Gower (1993)
9 BS 4778: Quality vocabulary. Br. Stand. Instn., London
10 BS ISO 8402: Quality management and quality assurance vocabulary. Br. Stand.
   Instn., London
11 Scrimshire D., *Foundrym.*, **87,** 137 (1994)
12 Scrimshire D., *Foundry International*, **19**(4), 146 (1996)
13 Hoyle D. *ISO 9000 Quality Systems Handbook*. 3rd Edn., Butterworth-Heinemann, Oxford (1998)
14 BSI Standards Catalogue, British Standards Institution, London (Annual)
15 ASTM Annual Book of Standards, American Society for Testing and materials,
   W. Conshohocken, Pa (Section 00 Index)
16 Villner, L. and Jansson, L., *Cast Metals Res. J.*, **1**(2), 32 (1965)
17 BS 1134 Assessment of surface texture, Br. Stand. Instn., London
18 Swing, E., *Trans. Am. Fndrym. Soc.*, **71,** 454 (1963)
19 Everest, A. B., *Br. Foundrym.*, **57,** 273 (1964)
20 Vogel A., *Metals and Materials*, **5**(1), 33 (1989)
21 Flemings M. C., in Vol 15, Materials Science and Technology Ed. R. W. Cahn,
   VCH Publishers Inc. New York (1991)
22 BS EN 10002. Tensile testing of metallic materials, Br. Stand. Instn., London
23 Armstrong, R. W., *J. Mech. Phys. Solids.*, **9,** 196 (1961)
24 BS EN 10003. Metallic materials. Brinell hardness test, Br. Stand. Instn.,
   London
25 BS EN 10109. Metallic materials. Hardness test, Br. Stand. Instn., London

26 BS EN ISO 6507. Metallic materials. Vickers hardness test, Br. Stand. Instn., London
27 BS 10045. Charpy impact test on metallic materials. Br. Stand. Instn., London
28 BS 131. Notched bar tests, Br. Stand. Instn., London
29 Gilbert, G. N. J., *Br. Foundrym.*, **54,** 5 (1961)
30 Angus H. T. *Cast Iron: Physical and Engineering Properties*, Butterworths, London (1976)
31 Laird G. II, Dogan O. and Bailey J. R., *Trans. Am. Fndrym. Soc.*, **103,** 165 (1995)
32 BS 1639. Methods for bend testing of metals, Br. Stand. Instn., London
33 BS 3683. Glossary of terms used in non-destructive testing. Br. Stand. Instn., London
34 BS EN 1330. Non-destructive testing. Terminology, Br. Stand. Instn., London
35 McMaster, R. C., *Non-Destructive Testing Handbook, Vols. 1 and 2*, Ronald Press, New York (1963)
36 ASTM Annual Book of Standards. Section 3 Metals Test Methods and Analytical Procedure: Vols 03–01 Metals – Mechanical Testing; 03–03 Nondestructive Testing (1998)
37 ASM Metals Handbook 9th Edn. Vol 17. Nondestructive Evaluation and Quality Control, ASM International, Metals Park, Ohio (1989)
38 Nondestructive Testing Handbook 2nd Edn. Vols 1–6 Amer. Soc. for Nondestructive Testing and ASM (1982–1989)
39 Hull, B. and John, V., *Non-destructive Testing*. Macmillan Education, Basingstoke (1988)
40 Halmshaw, R., *Non-destructive Testing*. Edward Arnold, London (1987)
41 BS EN 1371. Founding. Liquid penetrant inspection, Br. Stand. Instn. London
42 Starr, C., *Br. Foundrym.*, **58,** 143 (1965)
43 BS EN 1369. Founding. Magnetic particle inspection. Br. Stand. Instn., London
44 BS EN 444. Non-destructive testing. General principles for radiographic examination of metallic materials by X- and gamma-rays. Br. Stand. Instn., London
45 *Recommended Procedure for the Radiographic Flaw Detection of Steel Castings*, B.S.C.R.A., Sheffield (1967)
46 Schlieper F. *Trans. Am. Fndrym. Soc.*, **93,** 787 (1985)
47 Bossi R. H. and Georgeson G. E., *Trans. Am. Fndrym. Soc.*, **101,** 181 (1993)
48 BS EN 462. Non-destructive testing. Image quality of radiographs, Br. Stand. Instn., London
49 BS 2737. *Terminology of internal defects in castings as revealed by radiography*, Br. Stand. Instn., London
50 BS EN 583. Non-destructive testing. Ultrasonic examination. Br. Stand. Inst. London
51 BS 12223. Ultrasonic examination. Specification for calibration block No. 1 and BS EN 27963. Calibration block No. 2
52 Sully, A. H. and Lavender, J. D., *Br. Foundrym.*, **54,** 293 (1961)
53 Lavender, J. D. and Fuller, A. G., *Br. Foundrym.*, **58,** 54 (1965)
54 Abrahams, G. J. and Lavender, J. D., *Br. Foundrym.*, **58,** 66 (1965)
55 Dudding, B. P. and Jennett, W. J., *Control chart techniques*, BS 2564, Br. Stand. Instn, London
56 Fuller, A. G., *Br. Foundrym.*, **58,** 25 (1965)

57 BS 2635. Drafting specifications based on limiting the number of defectives permitted in small samples, Br. Stand. Instn., London

58 Huitson, A. and Keen, J., *Essentials of Quality Control*, Heinemann, London (1965)

59 Grant, E. L., *Statistical Quality Control*, 3rd Edn., McGraw-Hill, New York (1964)

60 Oakland J. S. *Statistical Process Control*, 4th Edn., Butterworth-Heinemann, Oxford (1999)

7

Casting design

The purpose of design is to achieve functional performance of structures at minimum cost in materials and manufacture. Both these factors are involved in the initial choice of casting as the means of production; the individual design is itself usually a compromise between a purely engineering concept based on service requirements and a shape lending itself to high quality, low cost production. The relative importance accorded to the two factors of function and cost naturally depends upon the class of product and the inspection standards to be applied, but both are always present to some degree. The concept of value analysis relates to the systematic weighting of all the factors in deciding between alternative processes and materials.

In the captive foundry attached to a large engineering organisation there is opportunity for a unified approach to casting design. More commonly, however, the design engineer is associated with the manufacturer of an end product and not with the foundry. Being concerned with a wide range of products he is seldom wholly conversant with all the features of the casting process. Many castings are nevertheless designed without foundry advice. This may even extend to the independent manufacture of pattern equipment, in which case, although a general knowledge of moulding techniques is available, no account can be taken of the particular facilities of the producer or of the casting method to be adopted. The role of the foundry in design is in these cases restricted to requests for modifications at a stage when shortage of time or reluctance to inconvenience a customer may preclude major alterations; the tendency is thus for the foundry to live with continuous difficulties which might readily have been eliminated at an earlier stage. Particular difficulties arise when a design originally intended for production by machining or weld fabrication is translated to the foundry without major modification. Shape, dimensional tolerances and material specification may then be wholly unsuitable.

Examining the process by which a casting comes to be used, five separate stages can be distinguished:

Preliminary stages:
$\begin{cases} \text{1. the decision to use a casting,} \\ \text{2. detailed design,} \\ \text{3. commercial contact: enquiries, quotations, order;} \end{cases}$

Production stages:
$\begin{cases} \text{4. manufacture of tooling.} \\ \text{5. founding and finishing.} \end{cases}$

If the function of the foundry is confined to the production stages, technical contact tends to be restricted to the investigation of complaints. Although publications are available to give guidance in the principles of casting design[1-7], the ideal is to achieve close contact between designer and producer during the preliminaries, so that the specific knowledge of the foundry industry about its own products can be used from the beginning. This enables comments to be made on draft designs, whilst pattern construction can be arranged to suit the proposed method of production: both these measures can be reflected in lower prices. General contacts on the commercial plane are no substitute for this direct technical cooperation.

CADCAM and rapid prototyping

The advantages of casting from the standpoint of design flexibility were stressed in the Introduction, and this situation has been reinforced by the development of rapid prototyping techniques. These are themselves made feasible by CAD facilities. The role of CADCAM in component design, the manufacture of tooling and the casting process itself is becoming more widely established as in other areas of engineering. Shape design data can be developed and stored in software suitable for further processing, and for transfer between supplier and customer.

CAD enables the 3D geometry of a component to be visualized and manipulated, the role of shape changes explored and the effects of loadings determined, with other forms of analysis and simulation, including behaviour in the casting process itself. Shapes can thus be optimized for both function and manufacture, including machining operations.

The stages from initial selection, through toolmaking to the production of finished castings, are often extended by the need for sample or prototype castings for proof machining and other forms of assessment; design changes may also be required. Lead times can be greatly lengthened by these activities. In some cases prototype tooling may be needed, where a production run rather than a single casting is called for.

CAD can itself reduce the need for prototyping by virtue of its simulation capabilities, but prototypes will continue to be required for quality appraisal and testing of a representative product.

Modern rapid prototyping techniques have greatly shortened the lead times involved. They are mainly based on the principle of layer manufacturing, in which solid models in plastics or other materials are generated directly from 3D CAD representations.

In stereolithography (SLA), to use one example, a horizontal platform is submerged in a bath containing a light-sensitive liquid polymer. The platform is raised through the bath until just covered by a thin liquid film, at which stage a laser beam controlled by the CAD output systematically scans the area representing an initial thin slice of the required 3D form. This cures the polymer film within the area covered, after which the platform is lowered by the thickness of another slice, and a new film of liquid is formed on the initial solidified layer. This is similarly cured, the laser beam again following a path controlled by the individual CAD output for that slice. The same layer sequence is repeated until an entire solid model can be separated from the surplus liquid. The build process is normally completed within a few hours.

Selective Laser Sintering (SLS) operates on a similar principle, but employs a polymer powder rather than a liquid. The successive layers are sintered by the scanning laser beam, the growing solid form being supported by the remaining loose powder bed. In Laminated Object Manufacture (LOM) the solid model is formed from laser cut sheets of paper, held together by adhesives. The above and other techniques are more fully described, and their capabilities characterized, in References 8 and 9.

The layer thicknesses vary with the process but are usually within the range 0.025–0.75 mm. The models produced can have minutely stepped surfaces associated with the incremental layer system, but these can be abraded or filled to provide a smooth finish. Complex shapes are feasible, with accuracies generally in the range 0.1–0.5 mm.

Originally aimed at the production of 3D conceptual models, the techniques soon became directly applied to full functional prototypes. For this purpose models generated from the 3D CAD data can be employed as patterns and coreboxes for casting by the normal route: both positive and negative formats are feasible. In the case of investment castings the plastic positive can be employed as an expendable pattern for the ceramic shell production of a one-off prototype casting, or as a master pattern to produce others. In a further development wax itself is used for the expendable pattern, which is built up using a machine operating on the principle of an inkjet printer, but spraying molten wax on the 3D layer principle. Figure 7.1 shows such a machine with a superimposed CAD design file and casting, and containing a similarly shaped wax pattern, just completed.

The successful use of several of these techniques to produce advanced and complex cast products was demonstrated in a major project described by Cox in Reference 10, in which important aspects of the 3D CAD modelling systems were also examined.

In yet a further development, the Direct Shell Sand (DSSRP) or Direct Croning Laser Sintering Process (DCLSP),[11,12] selective laser sintering is employed in conjunction with resin coated silica or zircon sand to shape both moulds and cores, i.e. the negative forms, directly from the positive CAD file, these being poured to produce prototype castings in any required alloy. The particular advantage claimed for this route is its emulation of the actual production process to be used, giving truer representation for a functional prototype. Other binders can also be employed, for example inorganic chemical types.

Other prototype routes based on the availability of CAD files employ direct linkages with CNC machining facilities to produce expendable

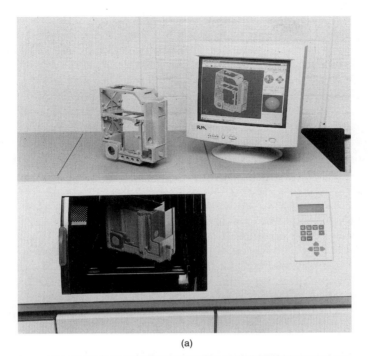

(a)

(b)

Figure 7.1 System for direct production of expendable wax investment casting patterns from CAD file, using wax spray layer deposition: (a) CAD file and casting, with similar wax pattern just completed, (b) other wax patterns produced by the same technique (courtesy of PI Castings Ltd.)

patterns, moulds or the cast alloy prototypes themselves, in each case by-passing the normal requirement for tooling provision. In the first of these categories, polystyrene patterns can be machined from solid blocks for use with lost foam and Replicast® processes; wax too can be shaped by cutting, although it remains difficult to achieve both machinability and investment casting qualities in the same wax. Mould and core production based on chemically bonded sands can be similarly achieved by direct machining from pre-hardened blocks of the material, again using the appropriate CAD data. This technique has been actively pursued by the Castings Development Centre for prototype and one-off production items. Lastly, cast alloy prototypes can be similarly machined from precast solid blocks or multi-purpose blanks; although these do not provide precise representation of the shape, structure and properties of the individual casting they may suffice for more limited purposes.

The subject thus presents a complex picture of intensive development of aids to design, prototyping and small quantity production. Bureau services are particularly useful for access to some of the more specialized techniques by smaller companies or others without dedicated facilities: such services can also extend to the full range of computerized design functions relating to cast products.

Where the possible use of castings is involved the work of the designer involves two main steps. Initially, the general characteristics of castings and of the foundry process need to be considered in relation to the whole range of metallurgical products, culminating in a decision based on technical and economic comparisons. The second and main stage of design involves determination of the overall shape and detailed features of the casting.

Initial considerations in design

Engineering structures may be produced from metal in the cast, wrought or sintered condition and the final shape may be developed by many different combinations of manufacturing process. Although the nature of the shaping process exercises a characteristic influence upon metallurgical structure and properties, the cost factor often predominates in process selection, since the engineering requirements can usually be attained by a number of alternative routes, provided that design measures are matched to the properties actually available. The comparative costs of materials and processes must, however, be established in relation not only to the technical requirements but to production factors. The most important of these is the quantity requirement as it affects production and tooling costs. This aspect will be further examined in the following section and in Chapter 9. The results of any such comparison will remain valid only for the particular circumstances considered.

The principal methods of shaping metals were briefly reviewed in the Introduction. Figure 7.2 gives a general picture of the basic operations and alternative paths to the production of a finished engineering structure. Each shaping operation offers compelling advantages within a limited field, but direct competition exists over a wide range of products. In certain cases the best results can be obtained by using composite construction, in which the particular advantages of separate processes can be combined to achieve the most effective overall design. Cast-welded assemblies offer a good example of complementary techniques and examples will be reviewed later in the chapter.

The distinctive characteristics of the foundry process and its products will now be summarised, after which casting design will be considered in more detail. Some of these characteristics are of importance primarily through their influence upon the economics of production, whilst others are mainly

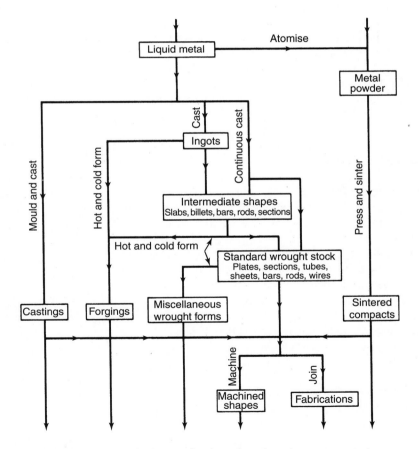

Figure 7.2 Major routes in the production of engineering components

of technical significance. In generalising, however, it must be remembered that wide variations exist within the diverse range of casting processes now available. Comparisons and contrasts between individual processes will be made elsewhere.

Economic characteristics

Economic considerations in the selection of castings can be summarised under two main headings:

1. *Tooling cost.* Pattern equipment can be cheap compared with form tools, giving a clear cost advantage for short production runs. Where quantity production is required, on the other hand, the cost of more sophisticated pattern equipment can be amortised over more units to reduce the cost of moulding, for example by the use of multiple impression moulds: similar considerations apply to dies.
2. *Process costs and flexibility.* Casting, being a direct process, is basically inexpensive to operate, although adaptable to widely different scales of capital investment. Not only does moulding offer a low cost route to elaborate form and intricate feature, but changes in design can be readily accommodated. This flexibility is reduced as pattern equipment becomes more permanent and complex in response to demands for large output from a single design, or for castings of exceptional quality.

It is important to note that the overall cost of castings varies widely with factors other than basic material cost, so that selling prices based solely upon casting weight can be fundamentally unsound and may either attract unremunerative orders or discourage the use of castings. This can be appreciated from Table 7.1, in which some of the main cost items are classified. The true cost depends not only upon weight of metal but upon the repercussions of the individual design, quality and quantity requirements upon manufacturing technique and costs. Realistic selling prices, and comparisons between casting processes and with other types of product, should thus be based upon the individual circumstances of each design of component.

Of the items in the table, pattern costs are of greatest significance where the quantity requirement for castings is small, but shrink to minor proportions under mass production conditions.

The foundry costs for an individual casting are not simply related to the weight or volume of cast metal but rather to the complexity of the design. Whether the shape is compact or spidery, solid or hollow, determines the mould volume, number of joints and cores and thus the cost of moulding. Casting yield and fettling costs are similarly affected by shape factors.

The relative importance of metal cost increases as other costs decline with increasing production and its impact is naturally greatest in the case of

Table 7.1 Some factors influencing casting costs

Circumstance	Tooling costs	Major effects on product costs	
		Foundry operational costs	*Material and melting costs*
A. Design characteristic	1. Alloy type and composition			Metal (intrinsic cost, losses and metallic yield) Melting (thermal properties and energy consumption)
	2. Basic dimensions	Pattern (size)	Moulding (mould volume) Finishing (casting surface area)	Metal (weight) Melting (weight)
	3. Shape	Pattern (complexity)	Moulding (complexity) Finishing (casting surface area and complexity)	Melting (casting yield)
	4. Quality standards	Pattern (type, material and construction)	Moulding (method and materials) Finishing (fettling and inspection standards) Rejection rate	Metal (raw materials) Melting (compositional tolerances)
B. Quantity requirement		Pattern (number, type, material and construction)	Moulding (production method)	

alloys of high intrinsic value. In a comparative survey of materials compiled by Sharp[13], metal cost for mass produced castings ranged from approximately 15 per cent of the total cost for malleable cast irons and carbon steels to 80 per cent and even higher for certain bronzes and nickel base alloys: in one example of a one-off casting, metal cost amounted to only 0.5 per cent of the total.

Technical characteristics

Technical considerations include the suitability of the casting process for particular types of form, dimensional and surface characteristics, and the metallurgical characteristics of cast alloys.

General form

Throughout a wide range of simple and complex shapes and sizes casting offers the greatest freedom of design and the most universally practicable means of forming. For simple shapes casting is in competition with closed die forging processes but as designs become more elaborate the shaping freedom allowed by moulding becomes decisive. This freedom is greatest in the case of sand castings, where sections can be disposed in virtually any desired pattern in relation to the applied load, whilst the weight range extends from a few grammes to several hundred tonnes. Although such castings are not highly accurate and surfaces are imperfect, specific need for precision can be met by local machining allowances. The freedom of shaping also offers a special advantage for alloys with poor machinability, however, in that design features can be incorporated in the original cast shape: this applies to many hard, tough or brittle alloys used for wear, heat and corrosion resistance.

More design restrictions exist in die casting and other specialised techniques but intricacy and precision then compensate for the loss of flexibility: machining can be greatly reduced or eliminated.

It should however be stated that the use of castings becomes less advantageous as forms become greatly extended and light in section. Excessively high surface area to volume ratios and tenuous, spidery frameworks create increasing casting difficulties. Such designs are difficult to run because of their high rates of solidification; they are also prone to contraction stress defects such as tearing and distortion. These shapes are also less economical to produce by casting because of the large volume of mould in relation to the weight of saleable metal, and the large amount of coring which they often entail. Casting difficulties vary greatly with the alloy: grey cast iron, for example, retains its casting capabilities under extremely adverse shape conditions.

It is in this field that the advantages of fabricated construction emerge, since a variety of standard wrought products in the form of sections, tubing, plate and sheet offer ready means of forming the thinner and more extended members of the structure. The particular advantages of fabricated weldments include lightness of construction and the possibility of differential composition, enabling expensive alloys to be confined to regions where their properties are strictly necessary. Castings are more suited to mass production, however, and for many complex shapes unit construction as a single piece casting would be expected to reduce relative manufacturing cost.

Weld fabrication is naturally restricted to those alloys which can be welded without metallurgical complications; in other cases alternative methods of joining, for example rivetting and bolting, are required if a single casting cannot be employed.

Within the general field of fabrication there is scope for the use of castings to form local complexes or junction pieces; this and other forms of composite construction will be separately discussed.

Other forms for which alternative processes may be preferred on shape grounds include those sufficiently simple to be produced by drop forging, upsetting, coining and allied processes of deformation, but in many such cases castings offer a feasible alternative and the decision will be made on a purely economic basis.

Dimensional and surface characteristics

Dimensional accuracy is one of the critical attributes of a casting: it influences the choice of process and determines whether local or general machining will be required. In general castings are not made to close tolerances in the engineering sense, machining allowances being incorporated for accurate finishing, although such allowances can often be reduced or omitted when using precision casting techniques. Even where machining is to be carried out, an understanding of dimensional behaviour is needed to ensure adequate depths of cut. The machining of castings will be considered more fully in a subsequent section.

Every engineering dimension deviates from its nominal value, tolerances being used to cover those variations expected in the normal course of production; tolerances should therefore conform to the capabilities of the process. For castings, the attainable accuracy depends partly on the process and partly on the nature of each dimension. The main sources of error are:

1. pattern errors,
2. errors in mould dimensions,
3. anomalous contraction and distortion on cooling,
4. finishing.

More detailed analysis of these sources reveals two separate possibilities for departure from nominal dimensions. Errors may be either statistical or systematic in their occurrence.

Each dimension is subject to an inherent variation due to minor changes in process variables. This results in a statistical spread of measurements about a mean value, depicted by the normal type of frequency distribution curve shown in Figure 7.3a. Assuming that the mean can be made to coincide with the nominal dimension, the characteristics of the distribution curve determine the tolerances feasible during a production run. The form of this curve is dependent largely on the procedures used in moulding and coremaking.

In the second type of deviation the mean or peak value is displaced in one direction from the target value in the manner shown in Figure 7.3b. The main causes of such deviations are contraction uncertainty and errors in pattern dimensions. The initial contraction allowance is normally based on a standard value for the alloy and on previous experience with castings of similar design; the predicted value can however be subject to an appreciable percentage error from plastic deformation resulting from hindrance to contraction. This type of error is approximately proportional to the size of the dimension.

The two types of dimensional variation can be represented in an expression for casting tolerances:

$$T = \pm(aD + b)$$

where D is the dimension and a and b are constants for particular casting conditions. Some typical quoted tolerances for sand castings in accordance with this type of relationship are shown in Table 7.2.

In the above expression the term aD, representing the contraction uncertainty, can be of considerable magnitude. Anomalies in contraction behaviour can, on occasion, be equivalent to a substantial proportion of the total normal allowance: given an original allowance of 2 per cent, for example, a dimensional error representing 1 per cent of the dimension can be encountered. Although tolerances cannot be designed to cover these exceptional cases, it will be noted that the normal tolerance allowed in

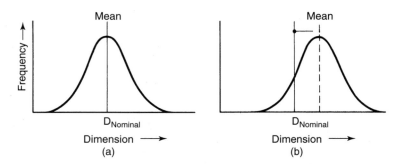

Figure 7.3 Nature of dimensional variation: (a) statistical spread of measurement, (b) systematic displacement from nominal dimension

Table 7.2 Dimensional tolerances for steel castings (after Refs 4 and 6, Courtesy of Castings Development Centre)

Greatest dimension of the casting	Less than 250 mm	250 to 1000 mm	Over 1000 mm
Normal	1.5+6D/1000	2.0+6D/1000	2.5+6D/1000
Reduced	1.0+5D/1000	1.5+5D/1000	2.0+5D/1000
Close	0.7+4D/1000	1.0+4D/1000	1.5+4D/1000

Typical tolerance values for special casting process (*q.v.* Chapter 9) can be expressed in similar form, e.g. $T = \pm0.13 + 5D/1000$ mm (investment castings); $T = \pm0.08 + 0.5D/1000$ mm (zinc base die castings, close tolerance, defined circumstances).

Table 7.2 provided for a possible error of 0.6 per cent from this cause, amounting to approximately 30 per cent of the normal contraction. On large dimensions the contraction uncertainty term accounts for the major part of the tolerance band.

Where castings are required in quantity, opportunity exists for progressive modification of the pattern equipment in the light of dimensional experience with the first castings; patterns may even be made slightly oversize with this adjustment in mind. Distinction must therefore be made between the initial accuracy obtainable on a single casting, in relation to the target dimensions, and the reproducibility of dimensions finally obtainable, given the rectification of initial systematic errors due to anomalous contraction.

The reproducibility of dimensions, mainly represented by the term b in the previously used expression and by the normal spread of the distribution curve (Figure 7.2a), is constant, for one particular process and dimensional circumstance and for one type of moulding equipment and procedure unless a specific error occurs (q.v. Chapter 5). The magnitude of this variation, and thus the ultimate accuracy attainable, is governed by the following factors:

1. Inaccuracies in the mutual location of pattern sections, mould parts and cores. The closest accuracy is achieved when a dimension is wholly dependent upon a single element of the pattern, contained within a single mould part, but falls away as the number of locating operations increases. Accuracy of alignment is determined by such factors as dowel, box pin and coreprint clearances; the precision casting processes are distinguished chiefly by their more accurately engineered location systems and by the smoothness and superior mutual fit of mould parts and cores.
2. Changes in mould shape. Discrepancies between mould cavity and pattern dimensions occur through temporary pattern distortion during ramming, pattern rapping, manual finishing, and mould sagging.

3. Changes in shape on casting. Errors result from swelling under metallostatic pressure and from irregular contraction, influenced by such variables as ramming density and pouring temperature.
4. Surface conditions. Dimensions are affected by initial roughness of the cast surface and by changes due to scaling and fettling.

Since these factors depend on the equipment and the individual circumstances of moulding and casting it is not possible to give general values for the dimensional variations to be expected. Systematic case studies of production castings have however been carried out and the effects of individual factors such as nature of pattern equipment, cores, contraction variation and parting lines determined[1,2]. As a rough guide dimensional tolerances are equivalent on average to half the contraction allowance, but the use of published recommendations for widely varying dimensions and circumstances can produce values as low as ±0.3 per cent D for a large dimension within a single element of a mould or exceeding 2 per cent D for small dimensions involving parting lines[1,2,4,6]. It can thus be seen that, whilst any formula or table of standard values must cater for a limited range of contingencies, the use either of a single percentage or of a single absolute value can never be more than a general guide to behaviour.

Villner[14] detailed the development of a unified system of tolerances for various alloys, based on the statistical analysis of 25 000 measurements in Swedish foundries. Typical general tolerances derived from this study are illustrated in Figure 7.4, which included analogous details of German specifications. The overall curves followed the general relation

$$T = k_n D^a$$

where D is the dimension, k_n a coefficient characteristic of the particular grade or stringency, and the exponent a a constant for the class of alloy ($a = 0.33$ for grey iron, steel and non-ferrous alloys; $a = 0.45$ for malleable and S.G. iron).

In a later contribution it was pointed out by Svensson and Villner[15] that tolerances designed to provide for both systematic and statistical errors were likely to be unnecessarily wide, and to be poorly related to the nominal size. This brings out the potential weakness of any general specification, namely that standard values cannot take account of all the circumstances contributing to the variation of every individual dimension.

Several specific mould and alloy circumstances were taken into account in other comprehensive studies of dimensional variation undertaken by the Institute of British Foundrymen, again based on the collection and analysis of data from industrial castings. Two reports dealt with dimensional tolerances in sand castings[16,17], for which regression equations were developed to enable the effects of prints, cores and similar factors to be allowed for. Further reports[18,19] adopted a similar approach for aluminium alloy gravity die and low-pressure die castings respectively. The significance of these extensive findings has since been further examined by Campbell[20] in the course of his wider considerations of casting accuracy, including problems of measurement and datums.

An important feature of rational systems of tolerances is the inclusion of grades or classes enabling different standards of stringency to be applied according to the circumstances of design and manufacture. Examples are seen in Figure 7.4: in the case of the German standards these were coupled with specific manufacturing techniques. Analogous grading is observed in Figure 7.5, which emerged from studies in the UK[21]. Further discussions, with international comparisons, are included in Reference 6. Other sources of typical tolerance values for varied casting processes and alloys are included in References 22 and 23.

The earlier international investigations of dimensional tolerances culminated in the introduction, in 1984, of the standard ISO 8062, followed by its British equivalent BS6615; these have been superseded by the 1994 and 1996 issues respectively[24,25]. The standards, which are identical, cover both dimensional tolerances and machining allowances.

Tolerances are placed in sixteen grades of stringency from CT1 to CT16, involving progressively increasing permissible limits. Values specified for each grade are provided for casting dimensions falling within sixteen

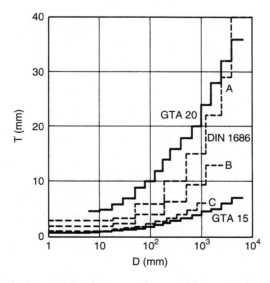

Figure 7.4 Typical general tolerances for grey iron, steel and non-ferrous castings. Tolerance width T as a function of basic dimension D (from Villner[14]) (courtesy of Institute of British Foundrymen); Swedish G.T.A. Grade 15 ($a = 0.33$, $K_n = 0.43$); Swedish G.T.A. Grade 20 ($a = 0.33$, $K_n = 2.2$); German D.I.N. 1686 A (hand moulded), B (machine moulded) and C (machine moulded in large numbers)

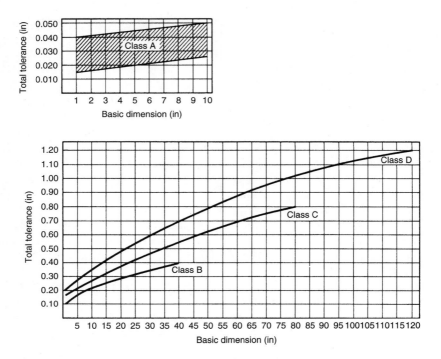

Figure 7.5 Recommended total tolerance values for basic dimensions of steel castings (from Reference 21). Typical fields of application: Precision Class A (range), shell moulded castings; Class B, metal patterns, large output; Class C, normal; Class D, simple equipment, short run production. Note: separate recommendations were made for subsidiary dimensions (see original Reference 21) (courtesy of CDC)

separate bands covering the range 0–10000 mm, the limits widening with increasing size of the dimension. Such aspects as mismatch across jointlines, draft taper and wall thickness are also positioned in relation to the tolerance bands for relevant dimensions.

Required machining allowances (RMA) are categorized on a similar principle, increasing over ten grades A–H, J and K and again catering for progressively greater allowances as the largest affected dimension increases over successive ranges to 10000 mm.

The tables of numerical tolerances and allowances corresponding to the various grades do not themselves relate to specific casting processes or circumstances, but provide the framework for agreement between customer and supplier in defining the formal requirements for a particular order. The second function of the standard is to offer process capability guidance, contained in informative annexes. These are based on industrial surveys and investigations of the kind previously referred to and indicate the grades normally expected from castings in designated categories.

The user is reminded of factors influencing accuracy, some of which have already been examined in the foregoing pages (see also Chapter 9). Apart from such aspects as type and condition of tooling, and complexity of design, important effects relate to production circumstances and to alloy type. An initial distinction is therefore made between long-series castings, where there is opportunity to adjust and optimize as previously discussed on page 373, and short-series castings, for which high-grade patterns may be unjustified and progressive modifications impracticable.

Separate tolerance capability tables are therefore given for each of the two categories, and each table provides normally expected grades for the main alloy types. Separate values are given for hand and machine moulded long series castings, and for clay and chemically bonded sands in the short series group.

The magnitudes of the tolerances associated with the full range of CT grades are indicated in the abbreviated Table 7.3a, whilst the typical influences of process conditions and alloy type on the normally expected grade designations are similarly summarized in Table 7.3b. Study of the standard itself is required for specific cases, but it is notable from the typical differences in normal gradings that the high mould density and pattern quality associated with machine moulding are the most potent factors determining casting accuracy. The chief distinction in the alloy field lies between the higher melting point alloys, represented by steel and cast irons, and the light alloys; most of the other alloy groups range over similar tolerance grades to steel, with copper in an intermediate position.

The overall curves follow the general relation

$$T = k_n D^a$$

The complexity of the dimensional problem emphasises the need to specify functional tolerances appropriate to each dimension and to avoid the use of unnecessarily close blanket tolerances covering all dimensions on the casting.

A further aspect of casting accuracy is *weight variation*: in special circumstances this may be of greater consequence than the dimensions. Although weight is affected by the degree of internal soundness, variations arise mainly from the influence of fluctuating dimensions upon volume: thus the percentage deviation from nominal weight is liable to exceed that for linear dimensions by a factor of up to 3 for small variations. The subject of weight tolerance is further discussed in Reference 6.

Associated with dimensional accuracy is *surface finish*, which also affects appearance, fatigue and corrosion resistance, and suitability for surface treatment. Surface characteristics are acquired from the original mould surface, metal flow behaviour, fettling and heat treatment.

Numerous surveys have been carried out on the surface roughness of castings, wide variations being encountered according to the process and

Table 7.3 Format of dimensional tolerance data presented in BS6615/ISO 8062. (a) Influence of casting dimension on tolerance; (b) Influence of process and alloy type on recommended grades

(a)

Basic casting dimension mm	Total casting tolerance mm					
	Tolerance grade CT					
	1	4	7	10	13	16
0–10	0.09	0.26	0.74	2	–	–
25–40	0.12	0.32	0.9	2.6	7	14
100–160	0.15	0.44	1.2	3.6	10	20
400–630	–	0.64	1.8	5	14	28
1600–2500	–	–	2.6	8	21	42
6300–10 000	–	–	–	11	32	64

Note: Special requirements apply in some cases eg. wall thickness dimensions, where the tolerances applied are one grade coarser than those tabulated. Other exceptions make reference to the full standard essential.

(b)

Production method		Tolerance grade CT casting material	
		*Ferrous	Light alloy
Long series	Sand cast hand moulded	11–14	9–12
	Sand cast machine moulded	8–12	7–9
Short series	Clay bonded hand moulded	13–15	11–13
	Chemically bonded hand moulded	12–14 (steel) 11–13 (cast irons)	10–12

Note: ferrous alloys – recommended values for steel and cast irons are identical, with the exception shown in the table.

These tables are adapted from BS6615:1996 with the permission of BSI under licence number 2000 SK/0049. The complete standard can be obtained by post from BSI customers services, 389, Chiswick High Road, London W4 4AL.

alloy[1,2]. Quoted values for sand castings are normally within the range 3–25 μm ($\approx$0.000 125–0.001 in), much finer finishes being attained in precision casting. Typical roughness ranges are illustrated in Figure 7.6. It should be emphasised, however, that the values quoted are for castings of normal finish. Regions with incipient metal penetration may show

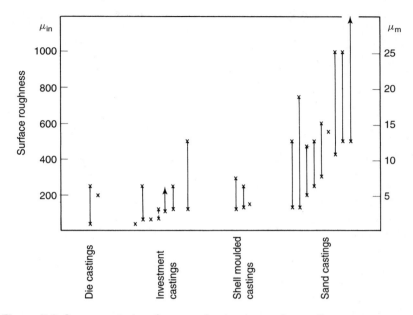

Figure 7.6 Some quoted surface roughness ranges for castings

irregularities as high as 0.5 mm (0.020 in) although surfaces as rough as this are usually dressed by grinding: in this case the final surface is governed by the fettling process. In other cases the final surface of an initially smooth casting may be determined by the abrasive used in shotblasting.

Metallurgical characteristics

Castings can be manufactured to virtually any required composition, although the best results are obtained by using alloys designed specifically round the requirements of the foundry process. Since many of the distinctive characteristics of cast metals are acquired during solidification, the most relevant comparison is with wrought materials in which the structures have been modified by mechanical deformation.

The influence of solidification on the structure of castings was considered in Chapter 2, where the effects of such variables as cooling rate on the nucleation and growth processes were examined. In some alloys, including many cast irons and non-ferrous alloys and some steels, solidification plays a leading role in establishing material characteristics. In other cases, for example the precipitation hardening light alloys, many high strength steels, and austempered ductile irons, the final structure and properties depend mainly upon phase changes in subsequent heat treatment, although the original structure can still exert a considerable influence. Thus the overall metallurgical situation relating to the use of cast materials might be represented thus:

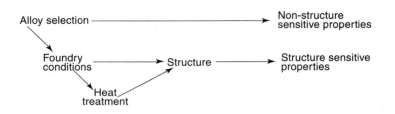

Structural characteristics affected by casting conditions include grain size and shape, substructure, solute and second phase distribution and the form and distribution of indigenous inclusions. Superimposed on the normal structure may be varying amounts of microporosity and segregation, although it is clear that comparison of cast with wrought materials should be based on sound products free from abnormal defects.

Structural comparisons generally show grain and substructures to be coarser in the cast material, especially in heavy sections, since the wrought materials have usually undergone recrystallisation in the solid state. This difference can persist after heat treatment, since the original structures influence the nucleation and growth conditions of the new crystals, giving finer grains and dispersions in the wrought material; voids are closed, giving high density and soundness. The original non-metallic inclusions and regions of alloy concentration also become deformed in working, so that the whole structure acquires a directional character associated with the forming process. By contrast, many cast structures are random in character, although directionality is seen in regions of columnar grain structure, as previously discussed in Chapter 2.

Certain structural features are associated particularly with cast materials; one of these is coring. Although this can be modified by homogenisation heat treatment, compositional differences are more effectively diminished where mechanical and thermal treatments are combined. Other structures frequently found in cast alloys are the eutectics, which are associated with particularly favourable foundry characteristics and which are present in many alloys produced by casting alone, for example the cast irons and aluminium-silicon alloys. These examples emphasise the greater compositional freedom in cast materials in the absence of the need for plasticity. Direct comparisons with wrought material are thus only relevant to certain groups of alloys, such as steels and light alloys, in which both castings and forgings are widely employed; the cast irons offer no real parallel since the cast structures are unique.

Arising from the above structural characteristics, certain general observations can be made concerning the properties of castings and their comparison with wrought products.

Castings have traditionally been accepted as being free from marked directionality of properties. Structural alignment exists in wholly columnar regions in some cast alloys but its effects are frequently modified by changes in subsequent cooling or heat treatment. Anisotropy of properties can nevertheless remain in some cases and examples of its control and utilisation are examined in Chapters 2 and 9.

Local variations in structure, brought about by differences in cooling rate and other solidification conditions, give rise to the general phenomenon of section sensitivity. In most types of alloy this arises from variations in grain size and substructure and the associated distribution of second phase constituents, solutes and microporosity. Brittle constituents and porosity exert a greater influence on ductility and tensile strength than on yield or proof stress and it is these properties, and related fatigue properties, that tend to be lowest in thick sections and central regions, due to the coarser structures and greater incidences of segregation and microshrinkage. The outer, more rapidly cooled regions, show greater strength and toughness; the benefits of refinement are also evident in the superior properties of chill castings and thin sections.

That section sensitivity can be attributed partly to connections between structure and soundness was demonstrated by Polich and Flemings[26] in work with low alloy steels. Dendrite arm spacing was found to exert a greatly reduced influence on properties in unidirectionally solidified as opposed to orthodox cast material. The reduced section sensitivity obtained under such conditions is illustrated in Figure 7.7.

A similar contrast can occur between alloys: the properties of aluminium bronzes are found to be less sensitive to structure and section than those of many other copper alloys because of their short freezing range and reduced tendency to shrinkage porosity. Marked sensitivity in the former group only occurs in the presence of gas[27].

Section sensitivity can be particularly marked in cast iron, in which cooling rate affects the proportions of cementite and graphite as well as the form and dispersion of the graphite itself. This is specifically recognised by relating properties to test bar diameter and to equivalent cast metal sections. The effect on tensile strength for various grades of grey cast iron is summarised in Figure 7.8: such data enable grades to be selected to suit the particular section thicknesses concerned.

Although section sensitivity is much less pronounced in other materials, design to a uniform section thickness is one means of minimising the variation of properties within a single casting. Whilst the effects discussed originate in casting, it must also be borne in mind that the influence of mass on properties is not confined to cast products, since many alloys, independently of the method of forming, are sensitive to cooling rate by virtue of transformations in heat treatment.

Independently of the influence of section thickness, a greater scatter of properties is liable to be encountered in cast than in wrought alloys: this can be partly attributed to the coarser cast structure relative to a given size of test specimen. The nature of this variation was examined by Scheuer[29] using data from many sources: coefficients of variation of 10–15 per cent

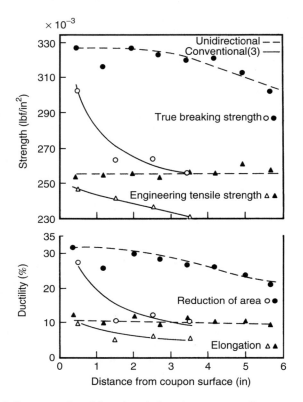

Figure 7.7 Influence of unidirectional freezing on tensile properties of a high strength cast steel (after Ahearn and Quigley,[28]) (courtesy of American Foundrymen's Society)

for tensile strength and 30–50 per cent for elongation were found to be common, with wider variations in certain cases.

The possibilities of heterogeneity and scatter must obviously be taken into account for the proper correlation of casting with test bar properties. It could be argued that since the best properties are usually found in the most highly stressed outer regions such differences will not be of great significance, but more systematic knowledge of property distribution is clearly desirable. Destructive examination and numerical modelling can both contribute to this form of assessment. An important aspect of progress in cast alloy technology is the reduction of scatter by refinement, close quality control and non-destructive testing techniques. In the manufacture of 'premium' castings property scatter is reduced both by increasing the general level of soundness and by maintaining high local cooling rates so as to produce finer and more uniform microstructures.

Direct comparisons between cast and wrought structures of precisely equivalent composition are not often reported, although specifications

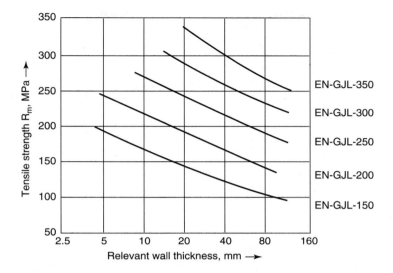

Figure 7.8 Examples of relationship between minimum values of tensile strength and relevant wall thickness, or ruling section, of simple shaped castings in various grades of grey cast iron (This extract is adapted from BSEN 1561:1997 with the permission of BSI under licence number 2000SK/0049. The complete standard can be obtained by post from BSI Customer Services, 389, Chiswick High Road, London W4 4AL)

frequently stipulate lower ductility values for castings. The directional influence of mechanical working on properties is exemplified in Figures 7.9 and 7.10. Deformation is seen to improve the ductility of a steel in the longitudinal direction but partly at the expense of that in the transverse direction. The same is true of impact and fatigue properties[31]; in the latter case the wrought material shows superior longitudinal fatigue properties in the absence of a notch (Figure 7.11) but is more notch sensitive than the cast equivalent. In the specific case of high temperature properties the finer wrought structure is no longer beneficial and cast structures of comparatively large grain size offer a positive advantage with respect to creep resistance.

Numerous examples have been quoted in the literature of the direct replacement of forgings by cast products in applications calling for high strength and reliability[32]. What is required in these circumstances is design for casting rather than direct simulation of forging or fabrication condition.

Apart from the contrasts between cast and other types of material, significant metallurgical differences exist between individual casting processes. The previously mentioned difference between sand and chill cast alloys, with their finer structure, is one example; a similar difference can exist between static and centrifugal castings and will be further discussed in Chapter 10.

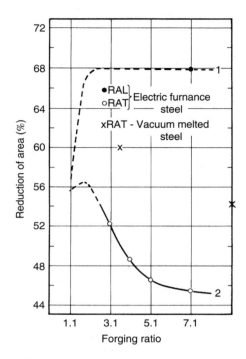

Figure 7.9 Effect of forging reduction on reduction of area transverse (R.A.T.) and reduction of area longitudinal (R.A.L.) (from 'Quality Requirements of Super-duty Steels[30], after Wells and Mehl) (courtesy of the Metallurgical Society of AIME)

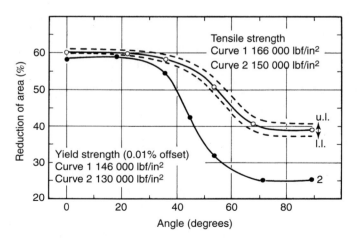

Figure 7.10 Relationship between reduction of area and angle to fibre direction in forging (from 'Quality Requirements of Super-duty Steels[30], after Wells and Mehl) (courtesy of the Metallurgical Society of AIME)

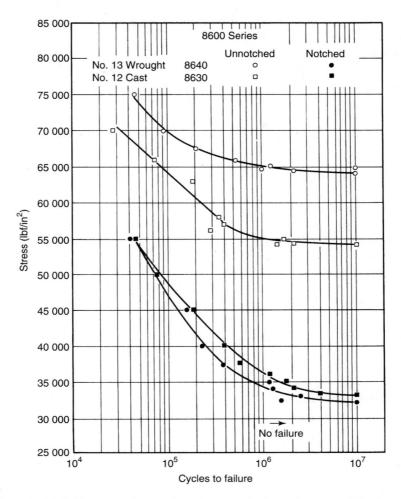

Figure 7.11 *S–N* curves for cast and wrought steels in the notched and unnotched conditions, when normalised and tempered to the same strength level (from Evans *et al*[31]) (courtesy of American Society for Testing and Materials)

Process and alloy selection

Casting processes

At an early stage in design and before proceeding to the detailed determination of shape, the casting process, the alloy and if possible a general outline of the proposed manufacturing method need to be established. One of the first considerations is whether a sand casting, a die casting or a casting made by some other specialised process will be required. The broad process groups and the principal factors influencing selection are listed in Table 7.4; the characteristics of the processes themselves will be examined in Chapters 8, 9 and 10.

Once the process has been decided in broad terms, the general approach to the shaping of the component must be considered and at this stage the possibility of using multiple construction and other supplementary techniques arises.

Cast–fabricated construction

Although casting is frequently in competition with other methods of forming, a wide field exists within which two or more techniques can be regarded as complementary and the best features of each exploited. A single casting may, for example, be broken down into smaller units to be subsequently fabricated by welding, bolting or other mechanical means; castings may be similarly employed in conjunction with wrought materials such as plate or rolled sections. These assembly techniques can be used either for economic reasons or to simplify the technical problems in the production of larger units.

An outstanding example of the introduction of castings into a fabricated assembly is the cast steel node joint employed in offshore oilrig construction

Table 7.4 Process selection

(a) Casting processes	(b) Factors in process selection
1. Sand casting, including core assembly, close tolerance and low pressure/vacuum variants	1. Suitability for product shape and size
2. Resin shell (Croning) and cold shell mould casting	2. Suitability for alloy
3. Investment casting and permanent pattern variants: plaster and ceramic mould casting	3. Accuracy and surface finish
4. Gravity die or permanent mould casting	4. Quantity requirements and overall cost of production
5. Low pressure die casting	
6. Pressure die casting; squeeze and slurry casting	
7. Centrifugal casting	

Figure 7.12 75 tonne cast steel node joint employed in a modern North Sea installation, Norwegian sector (courtesy of River Don Castings, now part of Sheffield Forgemasters Engineering Ltd.)

(Figure 7.12). The use of this component offers several vital advantages over the original construction based wholly on fabrication from wrought tubes. Firstly, welding operations are eliminated from the stress concentration zone at the intersection of the tubular members, providing a homogeneous composition and continuous microstructure in the crucial positions, and reducing residual stresses. Of still greater importance is the ability to incorporate generous fillet radii in the angular junctions, giving smooth contours and greatly reducing the stress concentration factor. The modified geometry, typified in a sectioned node in Figure 7.13, provides highly favourable fatigue characteristics as compared with the fully fabricated construction.

These advantages were confirmed in large scale fatigue tests. A comparison between the respective stress concentration factors as derived from practical stress analysis studies is summarized in Figure 7.14. Other examples of comparisons between cast and fabricated structures used in heavy duty offshore engineering were detailed by Marston in Reference 34.

The use of cast nodes to simplify fabrication and extend design capability is seen at the opposite end of the size scale in the car body space frame illustrated in Figure 7.15. Aluminium alloy extrusions are interconnected by die cast nodes weighing up to 1.8 kg. This advanced concept[35], employing 47 such nodes and robotic welding, replaced a conventional sheet steel body requiring some 300 spot-welded units. Weight saving and reduced tooling and production costs were combined with increased torsional stiffness. The

Figure 7.13 Section through a cast steel node joint, showing fillet radii and tapered sections (from Reference 33) (courtesy of the Institute of Materials)

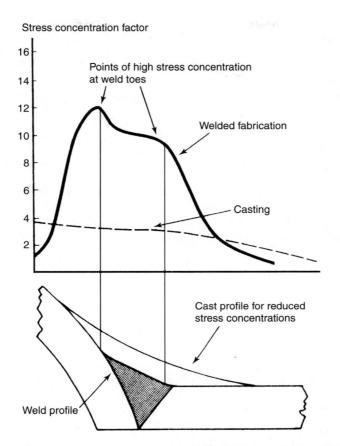

Figure 7.14 Stress concentration factors for cast and welded profiles (from Reference 34) (courtesy of the Institute of British Foundrymen)

development employs special vacuum die casting techniques and the use of a strong but ductile heat treated alloy.

Division of an extensive structure may be used to bring the overall dimensions or weight of the members within the capacity of the manufacturer, or to reduce the total mould volume required for a given weight of metal. More frequently, however, division is sought to simplify feeding problems or to remove the risk of tearing in conditions where a monolithic casting design would be subject to serious hindrance to contraction. Large flanges on cast pipework or long projections on otherwise compact forms are features which can usefully be produced separately and subsequently welded to the main body of the casting. For simple members of such structures wrought forms will often provide the most economic source of material.

Apart from simplification, structural division enables the best material to be utilised for each section of the design. It is thus possible to incorporate

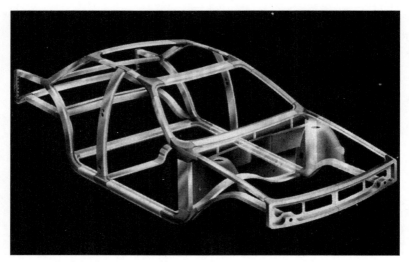

(a)

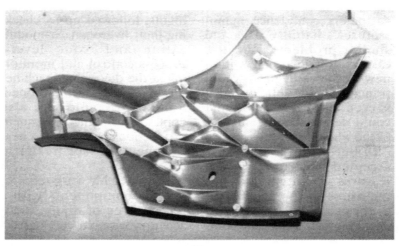

(b)

Figure 7.15 (a) Car body spaceframe incorporating aluminium alloy die cast nodes, (b) thin walled node casting (after Reference 35) (courtesy of Foundry International)

wear resisting alloys at critical points alone, or to use wrought sections where their particular properties can be best exploited. Solid drive shafts and spindles, difficult to produce as castings without centre line shrinkage, may be readily produced from bar or as forgings and embodied in a cast structure. Two further examples of cast–welded construction are illustrated in Figure 7.16. Other examples and illustrations were cited in Reference 36.

(a)

(b)

Figure 7.16 Examples of cast–welded construction: (a) steam chest assembly fabricated from eight low alloy steel castings and designed to operate at temperatures up to 550°C and pressures up to 17.2 MPa (2500 lbf/in^2) (courtesy of F. H. Lloyd & Co. Ltd.); (b) heavy steam chest, branch pipe and housing assembly fabricated from three alloy steel castings. Welds indicated by arrows (courtesy of British Steel Corporation Special Steels Division, Grimesthorpe Works)

Ideal conditions for cast–welded and other types of duplex construction are achieved if the foundry is able not only to influence the design, but to undertake the production of the finished assembly. This is particularly important where welding is involved, since weldability of the alloy and metallurgical effects involving the structure of the component castings must be taken into account.

Multiple construction is more commonly used in conjunction with steel than with cast iron and non-ferrous castings. The contraction characteristics of cast iron are such that there is less tendency for tearing and distortion to occur in extended castings, whilst the high fluidity and low shrinkage tendency facilitate the direct casting of thin members. Steel, on the other hand, has good weldability, giving an incentive to circumvent casting difficulties by division. The development of electroslag welding enables relatively thick cast sections to be joined.

The above emphasis on the use of castings in multiple construction needs to be seen as one useful approach to design, particularly suited to very large structures. The reverse possibility, that of combining an existing assembly to form a single cast component, has been successfully applied in the investment casting field, exploiting the capability of that process to produce shapes of great complexity: an example of such a casting is shown in Figure 7.17.

Apart from division and subsequent fabrication, several supplementary techniques are available to extend the design potential of cast products;

Figure 7.17 Investment cast gearbox housing for auxiliary aircraft engine and starter motor, in 20 Cr 10 Ni 3 W steel (courtesy of Cronite Precision Castings Ltd.)

they include the local introduction of special properties by means of inserts, bimetal casting and surface treatment. Such techniques may be employed either as a means of lowering product cost or to achieve otherwise unobtainable combinations of properties.

Inserts

Metal inserts, of different composition from that of the casting, may be incorporated permanently in the structure by placing in the mould before pouring. Although the ideal is to achieve interfacial adhesion through wetting, inserts should also be designed for mechanical interlocking with the parent casting, for example by undercut grooves or milling.

Inserts are widely used in die castings as local reinforcements for the relatively soft parent alloys. Steel or bronze bushes, for example, can be included to provide bearing surfaces; the technique can also be used to introduce preformed threaded holes or stud projections of such alloys as brass or steel. Machinable inserts can be embodied in sand castings of wear resistant alloys which are difficult to machine, enabling bores, keyways and similar features to be cut after casting. Figure 7.18 illustrates a series of crane wheels in austenitic manganese steel, provided with cast-in carbon steel bushes to carry the shafts. A similar example is seen in Figure 9.32a which illustrates a precision cast aluminium alloy torque converter component embodying a cast-in steel bush. A further interesting use of the insert technique is shown in Figure 7.19: in this case a turbine half ring has been produced from wrought steel blade sections embedded in cast steel shrouds.

Figure 7.18 Manganese steel crane wheels with cast-in carbon steel bushes (courtesy of Osborn-Hadfield Steelfounders Ltd.)

Figure 7.19 Turbine half ring with cast-in wrought blade sections (courtesy of Reyrolle-Parsons & Co. Ltd.)

The reverse of the insert procedure is seen in the technique illustrated in Figure 7.20, where an integral bimetal aircraft brake drum is produced by pouring cast iron into a rotating shell, previously fabricated from a mild steel pressing and ring. The shotblasted and preheated shell is coated with a compound to promote intermetallic adhesion, a strong permanent bond being achieved between the steel case and the centrifugally cast iron forming the braking surface. The same principle is used in the casting of white metal bearings into tinned and preheated shells.

Bimetal casting techniques

True bimetal castings of tubular form are produced by sequential pouring of separate alloys in centrifugal casting, a technique which will be more fully outlined in Chapter 10. The principle is also used in static casting, e.g. in the manufacture of rolls by the double-pour technique: a casting is poured in one alloy and the liquid core is subsequently displaced or diluted by a second molten alloy of different properties.

Conditions governing the bonding of two metals in bimetal structures were examined by Talanov *et al*[37], who emphasised the value of flushing the interface with additional metal as in burning-on or flow welding. Joint strength comparable to that of the parent metals was found to require heating of the interface above the solidus for a period of 5–10 s. This could be achieved by alternative combinations of preheating of the solid and superheating of the liquid. The latter is the more practicable: using a sufficiently high pouring temperature adhesion can be achieved with the solid surface at ambient temperature.

Figure 7.20 Bimetal aircraft brake drum (courtesy of Sheepbridge Alloy Castings Ltd.)

Surface treatment of castings

Many processes are used to develop special properties over the whole or part of the casting surface. Most of the techniques are not confined to castings and are too numerous to detail in the present text. They are commonly adopted either for enhanced appearance and corrosion resistance or to improve wear or fatigue resistance. The latter can be achieved by surface hardening and by generation of residual compressive stresses in the surface layers: shot peening, carburising and nitriding owe their beneficial effects to both these mechanisms. Polishing also contributes to fatigue resistance.

One modern development in this broad field warrants special mention, namely laser treatment. In this case electrical energy is converted to a high intensity coherent light beam, which produces a remote local heating effect on contact with the metal surface. Rapid heating of a thin layer is followed by self quenching by conduction into the bulk metal. The effect can be used to achieve martensitic transformations and hard cases on cast irons and steels of suitable composition, whilst superficial melting or glazing can be used to form rapidly solidified surface structures, or for the introduction of surface alloying elements. Metallurgical aspects of the process were examined in References 38 and 39 and a substantial review was presented in Reference 40. The laser surface alloying of ferrous alloys with chromium, nickel, boron and other elements by powder feed was the subject of the further review of Reference 41, which also examined the introduction of solid high melting point carbide particles to produce refined, hard composite layers.

The principal finishing processes used with castings are summarized in Table 7.5.

The surface treatments extend the effective range of application of the parent alloy, so that their use must be considered at the outset as an element in alloy selection. This choice must again be made on both technical and economic grounds, including the suitability of the treatment to the bulk material. Thus a decision must often be made between a cheaper material with a surface coating and a more expensive alloy with the intrinsic surface properties required.

Table 7.5 Surface treatments for castings

Primary purpose	Treatment
A. Wear resistance and improved mechanical properties, particularly fatigue resistance	1. Flame, induction and laser hardening 2. Hard surfacing by alloy weld deposition or plating 3. Carburising and nitriding 4. Local chilling (as applied for example to camshafts in cast iron) 5. Shot peening
B. Appearance and corrosion resistance	1. Electroplating 2. Tinning and galvanising 3. Painting, stove enamelling and associated pretreatments 4. Vitreous enamelling 5. Anodising, pickling, and chromate baths 6. Chemical and electrochemical polishing 7. Mechanical polishing and machine finishing

Property criteria in design

The design of castings may be based upon any of a wide range of cast metal properties, depending upon the service conditions of stress, temperature and environment. The first essential is to define these conditions and to select the right design criteria. Many of the key properties are structure sensitive and their relationship to the typical characteristics of cast structure was examined in Chapter 2.

As previously emphasised, the economic use of materials demands that the properties of a given alloy be considered in relation not only to function but to cost, since the strength or rigidity of a component, for example, may be achieved by alternative combinations of material properties and dimensional proportions. These criteria cannot be comprehensively treated in the present work but the properties will be reviewed and will be subsequently related to the main groups of cast alloys.

The choice of alloy is also governed partly by production requirements: foundry characteristics, machinability and weldability need to be considered in relation to the particular shape and application.

Mechanical properties

The most frequently specified property is *tensile strength*, although its significance in the design of static structures is less than that of the *yield stress* or *proof stress*, which marks the effective limit to the elastic range.

The other essential guide to elastic behaviour is *Young's Modulus*, the measure of stiffness, or deflection under a given load: this property determines the resistance of a structure to elastic distortion and eventual instability. In many designs rigidity is the critical factor governing dimensions, requiring section thicknesses which are more than adequate from a strength viewpoint. Young's Modulus is relatively insensitive to changes in cast structure, being primarily established by composition. In grey cast iron, however, the effective value varies markedly with the graphite content and decreases with an increase in the applied stress.

The tensile properties also give a guide to other properties which are not themselves frequently determined, for example the shear, torsional and compression strengths of the material; in each case there is a consistent relationship with tensile strength for a given class of alloy. Tensile strength will also be shown to be relevant to fatigue performance.

In calculating strength requirements, stress concentrations at notches must be taken into account; the elimination of stress raising features such as sharp corners is essential in deriving maximum strength from cast components. Notch effects have been previously discussed in Chapter 5; their predominant influence on strength and the vital importance of fillet radii were demonstrated by Caine in his major works on casting design[42,2].

In prescribing the mechanical properties of materials to be used in engineering designs, two major uncertainties arise, firstly in respect of the exact distribution of service stresses and secondly in respect of quality and the possible presence of defects or residual stresses.

Nominal properties quoted in specifications are usually those expected from sound cast test bars; the relationship of test bar to casting properties has been discussed in the previous chapter. To take account of possible errors in predicting the magnitude and distribution of stresses and to allow for the possibility of stress concentration at unseen defects, a safety factor is normally incorporated in the design. A design stress of approximately half the yield stress, i.e. a safety factor of 2, is commonly used where static stresses alone are involved, but much higher factors are employed for dynamic stresses and in other special circumstances. Safety factors can be greatly reduced where rigorous non-destructive testing is used to ensure quality; there is also less need for high factors where simple designs increase the accuracy of stress predictions.

Mechanical properties are sensitive to high strain rates and to cyclical stresses, either being capable of producing failure well below the ultimate tensile stress, especially in the presence of notches or surface imperfections; knowledge of tensile properties is of limited value under such circumstances. Notch brittle behaviour under shock loading arises from triaxial stresses, which can produce brittle fracture even in alloys showing ductility in tensile testing.

Notched bar impact properties derived from Charpy and Izod tests are relevant under these conditions, although such tests cannot be used to provide design stress data since results are dependent on notch geometry and specimen size, which cannot be directly related to other shapes and sections. Impact test results are of particular significance where low temperature service is envisaged and where the metal is susceptible to the ductile–brittle transition, commonly encountered in ferrous and other metals with the body-centred cubic crystal structure. In such cases impact tests may be used to determine the actual transition temperature or simply to provide comparative fracture energies at a standard temperature. Austenitic steels and non-ferrous alloys with the face-centred cubic structure are comparatively immune from low temperature embrittlement.

The design situation with respect to brittle fracture was radically transformed with the development of the fracture mechanics approach to the problem, eliminating the need for excessive safety factors to cater for the possible effects of unknown flaws in heavier sections.

The relevant material property in this case is the *fracture toughness*. This relates to static stresses and enables the fracture stress of components containing sharp flaws to be predicted. The plane strain fracture toughness of a material is determined under conditions involving triaxial stresses, and is expressed in terms of the critical value of the stress intensity factor K_{IC} at a crack tip which will cause crack propagation and fracture.

The special test required for this purpose is the COD or crack opening displacement determination, described in References 43 and 44. A sharp fatigue crack is first induced in a specimen of suitable thickness. The specimen is then statically loaded and the width of the crack continuously measured as the load is increased. The load producing crack instability or a fixed percentage extension is determined, and is independent of specimen geometry. Automatic pre-cracking and testing procedures[45] have greatly reduced the time required for these operations.

Using the basic relationship between fracture toughness, applied stress and crack length, it becomes possible to calculate the critical defect size that can be tolerated without crack growth leading to brittle fracture. The relationship is expressed in Figure 7.21. Knowledge of any two of the three factors enables the third to be determined. This relationship and other major aspects of mechanical properties in design were reviewed by Jackson in Reference 46, and more extensively by Jackson and co-authors Selby and Found in Reference 6.

Under fatigue conditions, involving periodic stress fluctuation, failure can occur at stress levels far below the normal yield stress. In these conditions the significant design criterion is the *fatigue limit* for the material as revealed by S–N curves of the type shown in Figure 7.22. The limiting fatigue stress is normally encountered after 10^6–10^7 reversals in ferrous alloys, being seen as a marked break in the slope of the S–N curve, but fatigue strength continues to fall for much longer periods in many non-ferrous alloys. This contrast between the two groups of materials is illustrated in Figure 7.22.

The fatigue limit bears a general relation to the ultimate tensile strength, the endurance ratio, or ratio of fatigue to tensile strength, being roughly constant for a particular material and mode of testing. The lowest ratios are encountered under conditions of complete reversal of the load to give alternate tensile and compressive stresses, higher values being obtained when the fluctuating stress retains the same sign throughout the cycle. Endurance ratios in the range 0.3–0.5 are common in ferrous and 0.15–0.35 in non-ferrous alloys. The fatigue–tensile strength relationships for various types of alloy are portrayed in Figure 7.23.

An exception to the normal role of tensile strength as a guide to fatigue behaviour occurs in the case of the austempered ductile cast irons. The maximum fatigue strength is in this case associated with those austempering temperatures which produce maximum elongation rather than maximum strength, a feature attributed to the distribution of retained austenite in the microstructure and its effect on crack propagation. Reference 47 provides a general review of this and other aspects of ADI properties.

Fatigue properties are particularly sensitive to non-metallic inclusions and to particles of intermetallic compounds. The lowest endurance ratios are thus associated with high strength alloys deriving their properties from precipitates. In view of the particular sensitivity of fatigue properties to

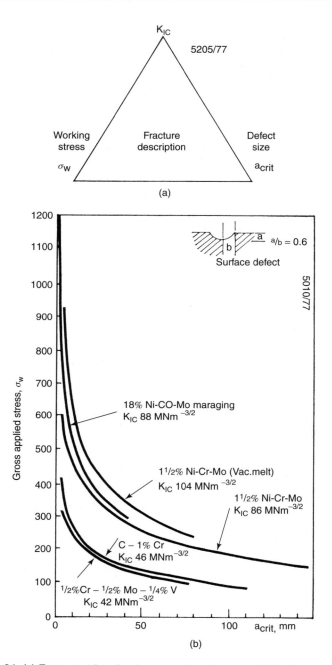

Figure 7.21 (a) Fracture triangle: the operating factors, (b) Basis for comparison of materials: variation of critical defect size with gross applied stress for steels of varying fracture toughness (from Jackson[46]) (courtesy of Institute of British Foundrymen)

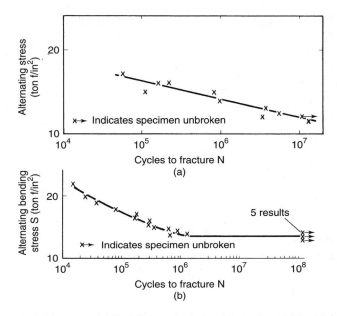

Figure 7.22 S–N curves: (a) for heat treated aluminium alloy, (b) for high carbon steel (from Sharp[13])

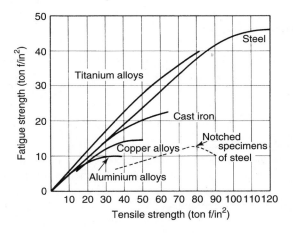

Figure 7.23 Approximate relationship between fatigue strength and tensile strength (fatigue limit based on 10^8 reversals) (from Sharp[13])

notch effects and surface or internal imperfections, a further substantial factor of safety is required, minimum factors of 2–3 being usual.

The influence of a notch on fatigue performance may be seen by reference back to Figure 7.11. Despite the initial superiority of the wrought material the notch exerts a proportionately greater effect than in the cast material.

This reflects a general characteristic of cast metals, which exhibit less notch sensitivity than their wrought counterparts, possibly owing to the stress raising features inherently available in cast structures.

Ductility. In all applications involving strength properties the ductility of the cast metal is significant, whether measured as elongation and reduction of area in tensile testing or as fracture energy in impact testing. The capacity for plastic deformation before fracture, whilst seldom exploited directly in casting design, provides an assurance against failure through unforeseen overstressing.

The level of ductility which can be usefully employed is problematical. As with other kinds of safety factor the need is least when high standards of inspection are adopted. Although high elongation values are a feature of many specifications for both cast and wrought materials, the need for such high values has been questioned, at least where simple tensile stresses are concerned. Caine[42] suggested that 1–2 per cent elongation is more than adequate for its accepted purpose, and Mercer[48] advanced the view that taking the 0.1 per cent proof stress as the design criterion for static loads, an elongation of 0.5 per cent is more than can be usefully employed in liquidating stress concentration at cracks in structures.

In some alloys, including the grey cast irons, ductility is virtually absent. This is recognised in the higher safety factors allowed, factors of 3–5 being usual even under conditions of static loading. Structures in grey cast iron are however normally designed to exploit the superior compression strength of the material.

Shear, bend, transverse, torsion and other mechanical test values may be variously specified by designers to evaluate material quality under conditions analogous to those in service, but are not normally used directly in design: stress calculations for all these conditions are normally based on tensile properties.

High temperature properties. The properties discussed in the foregoing section no longer provide an adequate guide to behaviour at high temperatures, for which creep data must be employed in conjunction with specific information concerning expansion, oxidation resistance and other relevant temperature dependent characteristics. Structural stability is also required to minimise deterioration of properties with time.

Creep, or continuous plastic deformation at stresses below the normal elastic limit, requires design on the basis of strain–time relationships determined for particular conditions of stress and temperature and illustrated in groups of creep curves as typified in Figure 7.24. Of special significance is the rate of secondary creep, the period during which strain hardening and recovery are in dynamic equilibrium and produce a linear section of the strain–time curve. The acceptable creep rate depends in some cases upon

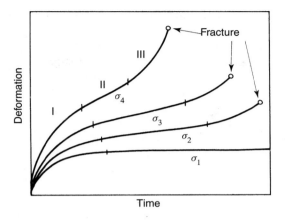

Figure 7.24 Group of creep curves for various stress levels at constant temperature. Curves show characteristic primary, secondary and tertiary stages

dimensional tolerances and clearances in relation to the expected life of the component. In many designs the applied stress remains substantially constant irrespective of dimensional changes resulting from creep: a suitable basis for design may then be found in knowledge of the constant stress required to produce a given small rate of deformation at the specified temperature. The design data may take the form of a family of stress–time curves for fixed amounts of strain, or a single criterion may be employed; one such criterion is the stress required to produce a deformation of 0.1 per cent in 1000 h, or a creep rate of 10^{-6} per hour, although much lower rates, for example, 10^{-9} per hour, may be relevant in some cases, as in steam turbine applications. For very high temperatures and where clearances are unimportant, design may be based on stress rupture rather than creep strain data; in this case plots of stress against time to fracture are employed.

In certain applications the effect of creep is to cause progressive relaxation of the applied stress by conversion of elastic to plastic deformation: in this case the significant relationship is that between stress and time for a condition of constant strain. These conditions occur in prestressed elements such as bolts and tie bars and can be simulated in creep relaxation tests.

Hardness can be used as a design criterion in certain situations demanding wear resistance, although indentation hardness values do not provide a measure of resistance to all types of wear. Bearing properties or behaviour in sliding friction cannot be directly related to hardness, since such metallographic features as the duplex structure in bearing metals or the graphite phase in cast iron predominate; alloy selection is thus on an empirical basis.

There is a better correlation between hardness and resistance to abrasive wear, whilst the property may also be specified for its influence on machinability, although in both cases the additional factor of work hardening must also be taken into account: this is illustrated by the exceptional resistance of the initially soft austenitic manganese steel to impact abrasion and by the work required in machining austenitic stainless steels. Abrasion resistance is however most commonly associated with the presence of hard intermetallic compounds, carbides and nitrides, or with alloys developing martensitic structures.

The need for hardness can be met in two alternative ways. Bulk mechanical properties can be adapted to surface requirements by choice of an intrinsically hard and therefore strong alloy; such alloys are frequently of low ductility. Alternatively the parent metal may be surface hardened by heat treatment, selective chilling, or through a change in surface composition, in which case alloy selection can be based primarily on the ideal bulk properties. The choice depends partly upon whether the casting will be subject to general stresses as well as local wear; intrinsically hard, brittle alloy castings find wide use as liner plates where they can be backed by tougher materials, but surface or work hardening alloys are preferred for structural applications.

In its additional role as a guide to strength properties, hardness is used primarily for control in heat treatment; strength requirements in design are specified directly in terms of the mechanical properties previously discussed.

Corrosion resistance

The corrosion resistance of cast alloys is relevant to a variety of service conditions; in the special circumstances of chemical plant it may be the primary consideration. The same applies to scaling resistance at high temperatures.

Since the corrosion resistance of an alloy depends on a wide range of environmental factors, no single recognised property can be used either to predict performance or to form a basis for design and specification. Alloy selection is therefore based primarily on empirical data for particular combinations of alloy and reagent, preferably obtained under physical conditions resembling those expected in service. Sources of such data will be referred to in the ensuing section. Specific tests may also be carried out on samples exposed to the real or simulated environment.

Although corrosion is primarily a surface phenomenon, the possibility of more profound effects must also be taken into account. These include intergranular corrosion and the combined influence of stress and corrosive attack. The possibility of stress corrosion, with drastic reduction in tensile and fatigue strengths, requires the avoidance of susceptible alloy–environment systems and the use of greatly increased safety factors.

As in the case of hardness, a choice must be made between an alloy with intrinsic corrosion resistance and an alloy selected primarily for its mechanical properties or low cost but deriving corrosion resistance from a protective coating. Under severely corrosive conditions the former approach is usually adopted, but surface treated castings are widely employed for resistance to atmospheric and aqueous corrosion.

Other functional properties

A wide range of additional properties may be used as a basis for alloy selection. Although service requirements are most commonly defined in terms of mechanical properties, corrosion resistance, and related attributes such as strength–weight ratio, selection may be based on a physical property such as electrical resistivity or a characteristic such as thermal shock resistance. Some of these criteria are listed in Table 7.6. Many of the physical properties are not markedly structure sensitive, being influenced mainly by composition, especially that of the alloy base.

Manufacturing properties

A major consideration in alloy selection is response to the various phases of manufacture. In the choice of a casting alloy foundry properties are naturally the foremost concern, but machinability and welding qualities may also be required.

Foundry properties include fluidity, solidification and feeding characteristics, and susceptibility to contraction and gas defects: these are governed mainly by the constitution and thermal properties of the alloy. Compositions

Table 7.6 Additional properties employed in casting design

A. *Physical properties*	B. *Special characteristics*
1. *Thermal properties*	(a) Thermal shock resistance
(a) Specific heat	(b) Damping capacity
(b) Thermal conductivity	(c) Dimensional stability
(c) Coefficient of thermal	(d) Bearing properties and
expansion	frictional characteristics
	(e) Hydraulic soundness
2. *Electrical and magnetic*	(f) Colour
properties	
(a) Resistivity	
(b) Electrical conductivity	
(c) Temperature coefficient	
of resistivity	
(d) Curie point	
(e) *B–H* curve characteristics	

have been developed to meet the broad needs of the casting process as well as the engineering criteria: one example is the frequent preference for alloys of near eutectic composition, which combine several favourable characteristics, whilst low melting point and freedom from hot shortness have been major influences in the development of the die casting alloys. Castings in other types of alloy are nevertheless employed where needed to achieve the engineering properties, as when using precipitation hardening systems. The influences of alloy constitution in founding have been examined in Chapters 1, 2, 3 and 5 and will not be pursued further in the present discussion.

Machinability determines the rate of stock removal attainable in machining operations and the costs incurred in power consumption and tool wear. The property cannot be defined in terms of a single criterion but is inversely related to the work of deformation. This is represented by the area enclosed beneath the stress–strain curve for the material and is thus diminished by low yield strength, low ductility and a low rate of work hardening. In general, therefore, soft, brittle materials are readily machined: they offer minimum resistance to the tool and produce small chips rather than bulky turnings.

There is no universal relationship between machinability and mechanical properties because of sensitivity to features of the microstructure, notably the mode of distribution of dissimilar phases and non-metallic inclusions: this is exemplified by the influence of lead and sulphur additions made to copper alloys and steels to confer free cutting quality. Frictional properties of the alloy are also involved.

The machinability of an alloy is not constant for all conditions, being partly dependent upon tool design and material, lubrication, and the feed and speed ranges employed. Comparative machinability indices based on a single factor such as power requirement or tool wear may however be derived by standardisation of all the other variables.

Weldability is the capacity of an alloy to give mechanically strong, defect-free welded joints, a valuable property in many castings, especially where cast–welded construction is envisaged. The property is a function both of composition and initial structure and can be measured by various forms of empirical test related to manufacturing and service conditions.

The most important aspect of weldability is freedom from embrittlement and cracking in the heat affected zone: this freedom is greatest in alloys which remain ductile throughout the heating cycle, absorbing thermal shock and differential expansion by plastic flow. Weldability is least in intrinsically brittle alloys and in alloys subject to martensitic transformation, although welding may still be possible using preheating and postheating techniques and filler metals of special composition.

Other aspects include susceptibility to contamination by oxygen, hydrogen and nitrogen, and the extent to which corrosion resistance is impaired by structural changes in the parent metal.

Response to heat treatment. In stress bearing applications it is the bulk mechanical properties that are significant. Where the development of these properties is dependent on cooling rate or heat treatment the material must respond to the cooling rates obtainable in the particular section thickness. Criteria such as hardenability reflect this type of requirement and Jominy end quench tests may therefore be specified as metallurgical controls. More fundamental data in the form of isothermal or continuous cooling transformation diagrams can provide an initial basis for material selection.

Sources of design data for cast alloys

Any attempt at concise summary of design data for casting alloys is beset by serious difficulties, due both to the very large number of alloys and the wide range of properties obtainable within each group by changes in alloy composition, foundry conditions and heat treatment. For specific design values, therefore, reference needs to be made to published sources of design data, including various handbooks treating the design of castings and the products of individual casting processes. Much of this information is classified by alloy family rather than by engineering function. Some sources of cast alloy data are listed in References 49–61, 13, 22 and 6, and can provide a general picture. However, since specific values and nomenclature become dated and new alloys emerge, it is essential to employ current issues of specifications and other literature for definitive data.

For comparative treatments of engineering materials which cut across conventional boundaries, broader works on materials selection need to be consulted. A classic contribution in this area, including consideration of numerous cast alloys as well as other materials, was that of Sharp[13] which also provided bibliographies on design issues. Systematic treatments employing sophisticated selection criteria were subsequently produced by Charles and co-authors[60], and by Ashby[61]. The concept frequently used is the performance index, based on criteria chosen as most pertinent to the required function. This aspect is extensively treated by Ashby: materials selection charts can be constructed to portray, in the form of area maps, combinations of two or more designated properties of alternative groups of alloys and other materials. One example is the plot of strength against density shown in Figure 7.25. Young's modulus, fracture toughness, density and cost parameters can be included in such treatments, along with more specialized thermal and electrical properties. Certain parameters already embody two properties, for example the specific modulus, or Young's modulus per unit of density. The main emphasis is on selection between

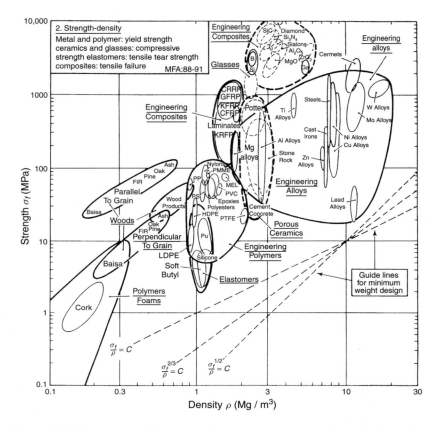

Figure within chart:

2. Strength-density
Metal and polymer: yield strength
ceramics and glasses: compressive
strength elastomers: tensile tear strength
composites: tensile failure   MFA:88-91

Figure 7.25 Typical materials selection chart, of strength plotted against density. The guide lines of constant σ_f/ρ, $\sigma_f^{2/3}/\rho$ and $\sigma_f^{1/2}/\rho$ are used in minimum weight, yield limited design (from Ashby[61]) (reprinted by permission of Butterworth-Heinemann (Publishers), a division of Reed Educational & Professional Publishing Ltd.)

rather than within particular groups of materials but the general principle can be valid for narrower comparisons.

The work by Charles et al[60] contains detailed considerations of the significance of particular properties and information on numerous individual alloys, including casting alloys, in relation to major engineering functions. These and associated publications have transformed the process of alloy selection into a systematic field of study for the designer using castings and other products.

More comprehensive summaries and comparisons of the properties and capabilities of cast alloys require the use of materials selection databases as typified in References 51 and 52, in conjunction with the casting specifications themselves. Electronic data and computerized selection processes are playing an increasing role in this field, with their potential for integration of both shape and material aspects of design.

The main British Standard specifications for casting alloys are listed in the Appendix, with indications of content: current issues are detailed in the BSI Standards Catalogue[49], with cross-references to links with ISO and other international standards. It should, however, be emphasized that some commercially available cast alloys, particularly those of recent development, are not represented in these specifications; properties of additional proprietary alloys can be found in manufacturers' literature.

Some of the physical properties of the alloy families may be roughly predicted by reference to the properties of the parent metals since such properties are not strongly structure sensitive. A number of the more important properties are given in Table 7.7. A large amount of data for individual alloys and for other physical properties is available in Reference 62. In the case of mechanical properties, most of which are structure sensitive, the properties of the parent metal offer little guidance to the performance of the alloy derivatives. Some generalisations can nevertheless be made with respect to the principal engineering characteristics of strength and wear resistance, strength–weight ratio, corrosion resistance and response to manufacturing operations. In making such generalisations, however, the importance of the economic aspects of alloy selection must not be overlooked.

Cost

Substantial quantities of castings, including some of considerable mass, are used under lightly stressed conditions. Alloy selection can in these cases be based largely upon low material cost and ease of production and grey cast iron emerges with a clear advantage over other alloys. In stressing the low cost, exceptional foundry properties and machinability of this material, however, it must be reiterated that it is also frequently adopted for its high compressive strength, damping capacity and resistance to thermal shock, each of which contributes to its traditionally strong position amongst the casting alloys.

The relative economic positions of other casting alloys are less well defined, being subject to the fluctuations of the metal market as well as to the particular value criteria employed. Selling prices are naturally influenced by the intrinsic cost of the main alloy constituents, although higher melting costs, as incurred for steel, and special casting techniques such as those needed for magnesium alloys also contribute to the effect of the alloy on the price structure.

A comprehensive examination of the relationship between function and cost of alloys was undertaken by Sharp[13]. Comparisons were given of a wide range of engineering materials and product types, various performance criteria being related to material and overall costs. A comparison was also given of the relative fabrication costs associated with the different products

Table 7.7 Physical properties of metals (after Reference 62)

Metal	Density g/cm^3	Melting point °C	Specific heat J/kg K	Coeff. of thermal expansion 10^{-6}/K	Thermal conductivity W/mK	Resistivity μΩ cm	Temp coeff. of resistivity 10^{-3}/K	Young's modulus GPa
Aluminium	2.70	660	917	23.5	238	2.67	4.5	70.6
Copper	8.96	1085	386.0	17.0	397	1.69	4.3	129.8
Iron	7.87	1536	456	12.1	78.2	10.1	6.5	211.4
Lead	11.68	327	129.8	29.0	34.9	20.6	4.2	16.1
Magnesium	1.74	649	1038	26.0	155.5	4.2	4.25	44.7
Nickel	8.9	1455	452	13.3	88.5	6.9	6.8	199.5
Tin	7.3	232	226	23.5	73.2	12.6	4.6	49.9
Zinc	7.14	420	394	31	119.5	5.96	4.2	104.5

and materials. The cost factor in materials selection has also been examined by Charles[60] and by Ashby[61].

In the following sections emphasis is upon comparisons of specific properties of the main groups of casting alloys. No attempt is made at systematic value analysis in relation to individual design criteria; for which reference should be made to the more specialised literature.

Mechanical properties

Modern developments in commercially available cast alloys have markedly influenced the picture with respect to mechanical properties. The most significant feature has been the emergence of the austempered ductile irons, in which property levels traditionally associated with the alloy steels can be achieved. Neither the austempering principle nor ADI itself is new but there had been only slow progress in exploitation. Different combinations of properties are included in BS EN 1564: tensile strengths exceeding 1000 MPa, for instance, can be combined with 7 per cent elongation and unnotched Charpy values of 80 J. In common with the cast steels, still higher strength and hardness levels can be attained at the expense of ductility and toughness, for example 1400 MPa tensile, 1100 MPa yield, 1 per cent elongation and over 400 HB, providing excellent wear resistance and potential applications in tooling. Particular advantages of ADI include low density, some 10 per cent less than that of steel, giving potential for thinner castings; other positive features are the low distortion associated with the heat treatment, and lubricating effects of the graphite phase. A full appraisal of these modern irons was presented by Harding[47].

Low alloy cast steels with tensile strengths above 1000 MPa and notched Charpy minima of 20 J are included in BS 3100; similar high strength steels, produced to BS 3146 as investment castings, combine 1000 MPa tensile and 880 MPa yield strengths with 9 per cent elongation and 40 J Izod minimum impact values. More specialized high alloy compositions offer further enhanced properties, the 17/4 PH precipitation hardening stainless steel giving 1300 MPa tensile and 1170 MPa yield strengths with 8 per cent elongation, whilst the ultra-high strength vacuum melted maraging steel VMA1B combines 1760–1930 MPa tensile and 1700 MPa yield strength minima with 5 per cent elongation and 8J Izod impact values.

The cast steels and ductile irons thus remain as the principal groups of high strength alloys for exacting loading conditions, particularly those also involving wear and fatigue in varied degrees: heavy duty gears, crankshafts, earthmoving equipment and safety critical steering and brake components for vehicles incorporate such requirements. Needs for exceptional toughness and high notched bar impact properties, including retention to low temperatures, can be met particularly well by appropriate cast steels: 3–4 per cent nickel steels remain suitable for service down to −60°C, whilst some austenitic chromium nickel steels can sustain minimum notched impact values of 40J to −196°C.

Further high strength cast materials include the precipitation hardening copper-beryllium alloy, with tensile strength approaching 1000 MPa. Other copper alloys, particularly the high tensile brasses and aluminium bronzes, have more modest strength levels, in the ranges up to 740 MPa tensile and 400 MPa 0.2 per cent proof stress, with elongation values between 10 and 20 per cent, although the lower cost of the ductile cast irons and carbon and low alloy steels makes these the normal choice unless some additional quality, for example corrosion resistance or an electrical property, is sought.

Table 7.8, from Reference 60, whilst not specifically concerned with cast materials, provides a useful yardstick against which the various cited strength values can be seen, particularly in relation to its four arbitrary strength categories and the broad alloy groups in each.

In most cases the tensile strength of the cast alloy gives a good indication of its behaviour under other types of stress, as for example in the case of fatigue strength, previously illustrated in Figure 7.23, although the significant exception of austempered ductile iron was also noted. A further exception is the compression strength of grey cast iron, which is approximately four times the tensile strength. This exerts an important influence upon design for maximum strength, bringing this low cost material into the category of the high strength alloys in those cases where the compression strength can be fully exploited.

The mechanical properties of the non-ferrous cast alloys are, with a few exceptions, much lower in absolute terms than those of the steels and ductile

Table 7.8 Strength ranges of metals and metal alloys (from Charles et al[60]) (reprinted by permission of Butterworth-Heinemann (Publishers), a division of Reed Educational & Professional Publishing Ltd.)

Low yield strength (0–250 MPa)	Annealed pure metals Mild steels Non-heat-treatable Al–Mg alloys
Medium yield strength (250–750 MPa)	Heat-treatable 2xxx/7xxx Al alloys High strength structural steels Engineering steels Commercially pure (CP) titanium Stainless steels
High yield strength (750–1500 MPa)	Titanium alloys Cu–2% Be precipitation hardened Medium-carbon low alloy steels High strength low alloy steels Precipitation hardened stainless steels
Ultra-high yield strength (>1500 MPa)	Maraging steels Patented wire Tool steels

irons, but the aluminium and magnesium alloys emerge as major contenders for use in design situations where density is a significant factor. To these can be added relatively modern zinc base alloys with much higher aluminium contents than those traditionally produced as pressure die castings. Campbell reviewed these with other advanced cast alloy developments[63], including the alloy containing 27 per cent Al, which provides a tensile strength of 425 MPa, with 0.2 per cent yield strength of 370 MPa, 2.5 per cent elongation and 10 J unnotched Charpy impact value, with a relative density of only 5.

The traditional strength-weight ratio gives some indication of the shift in the relative position of the lower density alloys under these circumstances. Table 7.9 is based on the simple ratio of tensile strength to density of selected alloys in the ferrous and low density non-ferrous groupings referred to above. This does not provide a ranking, since other compositions and heat treatments, and the need for other properties, give different strength values, but the convergence when introducing the density factor is demonstrated. More complex design criteria, as used to maximize stiffness–weight ratio for example, give proportionately greater emphasis to density (see References 60 and 47), further enhancing the relative advantage of the aluminium and magnesium alloys.

The role of the light alloys is assisted by their fluidity, which facilitates the production of the thin, hollow forms often required if advantage is to be taken of the ability to use light structural members carrying a minimum of surplus section thickness.

High temperature strength is roughly related to melting point, the most suitable casting alloys to withstand high temperature stresses being found principally amongst the high alloy austenitic steels and the nickel and cobalt

Table 7.9 Tensile strengths and strength–weight ratios of some casting alloys

Alloy type	Tensile strength MPa	Relative density approx	$Ratio = \dfrac{Tensile\ strength}{Relative\ density}$
Low alloy steel	850	7.85	108
Austempered ductile iron	850	7.1	120
Aluminium alloy (Al Si 7 Mg 0.6)	250	2.7	93
Magnesium alloy (Mg Al 8 Zn 1)	240	1.8	133
Zinc aluminium (ZP 27)	425	5.0	85

based alloys. The advanced creep resisting alloys are mostly the subject of proprietary specifications, although the British Standards include a number of heat resisting steels possessing good high temperature strength and oxidation resistance. For more lightly stressed applications heat resisting irons containing large amounts of chromium, silicon or nickel show good resistance to oxidation.

At more moderate temperatures there are applications for low alloy steels, cast irons and aluminium bronzes, whilst certain light alloys can be used at temperatures up to 250°C.

Wear resistance

The principal alloys for abrasive wear resistance are again provided by the cast iron and steel groups. Initial hardness values of up to 650 HB are a feature of the plain and alloy white cast irons, which derive their wear resistance mainly from free cementite. To this group may be added the martensitic white irons containing nickel and chromium. For heavy duty impact and gouging abrasion the work hardening austenitic manganese steel is the normal choice: its initial hardness of approximately 200 HB increases to over 500 HB after severe deformation of the surface.

For less extreme forms of wear, use is made of martensitic steels and of cast irons and steels suitable for nitriding and flame hardening. For gears, steels, cast irons and numerous bronzes are represented. More specialised wear resisting materials include the chilled irons used in roll manufacture and the copper based and tin based bearing alloys suitable for conditions of lubricated sliding friction, whilst the wear resistance of the hypereutectic aluminium–silicon alloys has been the prime consideration in their use in the motor vehicle engine.

Corrosion resistance

Many casting alloys can have their effective corrosion resistance extended by surface coatings, but under more rigorous conditions it becomes necessary for selection to be based upon intrinsic corrosion resistance. At this stage the emphasis moves away from the high strength constructional alloys. Nonferrous casting alloys based on copper and nickel assume great importance, together with certain aluminium alloys, although the ferrous metals are still well represented by the stainless steels and corrosion resisting cast irons.

For resistance to atmospheric corrosion, wide use is made of castings in stainless steel and in many of the copper and non-heat-treated aluminium alloys detailed in the British Standards. The copper alloys used in this field range from the less expensive general purpose brasses with their moderate corrosion resistance, fair strength and excellent machinability, to the wide range of bronzes which involve higher material costs. Of the alloys of aluminium, the binary alloys with magnesium are preferred although the aluminium–silicon alloys also possess useful corrosion resistance. Castings in most of these classes of alloy retain a good surface appearance after prolonged exposure to atmospheric conditions.

The copper alloy, nickel–copper alloy and stainless steel groups are frequently used for castings for hydraulic applications, for processing a wide range of chemicals and foodstuffs and for withstanding seawater corrosion; in the latter category the manganese and aluminium bronzes are outstanding. Under more exacting conditions, as in the handling of the mineral acids and strong alkalis, it becomes necessary to use the high alloy cast irons containing copper, chromium and silicon, and a range of proprietary nickel base alloys containing copper, chromium, silicon and molybdenum.

In general the least satisfactory groups of alloys from the point of view of corrosion resistance are the carbon and low alloy constructional steels and the magnesium base alloys, both, it will be noted, offering high strength–weight ratios. Surface treatments are however available in cases where an oil film is not present; most magnesium alloy castings, for example, are subjected to chromate bath treatment before use. Where intrinsic corrosion resistance needs to be combined with high strength, martensitic or precipitation hardening stainless steels and aluminium and manganese bronzes can be considered. In combinations of stress with corrosive conditions, however, the possibility of stress corrosion phenomena may introduce restrictions. Certain high tensile brasses, for example, carry specific warnings of susceptibility to season cracking under stress. Stainless steels and other alloys can be similarly affected in certain environments.

High temperature scaling resistance is conferred by alloying elements such as chromium and silicon and is an important characteristic of the heat resisting materials in the high alloy steel and nickel based alloy groups. Where scaling resistance is the predominant requirement, high chromium steels and irons can be employed; nickel-free materials are necessary to avoid high temperature embrittlement in sulphur contaminated atmospheres.

As in the case of the heat resisting alloys, many corrosion resisting alloys are not represented in the formal casting specification ranges but are the subject of proprietary specifications. Much data on the corrosion resistance of individual alloys in various environments is available in published manuals[64,65] and advice can be sought directly from manufacturers.

Other functional properties

It is not possible to itemise the numerous alloy compositions selected on service criteria other than those of stress, temperature and corrosive environment. Additional characteristics referred to in Table 7.6 included thermal and electrical properties; reference to Table 7.7 suggests the alloy groups within which certain physical properties are likely to be found. Some alloys whose selection may result wholly or partly from characteristics mentioned in Table 7.6 are listed in Table 7.10. As in the case of strength or stiffness, density can be an important factor in selection: choice between copper and

Table 7.10 Examples of cast metals and alloys employed for special properties

Characteristic	Alloy type
Thermal properties	
High thermal diffusivity	Aluminium alloys
Low thermal expansion	Iron–36% Ni
Electrical properties	
High electrical conductivity	Pure copper and aluminium
Electrical conductivity combined with wear resistance	Aluminium and tin bronzes
High magnetic permeability	Low carbon steel
High coercivity and remanence	Aluminium–nickel–cobalt alloys
Special characteristics	
Thermal shock resistance	Grey cast irons
Damping capacity	Grey cast irons
Bearing properties	Bronzes; tin base alloys
Hydraulic soundness with specific corrosion resistance	Leaded gunmetals; numerous ferrous and non-ferrous alloys in other groups

aluminium, for example, may depend on a need for minimum weight or minimum volume as well as on relative cost.

Manufacturing properties

Most casting alloys can be machined or welded if required, although there are major differences in response to these operations.

The relative machinabilities of various types of alloy may be approximately assessed from Table 7.11, although such a generalisation is subject to variation according to the exact nature of the alloy and the machining conditions. The relatively soft magnesium and aluminium alloys are most readily machined, but the majority of the cast irons, steels and copper alloys can be machined without serious difficulty given the appropriate technique. In many of the high strength alloys the greater part of the machining can frequently be carried out in the soft condition before heat treatment for full development of properties. This is particularly desirable in the case of the austempered ductile irons, which work harden on cutting due to the retained austenite in the microstructures.

The greatest difficulty in machining arises in the case of intrinsically hard, brittle materials such as the white cast irons, some of which are virtually unmachinable, and in alloys with great work hardening capacity, for example austenitic magnanese steel. In these wear resisting alloys it is

Table 7.11 Relative machinabilities of typical alloys

Alloy type	Rating %	Alloy type	Rating %
Magnesium	500–2 000	Low alloy steel	50–65
Aluminium	180–2 000	Manganese bronze	60
Free cutting brass	180–400	Aluminium bronze	60
Zinc	200	Gunmetal	60
Malleable cast iron	90–120	Nickel	55
Free cutting steel	100	18–8 free cutting stainless steel	45
Leaded phosphor bronze	100	Monel	40
Cast iron	50–80	Phosphor bronze	40
0.35% cast steel	70	18–8 stainless steel	25
Copper	70		

important for holes, slots and similar features to be cored out accurately in the original casting.

With respect to weldability the most responsive of the casting alloys are the low carbon and austenitic steels, which can be readily embodied in cast–weld assemblies. Weldability, however, diminishes with increasing carbon content and the high strength, low alloy steels require special precautions because of their susceptibility to martensitic transformations. Corrosion resistance can also be lost through structural change induced by welding. An example is weld decay due to carbide precipitation in austenitic stainless steels, necessitating the use of stabilised grades unless full postweld solution treatment is feasible.

Most other classes of casting alloy can be welded with suitable precautions, although in some cases with considerable difficulty. In the case of the cast irons, preheating and postheating techniques are essential to satisfactory welding. However, fabrication is not normally required in this case in view of the excellent founding qualities of the material. Of the nonferrous alloys, those based on aluminium and magnesium can be readily welded by methods employing inert gas shielding; preheating is again often required. In the heat treated alloys, treatment after welding enables the full properties of the alloy to be developed in the heat affected zone, but properties of the weld remain inferior to those of the parent metal. Most copper based alloys, including the high strength bronzes, can be welded if required, with the exception of those containing lead.

The foregoing review has been based on each of the major characteristics and properties governing the selection of cast alloys. The alloys have not, therefore, been considered under the headings of the parent metals. Table 7.12, however, summarises the main characteristics which

Table 7.12 Outstanding characteristics of casting alloy groups, as influences in selection

Alloy type	Main positive characteristics	Examples of applications emphasising main characteristics*
Cast iron (grey)	Low cost combined with appreciable hardness, tensile strength and rigidity; high compressive strength; high damping capacity and thermal shock resistance; excellent founding qualities for complex designs	Manhole cover; tunnel segment; lathe bed; i.c. cylinder block; brake drum; ingot mould; gear blank; piston ring
(ductile, malleable and special)	High yield and tensile strength with ductility; wear resistance; corrosion resistance; low or moderate cost	Heavy duty gear; digger teeth; crankshaft; agricultural implements; ball mill liner; pump and valve components for acid plant
Steel (carbon and low alloy)	High yield and tensile strength, stiffness and strength–weight ratio, combined with toughness and fatigue resistance, at moderate cost	Track link; aircraft undercarriage member; mill housing; die block; heavy duty gear blank; oilrig node.
(high alloy)	Corrosion resistance under a wide range of conditions; resistance to high temperature creep and oxidation; abrasion resistance	Water turbine runner; pump and valve components; gas turbine casing; radiant tube; tube support; carburising box; excavator bucket lip; rock crusher jaw
Copper alloy	Corrosion resistance, especially to seawater, combined with high strength if required; bearing properties; electrical properties	Marine propeller; hydraulic and steam pump and valve components; rolling mill bearing; switchgear contacts; gear blanks

(continued)

Table 7.12 (*continued*)

Alloy type	Main positive characteristics	Examples of applications emphasising main characteristics*
Aluminium alloy	High strength–weight ratio; useful corrosion resistance, especially to atmospheric corrosion; high thermal diffusivity; comprehensive range of cast products and alloys available at moderate cost	Clutch housing; automotive piston; i.c. cylinder head; exhaust manifold; marine fittings; beer cask
Magnesium alloy	High strength–weight ratio; low density; intricate sand and die castings available	Crankcase; transmission casing; binocular body
Nickel alloy	High corrosion resistance; strong resistance to high temperature creep and oxidation	Pump and valve components for chemical plant; gas turbine blade.
Zinc alloy	Pressure die cast forms give intricate components of reasonable strength and toughness by mass production at low cost	Radiator grill; door handle; carburettor body

*The selected examples, although typical, are neither fully nor exclusively representative of the alloy groups. Many properties and applications are common to several groups of alloys; the more specialised applications cannot be included in a brief review.

dominate the choice of the respective groups of alloys and gives examples of applications. The table illustrates the wide range of available characteristics and the overlap which permits alternative design solutions to many problems.

Physical design features

The detailed shape of a casting is important from the point of view both of engineering function and suitability for the casting process. Whilst some reference will be made to engineering aspects of shape, the principal object will be to consider the influence of shape features on the problems of casting manufacture. Although much of the following detailed discussion relates specifically to sand castings, many of the underlying principles are applicable to other casting processes, for which more specific points are dealt with in Chapter 9.

The most important general principle in determining shape is for the component to be designed from the outset as a casting, exploiting the particular advantages of the foundry process and adapting the individual features to take account of foundry production techniques. It is bad practice to use a design originally intended, for example, for weld fabrication, since the restriction imposed by the need to build up from standard, parallel and initially straight sections is absent; the foundry process, by contrast, is uniquely suited to the production of curved and streamlined forms with smooth changes of direction, both functional and pleasing in appearance. In fluid passages and similar applications these advantages can be overwhelming.

In designing for casting, although complexity, changes in section and coring should naturally be minimised, it is widely recognized that the capacity to cope with these conditions constitutes the greatest asset of the foundry industry in competition with other methods of shaping. Design for simplicity of foundry production, the subject of much of the present section, should not therefore be carried to a point at which these advantages are lost.

The principal aims in designing for casting are to achieve soundness, accuracy and freedom from residual stress at minimum cost. Design measures are thus concerned with ease of moulding, with problems of metal flow and feeding, and with minimising stresses generated in cooling; each of these aspects will be separately examined. Machining factors must also be considered, whilst the whole concept must be compatible with the engineering function.

It is advantageous if design measures directed towards both moulding and casting operations can be tailored to the orientation to be used at these stages: this factor will therefore be the first to be examined.

Orientation. Design to a particular orientation is only feasible if the general method of casting is selected before the detailed features of casting geometry are decided. The direction of moulding is governed largely by the need to select convenient parting lines and to obtain moulds of adequate strength. From the latter viewpoint, for instance, recesses involving projecting masses of sand are best located at the bottom, where the weight of the sand is directly supported, rather than at the top where it is suspended. On the other hand orientation must also meet the requirements of the casting stage and their repercussions on quality; from this point of view deep recesses are better positioned at the top, where the enclosed sand cods or cores can be more readily vented to atmosphere.

The most important of casting requirements, that of feeding, is best served by the positioning of the heaviest masses of metal uppermost to provide direct access for top feeder heads. Other points to be considered, however, include the desirability of important machined surfaces or extensive flat surfaces being placed either downwards or in the vertical plane to avoid scabbing and non-metallic inclusions.

Thus, orientation for casting must be selected as the best compromise between the often conflicting requirements of moulding, feeding, venting and other foundry needs. If orientation can be decided at an early stage, however, subsequent design measures can be in general conformity with this initial step. This is an example of the benefit to be derived from full collaboration between designer and producer.

Moulding factors in casting design

The pattern

Decisions are made at the patternmaking stage which determine the positions of parting lines and the general approach to moulding and coring. The pattern should ideally, therefore, be made with full knowledge of the intended production method. It requires moulding taper and must incorporate contraction allowances specific to the alloy (see Chapter 8); these may be supplemented by allowances based on the characteristic behaviour of certain types of design, derived from previous experience with castings of similar shape and including, for example, allowances for camber and other forms of distortion.

Although patternmaking is essentially a part of the production sequence, responsibility for provision of the equipment is often undertaken by the purchaser. Since the casting can be no better than the pattern from which it is produced, it is in such cases essential for the designer to concern himself with the quality and accuracy of the equipment to ensure that this is consistent with the standards expected from the casting.

Mould parting

To minimise moulding cost the aim must be to achieve a small mould volume and to employ a minimum number of joints, cores and loose pieces:

all design features should be consistent with these aims. Parting lines should be as few as possible and should preferably be completely flat. Boxed moulds are normally cast with the partings in a roughly horizontal plane, giving top access to runners and open feeder heads, but vertical partings are widely employed for smaller castings produced by automatic boxless block moulding; this is also the normal mode for gravity die casting.

Parting planes are usually associated with the maximum projected areas of the casting, giving minimum draft distances and relatively shallow mould parts. As with orientation the choice is subjective and depends in part on the equipment and skill available. Subsequent design measures, however, should conform by providing unimpeded draft and adequate taper to the parting surface. Examples of features designed on this principle are illustrated in Figures 7.26 and 7.27. Figure 7.26 shows a projecting boss modified to a *D* shape, thereby eliminating an undercut feature which would otherwise require treatment by coring, additional jointing or loose piece moulding. The alternative corner boss treatment illustrated in Figure 7.27c is again designed to provide unrestricted draft to the chosen parting line.

All faces normal to the parting require taper for easy pattern withdrawal. The minimum allowance depends on the moulding process and the shape of the individual casting, but taper in the range 1–3° is desirable for sand castings, the higher allowance being preferred for inner surfaces which tend to restrain the withdrawal of enclosed cods of sand. Much lower allowances are adequate in other cases: in shell moulding, for example, a taper of as little as 1 in 1000, or approximately 4′, may sometimes suffice.

Taper allowances of these magnitudes can be incorporated by the pattern-maker as a matter of course, but it is preferable for more generous taper to be embodied wherever possible in the basic design.

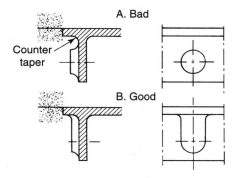

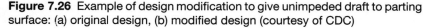

Figure 7.26 Example of design modification to give unimpeded draft to parting surface: (a) original design, (b) modified design (courtesy of CDC)

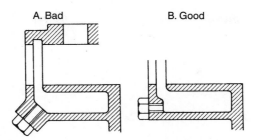

A. Bad   B. Good

Figure 7.27 Further example of design for unimpeded draft: (a) original design, (b) modified design (courtesy of CDC)

Cores

Design should minimise the need for cores, subject only to the point that an additional core may, by eliminating or simplifying a joint, produce a major reduction in moulding time. The need for coring is also partly governed by the equipment available. A pin lift moulding machine, for example, on which the mould is pushed upwards from the pattern, will not readily lift deep cods of sand from hollows in the pattern: separate cores are therefore required. A rollover machine may permit the production of the same features using integral mould projections.

Designs requiring cores should include provision for the maximum attainable coreprint area, with enough points of support for extended cores. These measures are needed to prevent core shift or rupture, to provide channels for gas escape and to facilitate the extraction of sand and reinforcing irons after casting. Where the design itself has insufficient apertures, it may be possible to provide temporary openings which can be plated over or plugged before the casting is put into service; there is otherwise no alternative to the extensive use of chaplets and studs.

Some limitations in moulding

One basic design limitation is the minimum mass of sand with which it is practicable to form a recess or aperture in a casting. Small cores and mould projections are exposed to erosive and hydrostatic forces during pouring, whilst in the case of the higher melting point alloys metal penetration or fusion may occur. Venting is also difficult where sand is closely confined. These problems are most acute where the enclosed mass of sand is small in relation to that of the surrounding metal. Difficulties associated with confined sand are exemplified in the pump impeller illustrated in Figure 7.28; such castings can in some cases be radically redesigned for multiple construction.

The minimum permissible thickness of an enclosed sand mass must depend partly upon the thickness of the surrounding metal sections. In the case of steel castings, for example, it is recommended that cores and

Figure 7.28 Pump impeller casting requiring thin core between shrouds (courtesy of Catton & Co. Ltd.)

isolated mould elements be at least twice this thickness[4]. To avoid fusion in small cylindrical cores, the recommended dimensional relationships are summarised in Figure 7.29, where it can be seen that a hole through an extended wall (represented by Case *a*) should be of a diameter at least equal to its length. Where the metal section is less than half the diameter of the core, on the other hand (Case *c*), the core diameter can be as little as one third of its length. Permissible lengths are reduced for blind holes where the core is supported at one end only.

Where the thickness of surrounding metal is small in relation to the core diameter (less than $D/3$), the principal factor restricting the ratio of length to cross-section of the core is distortion under hydrostatic forces. For cylindrical cores of a wide range of sizes, suggested limiting proportions of length to diameter are summarised in Figure 7.30 for metals in the two principal density groupings, namely steel and light alloys.

The minimum diameter normally used for cored holes in sand castings is in the range 4.5–6 mm ($\approx \frac{3}{16} - \frac{1}{14}$ in), although much smaller holes are feasible when using special casting processes (see Table 7.13). Holes of less than 25 mm diameter, however, can frequently be produced more economically and positioned more accurately by drilling from the solid.

All casting features should be designed to provide moulds free from sharp corners and feathered edges of sand. Not only are these fragile and susceptible to metal penetration but angular recesses form potent sources of stress concentration and puncture defects in the finished casting.

Design considerations at the casting stage

The importance of design at the casting stage lies in its impact upon problems of metal flow, feeding and subsequent cooling, which collectively

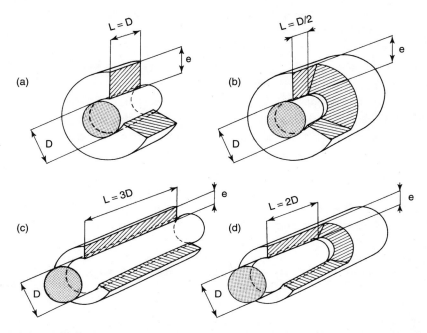

Figure 7.29 Recommended dimensional relationships for cores to avoid fusion. (a) Cylindrical hole with D less than $2e$, Maximum length $= D$; (b) blind hole with D less than $2e$, maximum length $= D/2$; (c) cylindrical hole with $2e \leq D \leq 3e$, maximum length $= 3D$; (d) blind hole with $2e \leq D \leq 3e$, maximum length $= 2D$ (courtesy of CDC)

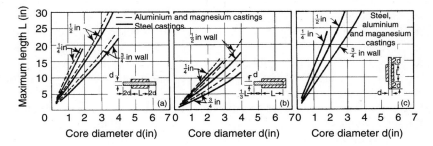

Figure 7.30 Recommended limiting dimensional relationships for cores for adequate dimensional stability; (a) core anchored firmly at both ends along mould parting, (b) core anchored at one end only, (c) core anchored at both ends but vertical to mould parting (courtesy of American Society for Metals)

determine the soundness and dimensional stability of the casting. The design features of greatest significance in these respects are section thickness and the general disposition of the various casting members in relation to the whole. These features are determined with the multiple aim of combining

Table 7.13 Recommendations for minimum diameters of cored holes in castings*

Product	Hole diameter
Sand castings	4.8–6.3 mm ($\frac{3}{16}-\frac{1}{4}$ in)
Gravity die castings	4.8–6.3 mm ($\frac{3}{16}-\frac{1}{4}$ in)
Shell moulded castings	3.2–6.3 mm ($\frac{1}{8}-\frac{1}{4}$ in)
Pressure die castings: Cu base	4.8 mm ($\frac{3}{16}$ in)
Al base	2.4 mm ($\frac{3}{32}$ in)
Mg base	2.4 mm ($\frac{3}{32}$ in)
Zn base	0.8 mm ($\frac{1}{32}$ in)
Plaster moulded castings	12.7 mm (0.50 in)
Investment castings	0.5–1.3 mm (0.02–0.05 in)

*Data from Reference 66: Cast Metals Handbook, 4th ed., American Foundrymen's Society (1962).

strength with lightness, minimising the task of feeding whilst providing adequate channels for metal flow, and avoiding serious stresses during cooling. The last two points will now be separately considered.

Metal flow and feeding

Metal section is normally kept as light as possible consistent with the required strength and rigidity and with metal flow, although in certain cases thickness may be locally increased to provide conditions for directional solidification. In the case of malleable cast iron, light section is also needed to ensure formation of the white iron structure in the first instance. Mould filling must be achieved without excessive superheat which may lead to coarse microstructures and defects.

The minimum feasible section thickness for metal flow depends on alloy fluidity, extent of section, and moulding process and material. In sand casting, wall thicknesses of less than 3 mm ($\approx\frac{1}{8}$ in) are unusual, whilst for less fluid alloys or for extensive plate sections, a much greater minimum thickness may be required. Solidification proceeds most rapidly in green-sand moulds, more slowly in drysand moulds and more slowly still in shell moulds, where flow is further assisted by smooth surfaces and high permeability. Under the special conditions of investment and die casting, where flow is assisted by hot moulds or by pressure, sections as thin as 0.4 mm (≈ 0.015 in) can be attained over short distances.

Published recommendations for minimum wall thickness are normally based upon feasibility for routine production rather than the limits of technical possibility. Table 7.14 gives some values for sand castings in the main groups of casting alloys; other original recommendations differ only in minor respects. For comparison purposes, selected values are shown for pressure die castings and other special products.

The dependence of the attainable minimum upon the general dimensions of the section is illustrated in Figure 7.31, which shows the recommended relationship between minimum thickness and length of section for steel castings: the lowest value of 6 mm ($\approx$0.25 in) or, exceptionally, 4 mm ($\approx$0.15 in), is seen to increase progressively to 50 mm (2.0 in) for very large dimensions.

For grey cast iron, 4.5 mm ($\approx$0.16–0.20 in) is the normally recommended minimum for the general run of engineering castings[3]; in this section-sensitive material the possibility of chilling rather than running capability is the limiting factor. Castings of much lighter section can however be produced for special thin wall applications in high phosphorus irons of exceptional fluidity, thicknesses as low as 1.5 mm ($\approx$0.06 in) having been achieved. The manufacture of thin walled castings in grey iron naturally depends on precision mould production as well as running requirements: the successful production of such castings has been an important development in the design of automobile cylinder blocks and heads.

A similar problem to that of thin sections is presented by sharp corners and knife edges, which need to be avoided whenever possible since they are both difficult to run and susceptible to stress concentrations at small defects.

Table 7.14 Recommendations for minimum wall thickness of cast metal*

Product	Wall thickness
Sand castings: Steel	4.8–12.7 mm ($\frac{3}{16}$–$\frac{1}{2}$ in)
Grey cast iron	3.2–6.3 mm ($\frac{1}{8}$–$\frac{1}{4}$ in)
Malleable cast iron	3.2 mm ($\frac{1}{8}$ in)
Copper alloy	2.4 mm ($\frac{3}{32}$ in)
Aluminium alloy	3.2–4.8 mm ($\frac{1}{8}$–$\frac{3}{16}$ in)
Magnesium alloy	4.0 mm ($\frac{5}{32}$ in)
Gravity die castings: Grey cast iron	4.8 mm ($\frac{3}{16}$ in)
Aluminium alloy	3.2 mm ($\frac{1}{8}$ in)
Pressure die castings: Copper alloy	1.5–2.5 mm (0.060–0.100 in)
Aluminium alloy	1.1–1.9 mm (0.045–0.075 in)
Magnesium alloy	1.3–2.0 mm (0.050–0.080 in)
Zinc alloy	0.4–1.1 mm (0.015–0.045 in)
Plaster moulded castings	1.0–1.5 mm (0.040–0.060 in)
Investment castings	0.6–1.3 mm (0.025–0.050 in)

*Data from Reference 4, courtesy of Castings Development Centre

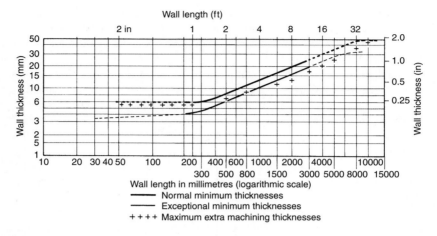

Figure 7.31 Recommended relationship between length of section and minimum wall thickness for steel castings (courtesy of CDC)

Caine[2] suggested that external corners be provided with radii of 10–20 per cent of section thickness. Where sharp external edges are definitely required, these are best obtained by provision of a beading for subsequent removal by machining or grinding.

No single principle can be laid down for the treatment of section thickness to assist feeding, which may be promoted either by uniformity or progressive section change according to the nature of the design. Problems of isolated heavy sections, remote from the primary feeder head locations and inconveniently placed for direct feeding, are naturally minimised where general uniformity of section can be achieved. The even cooling rate is also beneficial in encouraging a homogeneous structure throughout the casting. Carefully controlled differences in section thickness can, on the other hand, be used in conjunction with other measures of the kind discussed in Chapter 3 to produce directional solidification and a high standard of internal soundness. Basic design is in this case used to achieve results equivalent to those obtainable from temporary padding but without the need for extensive metal removal in fettling or machining. The incentive for a designer to match section thickness to feeding requirements must naturally depend upon pricing policies which do not penalise additional weight where this is necessary.

An example of the redesign of a component to induce directional solidification is illustrated in Figure 7.32. In the modified design the progressive increase in section towards the hub enabled a single feeder head to be employed where two had previously been required: hot tearing in the intermediate section was also eliminated.

Figure 7.32 Redesign of cast component for directional solidification; (a) original design, (b) redesign (courtesy of American Society for Metals)

Special design problems are presented by intersections of casting members. These can be treated in alternative ways, previously referred to in Chapter 3, but the most radical approach is to reduce or eliminate local hot spot formation by design. Knowledge of the behaviour of intersections and recommendations for their basic design owe much to the extensive early investigations of Briggs, Gezelius and Donaldson[5]. The elimination of hot spots is sought by measures to minimise mass and to obtain the largest possible included angle between the members. Figure 7.33 illustrates alternative approaches to the design of the three fundamental types of intersection, namely the L, T and X sections. The methods include the use of central lightening cores, feasible for intersections of no great depth, and the radical redesign of the more complex junctions to form simpler intersections which can be made fully sound either by design alone or with the aid of chills.

Sharp angles at intersections produce local hot spots as well as providing sources of stress concentration: they should therefore be eliminated by

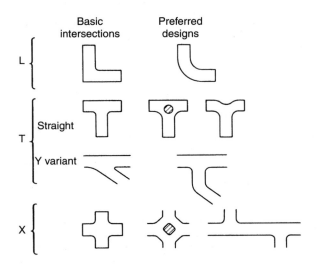

Figure 7.33 Design of intersections: some techniques for elimination of hot spots. (Refer also to Table 7.15)

fillet radii. Since these increase the mass, however, the choice of radius represents a compromise; the magnitude of fillet radii will be discussed in the following section.

Cooling: avoidance of stresses and stress defects

The hot tears, cracks, distortion and residual stresses which can result from restraint of free contraction were fully examined in Chapter 5, where it was shown that design could be a major factor in stress generation through the interaction of separate members when subject to varying rates of cooling. From this point of view uniformity of section thickness is ideal, since contraction of the members is then synchronised and the whole casting enters the elastic region at the same time. There can thus be conflict between requirements for feeding and stress-free cooling, progressive rather than uniform cooling being generally preferred for the former purpose.

The principal factor influencing stress generation is the relative disposition of the component members of the casting. A rigid framework will, if exposed to a range of cooling rates, be subject to much higher stresses than an inherently flexible structure. This point is illustrated by the well known example of a spoked wheel, stages in the evolution of a satisfactory design being shown in Figure 7.34. The first essential in this case is to maintain the correct relationship between the sections of spokes and rim in

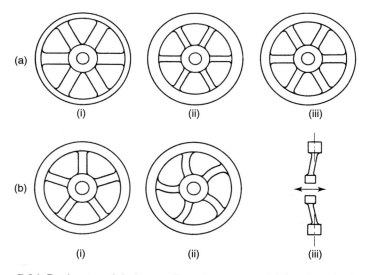

Figure 7.34 Design to minimise cooling stresses: example of spoked wheel. (a) Section thickness: (i) rim susceptible to hot tearing or to residual compressive stress and buckling; spokes to tearing or residual tensile stress; (ii) spokes susceptible to tearing or residual compressive stress; rim to residual tensile stress and flattening; (iii) balanced section thickness for uniform cooling. (b) Flexibility: (i) odd number of spokes; (ii) curved spokes; (iii) axially offset hub

the interest of uniform cooling, thus avoiding differential contraction and buckling distortion (Figure 7.34a). A more flexible design (Figure 7.34b) is achieved by using an odd number of spokes to prevent the behaviour of opposite pairs as rigid struts or ties; curved spokes represent a further improvement in the same direction. Similar flexibility is provided by longitudinal displacement of the hub of a wheel; the departure from a rigid plane structure enables spoke or web stresses to be relieved by small axial movement of the hub relative to the rim.

A further example of structural design based on a similar principle is illustrated in Figure 7.35: in this case continuous strut or tie effects are avoided by staggered construction. It must nevertheless be appreciated that rigidity is a prime functional requirement in many designs, in which case the opposite characteristic of flexibility for manufacture can only be sought as a compromise.

Factors contributing to internal stresses include high elastic modulus and high coefficient of thermal expansion; the tendency is diminished, on the other hand, in alloys of high temperature diffusivity, since temperature differentials are lower in these materials.

Apart from the framework as a whole, numerous individual features of casting design can increase the resistance to cooling stresses and tearing. Generally speaking, features which contribute to strength in the engineering sense provide a corresponding strengthening effect during the critical period of cooling. The most important single measure is the provision of adequate fillet radii in corners to reduce local stress concentrations. Caine[2], using data from many sources, considered the magnitude of stress concentrations at corners and junctions for various geometrical conditions and modes of stressing; fillet radius was shown to be crucial in all cases. Figure 7.36 illustrates a typical relationship between the stress concentration factor K_t and the ratio of fillet radius to section thickness, in this case for an L junction: in other cases too the stress concentration factor becomes quite low when the fillet radius attains a similar magnitude to the section thickness. Fillet radii should if possible, therefore, equal or exceed the section thickness, although such a rule can obviously not be applied in the case of very thick

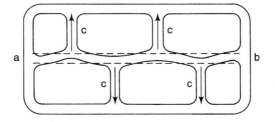

Figure 7.35 Design for grid casting using staggered construction (courtesy of CDC)

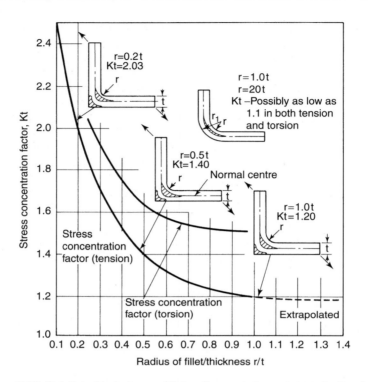

Figure 7.36 Relationship between fillet radius and stress concentration factor for L junctions (from Caine[2] after Roark *et al* and Lipe and Johnson) (courtesy of American Foundrymen's Society)

sections without aggravating the problem of feeding. These considerations are equally important in relation to service stresses, especially under fatigue conditions, as discussed more fully earlier in this chapter.

Specific recommendations for the optimum proportioning of radii at intersections have been made in the literature[2–6]. For both L and T sections the suggested radii range from the section thickness E to E/3 according to the magnitude of this thickness, provided that the ratio of the thicknesses of the sections, E/e, is less than 1.5. Where this ratio is exceeded gradually tapered junctions are recommended. For the L section, however, the doubly radiused section previously illustrated in Figures 7.33 and 7.36 is much to be preferred. Typical recommendations for L and T sections are summarized in Table 7.15.

Prevention of hot tears and general stiffening against distortion can be assisted by additional ribs, brackets or tie bars; the typical use of ribs is illustrated in Figure 7.37. The thickness of such reinforcements has been recommended not to exceed two-thirds that of the parent section[4] and a parabolic shape can be used to avoid stress concentration at the end of the

Table 7.15 Typical recommendations for design of L and T sections

Circumstances		L section		T section
Relative section thickness	Thickness magnitude mm	Right angled 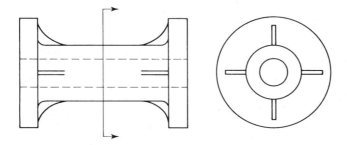	Doubly radiused	
Sections similar $E/e = 1 - 1.5$	$E > 30$ $E = 10 - 30$ $E < 10$	$r = E/3$ $r = 10$ $r = E$	$r = (E+e)/2$ $R = E+e$	$r = E/3$ $r = 10$ $r = E$
Sections markedly unequal $E/e > 1.5$	$E > 30$	Taper 15% $r = E/3$	$r = (E+e)/2$	Taper 7.5% $r = E/3$
	$E = 10 - 30$ $E < 10$	$r = 10$ $r = E$	$R = E+e$	$r = 10$ $r = E$

Figure 7.37 Typical use of ribs for reinforcement

rib when stressed in tension[2]. Ribs positioned on opposite faces of plate sections should be staggered to minimise hot spot formation. Very thin ribs, on the other hand, accelerate heat extraction and assist the development of a solid layer.

Intersection effects introduced by ribs and brackets can be reduced by coring out at the root of the reinforcing member as illustrated in Figure 7.38: this makes little difference to strength and avoids a triple intersection.

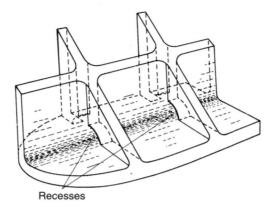

Recesses

Figure 7.38 Design of rib with cored aperture at root to avoid triple intersection (courtesy of CDC)

Machining factors in casting design

A casting may need to be machined for any of three related reasons. The attainable accuracy of the as-cast dimensions may be inadequate or the surfaces insufficiently smooth for functional requirements, or machining may be used to shape design features too intricate for production as part of the cast form. These purposes can frequently be served by confining machining to important surfaces or to those embodied in critical dimensions, but other castings are used in the fully machined condition. Either selective or general machining allowances may therefore be required.

Allowances should be of a depth to ensure substantial stock removal and to bottom minor surface irregularities; inadequate allowances can produce uneven surfaces, tool chatter and excessive tool wear. Each allowance must be sufficient to give the required depth of cut when the normal dimensional variation produces the most unfavourable condition within the accepted tolerances. This requires larger allowances where the overall dependent dimension is large, because of the increasing significance of contraction uncertainty. This circumstance is recognised in one system for the determination of machining allowances for steel castings (Table 7.16): this included a variable element representing approximately 0.5 per cent of the dimension, so allowing for possible departure from the expected 2 per cent linear contraction.

Recommendations for machining allowances depend also upon the nature of the alloy and the situation of the surfaces concerned. Table 7.17 gives comparative allowances for sand castings in the main groups of casting alloys, from which it can be seen that much lighter allowances are acceptable for the softer non-ferrous alloys. The table gives separate values for bore and upper surfaces: maximum allowances are used in the latter case since the top surfaces of the casting are the frequent site of inclusions

Table 7.16 Machining allowances for steel castings*

Greatest dimension (D)	<250 mm	250–1 000 mm	>1 000 mm
Minimum allowance (mm)	3	3	3
Maximum allowance (mm)	$4 + \frac{6D}{100}$	$4.5 + \frac{5D}{1\,000}$	$5 + \frac{4D}{1\,000}$

*From Reference 4, courtesy of CDC.

Table 7.17 Machining allowances for various casting alloys and locations*

	Dimension (mm)														
	<150			150–300			300–500			500–900			900–1 500		
Alloy group	N	B	T	N	B	T	N	B	T	N	B	T	N	B	T
Non-ferrous	**1.6**	2.4	2.4	**1.6**	2.4	3.2	**2.4**	3.2	3.2	**3.2**	3.2	4.0	**3.2**	4.0	4.8
Cast iron	**2.4**	3.2	4.8	**3.2**	3.2	6.4	**4.0**	4.8	6.4	**4.8**	6.4	6.4	**4.8**	8.0	8.0
Steel	**3.2**	3.2	6.4	**4.8**	6.4	6.4	**6.4**	6.4	8.0	**6.4**	7.1	9.5	**6.4**	8.0	12.7

*Adapted from Reference 66.
Locations: N: Normal surfaces, general allowance
         B: Bore
         T: Topmost surface

and other minor defects. This again emphasises the importance of early consideration of the orientation to be used in casting.

Much smaller allowances, in the range 0.8–1.6 mm ($\frac{1}{32} - \frac{1}{16}$ in) are commonly employed on die castings and other special cast products: in some cases a grinding allowance alone may suffice. Even on sand castings the normal allowances may be reduced under well tried mass production conditions.

More recent recommendations for machining allowances are included in the published standard ISO 8062[24] and its British counterpart BS6615[25] Required allowances corresponding to ten grades of increasing magnitude are tabulated for cast dimensions in increasing size bands and, as in the case of dimensional tolerances, guidance is given as to the most appropriate grades for specific casting processes and alloys.

Tables 7.18a and b have been abridged from the standard itself to give a picture of the main trends but the full standard needs to be consulted for definitive data.

Other points to be considered in designing for machining include the provision of chucking surfaces or lugs on shapes which would otherwise be difficult to lift and hold in the machine and the elimination whenever possible of intermittent cuts in favour of a steady cut through solid metal.

Table 7.18 Format of machining allowance data presented in BS6615/ISO 8062. (a) Some allowance grades according to magnitude of dimension; (b) influence of process conditions and alloy type on allowance grade

(a)

Largest dimension mm	Recommended machining allowance RMA mm					
	Grade					
	A	C	E	G	H	K
0–40	0.1	0.2	0.4	0.5	0.7	1.4
100–160	0.3	0.5	1.1	2.8	4	8
400–630	0.5	1.1	2.2	4	6	12
1600–2500	0.8	1.6	3.2	6	9	18
6300–10 000	1.1	2.2	4.5	9	12	24

Note: The very small allowances in the first two grades A and B are for special cases only, involving series production with pattern, casting and machining procedures agreed between customer and foundry.

(b)

Production method	RMA grades			
	Casting material			
	Steel	Cast irons	Zn base	light alloy
Sand cast hand moulded	G-K	F-H	F-H	F-H
Sand cast machine moulded and shell moulded	F-H	E-G	E-G	E-G
Metallic permanent mould gravity and low pressure die casting	–	D-F	D-F	D-F
Pressure die casting	–	–	B-D	B-D
Investment casting	E	E	E	E

The above tables are adapted from BS 6615: 1996 with the permission of BSI under licence number 2000 SK/0049. The complete standard can be obtained by post from BSI Customer Services, 389, Chiswick High Road, London W4 4AL.

In machining castings one of the most important points concerns the correct use of datum surfaces, whether as reference planes or as locations for jigs and fixtures, to ensure the correct amount of stock removal from the various machined faces: the allowances must be established relative to such surfaces. Serious errors can otherwise arise from an apparently harmless dimensional variation. This illustrates the importance of specifying reference surfaces which are not inherently susceptible to dimensional errors: moulded rather than cored surfaces should be used for this purpose. In the absence of full marking out, template checks should preferably precede machining. Richards[67] discussed this problem with particular reference to precision castings, stressing the need to split errors by providing initially machined location faces in the interest, for example, of eventual dynamic balancing. A comprehensive treatment of datums and tooling points for machining has since been undertaken by Campbell[20].

For more comprehensive and detailed treatments of the physical design features of castings reference should be made to the specialised works listed as References 1–4 and 6.

Engineering aspects of casting geometry

The principles of design for engineering performance are outside the scope of the present work, although certain general points relating to castings have been discussed in this and earlier chapters.

The avoidance of stress concentrations due to notch effects, examined in Chapter 5, is one of the most important principles in casting design; Caine[42] expressed the view that the use of generous fillet radii would do more to increase the service strength and promote the use of castings than any half dozen new processes. The determination of suitable fillet radii has already been reviewed in the context of measures designed to secure stress-free cooling. The use of smooth, continuous contours and overall streamlining, well adapted to production by casting, conforms to the same principle. This aspect assumes even greater importance where fatigue resistance is of paramount concern, as is the case of the oilrig node, which demonstrated the clear advantage of the cast design alternative in this respect.

To achieve light, strong and rigid construction the weight of metal should be disposed in positions where the properties are fully utilised in carrying stresses: this entails the use of expanded rather than solid sections. Figure 7.39 illustrates some alternative cast sections which can be employed to sustain bending stresses in a lever casting: of these, the hollow forms provide much the highest strength and stiffness–weight ratios since they carry little material in the region of the neutral axis. This also simplifies the task of feeding. The superiority of the tubular and U sections over the oval and I beam sections also extends to fatigue conditions.

Although closed sections give maximum resistance to torsional stresses, open sided sections are to be preferred from a moulding standpoint and can include corrugated or 'sinewave' constructions for the stiffening of extended webs, avoiding the problem of intersections as introduced by

	Maximum fibre stress when loaded to 21 000 lbf (93.4 kN) about XX		Loading capacity about XX for maximum fibre stress of 50 000 lbf/in² (345 MN/m²)	
	lbf/in²	MN/m²	lbf	kN
9 in / 2 in / ¾ in / ¼ in r / Kt = 1.8 / Oval	102 000*	703*	10 000*	44.5*
4⅜ in / ⁷⁄₁₆ in / ⁹⁄₁₆ in / 2¾ in / I-beam	34 000	234	31 000	137.9
2¾ in / 4⅜ in / ³⁄₁₆ in / ⁵⁄₁₆ in / Tubular	37 000	255	28 000	124.5
⅝ in / 2⅜ in / 4⅜ in / ⁵⁄₁₆ in / ⅝ in / 1¼ in / 2¾ in / U-section	45 000	310	23 000	102.3

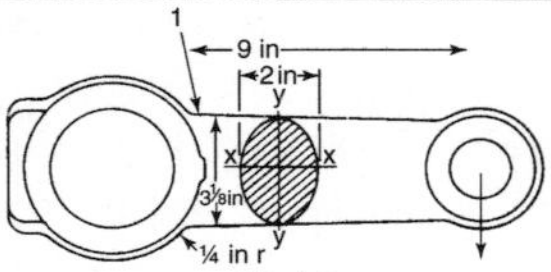

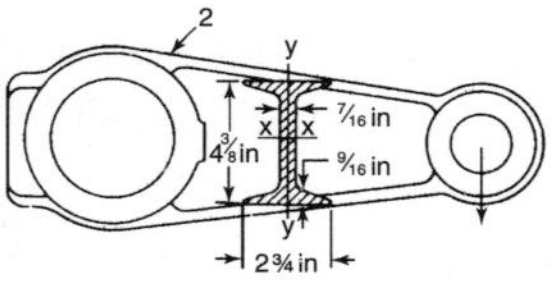

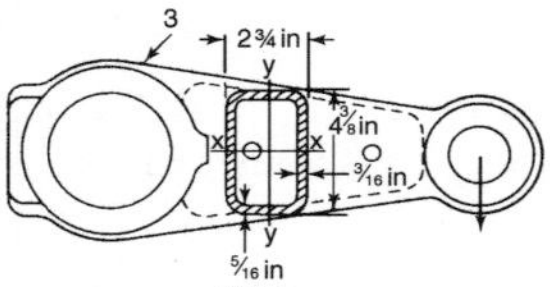

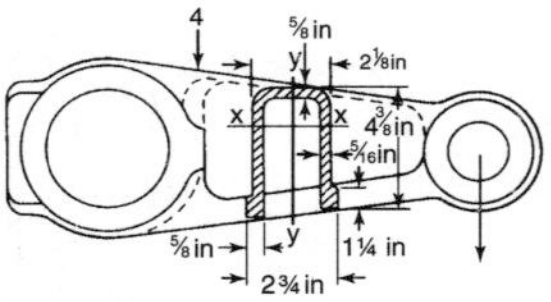

* Values for oval section design do not take account of additional stress concentration (factor $K_T = 1.8$) at junction with hub.

Figure 7.39 Design characteristics of alternative cross-sections for lever casting (after Caine[2]) (courtesy of American Foundrymen's Society)

ribs. The characteristics of further designs of cross-section in relation to the stress pattern in a single member are summarised in Figure 7.40.

Design for maximum strength must also take account of the particular combination of mechanical properties of the alloy concerned. A good example of this principle is the bracket illustrated in Figure 7.41, designed to utilise the superior compression strength of cast iron as distinct from its much lower tensile strength.

Specific guidance to design calculations for some standard cast sections has been given for ferrous materials in the works by Caine[2] and Angus[54]; the former author also provides a guide to calculations for irregular sections.

For more general information of this kind reference needs to be made to specialised works on engineering design.

The design of most components is based on the principle of functional performance at minimum cost. Notwithstanding the virtues of lightness and minimum consumption of material, total production cost is not simply related to casting weight. Material consumption has the strongest influence

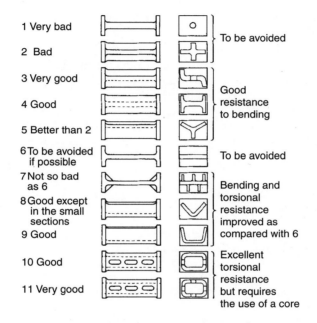

Figure 7.40 Summarised characteristics of various designs of cross-section for casting members (courtesy of CDC)

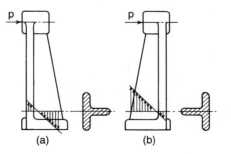

Figure 7.41 Design of bracket to exploit compression strength of cast iron: (a) preferred, (b) unsuitable design (from Shestopol[68]) (courtesy of Institute of British Foundrymen)

on overall cost in the case of intrinsically expensive alloys, but in many cases it is economically more important to reduce manufacturing cost than to minimise weight. This is accomplished by eliminating unnecessary coring and by designing to section thicknesses suitable for running and feeding: thickness may in these cases be determined not by strength or stiffness but by casting considerations, taking into account the fluidity and other foundry characteristics of the alloy and the advantage of uniform cooling in preventing residual stresses. Similarly, the costs of coring minor holes, slots or grooves should be weighed against the costs and relative accuracy of machining.

Lightness and high strength to weight ratio are the greatest assets in transport applications, where payload economics and power to weight ratio may outweigh initial cost considerations: design for minimum weight is however the object in only a minority of cases.

Both cases emphasise the importance of basing price structures on unit costs of production rather than upon weight alone: commercial policy can thus provide incentives to the designer to minimise the true cost of the final product.

References

1 *Casting Design Handbook*, Am. Soc. Metals, Metals Park, Ohio (1962)
2 Caine, J. B., *Design of Ferrous Castings*, Am. Fndrym. Soc., Des Plaines, Ill. (1963)
3 *Practical Guide to the Design of Grey Iron Castings*, C.I.F.A., London (1956), (Centre Technique des Industries de la Fonderie, Paris)
4 *Practical Guide to the Design of Steel Castings*, B.S.C.R.A., Sheffield (1958), (Centre Technique des Industries de la Fonderie, Paris)
5 Briggs, C. W., Gezelius, R. A. and Donaldson, A. D., *Trans. Am. Fndrym. Ass.*, **46**, 605 (1938)
6 Jackson, W. J., Ed. Steel Castings Design, Properties and Applications. SCRATA, Sheffield (1983)
7 ASM Handbook 10th Edn. vol. 20, Materials Selection and Design, ASM International, Metals Park, Ohio (1997)
8 Dickens, P. M., *Foundry International*, **16**, 1, 215 (1993)
9 Dickens, P. M., *Foundry International*, **18**, 3, 119 (1995)
10 Cox, G. M. A., *Foundrym.*, **89**, 331 (1996)
11 Elliott, W. O. and Kerns, W.J., *Trans. Am. Fndrym. Soc.*, **103**, 323 (1995)
12 Beaudolin, R., Carey, P. and Sorovetz, T., *Modern Casting*, **87**, Nov., 35 (1997)
13 Sharp, H. G., *Engineering Materials: Selection and Value Analysis*, Heywood-Iliffe, London (1966)
14 Villner, L., *Br. Foundrym.*, **62**, 488 (1969)
15 Svensson, I. and Villner, L., *Br. Foundrym.*, **67**, 277 (1974)
16 First Report of Sub-committee TS 71, *Br. Foundrym.*, **62**, 179 (1969)
17 Second Report of Sub-committee TS 71, *Br. Foundrym.*, **64**, 364 (1971)
18 Third Report of Working Group P4. *Br. Foundrym.*, **69**, 53 (1976)

19 Fourth Report of Working Group P4. *Br. Foundrym.*, **72**, 46 (1979)
20 Campbell, J., *Castings*, Butterworth-Heinemann, Oxford (1991)
21 *Quality Standards for Steel Castings*, Steel Castings Development Comm., Sheffield (1966)
22 ASM Metals Handbook, 9th Edn. Vol 15, Casting, ASM International, Metals Park, Ohio (1998)
23 Steel Castings Handbook, 6th Edn. SFSA, Cleveland (1995)
24 ISO 8062 Castings-System of dimensional tolerances and machining allowances. International Organisation for Standardization (1994)
25 BS 6615 Dimensional tolerances for metal and metal alloy castings. *Br. Stand. Instn.*, London (1995)
26 Polich, R. F. and Flemings, M. C., *Trans. Am. Fndrym. Soc.*, **73**, 28 (1965)
27 Report of Sub-Committee T.S. 63, Institute of British Foundrymen (1971)
28 Ahearn, P. J. and Quigley, F. C., *Trans. Am. Fndrym. Soc.*, **72**, 435 (1964)
29 Murphy, A. J. (Ed.), *Non-Ferrous Foundry Metallurgy*, Pergamon Press, London (1954)
30 *Quality Requirements of Super-Duty Steels*, Ed. R. W. Lindsay, Interscience, New York, London (1959)
31 Evans, E. B., Ebert, L. J. and Briggs, C. W., *Proc. A.S.T.M.*, **56**, 979 (1956)
32 Harris, H. H., *Trans. Am. Fndrym. Soc.*, **68**, 782 (1960)
33 Armitage, R., Solidification Technology in the Foundry and Casthouse. Book 273. The Metals Society, London, 385 (1983)
34 Marston, G. J., *Foundrym.*, **83**, 108 (1990)
35 Woelke, G., *Foundry International*, **17**, 4, S6 (1994)
36 The Incorporation of Castings in Welded Assemblies: Symposium of the Institute of Welding, *Fndry Trade J.*, **120**, pp. 325 and 395 (1966)
37 Talanov, P. I., Vlasevina, L. K. and Sysoev, S. O., *Russ. Cast. Prod.*, 506 (1966)
38 Kear, B. H., Breinan, E. M. and Greenwald, L. E., *Metals Technology*, **6**, 4, 121 (1979)
39 Trafford, D. N. H., Bell, T., Megaw, J. H. P. C. and Brandsen, A. S., *Metals Technology*, **10**, 2, 69 (1983)
40 Donaldson, E. G., *Br. Foundrym.*, **79**, 262 (1986)
41 Bamberger, M., *Internat. Mat. Rev.*, **43**, 5, 189 (1998)
42 Caine, J. B., *Trans. Am. Fndrym. Soc.*, **69**, 259 (1961)
43 BS 7448: Fracture mechanics toughness tests, *Br. Stand. Instn.*, London. (Part 1, Method for determination of K_{IC} critical CTOD and critical J values of metallic materials)
44 BS EN ISO 12737 Metallic materials. Determination of plane strain fracture toughness
45 Armsworth, R., *Materials World*, **6**, 2, 86 (1998)
46 Jackson, W. J., *Br. Foundrym.*, **70**, 191 (1977)
47 Harding, R. A., *Foundrym.*, **86**, 197 (1993)
48 Mercer, R., *Br. Foundrym.*, **54**, 410 (1961)
49 BSI Standards Catalogue (Annual issue). British Standards Institution, London.
50 ASTM Annual Book of Standards. American Society for Testing and Materials. Section 00, Index; Section 1 Vol 01.02 Ferrous Castings; Section 2 Vol 2.01 Copper and Copper Alloys; Vol 2.04 Nonferrous Metals etc.
51 *The Materials Selector*, 2nd Edn. Waterman, N. A. and Ashby, M. F., Eds. Chapman and Hall, London (1997)

52 *ASM Handbook*, 10th Edn. Vol. 1 Properties and Selection: Irons, Steels and High-Performance Alloys; Vol. 2 Non-ferrous Alloys (1990–1992)
53 *Steel Castings Handbook*, 6th Edn. SFSA, Cleveland (1995) ASM International, Materials Park, Ohio (1995)
54 Angus, H. T., *Cast Iron: Physical and Engineering Properties*, Butterworths, London (1976)
55 Elliott, R., *Cast Iron Technology*. Butterworth-Heinemann, Oxford (1988)
56 *Smithells Light Metals Handbook*, Ed. Brandes, E. A. and Brook, G. B., Butterworth-Heinemann, Oxford (1998)
57 British and European Aluminium Casting Alloys. Assoc. Light-Alloy Refiners Ltd. (1996)
58 West, D. R. F., *The Selection and Use of Copper-base Alloys*. OU Press, Oxford (1979)
59 *The Metals Black Book*, Vol. 1. 2nd Edn. Ferrous Metals (1995); *The Metals Red Book*, Vol. 2 Non-ferrous Metals (1993). Ed. Bringas, J. E., Casti. Pub. Inc., Edmonton, Alberta.
60 Charles, J. A., Crane, F. A. A. and Furness, J., *Selection and Use of Engineering Materials*, 3rd Edn, Butterworth-Heinemann (1997)
61 Ashby, M., *Materials Selection in Mechanical Design*, 2nd Edn, Butterworth-Heinemann (1999)
62 *Smithells Metals Reference Book*, 7th Edn, Brandes, E. A. and Brook, G. B., Eds., Butterworth-Heinemann, Oxford (1998)
63 Campbell, J., *Cast Metals*, **4**, 3, 129 (1991)
64 Schreir, L. L., Jarman, R. A. and Burstein, G. T., *Corrosion*, 3rd Edn, Butterworth-Heinemann, Oxford (1994)
65 *ASM Metals Handbook*, 9th Edn, vol. 13, Corrosion, ASM International, Metals Park, Ohio (1988)
66 *Cast Metals Handbook, 4th Ed.*, Am. Fndrym. Soc., Des Plaines, Ill. (1957)
67 Richards, G., *Br. Foundrym.*, **72**, 162 (1979)
68 Shestopol, V. M., *Proc. Inst. Br. Foundrym.*, **39**, B51 (1945–6)

8

Production techniques 1
The manufacture of sand castings

Although many process variations are used in metal founding, the production of castings from solid patterns in rammed refractory moulds still accounts for the greater part of the industry's output. The present chapter, a review of the essential stages in founding, is based largely on this process, although much of the subject matter is applicable to a wider range of casting processes. More specialized casting techniques will be reviewed in the ensuing chapters.

Sand castings, a term used loosely to include other castings made in refractory moulds, range in weight from a few grams to several hundred tonnes, thus covering almost the entire size range for cast parts; virtually all types of casting alloy are produced. The reason for the dominant position of the process lies mainly in its great flexibility in relation both to design and to production facilities. In these respects it epitomizes the qualities attributed to casting processes as a whole and discussed in the Introduction and Chapter 7. Almost unlimited freedom of shaping is combined with low capital and operating costs, whilst the process is suitable for any quantity of components. Thus, although other casting processes may excel in individual respects, sand casting is uniquely versatile in relation to weight, composition, shape and quantity. Specific design characteristics of sand castings are examined in Chapter 7.

The manufacture of a sand casting can be considered in three main stages, namely the production of the mould, melting and casting, and finishing operations: these will be the subjects of separate sections of the chapter.

Part 1. Mould Production

The first major stage in founding is the production of the mould, with its impression of the casting and its planned provision for metal flow and feeding. For this purpose a pattern is required, together with foundry equipment ranging from moulding and coremaking machines to moulding boxes, tackle and hand tools. Following a survey of this equipment the basic techniques of moulding and core production will be reviewed.

It should be emphasised that the moulding procedure for a particular casting is largely determined by the means chosen at the outset for pattern removal and embodied in the construction of the pattern. For this reason, decisions as to the entire manufacturing technique should be taken at the earliest stage, including consideration of the orientation of the casting for gating and feeding as well as for pattern withdrawal. A rational choice can then be made of the system of parting lines, cores and other features to achieve overall economy in manufacture.

A further decision which greatly influences production costs concerns the size of mould unit to be adopted and the arrangement of patterns in relation to mould dimensions. The object should be to employ the largest size of mould that can be conveniently handled by the plant, and to achieve intensive use of mould space through a high packing density of castings. This reduces the number of moulding operations and minimizes sand consumption. These objectives are achieved by the use, whenever possible, of multiple-casting moulds in which two or more patterns are grouped round a common sprue or feeding system, or in which mould parts are stacked to produce superimposed layers of castings. These techniques are illustrated in Figure 8.1. In the case of stack moulding it is sometimes possible for the back of one mould part to form the face of the next as shown in Figure 8.1c, virtually halving the number of mould parts required. Where such parts carry impressions on both faces they may be produced by machine squeezing between two pattern plates, a principle employed in some of the automatic boxless block moulding systems. A further advantage of such systems is the possibility of improved casting yields through the sharing of feeder heads and gating systems (q.v. Chapter 3).

It can also be advantageous in certain cases to combine components into larger units for casting purposes. A black casting may thus be designed for the production of a number of machined parts. Single rings may be parted from a cast cylindrical bush, or a half round component may be tied to a similar component to provide more symmetrical cooling and avoid distortion; all such measures need to be determined before the pattern is constructed.

The nature of the moulding practice is closely related to the moulding material employed. Various classes of binder were examined in Chapter 4, as were the characteristics of green and drysand practice and various mould and core hardening reactions. These considerations influence the choice between moulding box and block moulding, for example, and thus affect the pattern requirements from the outset.

Equipment for moulding

1. Pattern equipment

The nature of the pattern equipment depends in the first instance upon quantity requirements and the intended method of production. Patterns

(a)

(b)

Figure 8.1 Grouping of castings for production: (a) Steel valve cover castings produced two-in-a-box using a common sprue (courtesy of Catton & Co. Ltd.): (b) stack moulded cast iron piston rings (courtesy of Stockton Casting Co. Ltd.): (c) (overleaf) arrangement of stacked moulds to employ both faces of mould parts

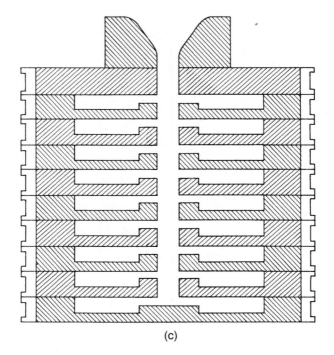

(c)

Figure 8.1 (*continued*)

can range from simple wooden equipment, requiring skilled handling and intended only for short life, to elaborate metal plate assemblies, complete with gating and feeding systems, suitable for mass production by machine moulding; similar considerations apply to coreboxes. The governing factor is the permissible first cost in relation to the output envisaged.

Although it is generally recognized that pattern equipment should be designed to suit the founder's own method of production, castings are frequently ordered from previously existing equipment. This can place a restriction on moulding and casting methods; a further disadvantage is the occasional survival of an insufficiently durable pattern, originally intended for the production of a few castings, into bulk production. In such cases the cost of pattern modification or replacement represents a valuable investment in terms of casting quality.

The basic types of pattern will be considered at a later stage in relation to moulding technique. There has been a major shift towards more refined equipment, accompanying a widespread transition from manual skill to engineering technique as the principal element in moulding. Standardization and mass production, coupled with fewer available skilled craftsmen, requires that mould production be 'pre-engineered' at the patternmaking stage. Patterns are accordingly designed with less reliance on the sculpture of difficult stepped joints and use of loose pieces, but with correspondingly greater use of cores in conjunction with design measures for simplification of moulding. Moulding skill can then be deployed on heavier castings and on jobbing work required in small quantities.

A further feature of mounted patterns as used in machine production is the inclusion of integral gating and feeding systems, giving complete standardization of casting method and avoiding the loss of mould quality which accompanies manual cutting.

Manufacture of pattern equipment may be undertaken either by the foundry or by a specialist concern able to supply a comprehensive range of modern tool manufacturing techniques. Materials used include wood, plastics or resins, cast iron, brass, white metals and light alloys. Composite construction, with harder materials used as inserts at critical wearing points, can be used in some cases: the standard of materials and construction depends on the need for durability and dimensional stability.

Important features are the allowances made for contraction and draft. Examples of standard contraction allowances are given in Table 8.1, but it is the practice to vary these in the light of experience with particular types of dimension. The various factors influencing contraction behaviour, including mould constraints, have been examined in great detail by Campbell[1] with a view to the production of calibration curves for the more accurate prediction of dimensional behaviour in individual circumstances. Draft taper, incorporated on faces normal to the intended parting plane, is usually in the range $\frac{1}{2}$–3 degrees for sand castings, although it is preferable to design castings incorporating more generous natural taper.

Moulding operations are aided by a colour coding system indicating the functions of the patterns, for example coreprints and machined surfaces. Typical colours used for some of the principal features are as follows:

As-cast surface–main body of pattern:	Red/orange
Machined surface:	Yellow
Coreprint:	Black
Loose piece seating indication:	Green

Table 8.1 Typical contraction allowances for castings

Alloy type	Allowance %	Equivalent ratio	Equivalent, in per ft
Steel	1.6	1 in 64	$\frac{3}{16}$
Cast iron	0.8	1 in 120	$\frac{3}{32}$
Brass	1.4	1 in 70	$\frac{11}{64}$
Aluminium	1.3	1 in 77	$\frac{5}{32}$

In loose pattern moulding this system can be extended to include indications for the positioning of heads and chills, thus using the pattern as a vehicle for method instructions to the shop floor.

2. Moulding boxes

Moulding boxes and flasks are selected for minimum sand consumption and production time per unit weight of casting. This aim can often be realized by grouping and stacking castings as previously described.

The shapes and sizes of box parts selected to form a stock must depend on the type of product, but much is to be said for standardization on a few sizes, with a high degree of interchangeability of parts. This is particularly necessary for machine production, for which length and breadth can be fully standardized, leaving only depth to be varied according to need.

Boxes can be of cast or fabricated construction in steel, cast iron or light alloy. Robust construction is required to meet the rough handling to which they are often exposed. For small castings, units fabricated from rolled steel section find wide use, being both strong and light. Heavy construction is needed for large moulds to give rigidity in lifting; for the largest boxes sectional construction can be adopted, side and end elements being bolted together as required, reducing the number of parts which need be held in stock.

One of the most important features is the location system for accurate alignment of the mould parts. This is commonly based on steel pins riding in holes through projecting lugs. The holes should be jig drilled and the boxes thereafter carefully maintained with respect to the accuracy of the pin-centres dimension. Many boxes are provided with wear resisting hardened steel bushes in the pinholes; the use of one round and one elongated hole enables accurate location to be maintained even with limited box distortion. This system can be combined with loose pins which can be renewed as soon as wear is detected: these must however always be left in position in closing to avoid movement when clamping.

Other features include fixed or detachable bars, often fitted to large boxes to reinforce the sand mass. Handles and crane lifting attachments, trunnions for turnover, clamping bars and sand retaining flanges are also fitted according to the size and purpose of the equipment.

Moulds may also be produced using the *snapflask* or similar systems in which the box is used only as a frame for ramming purposes. The flask is hinged at its corner, being opened after ramming to leave a block mould bound by steel bands placed in the flask beforehand. This method, suitable for the mass production of small moulds, dispenses with the need for a large box stock. In certain cases boxes are dispensed with altogether as in floor moulding, high pressure block moulding, and core assembly production; these techniques will be further referred to at a later stage.

Intensive box utilization is desirable in view of the appreciable capital value represented by the box stock: one object of production planning is to achieve rapid box turnround by minimizing delays between moulding, casting and knockout.

3. Ancillary equipment and tools

Apart from the moulding box, tackle is required for mould strengthening and core support. Reinforcing grids for larger moulds are usually cast to the required shape and size in an open sand bed in the foundry, whilst general stocks of irons, nails and sprigs need to be maintained, with studs and chaplets for core support. Tools for hand moulding include the following:

Ramming:	peg and flat rammers; pneumatic rammers for bench or floor work;
Cutting and finishing:	trowels, gate knives, cleaners;
Venting:	vent wires of various diameters.

Moulding and coremaking machines are vital items in modern foundries: these will be considered later in relation to the ramming of moulds and cores.

With increasing demand for precision in casting, patterns, coreboxes, driers, moulding boxes and machines must be manufactured and maintained to the highest standards. This minimizes the need for hand finishing, patching and rubbing of moulds and cores or for subsequent sculpture in the fettling shop due to imperfect or worn equipment.

Moulding techniques

1. Pattern utilization

Moulding methods range from manual operations with the simplest forms of pattern to highly mechanized operations for mass production; in the most modern developments a high degree of automation is achieved. The moulding operations themselves are closely bound up with the design of pattern equipment, especially the arrangement of mould parts, cores and other devices enabling the pattern to be withdrawn. These aspects will now be further examined.

Loose pattern practice

In the simplest form of loose pattern moulding the parting line is formed during the sequence, the mould parts being rammed directly one upon the other with moisture resistant parting powder used for separation. The operational steps are summarized in Figure 8.2. A single flat parting plane can be used in many cases but for more complex designs stepped joints or additional partings may be required.

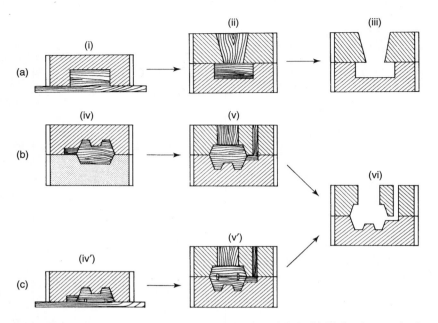

Figure 8.2 Basic sequences in loose pattern moulding. (a) Plain shape: single parting formed by flat surface of pattern; (i) bottom part rammed on flat board, (ii) top part rammed on bottom part after turnover, (iii) complete mould after separation, pattern withdrawal and closing. (b) Oddside employed to bed one-piece pattern to parting surface; (iv) bottom part rammed on oddside, (v) top part rammed on bottom part after turnover and removal of oddside, (vi) complete mould after separation, pattern withdrawal and closing. (c) Split pattern; (iv') bottom part rammed from half pattern on flat board, (v') top part rammed on bottom part after turnover and positioning of second half pattern, (vi) complete mould after separation, pattern withdrawal and closing

In floor or pit moulding a different procedure needs to be adopted since the mould cannot be turned over. For the making of the drag or bottom part, normal ramming against the pattern is not possible; instead, the pattern is bedded on to a previously rammed sand base, laid over a coke bed for venting.

One method of accommodating features preventing straight draw to a single parting plane is the provision of loose pieces on the pattern, suitable for separate lateral withdrawal into the mould cavity after removal of the main body of the pattern. The most widely used method of all, however, is the introduction of a core: the various functions of cores will be separately examined. The common alternative methods of providing for features impeding direct withdrawal are summarized in Figure 8.3. An interesting additional solution to this problem is offered by the full mould casting process: this utilizes an expendable loose pattern piece which is left in the mould and displaced by the metal on casting.

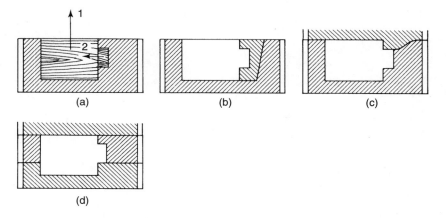

Figure 8.3 Accommodation of undercut features in moulding: (a) pattern with loose piece, showing direction and sequence of draw, (b) core, (c) stepped joint, (d) additional joint involving extra mould part

Oddsides.   Where a pattern has a flat surface suitable to form a parting plane (as in Figure 8.2a), ramming of the first mould part can be carried out on any flat surface. Where the parting plane intersects the casting, however, the parting surface must be formed initially by bedding the pattern in moulding sand to form a base for ramming the first section. This initial support is termed the oddside (see Figure 8.2b). This may be made in dried sand or coresand for repeated use, whilst for long production runs a permanent oddside can be made in plaster. The device saves moulding time.

Split patterns.   Where a flat parting plane intersects the pattern, moulding is simplified by providing a split pattern and dispensing with the oddside. The first mould part can then be rammed from a half pattern on a flat surface and the second rammed on the first with the pattern halves mutually located by dowels. This sequence is illustrated in Figure 8.2c.

Mounted pattern practice

Moulding operations are greatly simplified by the provision of split patterns mounted on plates or boards. This obviates the need to form a parting surface during moulding: each half pattern is mounted on its own flat or contoured plate and the mould parts can be made independently of one another, whether by hand or on moulding machines.

One of the major advantages is the mounting of integral gating and feeding systems with the patterns. The accurate implementation of method decisions is thus ensured from the outset and the degree of skill required in moulding reduced. A typical assembly is illustrated in Figure 8.4.

Figure 8.4 Mounted pattern assembly for machine moulding

The mounted pattern principle can be used in conventional moulding box practice or in various block mould or core assembly techniques. In all cases provision is required for accurate mutual alignment of the mould parts. Where moulding boxes are being used pattern plates can be provided with pins for box location during moulding, the pattern positions being marked out relative to the pins. In machine moulding the pin locators are mounted on the machine, in which case both pattern plate and box part are located independently: the plate may slip over the pins or may sit in a recess on the machine table.

Matchplates. Normal pattern plate practice requires separate plates for top and bottom part production. A matchplate pattern enables a complete mould to be produced in a single operational sequence. In this system the pattern elements forming the cope and drag are mounted on opposite faces of the same plate. The plate is then sandwiched between the boxes for the successive production of each mould part, the whole assembly being aligned with long pins: the parts are readily realigned on removal of the matchplate. The sequence is illustrated in Figure 8.5.

Matchplates and conventional plates for long production runs are frequently produced as single units with integrally cast pattern elements. Using techniques of plaster moulding and pressure casting, accurate units can be produced in aluminium alloy, embodying flat or irregular parting surfaces with equal facility.

Block mould and core assembly techniques

Moulding from mounted patterns is frequently carried out in conjunction with block mould systems requiring no moulding boxes in the normal sense.

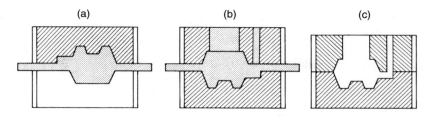

Figure 8.5 Hand or machine moulding from matchplate pattern: (a) bottom part rammed with matchplate supported on empty boxpart, (b) top part rammed after turnover, or separately supported on empty boxpart, (c) complete mould after pattern withdrawal and closing

Temporary boxes or frames are employed to contain the moulding material during compaction, but these are removed before closing and casting, leaving boxless mould parts analogous to cores. (see Figure 8.25e).

Core assembly moulds have themselves had a long history in the production of high quality castings, relatively small and intricate parts being produced in sands which develop high strength on hardening. The quality of such cores has been enhanced by the availability of the modern binder systems which permit setting in the box. This has provided the basis for the manufacture of complex castings in which close accuracy is achieved by the use of mould units built from clusters of numerous individual cores. These are pre-assembled in rigid bolted and banded packages for insertion into the main mould, virtually all cast surfaces being formed by the cores.

The ease of setting also enables the boxless assembly principle to be employed for very large moulds: the cement sand process was an early example of the use of block moulds even for very heavy castings. In all these cases mutual alignment of the mould parts requires integrally moulded location devices, of which one form is featured in the illustration shown.

In snapflask and other systems which similarly employ moulding boxes only during the moulding operation, a complete mould can be produced from mounted patterns in a single operational sequence. The block mould principle also extends to high pressure greensand, being employed in automatic moulding systems of which a typical example will shortly be described. In this case the sand blocks are mutually supported in the form of a continuous strand to the casting point.

Direct mould cavity generation

To the numerous techniques which shape the mould against pattern surfaces must now be added the ability to cut the cavity from solid by CNC machining. Milling and other operations, as employed in component or tool manufacture, can be applied directly to pre-formed blocks of chemically bonded sand. This offers a rapid and low cost route to the production of moulds for prototypes or small numbers of castings without the need for custom built tooling.

Strickle moulding

Strickle moulding too differs from other practices in that the customary solid pattern is dispensed with in favour of simple boards cut to the profile of the casting. By mounting such a board on a central spindle the mould for a casting of circular form can be struck up by rotation (Figure 8.6). A similar principle can be used for other shapes of uniform profile by providing guides to the path of the strickle board. A related practice uses skeleton patterns: the full form of a solid pattern is in this case obtained by infilling and shaping with moulding material. Strickle boards and skeleton patterns can be supplemented by solid pattern elements, for example gear tooth segments, to form features of irregular or complex shape.

The use of strickles is also traditionally associated with loam practice, in which the main body of the mould is constructed from bricks and reinforced with grids and irons; a wet and highly plastic moulding material is then sleeked on as a comparatively thin facing, the whole being consolidated by drying. Strickle techniques are, however, adaptable to the use of modern types of conventional moulding material, which have largely displaced the earlier loam practice.

In all these techniques the cost of pattern equipment is minimized at the expense of increased use of skilled labour in moulding.

2. Mould compaction

The object in moulding is to produce accurate parts able to withstand lifting and handling and to contain the hydraulic pressures in casting. Mould strength depends partly upon effective compaction and partly, in the case of heavier castings, upon mechanical support derived from box bars and metal reinforcements. The general aim in compaction is to bring the bulk density

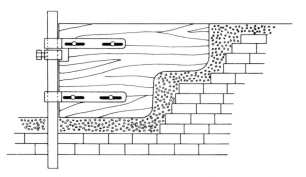

Figure 8.6 Strickle moulding system for casting of regular sections

of the moulding material from its 'loose' value of about $0.6–0.75\,g/cm^3$ towards a value when fully packed of about $1.6–1.8\,g/cm^3$. The influence on the properties of moulding sand is demonstrated in Figure 8.7, showing the direct relationship between strength properties and bulk density. The packing density is never uniform throughout the mould volume: its distribution depends upon the technique of compaction and is markedly different for the various methods of hand and machine ramming.

Hand ramming is adopted for jobbing production where small quantity requirements or unsuitable size preclude the use of moulding machines. Intermediate between hand and full machine moulding is the production of light moulds on simple pattern draw machines where ramming is still by hand, although machine type pattern assemblies are employed and the mould is stripped by a hand operated machine action.

The traditional method of compaction with peg rammers relies heavily upon manual skill for the production of dense moulds. The natural tendency in ramming progressively up the depth of a mould is to produce layers of varying density, giving rise to soft zones on vertical surfaces. The typical bulk density distribution within such a mould is shown in Figure 8.8, portraying the results of hardness surveys on moulds compacted by hand and machine techniques. Dense mould surfaces can however be achieved by careful hand ramming: since the rammer is selective, individual attention can be given to confined pockets such as give rise to difficulties in machine compaction. Sand flowability is less important than in machine moulding, since the aim is to localize the effect of each rammer blow. The tendency is thus to use sands of high green strength, in some cases based on naturally bonded sands.

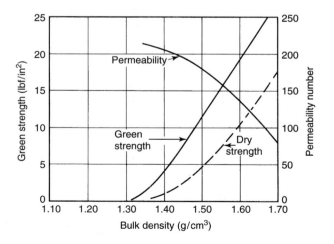

Figure 8.7 Influence of bulk density on strength and permeability of moulding sand (from Davies[2]) (courtesy of British Steel Corporation)

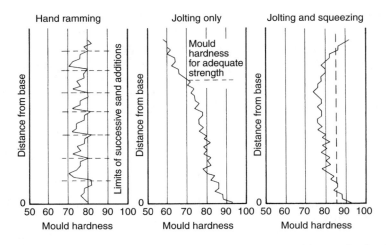

Figure 8.8 Influence of ramming technique on mould hardness distribution (from Davies[2]). Superimposed dotted line approximates to distribution obtained in similar survey by Stoch and Lambert for simultaneous jolt–squeeze (Reference 3) (courtesy of British Steel Corporation)

The use of pneumatic tools in hand moulding does not introduce any major change in mould characteristics, although production rates can be greatly increased and labour reduced.

Machine moulding

Machine moulding enables moulds to be produced in quantity at high rates. Excluding the simplest types of hand operated pattern draw machine, the principal feature of machine moulding is the use of power operated mechanisms for mould compaction; this mechanization can extend in varying degrees to pattern withdrawal and mould part manipulation.

Machine compaction is distinguished from hand ramming by the reliance placed upon flowability of the moulding material to achieve the required packing density: with the exception of the sandslinger, energy is applied to the whole area of the mould simultaneously, producing uneven packing unless considerable flow can occur. Machine moulding sands are thus selected for high flowability rather than maximum green strength.

The actual compaction is effected by actions of jolt, squeeze, vibration, blow/shoot, vacuum, and impact on the sand, either singly or in various combinations depending on the type of machine selected; characteristics of some of the typical systems will be reviewed.

Jolt and squeeze compaction

Many moulding machines rely on jolting and squeezing, either alone or in combination: the characteristics of the mould differ according to the action. In the former case the machine table carrying the pattern assembly is repeatedly jolted against a stop, the depth of travel being termed the stroke of the machine. The action may take the form of a plain jolt against a fixed anvil forming the main body of the machine, the shock being transmitted to the foundation, or a shockless jolt principle may be employed in which the jolt takes place against a spring supported floating anvil, the shock being largely absorbed within the box assembly. Jolting produces a mould density distribution giving maximum hardness against the pattern surfaces. This type of action is used on a wide range of machines, including those of largest capacity.

In squeezing, pressure is applied to the upper surface of the moulding material through a squeeze head. In this case the distribution of mould density is reversed, with a general tendency towards maximum hardness adjacent to the squeeze head. These effects are illustrated in Figure 8.8.

Combinations of jolting and squeezing can be employed with great advantage, using either successive or simultaneous application of the two actions: the latter provides the most effective and uniform compaction[4,5]. The shockless jolt machine uses this principle. Jolt squeeze machines such as that typified in Figure 8.9 became the workhorses of the mechanized industry and the principle is still used, although many machines have been replaced by types incorporating more sophisticated compaction systems.

Squeeze moulding is an attractive system on grounds both of speed and silence, but its limitations when working with patterns of appreciable depth were demonstrated in several original investigations[4–10]. The effect of frictional forces in impeding the free flow of sand is illustrated in Figure 8.10. Higher squeeze pressures and specially designed squeeze heads were subsequently introduced. Pressures in the range 0.7–2.8 MPa (100–400 lbf/in^2) and much higher were investigated by Dietert[8], Kremnev et al.[10] and Nicholas[4,12] but it was established that most of the advantage was achieved by around the 1.4 MPa (200 lbf/in^2) level: the beneficial influences on mould hardness and dimensional accuracy are shown in Figure 8.11.

To overcome the problem of disparities in the depth of sand below the squeeze head, differential thrust was introduced using individually contoured squeeze heads, flexible diaphragms, and heads embodying separate squeeze pistons or feet as illustrated in Figure 8.12. Using these techniques less extreme pressures are required, since an element of lateral thrust is achieved, giving high mould hardness even on deep vertical surfaces[6,7]. Such features have been adopted on numerous modern machines, often in combination with other aids to compaction. One widely adopted machine incorporating multi-piston squeeze is shown in Figure 8.13: in this case the squeeze, action is used for final compaction, following pre-compaction by through flow of compressed air. The machine is available with alternative combinations of compressed air and vacuum compaction in conjunction with

Figure 8.9 Jolt–squeeze pin lift moulding machine suitable for automatic or manual operation (Osborn Ram-Jolt Machine) (courtesy of Efco. Ltd.)

the high pressure piston squeeze facility; one version incorporates all the alternative modes in a single machine.

High pressure moulding requires a high standard of pattern construction and strong, rigid moulding boxes to withstand distortion. Appreciable deflection under pressure can cause springback and fouling of the pattern on withdrawal[11,12].

Mention should be made of one further technique of squeeze moulding, the *downsand frame* principle. Using this system box bars can be employed in conjunction with squeeze compaction and a high bulk density achieved against the pattern. The essential feature is the provision of a frame between moulding box and pattern plate (Figure 8.14). The pattern assembly is free to move up within the frame on squeezing, compressing the sand against

(a)

Mould hardness survey

	91	92	97	98	98	95	93	

Top

81	86	86	94	94	95	89	89	85
82	78	87	95	93	98	90	82	87
68	68	53				63	73	70
60	61	51				57	60	56
48	46	45				54	55	60
36	33	32				48	52	48
30	25	28				34	40	37
20	24	8				25	40	29

Top of pattern

			89	92	95			

(b)   Parting

Figure 8.10 Influence of friction on sand flow in squeeze compaction: (a) deformation of sand tracer grid, (b) hardness survey of mould section. Flat squeeze: pattern 3 in × 6 in: initial sand height 12 in, density 50 lb/ft^3; final 8 in, 65 lb/ft^3: density non-uniform, soft spots at pattern corners (from Bodsworth et al.[6]) (courtesy of Institute of British Foundrymen)

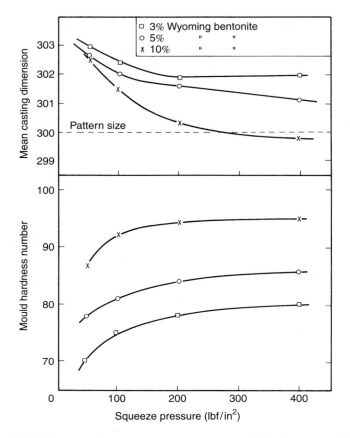

Figure 8.11 Influence of squeeze pressure on mould hardness and dimensional accuracy of castings (after Nicholas and Roberts[4]) (courtesy of British Foundryman)

the pattern until the plate lies flush with the bottom of the moulding box: the initial depth of frame is selected according to the degree of compression expected.

Vibration. The use of vibration in machine moulding was originally confined to the subsidiary role of assisting pattern withdrawal, but it later came to be widely employed in the compaction of chemically bonded sands requiring less energy than clay bonded greensands. These are typically dispensed directly from continuous mixers of types examined and illustrated in Chapter 4, consolidation being assisted by simple compaction tables. Vibration is also used in conjunction with unbonded sand systems of the types reviewed in Chapter 10, and performs a central function in the modern *vibratory squeeze* moulding machine. This employs a vibratory

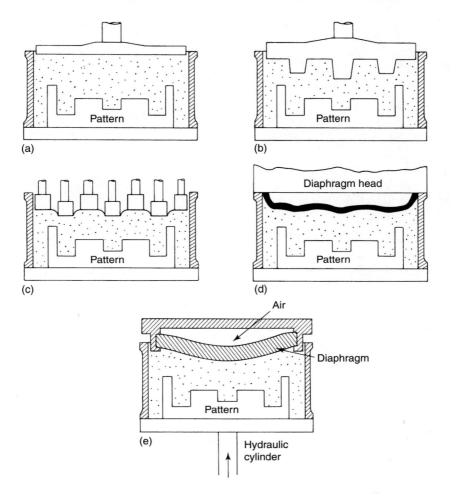

Figure 8.12 Designs of squeeze head (from Stoch and Lambert[17]): (a) conventional squeeze, (b) profile squeeze head, (c) equalizing piston squeeze, (d) diaphragm squeeze, (e) high pressure hydrapneumatic squeeze (courtesy of Institute of British Foundrymen)

pre-compaction stage on entry of the sand. The vibration then continues whilst a multi-piston contour squeeze head is brought down to intensify the effect. Final consolidation is achieved with differential penetration of the individual pistons under a further increase in pressure.

Blow-squeeze moulding

The long established and highly successful boxless block mould production system represented the first major application of compressed air blowing to

Figure 8.13 Example of a modern multiple action moulding machine: pre-compaction with through-flow of compressed air is followed by final compaction using multi-piston high pressure squeeze (AIRPRESS plus 2000 (courtesy of Künkel-Wagner))

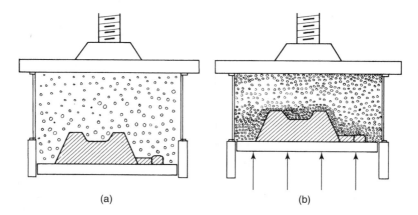

(a)   (b)

Figure 8.14 Downsand frame principle in squeeze compaction

mould compaction and embodied many new features. The original type of machine employs a blow-squeeze sequence and the vertically jointed mould parts carry pattern impressions on both faces. The blocks are compacted within a chamber the ends of which are formed by the two pattern plates, mounted on squeeze and counterpressure plates.

The moulding sequence is shown in simplified form in Figure 8.15. Sand from an overhead hopper is blown into the chamber through a slot in its ceiling and the compacted block is further consolidated by the application of the squeeze pressure. The blocks are pushed successively out of the chamber to form a continuous strand, which emerges on to an internal conveyor track for pouring and cooling.

Modern machines also have provision for automatic core placement in the interval between successive arrivals of mould blocks to join the strand. Each core is transferred into the back face of the last block, via a shaped holder or "mask" which has been presented to the operator at an opening in the machine casing. Feed to the mask can also be automated. Numerous other automatic control features can be incorporated.

This type of system was originally designed for the quantity production of small castings at rates of several hundred block moulds per hour, but block dimensions approaching one metre later became available. A large machine of this type is illustrated in Figure 8.16.

The already high production rates of such machines are enhanced by the use of both block faces, virtually doubling the output of castings as compared with cope-and-drag moulding. Precise alignment of all the movements, coupled with the high density moulds, ensures close product accuracy.

Other types of high pressure boxless machine produce horizontally parted moulds, using similarly rapid automatic production sequences but retaining the single cope-and-drag mould concept, which is often more

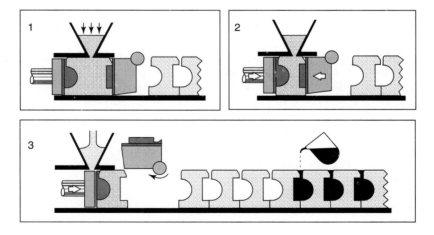

Figure 8.15 Schematic arrangement of vertical joint boxless high pressure moulding facility. Basic sequence as shown: 1. Sand fill and shoot from overhead hopper, 2. Squeeze from both faces, 3. Right hand pattern plate withdrawn and swung away; block mould pushed to join strand for casting (Disamatic principle) (courtesy of Georg Fischer Disa Ltd.)

Figure 8.16 Automatic high pressure moulding machine producing vertically jointed boxless greensand moulds. Aperture for core insertion is facing the operator (Disamatic 2070) (courtesy of George Fischer Disa Ltd.)

convenient for the gating and feeding of complex shapes and multiple arrangements of castings.

Variations of the successive blow-squeeze compaction sequence are also used in several moulding machine designs employing moulding boxes. Again air is used for pre-compaction, followed by squeeze from a flat, contoured or multi-piston head. Depending on the system used, venting of the pattern and pattern plate can be required to optimize the density distribution in the compacted sand.

Impact moulding

A major phase of modern moulding machine development came with the use of compressed air release or explosive gas combustion to propel and compact the sand. Prominent amongst several advantages is the favourable density distribution in the compacted mould, where maximum density is achieved at the pattern face as in jolting, but without the associated noise. Permeability increases with distance from the mould face.

The dynamic loading condition assists compaction in deep pattern recesses. Various design refinements have enhanced this benefit by selective distribution of the incoming air and sand, enabling patterns to be located near to the box walls without the danger of inadequate compaction in the resulting narrow spaces; the excellent lifting qualities of the compacted sand enable casting features requiring thin mould ribs and projections to be formed without the introduction of cores. Successful production of mould ribs as thin as 3 mm with height to width ratios of 5:1 has been reported[13] Fluidization of the sand under the rapid flow of air or gas reduces friction losses and so provides higher mould edge strength than in other compaction modes: these aspects were examined in some detail in Reference 14.

Air for each operation is admitted through a valve from a pressurized reservoir forming part of the machine, and compaction density can be controlled through variations in the air pressure; in some cases two successive pressure waves can be employed to optimize the packing, with or without an intermediate sand fill.

As in the case of jolting, impact or impulse compaction is used in conjunction with squeeze, applied in modes already referred to and represented in Figure 8.12, using plates, pistons or diaphragms. Some of these systems are designed to provide differential pressures tailored to the needs of specific pattern plates.

The action of one machine employing air impulse or "dynamic compaction" is illustrated schematically in Figure 8.17(a). In this case the air impulse can be used alone, whether as a single or two-phase operation, or can be followed by a mechanical stage involving application of a flexible squeeze pad as shown in Figure 8.17(b). Again, the respective actions can be applied in several alternative modes to suit the pattern characteristics.

Vacuum squeeze. The use of air pressure difference to generate kinetic energy for mould compaction is also seen in vacuum squeeze machines, in which initial compaction is achieved by admission of moulding sand to the previously evacuated mould container, followed by consolidation using multi-piston squeeze under continued suction; no vents are required in this system, since little air needs to be exhausted during the filling and compaction stages: this is one of the alternative modes available in the machine previously illustrated in Figure 8.13.

Sandslinger compaction. Sandslinger practice bears some resemblance to those techniques involving dynamic impaction of sand against the pattern. In this case a sand stream is ejected at high velocity from a belt fed impeller which is traversed systematically over the area of the moulding box. Moulds of high and uniform density are achieved, and the system provided a staple technique for the rapid production of large moulds using drysand practice. The system was largely superseded with the adoption of cold setting no-bake sands as previously discussed, the sand being dispensed directly from continuous mixers.

Reviews of many modern moulding machine designs, with commercial examples, include those by McCombe[15] and by the Institute of British Foundrymen[16]. Apart from varied methods of compaction, other features

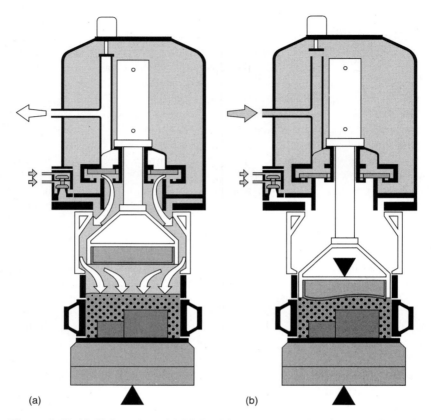

Figure 8.17 (a) Air impulse and (b) flexible squeeze pad actions employed in one moulding machine (6FD ComPac 540) (courtesy of Georg Fischer Disa Ltd.)

include the arrangements for pattern draw and changes, mould part handling and systems for the integration of the moulding operation into the wider foundry production sequence.

Comprehensively designed layouts are required to exploit the rapid rates of production achieved with modern moulding systems, whether based on moulding machines or direct sand feed from continuous mixers. Facilities for pattern draw, rollover, core setting and closing are incorporated in linear or closed loop conveyor systems, usually with provision for transfer to separate casting and cooling lines, and ultimately to casting separation and sand reclamation. Devices for rapid pattern change within the operating cycle of the moulding machine are a key feature supporting maximum production in conjunction with such layouts. Typical principles are illustrated in Figure 8.18. In some cases compact rotary or carousel units indexing between successive stations can be used to effect the main mould part production, especially in the field of boxless cold set moulding.

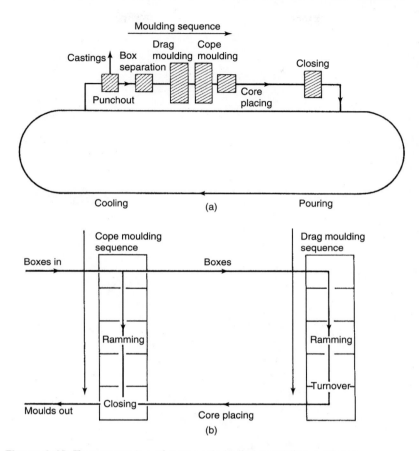

Figure 8.18 Two examples of automatic in-line moulding systems

Moulding machine manufacturers themselves provide full plant layouts embodying all the features required for the production of castings. An automatic moulding unit with an integral conveyor system for the emerging boxless moulds is illustrated in Figure 8.19. The continuous mould strand, supported at the sides, is transported smoothly through the casting and cooling zones on the walking beam principle. Figure 8.20 shows a similar unit incorporated into a closed loop with a rotary cooling drum, in which the sand and castings are separated before the sand is returned to the preparation plant which completes the cycle.

Analogous layouts are provided by the manufacturers of machines for boxed mould production.

Metallic reinforcement of moulds

Moulds for small castings normally derive adequate strength from the moulding material and box. As the overall dimensions increase, however,

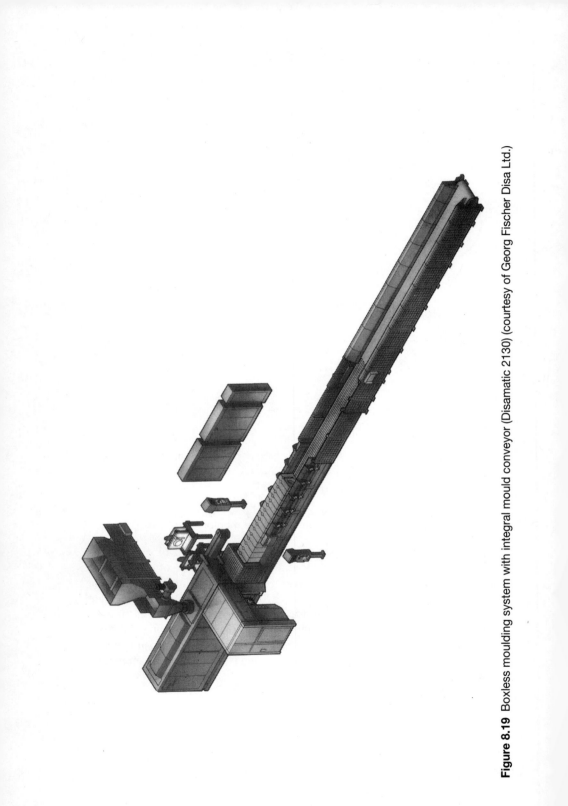

Figure 8.19 Boxless moulding system with integral mould conveyor (Disamatic 2130) (courtesy of Georg Fischer Disa Ltd.)

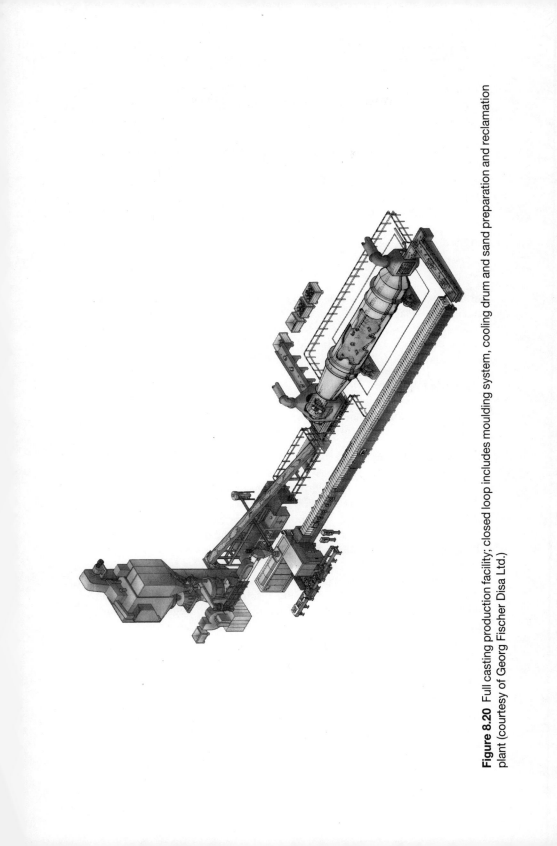

Figure 8.20 Full casting production facility; closed loop includes moulding system, cooling drum and sand preparation and reclamation plant (courtesy of Georg Fischer Disa Ltd.)

additional reinforcement is needed in the shape of box bars, lifters and nails. Integral or temporarily attached box bars give intermediate support within the box and can be used as anchorages for lifters embedded in the sand to provide a skeleton structure. Lifters are needed to reinforce deep cods of sand extending into pattern recesses, where they assist pattern withdrawal. The sand can be strengthened during ramming with preformed grids, rings and straight or bent irons to resist distortion under metallostatic pressure. Sprigs and clout nails can be used to reinforce the mould surface locally against scabbing and erosion: in general, however, there is a greatly reduced need for such reinforcements in the era of high pressure moulding systems and hard cold-set sands.

Venting

During casting a large volume of gas needs to be exhausted to atmosphere: the venting system is thus an important feature. Two separate forms of ventilation are required, for gas generated within the moulding material and air displaced from the mould cavity. Mould gases arise from three sources, namely.

1. evaporation of free moisture,
2. evolution of combined water and decomposition of organic binders and additives to form steam, hydrocarbons, CO and CO_2,
3. expansion of air present in the pore spaces of the sand.

A rough calculation shows that each 1% of free moisture in 1 litre of moulding sand produces about 20 litres of steam, so that the volume of gas from this source is particularly large in greensand moulds.

Dispersal depends primarily upon the permeability of the moulding material but it is often necessary to supplement this by general venting using a wire. The mould is systematically pierced from the outside surface to within a short distance of the pattern, forming local channels for the accumulation and escape of gas.

The problem of gas dispersal depends on the situation of the sand mass. As indicated in Figure 8.21, enclosure diminishes the sector available for gas escape: the problem is greatest in certain cores with egress only through tightly restricted coreprint areas. Central vent channels are required in cores in confined zones; in large moulds where the outside surface is distant from the casting, coke beds may be used as intermediate sinks for gas, which can be conducted to outside atmosphere through pipes. This is especially applicable to the bottom sections of moulds for plates and other castings of large area.

Typical arrangements of general vents are illustrated in Figure 8.22: such vents terminate short of the mould cavity to avoid blockage by metal on casting. In drysand practice the venting system assists the rapid elimination of steam generated during drying.

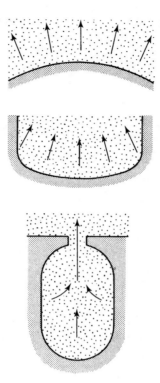

Figure 8.21 Gas escape as affected by mould geometry

Escape of air from the mould cavity must be ensured by free ventilation to outside atmosphere; few types of mould are otherwise sufficiently permeable to avoid back pressure on pouring. Natural vents exist in open feeder heads or risers but air pockets in corners or upward projections need separate 'whistler' vents formed by driving a wire from the mould cavity to the outside surface. In some cases the parting plane provides a convenient path, scratches being made from the extremes of the mould cavity. Whistler vents are illustrated in Figure 8.23.

Pattern draw

On completion of ramming the mould is ready for pattern withdrawal. In hand moulding the mould part is usually turned over after ramming, for which a crane, hoist, or simple turnover unit can be used. The pattern is rapped to provide clearance and lifted out by hand or crane: care is needed to avoid dimensional errors from excessive rap. Machine draw systems have already been discussed: the initial clearance is normally provided by low amplitude vibration, giving greater dimensional consistency.

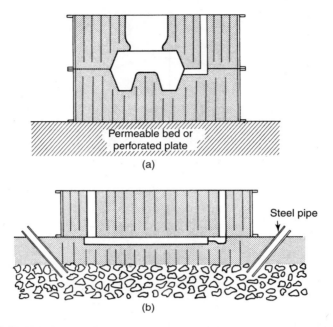

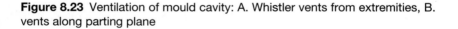

Figure 8.22 Ventilation of mould aggregate: (a) vents to sand surfaces, (b) additional vents to coke bed

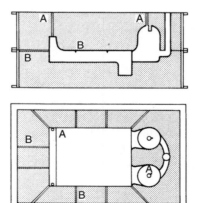

Figure 8.23 Ventilation of mould cavity: A. Whistler vents from extremities, B. vents along parting plane

Mould hardening

Many moulds are hardened before closing and casting, whether by drying or chemical reaction: factors influencing the choice between this and greensand practice were reviewed in Chapter 4. The introduction of high pressure moulding machines enhanced the capabilities of the greensand process, but the use of dry sand, even for heavy castings, became widely replaced by cold setting chemically bonded sands.

The hardening of cold-set sands has been discussed elsewhere. Where moulds do need to be dried the operation may involve complete stoving of the mould parts or may be confined to surface drying with portable equipment.

Stove drying

Stoving of moulds is carried out in the temperature range 200–400°C, above which loss of combined water is detrimental to the strength properties of many clay binders. Stoves are normally fuel fired and can embody recirculation: a proportion of hot gases from the drying chamber is drawn off and fed back through the combustion zone, giving increased thermal efficiency and greater temperature uniformity. Drying is also assisted by more vigorous circulation and a higher rate of gas flow past the mould surface.

The important factors in mould drying were investigated in corporate researches[3,18,19]. Apart from the initial moisture content the most important single factor governing drying time is temperature, which should be as high as is consistent with preserving the properties of the sand. The other major factor is the area of free drying surface, emphasizing the importance of the mode of stacking of the load to provide gas flow at the individual mould surfaces. Other factors such as sand permeability are of secondary importance.

Despite the overriding importance of temperature, investigations showed that the most economic system of stove operation could be achieved by intermittent firing: the heat supply is shut off before the moulds are fully dried to utilize heat stored in heavy moulding boxes and in the stove refractories and structure.

Surface drying

Equipment for surface drying can range from manually operated gas torches to various forms of drying hood designed to blow hot gases from the combustion of gas, oil or coke over the mould surface. Alternatively, the gases may be blown through the closed mould down the feeder head openings. Electric mould driers are also available, producing fan blown hot air. Using these systems the mould may be wholly or only partially dried. Where skin drying alone is sought, this can be achieved by the use of inflammable mould coatings, supplemented by gas torch. In certain cases air drying alone may be sufficient. The Randupson process of cement sand moulding relies on a period of exposure to normal shop atmospheres for the evaporation of excess moisture.

Since moisture evaporated on heating can diffuse through the sand in all directions, the intermediate stages of forced drying may be characterized by heavy local concentrations of moisture due to condensation in cold regions remote from the heating surface. This behaviour is illustrated in

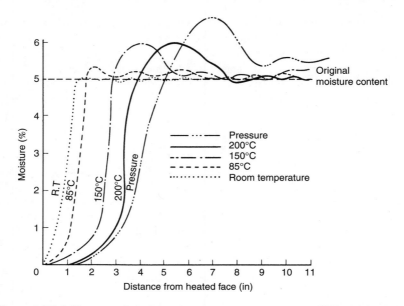

Figure 8.24 Influence of drying temperature on moisture distribution (from Reference 24) (courtesy of Institute of British Foundrymen)

Figure 8.24, which shows curves of moisture distribution after drying at various temperatures. In air drying at shop temperature the moisture moves only towards the drying surface but at elevated temperatures movement occurs in both directions and the condensation zone appears: it was shown[25] that the moisture content in such regions can attain a level as high as 60% above the original. If moulds are only partially dried, therefore, they should be cast with minimum delay because of the possibility of 'strike back'.

Reabsorption can occur either from still wet regions of the mould or from humid atmospheres. Moisture content tends to return to a level within the range 0.5 to 0.8%, most of the reabsorption occurring within the first 24 h after drying. This causes some loss of strength, which can however be recovered on briefly reheating.

Cores and coremaking

A core is essentially a mould component of sufficient strength to be handled as an independent unit without the support of a moulding box. Cores are most commonly used as inserts in moulds to form design features which would be difficult or impossible to produce by direct moulding, although the term can also be used to describe block mould parts for which coremaking materials and techniques are employed. The main functions of cores are summarized in Figure 8.25. For castings of hollow form they provide the means of forming the main internal cavities (Figure 8.25a). In the case of external undercut features of the type illustrated in Figure 8.25b, cores are frequently preferred to the introduction of loose pattern pieces or additional or stepped parting lines to achieve pattern draw. In the case of deep recesses (Figure 8.25c), whilst there is no direct obstacle to pattern draw, clean separation is impeded by friction combined with the low strength of moulding sand in tension. The point at which an integral sand cod needs to be replaced by a separate core as shown in the diagram depends partly on factors such as the type of pattern draw action and the amount of lifter reinforcement.

The two principal examples of cores as replacements for normal mould parts are illustrated in Figure 8.25d and e. In the first case an economy in production cost is effected by use of a weighted cover core to close the mould cavity, thereby dispensing with a conventional top part. Block mould or full core assembly techniques have been previously discussed: their use may be warranted by special requirements for small and intricate castings or for moulds of very high strength such as are required in the centrifugal casting of shaped components.

The design of a core must incorporate coreprints or projections locating in corresponding mould recesses to support the core in the correct position. The number and positions of these prints should preferably give firm

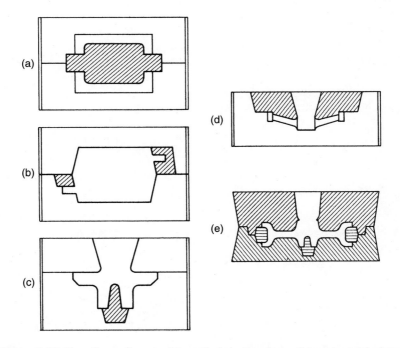

Figure 8.25 Functions of cores: (a) cavity in hollow form, (b) undercut features, (c) deep recess, (d) cover core, (e) core assembly or all-core mould

support without the use of studs or chaplets, and should provide adequate apertures for removal of the core from the solidified casting.

Coreboxes

As in the case of patterns, coreboxes range from simple wooden structures to precision metal assemblies capable of long life under exacting conditions. The simplest design is the open sided type producing flat walled cores or half cores which can be turned out directly on to a flat plate (Figure 8.26a); many cores are designed in separate halves for subsequent joining to enable this simple method to be adopted. In some cases the box is also made to split along a plane normal to the flat surface: the box can then be parted laterally to leave the core standing on the base plate (Figure 8.26b).

One piece cores without adequate flat surfaces are normally produced in split and dowelled coreboxes, of which one half is removed to enable the core to be extracted normally to the parting plane (Figure 8.26c). In this case a green core is transferred directly from the remaining box half on to a carrier or drier or a supporting bed of loose sand, although this is unnecessary for cores which have been wholly or partially hardened in the box. Extraction of cores with more complex features may involve the dismantling of multisection coreboxes and the withdrawal of loose pieces such as pins and sliding blocks.

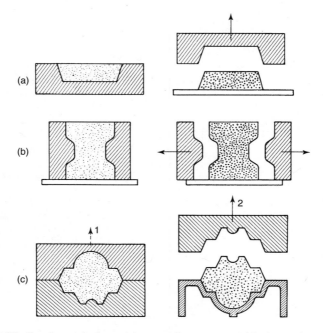

Figure 8.26 Corebox design: (a) one piece open sided corebox, (b) split corebox with vertical joint, (c) split corebox: core turned out on to carrier

More specialized corebox features are needed for coreblowing, hot box and other techniques: these include blowholes, vents and ejectors and will be referred to again in later sections.

Where small cores are required in large quantity, production times can be greatly reduced by the use of multigang coreboxes with several core impressions. These enable a number of cores to be rammed or blown simultaneously, reducing the time involved in stripping and reassembly.

Compaction

Cores may be compacted by hand ramming, jolting, squeezing or strickle operations analogous to those in moulding. In addition, standard core sections can be produced by extrusion of sand through an orifice using a simple machine. Most distinctive of all the processes, however, is coreblowing, also embracing core shooting, widely employed for the quantity production of small and medium sized cores.

Little need be said of those operations paralleled in the moulding field. The principal difference is that most coremaking is based on sand mixtures of comparatively low green strength, requiring little ramming. Even for very large cores, the use of stiff mixtures to prevent sagging has given way to methods in which partial or complete hardening is accomplished within the corebox, using setting reactions of the types discussed in Chapter 4.

Coremaking machines

Early machines designed specifically for the production of cores were based on the principle of coreblowing, using compressed air, normally at the standard mains pressure of $560–700\,kN/m^2$ ($80–100\,lbf/in^2$) to blow the sand into the corebox at high velocity, the air being exhausted through vents. Machines took several forms, ranging from bench models producing small cores to heavy units with sand capacities of several hundred kilograms.

In the simplest type of bench blower a separate sand cartridge, a manually filled tube, is inserted between blowing head and corebox, the whole being clamped together whilst the main air valve is opened to propel the sand slug into the corebox.

Heavier machines are of two types. The first embodies a sand reservoir into which air is introduced directly at mains pressure through a quick acting valve. This type may also incorporate agitator blades to avoid channelling in the sand chamber. The basic concept is illustrated in Figure 8.27. Such machines inject a fluidized mixture of air and sand, the sand being in effect filtered out by the vented corebox: compressed air consumption and corebox wear can consequently be heavy.

The second major type, which came to be widely employed, is based on the coreshooting principle illustrated in Figure 8.28. In this case blowing is accomplished by the admission of a fixed volume of compressed air from a separate machine reservoir to the sand chamber through a large capacity

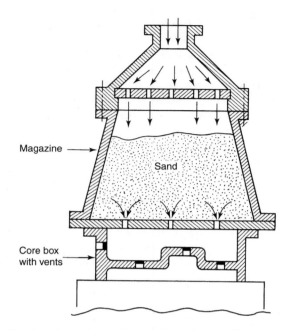

Figure 8.27 Section through coreblower (from Gardner[20]) (courtesy of Institute of British Foundrymen)

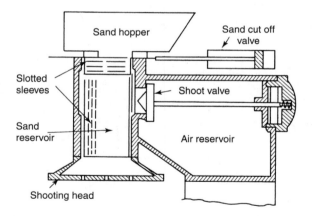

Figure 8.28 Section through coreshooter (from Gardner[20]) (courtesy of Institute of British Foundrymen)

quick acting valve; the valve to the main air line is meanwhile kept closed. Effective sand propulsion results from the rapid pressure build-up in the blowing chamber, air consumption and corebox wear being greatly reduced.

The larger machines incorporate automatic clamping devices to hold the corebox closed and maintain a seal between blowhead and box, the air

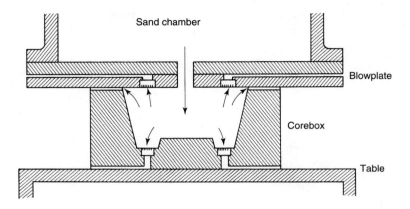

Figure 8.29 Positions for vents in coreblowing

escaping through vents on the principle illustrated in Figure 8.29. Venting requirements are minimized when the coreshooting system is adopted because of the reduced volume of air employed.

Shooting assumed great importance in the development of modern automatic and semi-automatic coremaking equipment. For rapid production fully automatic coreshooters are available, with inbuilt provision for cold box gas hardening (the options for core hardening based on the binders reviewed in Chapter 4 are the subject of the following section). The typical machine in this case completes the entire cycle, including purging of the reactive gas and core ejection.

In a further type of automatic machine, illustrated in Figure 8.30, the cold box sand is effectively extruded into the corebox by the instant admission of a predetermined volume of compressed air, which impacts on the top of a column of sand previously packed into the extrusion tube. The resulting shock wave extrudes the sand into the corebox, after which the sand chamber is vented to atmospheric pressure through a special valve system ready for the next operation. High and uniform densities are achieved, with little corebox wear; fully automatic ancillary facilities for clamping, corebox movements and core ejection are again provided.

Apart from these examples of advanced automatic machines, very high production rates are also attainable from simple coreblowing equipment by the simultaneous working of a number of coreboxes to achieve continuous operation of the blowing unit. Multi-station working, whether by operators or automatic devices, enables gassing, stripping and box reassembly to proceed separately in phase with blowing; boxes can be indexed between the stages using carousel and similar systems.

Core reinforcement and venting

Cores of appreciable size or fragile design may need reinforcement with core irons or metal grids, although this is less necessary with the strong

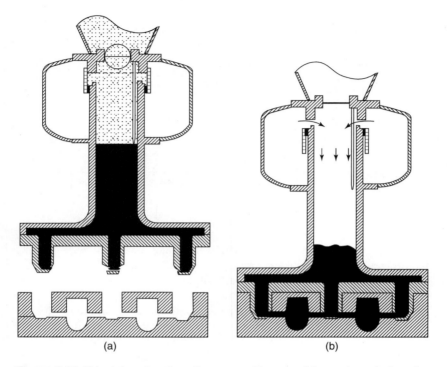

Figure 8.30 Principle of automatic coremaking machine using air impulse extrusion. (a) sand fill and (b) extrusion stages as illustrated are followed by exhaust and self cleaning of chamber. (6FD Core 300, DISA X-trude process®) (courtesy of Georg Fischer Disa Ltd.)

selfhardening binders now being widely employed. In larger cores the reinforcement provides an anchoring point for lifting staples used for handling or securing in the mould.

Small cores can frequently be vented with pierced holes leading out through the print surface. Alternatively woven nylon tubing can be incorporated during coremaking. Large cores may be moulded round pierced metal pipes which serve both to strengthen and as conduits for gases. Use is also made of coke centres, with vent tubes to exhaust the gas through the coreprint. Vents must be conducted to outside atmosphere, if necessary through communicating mould vents, whether pierced or scratched along the joint line. Examples of core vents are illustrated in Figure 8.31.

Core hardening and associated process variations

Although green cores are used in special situations, most cores need to be hardened to comparatively high strengths to meet handling and casting conditions. A range of alternative hardening techniques has given rise to

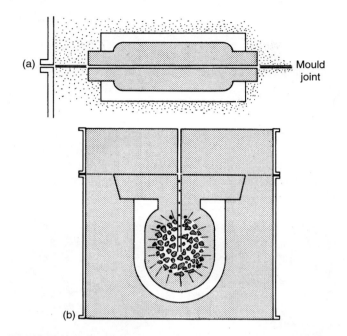

Figure 8.31 Core venting: (a) central vent in small core, continued to outside along mould joint or through vent in body of mould, (b) coke centre in heavy core, with egress through print

important variations in the entire coremaking process. Traditional practice involves separate hardening after the core has been stripped in the green state, but this has been mainly superseded by hardening within the corebox, made possible by the versatile range of chemical systems as previously reviewed in Chapter 4.

Hardening within the box

In-box hardening is associated with organic or silicate based binders, designed to react at rates suited to the selected process conditions. Distinctive systems are the hot and warm box processes, in which heat is the major hardening factor, and the cold box processes, in which hardening depends solely on reagents or catalysts, introduced either as liquids on mixing or by passage of gas or vapour through the compacted core.

All the processes enable precise, intricate cores to be produced. It must, however, be borne in mind that the extraction of an already hardened core requires a high standard of corebox construction and finish; the processes cannot be indiscriminately applied to worn or damaged equipment. In the case of the hot and warm box processes metal equipment is essential; metal, resin or wood can be variously used with the cold box systems depending on other circumstances. The boxes may need to incorporate vents and ejectors to suit particular modes of production.

The hot box process. The use of heated coreboxes was first associated with the production of shell moulded cores from dry resin bonded mixtures. The high accuracy of such cores led to their extensive use outside the field of shell mould castings (qv Chapter 9), in orthodox sand casting practice. Further advantages can be readily appreciated from the example shown in Figure 8.32: ready venting is combined with ease of handling.

The hot box principle was later extended to the use of sand mixtures based on liquid resin binders of the types discussed in the earlier chapter. The corebox, of steel, cast iron or aluminium and provided with vents

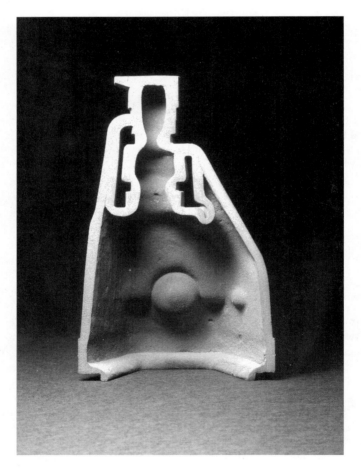

Figure 8.32 Section through a hollow shell core (courtesy of Hepworth Minerals and Chemicals Ltd.)

and ejectors as required, is maintained at a temperature usually within the ranges 230°–260°C. The sand mixture, formulated to be slow setting at room temperature, is blown into the heated box and curing proceeds at a rate sufficient for ejection within a fraction of a minute; the cure is completed by residual heat in the core. Stripping is assisted by periodic treatment with silicone parting fluid.

The corebox temperature is commonly maintained by integral gas or electric heating elements, but stove cycling an also be employed to accelerate the full curing of cores of thicker section after extraction. To counter the effect of contact with the hot coreboxes water cooling can be applied to the core machine blowplate.

The fast cure, rapid box turn-round and high strength of the cured sand render the process highly suitable for the mass production of cores, and especially for those of thin section and complex design. It can be applied to cores of section up to 50–70 mm and multi-station working can be used to increase production rates or to lengthen the time available for the curing of thicker sections. For thick and heavy cores, however, through curing becomes more difficult and preference may be given to shell or cold box production.

The need for metal boxes makes the process uneconomic for jobbing work, for which the various cold box systems or traditional stoved oil or resin sands are more suitable.

The warm box process. The equipment and general approach employed in warm box coremaking are very similar, but a more reactive binder system enables corebox temperatures in the range 150–190°C to be employed whilst maintaining short cure times. The cores feature very high strength and hardness with modest binder additions.

Cold box processes. Numerous cold box options are available and the main differences lie in the chemical binder and hardening systems as discussed in Chapter 4, together with the sand mixing and dispensing practice employed for both cores and block moulds. Setting time is the main production parameter for those processes employing liquid hardeners, being controlled mainly by the amount of the addition. The time is, however, also sensitive to variations in sand temperature, so that sand heating, sand cooling or addition adjustment may be required to maintain the necessary control.

Corebox materials and construction relate primarily to the quantity requirement and the method of compaction: wood is often satisfactory, particularly when resin coated. The processes are adaptable both to jobbing and quantity production, for which resin or metal coreboxes are preferred.

Gas and vapour cure processes require provision for diffusion of the gaseous phase through the compacted core or mould part. Various means for the generation and application of vapour catalysts, including determination of the required amounts and flow rates, are examined in Reference 21. The vapour, mixed with air or a carrier gas such as carbon dioxide or nitrogen, can be passed directly into the corebox through a venting system designed to achieve uniform distribution and avoid dead zones. High capacity vents and low vapour concentrations are considered most favourable for efficient gassing. Alternatively the gassing operation may be performed in a separate sealed cabinet after compaction. Some core machines are provided with integral gas generators suitable for fully automatic blow/shoot – gas sequences delivering hardened cores.

When using organic vapours or SO_2 hardening agents the gassing system must be effectively sealed and the residual gas collected to prevent health and pollution hazards: exhaust gases pass to wet scrubbers, although in some cases recirculating systems are employed for useful recovery of unused reagent.

In the CO_2 sodium silicate process the gas hardening can be accomplished either in the box or after stripping: methods of gassing in the corebox are illustrated in Figure 8.33(a–c) and, in the mould, f. This gas, at a pressure of around 140 kN/m^2 (20 lbf/in^2) is dispensed from banks of cylinders or from a bulk liquid container. Again the treatment can also be applied as a step in a coreblowing sequence. Some of the important factors in the gassing conditions were illustrated in Chapter 4. Theoretical gas consumption represents 0.2–0.5% by weight of the sand mixture, although this should be expected to be at least doubled under practical conditions.

Again, special coreboxes are not required, although the maximum permissible draft and a high standard of finish are desirable to assist stripping. Capital cost too is low. Classic works dealing with the process are those by Nicholas[22] and Sarkar[23]; production aspects are also included in other references cited in Chapter 4.

The gas hardening cold box processes are particularly suitable for high volume core production and especially for blown cores, where the rapid gas cure systems can be readily integrated with the core machine cycle or built into the machine itself. Tooling requirements are, by contrast, less exacting than for the high temperature processes, so that a wide range of cores needed in modest numbers can also be economically produced.

The benefits of cold box processes involving hardening within the box are well illustrated by reference to the Francis runner casting illustrated in Figure 8.34. The inter-vane cores, of complex contours with little flat surface, can be accurately formed and hardened without the need for high green strength, heavy reinforcement, or elaborate carriers such as were once required for traditional stoved cores of this type.

Separate hardening

Traditional coremaking practice, requiring thermal hardening of drying oil or synthetic resin sands, employs batch or continuous corestoves

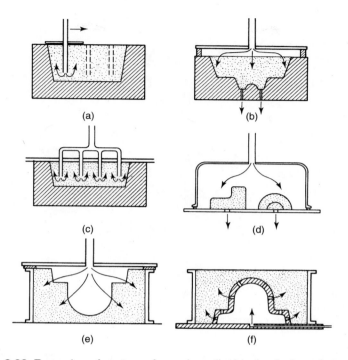

Figure 8.33 Examples of systems for carbon dioxide hardening of cores and mould parts: (a) progressive treatment using single probe, (b) cover board or hood, (c) multiple probe and manifold, (d) hood over previously stripped cores, (e) treatment of mould after pattern draw, (f) passage of gas through hollow pattern

operating within the temperature range 180–250°C. An essential feature is the provision for air circulation and ventilation to maintain temperature uniformity, to supply oxygen for certain of the hardening reactions, and to eliminate moisture and acrid fume generated on baking. Large cores require prolonged time at temperature to avoid underbaking in the centre.

Whole or split cores with flat surfaces are carried on perforated plates, but more complex shapes require support from carriers or sand beds: carriers may be of cast iron, light alloy or plaster. This need for support, and indeed for green strength, partly explains the strong preference for modern coremaking materials and systems, although baked core practice still retains a limited role for simple cores in restricted circumstances: rapid production by hand or machine remains possible.

Separate hardening after stripping is also used for some CO_2–sodium silicate cores and moulds. Gassing is in such cases carried out using hoods or chambers as shown in Figure 8.33d and e.

Final tasks in core production are the building of core sub-assemblies, including the jointing of split cores, and the application of surface coatings

Figure 8.34 Francis runner casting for water turbine, illustrating typical application of cold setting cores to form water passages

where needed. Jointing is carried out with adhesives, aided by templates and fixtures for alignment. The adhesives employed can be of hot melt, thermosetting or cold curing types: together with application techniques and potential problems these were fully examined in Reference 24. Refractory coatings for cores and moulds were reviewed in Chapter 4, including characteristics of the alternative water and spirit based types and the potential for dry coatings.

Coremaking operations have traditionally been conducted on a separate hand or machine production line, especially since much of the work involves relatively light loads with the need for rapid handling: pendulum and band conveyors are frequently featured. Such production can, however, be readily adapted to robotic and other automatic handling systems, which can be applied to individual operations, for example extraction and coating, as well as featuring in fully integrated production units. In some of the latter there can be close linkage with mould production lines, where cores are produced and ejected directly into the mould itself without intermediate handling.

Closing

Closing represents the final stage in mould production, being usually carried out either on the casting floor or on conveyors adjacent to the moulding machines. Principal requirements are the accurate mutual alignment of mould parts and cores, followed by clamping and weighting against metallostatic pressure.

Where precise moulding boxes and pattern equipment are used closing causes little difficulty, but in other cases painstaking checks with calipers, templates and thickness pieces may be required to avoid dimensional errors. Most cores derive support from their coreprints alone but in some cases studs or chaplets must be fitted. Heavier cores depending on a single print or suspended in the top part need to be secured by wires attached to staples moulded into the core. Before final closing the mould requires systematic cleaning by suction or compressed air.

Mould assembly includes insertion of insulating or exothermic sleeves used as feeder head liners and the positioning of runner bushes, feeder head extensions and tundishes for pouring.

The requirement for weighting to counter the tendency of the top part to lift can be determined from the expression:

$$W = \rho hA \text{ (kg)}$$

where ρ = density of metal (kg/m^3)

h = metallostatic head (m)

A = horizontal projected area of casting (m^2)

In the foregoing section the production of moulds has been considered principally in relation to sand casting techniques. With minor variations, similar principles are employed in the manufacture of moulds in other refractories as referred to in Chapter 4. Moulding practice was treated in detail in a number of classical published works[25-29].

The overall picture of mould production is thus one of great diversity both in techniques and in the organization of operations. This arises from the versatility of the moulding process in relation to a great variety of circumstances of design and quantity requirements. The contrast between jobbing and mass production was touched on in the Introduction and further illustrated in the techniques examined in the present chapter. Thus, whilst skilled manual techniques are still required for heavy castings of complex design, moulding is increasingly organized in mechanized and automated systems for flow production along the lines discussed.

A number of further casting techniques employ sand as the mould material but embody special features which are sufficiently distinctive for them to be regarded as separate processes: these are treated in subsequent chapters. The present chapter will continue with the mainstream production sequence through melting, casting and finishing stages.

Part 2. Melting and Casting

The wide variation in basic requirements for metal melting is illustrated in Table 8.2, which compares the thermal properties of those metals on

Table 8.2 Properties and heat requirements for metal melting

Metal	Melting point t_m °C	Mean specific heat 20–t_m °C c_p J/g deg C	Latent heat of fusion L J/g	Total heat requirements for melting $L + c_p(t_m - 20)$			
				For 1 kg kJ	For 1 litre kJ	For 1 lb BthU	For 1 in³ BthU
Iron	1536	0.590	272	1168	9228	502	141
Nickel	1455	0.548	302	1084	9651	466	149
Copper	1085	0.440	205	674	6067	290	93
Aluminium	660	0.992	388	1022	2759	439	43
Magnesium	649	1.210	558	1122	1909	482	29
Zinc	420	0.423	110	285	2018	122	32
Lead	327	0.142	23.9	67	783	29	12
Tin	232	0.243	59.5	113	825	49	13

Based on data from Reference 30

which the volume production of commercial casting alloys is principally based.

Apart from the type of alloy being produced, selection of equipment for melting and casting must depend upon economic factors. These include the balance between cost and quality requirements for particular classes of casting, and the nature of the demand for molten metal as determined by the pattern of production. This demand may vary widely: a regular flow of small quantities of metal may be needed to sustain a continuous production line, or much larger batches may be required for the intermittent production of heavy castings. Furnace capacities and melting rates must thus be related not simply to total output but to the whole pattern of production. Melting units and casting equipment range in capacity from a few kilograms to many tonnes, although the supply of metal to the casting floor is often regulated independently of the melting units themselves by the use of holding furnaces of suitable capacity. This practice gives an additional degree of freedom in the melting operations.

A further factor in melting and casting practice is the availability of facilities for the bulk transport of certain molten alloys from outside sources in insulated containers.

Melting furnaces

Much foundry melting is carried out on a batch basis from cold charges, the notable exception being the continuously melting cupola widely used in cast iron production. Intermediate between batch and continuous melting is the semicontinuous practice in which a 'heel' or proportion of liquid metal is retained in the furnace to provide a bath for the absorption of the succeeding charge of solid materials. In certain cases duplex practice is used, involving separate furnaces for successive stages: the use of a holding furnace to provide a reservoir of molten metal at the pouring temperature is the most important example.

Although melting furnaces must be designed for effective heat transfer to the charge, thermal efficiency is not the principal criterion in the choice of equipment and practice. The overall economics of melting operations depend upon many factors, including capital depreciation and degree of utilization; the operating costs themselves include maintenance and labour as well as fuel and power.

Metallurgical factors in the choice of melting facilities relate to the tendency of the charge to react with its surroundings, affecting compositional control, impurity level and metallic yield. Compositional effects must be judged in relation to casting quality requirements, whilst the economic importance of minimizing melting losses depends on the intrinsic value of the metals concerned: losses are, in general, more critical in the case of highly priced non-ferrous alloys.

To summarise, therefore, furnace selection is determined by cost, metal quality, production requirements and alloy type; flexibility for a wide range of conditions is itself an advantage for many types of production.

It is not intended to describe the many individual types of melting unit which may be found in the literature. The principal types of unit used in the foundry industry are classified in Table 8.3 and some of the main operating characteristics will be summarized.

It will be seen that the furnaces have been classified in two main groups, based on fuel combustion and electrical heating. Within each of these categories the principal distinction, apart from the special case of the cupola, is between hearth and crucible types: the fundamental difference in heat flow conditions is illustrated schematically in Figure 8.35. Important characteristics of these main design variations and heat sources will be separately considered.

Design

In general the closest control of composition, minimum losses and highest standards of cleanliness are achieved in those furnaces in which the charge is isolated from the combustion processes and their attendant dangers of contamination with gaseous elements and sulphur. In this respect crucible furnaces offer an advantage over hearth furnaces, whether of the static or rotary type, especially since the modern crucible is highly impermeable to contaminants. This comparison is to some extent also valid even in the case of electric melting, since electric arc melted metal is subject to contamination with atmospheric gases absorbed under the influence of the arc. Induction melting by contrast is analogous to other crucible processes. Where slag–metal reaction is sought, on the other hand, the

Table 8.3 Foundry Melting Furnaces

Energy	Basic type (see Figure 8.35)			Furnace	Usual means of heating	Main fields of application
I. Fuel fired	Shaft			Cupola	Coke. Charge in direct contact with fuel. Continuous melting	Cast iron
	Hearth			Reverbatory (air) Rotary (rotating or rocking)	Gas; oil Gas; oil Gas; oil	Non-ferrous alloys; cast irons, Non-ferrous alloys; cast iron, esp. malleable and special. Duplex holding
	Crucible			Crucible Lift out or pit type Tilting Bale out	Gas; oil Gas; oil Gas; oil	Most alloys, except steel Light castings, especially die castings
II. Electric		Arc	Hearth	Direct arc Indirect arc (rocking)	Arc to charge Radiant arc	Steel; cast iron Non-ferrous alloys; high alloy steel and special irons
		Resistance	Crucible	Resistor (static or rocking) Resistance	Radiant resistor rod Elements (shroud or immersion)	Steel; cast iron; copper alloys Non-ferrous alloys, especially holding for die casting
		Induction		Coreless induction	Medium frequency induction	Steel, esp. alloy and small tonnage; cast irons; Ni base
			Melting channel	Cored induction	Mains frequency induction	Non-ferrous alloys; holding for die and light castings

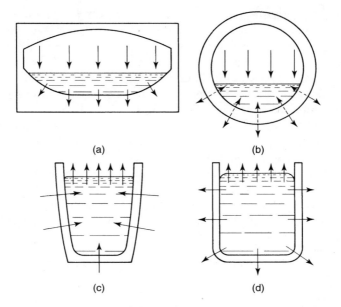

Figure 8.35 Heat flow to charge in melting furnaces (schematic): (a) hearth, (b) rotary, (c) crucible, (d) induction or immersion heated bath

thermal conditions in hearth furnaces are to be preferred since the top heating effect promotes refining. In the special case of cast iron, direct contact between fuel and charge can be a positive advantage, since the coke exercises the dual function of fuel and carburising agent: in this respect the cupola is unique.

Apart from the question of melting conditions, there is a size limitation on crucible furnaces. For tonnage melting a hearth, with a substantial thickness of refractory supported by a metal shell, is required. Conversely the crucible is a highly convenient and flexible unit for handling small quantities of metal.

Heat sources

Provision of heat for melting and superheating has undergone a shift away from solid and liquid fuels to successively more controllable forms of heating with gas and electric power, which offer many technical advantages, whilst environmental factors are also increasingly important in the choice of melting plant and practice.

The coke fired cupola, in which the coke serves as carburizing agent as well as fuel, retains a major role in cast iron production, especially where very large scale units can be employed. Developments in the field of hot blast operation, the use of basic linings and slags, divided blast, water cooling and oxygen injection and enrichment have provided closer control and increased thermal efficiency, as well as facilitating the use of cheaper raw material in the shape of steel scrap. These and other trends were reviewed in References 31–33. Both gas firing and electric power have also been employed in conjunction with the cupola; the cokeless cupola, relying entirely on gas or oil firing, has received much attention and some use on a production scale[34]. Plasma torches have provided a further source of energy for cupola melting.

Despite the continued success of the cupola, electric induction furnaces have increased in importance in both the melting and holding of cast irons as in other foundry applications. In the steel castings field the cupola-converter combination, with its high metallic losses and poor control, has long been superseded by the more flexible electric arc and induction furnaces.

A more complex picture exists in relation to non-ferrous metal melting: the general pattern of furnaces and practice in this field has, however, been well illustrated in group reviews in References 35–37.

Fuel fired hearth furnaces, a term used broadly in the present context to include rotary as well as static and tilting reverbatory furnaces, have traditionally been used for medium and high tonnage melting of a wide range of alloys, including those based on copper and aluminium as well as malleable cast irons; earlier fuels used included coke and coal. This type is now represented in modern gas or oil fired tilting reverbatory drum furnaces, as used for aluminium bulk melting and other non-ferrous applications. Advanced burner technology gives high combusion temperatures and efficiency: nozzle mix, recuperative and regenerative burner systems are variously employed and have been described in detail by Ahearne[38] and McMann[39]. Heat is extracted from the flue gases to achieve air preheat temperatures exceeding 600°C. Modern regenerative burners employ packed ceramic heat storage beds, with twin burners firing alternately for 2–5 minutes. Separate holding furnaces are usually employed, electrical resistance types being favoured for aluminium.

A further important development in aluminium melting and holding has been the shaft or tower furnace incorporating the thermally efficient counterflow principle, as seen in simple form in the cupola. The basic concept, which was already established in the 1960s[40], is particularly suited to the continuous melting of a single composition. The modern system, fully detailed in Reference 41, incorporates a holding stage, using a two-chamber system with separate burners for the melting and holding sections. The melting chamber incorporates the shaft or tower, into which solid charge material is fed, where it is preheated by gases from both melting and holding burners. Molten aluminium flows from the melting zone into the holding chamber. Both melting losses and emissions are very low and the exit gas temperature is in the range 150–300°C. Control of charging rate and firing is maintained by sensors monitoring exhaust gas temperature and metal level in the holder.

Whilst reverbatory and shaft furnaces are available down to capacities below one tonne, the most flexible and widely used fuel burning unit in the lower weight ranges is the gas or oil fired crucible furnace. Lift-out units of up to 150 kg capacity, and tilting units accepting charges exceeding 500 kg, are employed for most types of alloy where small quantity, varied composition and high quality are needed. Unlike hearth furnaces, in which top burners are employed, heat transfer through the crucible wall requires refractories of high thermal conductivity, based on graphite or silicon carbide. Bale-out furnaces are used primarily as melting and holding units for the production of light castings, especially die castings, for which they are provided with temperature controlled automatic burners. Some of the units in this category employ firing through refractory immersion tubes submerged in the molten bath, so cannot be described as crucible furnaces in the normal sense.

The casting industry as a whole has witnessed a marked increase in the importance of electric furnaces, not only because of the high degree of control and flexibility of operation, but because of the cleanliness and reduced levels of pollution in the absence of fuel combustion. Arc and induction furnaces account for most of the tonnage produced, but resistance furnaces are also widely used, including some operating on the immersion heating principle.

Arc furnaces are analogous to fuel fired hearth furnaces in that they provide the conditions for active refining as well as for straightforward melting. Direct arc furnaces are widely used in the production of steel castings, although induction melting has grown in relative importance with the availability of larger units and enhanced melting rates: Svoboda and Griffith[42] have critically reviewed comparative characteristics of the two types in relation to investment decisions for new plant.

Modern arc furnaces incorporate water cooled panels to replace the refractory lined wall sections formerly employed. Other developments include assisted melting with oxy-fuel burners to increase metal throughput[43], and carbon injection into steel baths to deliver chemical reaction heating during oxygen lancing. Furnace utilization is also enhanced by the use of separate secondary refining units for less power-intensive process stages. Ladles, ladle furnaces and various custom designed vessels contribute to greater availability of the prime melters.

Induction furnaces offer some parallels with fuel fired crucible furnaces, the charge being similarly isolated from external heat sources; any slag or flux derives heat solely from the metal charge. These furnaces can be of either channel or coreless type: their respective electrical features and main operating characteristics have been extensively reviewed in the literature[44,45].

Cored channel furnaces were long-establishing for non-ferrous melting and became widely adopted as holding units for cast iron, especially in duplex practice, with primary melting carried out in the cupola; they also came to be used for melting. The heating effect is generated within the induction loop or channel connected to the main bath, the molten metal circulating through the system under electromagnetic and convective forces. A molten heel incorporating the loop needs to be continuously maintained, but power utilization in the furnace is highly efficient. Quickly replaceable loops enable the refractory linings, which are subject to intensive local wear and in some cases to dross build-up, to be readily cleaned or renewed.

The outstanding advance in electric melting, however, came in the field of the coreless induction furnace, with the introduction of the medium frequency unit powered through the solid state inverter. This enabled high power density to be maintained throughout melting by automatic variation of the output frequency. The inverter dispensed with the capacitor switching which had previously been required to maintain a suitable circuit balance and power factor as the geometry and electrical characteristics of the charge alter during successive phases of the melting cycle. An expanded treatment of the electrical parameters is given in Reference 46.

The medium frequency coreless furnace provides great flexibility, operating without the need for the molten heel required in both coreless and channel mains frequency furnaces; it combines rapid melting with adequate stirring for the absorption of swarf and other finely divided charge materials. Small capacity furnaces are capable of the same output as earlier induction units of much larger size, and efficient heating can be maintained in partly emptied baths, a valuable feature when dispensing metal to the casting floor. The furnace has been widely adopted in preference to the mains frequency coreless units which had become important in cast iron melting and swarf recovery. It is used in both ferrous and non-ferrous founding, having displaced mains frequency systems in most new installations. As with earlier coreless furnaces, a spiral induction coil surrounds a plain cylindrical melting chamber within a tilting box body.

The frequency of induction melting units is selected on the basis of both size and function. Effective current penetration of the charge increases with decreasing frequency, so that the larger the furnace the lower the optimum value. The stirring effect too increases with decreasing frequency for a furnace of a given capacity: powerful stirring was a particular feature of mains frequency furnaces. Medium frequency coreless furnaces mostly fall within the range 200–3000 Hz. 200 Hz is a typical level for cast iron, providing strong stirring; tonnage capacity steel furnaces employ 1000 Hz, with 3000 Hz for small units in the 100 kg range. Copper alloy melting uses intermediate frequencies within the same range[44].

Most electric melting furnaces operate on the tilting principle, but coreless induction melting is also carried out using free-standing crucibles, which provide a flexible system for the fast melting of small batches. The furnaces used operate on the push-out, drop coil or lift-off coil principle, enabling the crucible to be separately handled for pouring: crucibles can be readily interchanged for the production of a variety of alloys.

Smaller coreless induction furnaces can be lifted bodily for pouring, whilst the rollover principle is employed for the filling of individual investment casting moulds: the mould is in this case clamped to the furnace body, which is provided with trunnions for rapid inversion.

Resistance furnaces provide the remaining substantial group of electrically heated molten metal units, which can be of either bath or crucible type. Radiant roof furnaces are used as holding units for aluminum alloys, often in conjunction with prior gas or induction melting, although some are also used for melting. Heat is generated in suspended silicon carbide rods or in wire elements sheathed in radiant tubes.

Resistance crucible furnaces are used for aluminium and zinc melting and employ varied and ingenious element designs[47]; bale-out units provide for both melting and holding of metal required in frequent small quantities for dispensing to die casting machines. Immersion heating elements are frequently employed in this type of application.

The operation of electric furnaces on any significant scale requires attention to the pattern of power consumption by the plant as a whole, with a view to minimizing total energy costs. This arises from the application of supply tariffs based not on power consumption alone but also on maximum demand over a given working period. The aim is to spread the electrical load by staggering periods of high demand from individual items of plant, and to shift heavy loads away from anticipated peak times. Load management equipment can range from simple monitors suitable for operating alarm devices, enabling decisions to be reached on a temporary interruption of part of the supply, to comprehensive systems in which graduated automatic load shedding is introduced on a predetermined basis. Night melting in conjunction with holding furnaces for daytime casting operations is one system offering economies in suitable circumstances.

Further developments

Given the substantial advances achieved in furnace design, efficiency and melting rate, perhaps the greatest scope for progress lies in the extension of controls based upon plant instrumentation. Modern pyrometers, pressure gauges, flow meters, analytical systems and metal level sensors can provide a basis for the establishment of various automatic linkages: a typical example is the coupling of temperature and energy input to optimize combustion, melting rate, metal quality and costs. Such linkages between crucial parameters seems likely to play an increasing part in maximizing the efficiency of future melting operations.

Melting practice

The aim in melting is to achieve close control of metal composition, with low melting losses and the avoidance of gas contamination and non-metallic inclusions. The choice of practice is governed partly by the composition and quality requirements of the alloy and partly by the nature of the materials forming the charge. Two broad types of practice can be distinguished. In many cases conditions are maintained as inert as practicable, the charge being of virtually identical composition to that required of the liquid metal: the system is one of rapid melting with minor corrections for unavoidable losses. In some cases, however, the final composition of the melt must be developed by controlled refining using slag–metal and gas–metal reactions. The former practice of simple remelting is often preferred in the batch melting of small charges, whereas large melts justify chemical control; the nature of the alloy is, however, the main factor influencing the type of practice.

The charge

The furnace charge may consist of pre-alloyed pig or 'ingot', virgin metals and 'hardener' alloys, scrap from outside sources or from internal fettling and machine shops, or any mixture of these materials. Externally purchased pig, ingot and alloy additions are normally supplied to certified analysis, but certain materials, especially scrap, require sorting and analytical control for charges to be calculated with accuracy.

The task of melting depends on the state of division as well as the composition of the charge. Large pieces, for example pigs and heavy scrap, have a small surface area and are therefore least susceptible to melting losses and contamination. Finely divided bulky materials such as swarf or turnings are much less satisfactory: they are most readily absorbed by feeding directly into a liquid bath but this introduces dangers of gas contamination, a problem previously considered in Chapter 5. It is nevertheless economically desirable that these materials be returned to the production cycle at the earliest stage, if necessary by premelting and casting in denser form: this also assists in maintaining accurate control of composition, although double melting losses are then incurred. The physical problem of swarf recovery in cast iron founding has been approached through various techniques of canning and briquetting. Work has also been carried out with systems of injection of borings directly into the melting zone of the cupola. The value of induction furnaces for the direct recovery of cast iron borings has previously been mentioned.

The processing and refining of these more inconvenient returns may be uneconomic for the founder as compared with disposal to specialists for the production of secondary ingots of controlled composition. Where extensive refining is undertaken in the foundry, however, the opposite principle may be adopted, the foundry not only processing its own returns but combing the scrap market for further sources of cheap raw material. The former practice is common in the non-ferrous foundry industry, particularly the light alloy industry, whilst ferrous alloy founders utilize a major proportion of returned scrap in crude form.

Efficient materials utilization in furnace charges is facilitated by computerized control systems. These are designed to achieve minimum cost charges and the subsequent adjustment of bath composition to the required specification.

Melting conditions

Apart from effects inherent in the type of melting unit, conditions can be varied in respect of the time–temperature sequence, the use of slags and fluxes, and the atmosphere in contact with the charge. The furnace atmosphere can contain water vapour, CO, CO_2 and SO_2 as products of fuel combustion as well as the normal atmospheric gases.

The most common reaction is between melt constituents and oxygen present in the atmosphere and refractory linings, or in certain cases deliberately introduced by injection or as a constituent of slags and fluxes. In some cases, for example iron, copper and nickel, oxygen has appreciable solubility in the melt, where it can take part in reactions with solute elements: substantial amounts of surface oxide are formed only after saturation of the melt. In other cases such as aluminium and magnesium, stable oxides are produced at the surface and solubility in the melt is negligible.

Reactions are governed by the oxygen affinities of the elements present in the system, conveniently represented by Ellingham diagrams in which the standard free energies of formation of the oxides are plotted against absolute temperature. A typical diagram is shown in Figure 8.36. Those elements with strongly negative free energies of oxide formation, which appear low down the diagram, oxidize preferentially to those above and form stable oxide products.

The decrease in free energy, while representing the driving force for a reaction, gives no indication of the rate at which it will occur since this depends upon kinetic factors. In a heterogeneous reaction system involving metal–slag or metal–gas contact, the overall reaction rate normally depends upon the rate of transport of the reacting elements to, and products from, the interface: the degree of agitation and mixing is thus critical. Quantitative predictions are difficult, partly on this account and partly because of the complexity of systems containing many elements. Free energy data are nevertheless useful in predicting the probable reactions in a particular system.

Surface phenomena in the oxidation of liquid metals were examined in a major review by Drouzy and Mascré[48]. There are many similarities with the oxidation of solid metals although much less information is available. In general the reactions follow parabolic rate laws:

$$m^2 = Kt$$

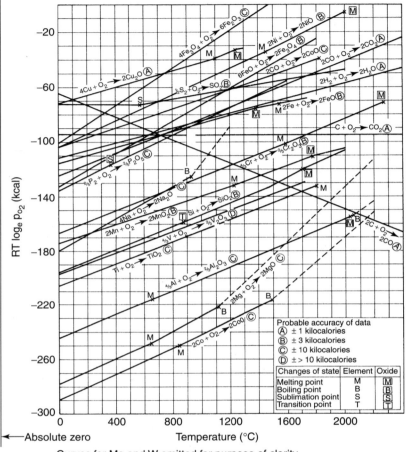

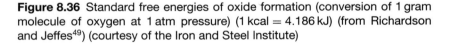

Curves for Mo and W omitted for purpose of clarity

Figure 8.36 Standard free energies of oxide formation (conversion of 1 gram molecule of oxygen at 1 atm pressure) (1 kcal = 4.186 kJ) (from Richardson and Jeffes[49]) (courtesy of the Iron and Steel Institute)

but linear laws are also found:

$$m = Kt$$

As in the case of metal–mould reactions (q.v. Chapter 5) preferential oxidation is an important factor and oxidation can again be inhibited by small amounts of certain elements. These analogies are seen in the heavy oxidation of aluminium–magnesium alloys, inhibited by small amounts of beryllium, and in the effect of aluminium in reducing the oxidation of zinc in brass.

The relationship between the various elements and oxygen is thus in most cases the predominant factor in melting. It affects compositional control in that losses of elements proceed at different rates. It affects overall melting losses and their economic consequences. Used to positive effect it provides a means of refining by selective oxidation of impurities. Melting conditions are thus designed either to suppress or to encourage oxidation according to need.

Where the process is one of remelting pre-alloyed ingot material or clean, heavy scrap, the principal concern is to avoid melting losses and gas contamination. Fast melting with minimum disturbance is therefore the aim and crucible type furnaces are most suitable, although melt protection may be obtained in hearth furnaces by use of a flux cover. This 'inert' practice is typified in the melting of aluminium alloys. Non-oxidizing fluxes are used, with only mildly oxidizing furnace atmospheres so as to minimize losses; oxidizing conditions in any case offer no protection against hydrogen absorption in these or in magnesium alloys. Degassing using a scavenging gas technique, together with any special grain refining treatment, is commonly carried out before casting. Reference has already been made in earlier chapters to the technique of filtration which can be employed to eliminate non-metallic inclusions from these alloys. Such treatment can in some cases be combined with degassing measures.

Magnesium alloys are susceptible to rapid oxidation and are normally melted in iron crucibles under a protective flux cover. After degassing, the melt is cleansed of suspended inclusions and residual melting flux using an 'inspissated' flux. This chloride–fluoride flux is stirred into the melt and coats the crucible and the metal surface: inclusions and residues are immobilized by adhering to the bottom and sides of the crucible on pouring. The elimination of deliquescent flux residues is vital because of the danger of corrosion and surface inclusions in the castings. Degassing can be carried out by similar techniques to those used for aluminium alloys: chlorine forms magnesium chloride which itself behaves as a flux. The special techniques used in melting magnesium alloys were treated in detail in works by Emley[50,51] and by Brace and Allen[52].

Most copper alloy melting is nowadays carried out under positively oxidizing conditions with subsequent deoxidation. In the case of the brasses, however, heavy oxidation and consequent zinc loss must be avoided; gas absorption is in any case not a problem in these alloys. The bronzes and gunmetals, by contrast, are susceptible to gas porosity both from hydrogen alone and from steam reaction. The usual approach, therefore, is to maintain a low hydrogen content by melting under fluxes containing an excess of unstable oxides and by using an oxidizing furnace atmosphere. The relationship between dissolved oxygen and hydrogen was fully examined in Chapter 5, where it was shown that an increase in oxygen content could be used as a means of eliminating other solutes by overthrow of existing melt equilibria (see Figure 5.10). Final deoxidation is based on a similar but opposite principle. In copper alloy melting phosphor–copper, lithium, silicon and various other elements are effective as deoxidants.

An alternative technique for such alloys is to minimize melting losses by maintaining neutral or reducing conditions and then to degas using a scavenging gas as employed for light alloys; this method is not however so widely adopted. In deciding the practice for this group of alloys it must be remembered that the oxidation–deoxidation sequence becomes less effective with increasing initial contents of elements such as phosphorus and zinc, which are themselves oxidized in preference to hydrogen.

The oxidation–deoxidation system is used in the melting of a wide range of materials, including steels and nickel base alloys. In many cases the level of oxidation is low, as for example in the high frequency remelting of heavy charge materials without the use of oxidizing slags. A limited carbon boil may, however, be induced for hydrogen removal, using additions of iron or manganese ore or nickel oxide.

Where more positive refining is sought for the removal of unwanted elements present in the charge, strongly oxidizing fluid slags and maximum agitation are employed: a vigorous carbon boil or gaseous oxygen injection can greatly accelerate the oxidation process. Hearth processes are preferred in these cases to accelerate the rate of slag–metal reaction: melting losses are generally higher and full deoxidisation is required at the final stage. This type of refining is widely used for carbon removal and for reduction of the phosphorus content of cast steel: slag–metal and gas–metal reactions have been most widely studied in this field[53,54]. A further development in refining is the injection of solids into the bath using a powder dispenser and a carrier gas. Such a technique is employed with lime for dephosphorisation in converter steelmaking and was adopted in the SCRATA fumeless refining process, based on the injection of iron oxides in the electric arc furnace.

To summarise, the common basis of a wide range of melting practice is the control of solute content through relationships of the type shown in Figure 5.10. Refinement is often sought by increasing the oxygen content to eliminate hydrogen, carbon and metallic impurities, after which similar relationships are employed in the reverse direction to remove excess oxygen. In rarer cases oxygen is attacked from the outset by reducing elements such as hydrogen and carbon, the subsequent problem being to remove hydrogen by inert gas or vacuum degassing.

The role of vacuum melting was discussed in Chapter 5. Like inert gas melting it offers clean, neutral conditions for melting a charge with a minimum of chemical change. Under certain conditions, on the other hand, it provides a means of active refining through pressure dependent reactions. These include the preferential evaporation of impurities with high vapour pressures, the extraction of dissolved gases, and the elimination of elements such as carbon, oxygen and hydrogen by the formation of compound gases.

Care is required to ensure that similar reactions do not upset the required composition of the alloy.

Melting losses

One of the most important aspects of melting practice is the level of melting loss, which has a strong influence on the economics of production. Sources of loss include oxidation or volatilization of metallic constituents during melting and the entrapment of liquid metal in dross or slag removed from the furnace or skimmed before pouring. Subsequent spillage is a further source of loss, although not strictly a melting loss. A less evident cost is incurred by the purchase and weighing of non-metallics included in charge materials: although the metal loss is only notional its cost is real.

Melting loss is important in respect of its influence on the overall metal yield, but compositional control is also affected since there may be preferential loss of elements with high oxygen affinity or vapour pressure: there is thus, to quote two examples, a tendency for magnesium to be lost from aluminium alloys and zinc from copper alloys.

The position of melting losses in relation to the whole pattern of metal utilization in the foundry may be seen by reference to Figure 3.7 in Chapter 3. It is evident that the melting loss incurred on the furnace charges represents a still higher percentage loss on the weight of castings sold, owing to repeated losses on that part of the metal recirculated as internal scrap. Melting losses are most serious in alloys of high intrinsic value, where the cost may exceed that of labour, fuel and maintenance combined. This point is illustrated by the examples in Figure 8.37.

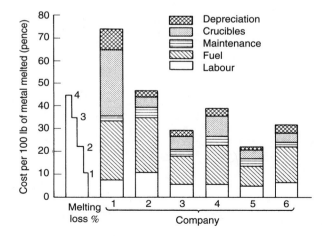

Figure 8.37 Cost of melting losses in copper alloys relative to other operating costs: oil fired crucible furnace (from Reference 55) (courtesy of Institute of British Foundrymen)

Melting losses in copper base alloys, light alloys and cast iron were the subject of past industrial surveys and bibliographies prepared by the Institute of British Foundrymen[55–57]. A notable feature of these studies was the wide variations arising from differences in plant, raw materials and melting practice. A number of clear conclusions, however, emerged and still remain valid.

Table 8.4 summarises the losses encountered in the various groups of alloys investigated. It is evident that the composition of the alloy has a major bearing on the level of melting loss. In copper base alloys, most results fall in the range 1 to 4%, the greater losses being encountered in those containing zinc and aluminium in contrast to the more stable bronzes. The highest losses in the aluminium alloys are met in the aluminium–magnesium series, whilst the aluminium–zinc alloys are subject to exceptionally low loss. A similar type of survey was carried out in over 2000 foundries in

Table 8.4 Melting losses in foundry alloys*

Alloy type	Range, omitting exceptional all-swarf melts %	Peak frequency range or typical value %	
Phosphor bronze	1.0–6.7	1–3	
Gunmetal	1.2–6.6	2–4	See footnote §
High tensile brass	3.0–5.0	3–5	
Aluminium bronze	1.0–5.0[†]	4–5	
All copper alloys	1.0–6.7	3–4	
Cast iron	Gain–6.6[‡] (0.2)	4	
Al–Si	0.2–9.7	1–3	
Al–Mg	0.4–10.7	{ 1–2	See footnote **
		6–10	
Al–Cu–Si	0.3–8.5	1–2	
Al–Zn	0.3–4.0	<1	
Al–Cu–Mg	1.7–3.5[†]	2–3	
All aluminium alloys	0.2–10.7	≈2	
Mg alloys (Zr free)	0–6.0	1–2	
Mg alloys (Zr bearing)	0.2–10.7	5–10	

*Based on References 55–57.
[†]Additional values in the range 12 to 14% were reported for all-swarf and contaminated scrap melts.
[‡] Up to 5% additional loss from swarf. Air furnace melting loss ≈5–7.5% (Ref. 59).
§Comparative values from survey of 2000 foundries in U.S.A.; range 0.5 to 12.5%; mean 3.3% (Ref. 58).
**Comparative values from source as §: range 0.75 to 10.0%; mean 3.4%

the U.S.A.[58]: some of the relevant findings are appended to Table 8.4. This reference also provides valuable data on total metal losses and on rejection rates for castings in various alloy groupings and commercial categories.

Apart from alloy composition, oxidation loss is a direct function of the time of exposure and of the surface area to volume ratio of the charge. High speed melting can thus itself be a major factor in reducing melting losses, especially when coupled with rapid and progressive movement of charged material into, through and out of the actual melting zone; the counterflow principle has already been exemplified in the previously mentioned tower furnace. The great progress in the provision of intense heat sources, including advanced burner design and high power density electric melting facilities, has also been emphasized.

The form of the melting stock is a major factor determining the level of loss: there is a general increase with the proportion of scrap, the highest losses of all being associated with heavily contaminated material containing large proportions of swarf and fines. Where clean, heavy scrap alone is used, however, no distinction should arise as compared with new ingot material. The use of shaped ingots in conjunction with crucible melting represents an interesting extension of the same principle.

The combination of a compact charge with rapid melting and casting is further pursued with single shot systems such as the self tapping induction furnace, as employed for the production of investment castings and shown in Figure 8.38. Pre-programmed power input enables the required superheat to be developed before final melting occurs at the tapping nozzle. The lost crucible induction melting concept embodies a similar single shot system, illustrated in Figure 8.39. Various aspects of these and other single shot melting systems have been the subject of a review by Hayes et al[62].

A further significant factor is furnace design, the survey findings being consistent with points previously made: rotary and reverbatory furnaces, with their more pronounced atmospheric contact and constant renewal of the metal surface, produce higher losses than crucible furnaces with their small bath area, although losses can be greatly reduced in all cases with

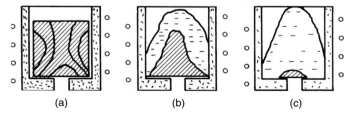

(a)   (b)   (c)

Figure 8.38 High speed self-tapping induction furnace for melting of single charge slugs. (a), (b) and (c) show progressive melting towards tapping nozzle (from Reference 60) (courtesy of Foundry Trade Journal)

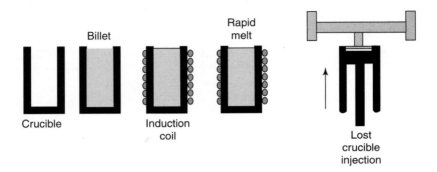

Figure 8.39 Principle of rapid induction melting lost crucible system (RIMLOC) (after Bird and Rickards[61]) (courtesy of Institute of British Foundrymen)

flux protection. In the melting of steel, losses as low as 1% in induction melting contrast with levels of 3–5% in electric arc practice. Similarly, the range encountered in the cupola melting of cast iron is well below the $5–7\frac{1}{2}\%$ obtained in the air furnace.

Since oxidation losses are directly related to time of exposure, melting rate is important and rapid melting offers one means of attacking the problem. Other savings can be made by avoiding excessive metal temperatures and unnecessary transfers. In all investigations, care and close control of melting conditions have been found to be crucial in achieving a low loss level.

Special melt treatments

The final stage of melting often includes treatment of the molten alloy for further purification or to influence the metallographic structure of the casting. Such treatments usually involve reagents added to the furnace bath or ladle by plunging, injection and similar techniques. Several examples have been discussed previously and a number are listed in Table 8.5.

Since furnace conditions are in most cases mildly oxidizing, final deoxidation is needed for those alloys in which oxygen absorbed by the melt forms embrittling constituents or reacts with other solutes, notably carbon, sulphur and hydrogen, to precipitate compound gases during freezing (see Chapter 5). Deoxidant additions are used to some degree in a wide range of iron, copper and nickel alloys. Suitable reagents can be deduced from the Ellingham diagram. Those elements low down the diagram form stable oxides and combine with available oxygen at the expense of elements above.

Treatments for removal of dissolved gases other than oxygen were fully discussed in Chapter 5; such treatments are most widely applied in light alloy casting production. Other chemical refining techniques include desulphurisation of ferrous alloys.

Table 8.5 Melt treatments

Treatment	Main fields of application	Typical reagents used
Deoxidation	Steel	Mn, Si, Al, Ca, C at low
	Copper base	P, Si, Li pressure
	Nickel base	Mg, Si
Degassing	Aluminium base	N_2, Cl_2
	Magnesium base	Cl_2
	Copper base	N_2
	Nickel base	O_2
Desulphurisation	Steel	Soda ash
	Cast iron	Soda ash, calcium carbide
Grain and constituent refinement	Aluminium base	Ti, B, Nb, Zr
	Magnesium base	C, Zr
	Steel	Al, Ti
Refinement and eutectic modification	Cast iron (inoculation)	Si, C
	Cast iron (S.G. iron structure)	Mg, Ce
	Aluminium–silicon	Na, Sr, P

Most other final treatments are applied for control of metallographic structure, often with grain refinement as the object: this topic was treated in some detail in Chapter 2. Grain refining treatments are extensively used in the casting of light alloys and involve additions of elements such as titanium and boron at the final stage in master-alloy form. A further melt treatment applied to light alloys is the modification of the eutectic structure in the aluminium-silicon alloys, using small additions of metallic sodium or treatment with sodium salts; the hyper-eutectic alloys in the same system are refined with the aid of phosphorus.

Whilst melt treatments with reactive substances can often be carried out by simple additions to the furnace bath, stream or ladle, special equipment and techniques are required for the efficient application of certain treatments, as will be shown by some examples.

Nodularizing and inoculation in cast irons.   To produce the spheroidal graphite structure either magnesium or cerium additions can be used, in conjunction with a low sulphur iron, although magnesium has been universally adopted by the industry and residual contents in the range 0.03–0.06% are required to achieve the structure. Various methods have been developed to overcome the special problems arising from the low density, low boiling point and high reactivity of the element, which can cause violent reactions, oxide fume and poor recovery of the magnesium content. Numerous reviews of techniques and equipment have been published, including the classic work by Karsay[63]; later surveys include those in References 64 and 65.

One of the simpler methods uses the "sandwich" principle, in which the magnesium-containing alloy addition is covered with a protective layer of solid plate scrap in a pocket in the bottom of an open ladle: the submerged alloy is only released into the melt when the ladle is substantially filled. A ladle of a type embodying a cover with an integral tundish is illustrated in Figure 8.40; these too are widely used[66], whilst other methods employ plunger bells, custom designed pressure vessels and pivoted converters. Some of these are suitable for the use of pure magnesium, which is cheaper than the magnesium ferrosilicon alloys, preferred in the simpler systems to moderate the vigour of the reaction.

In a special category is the mould based treatment in which a bonded nodularizing powder is placed in a reaction chamber within the gating

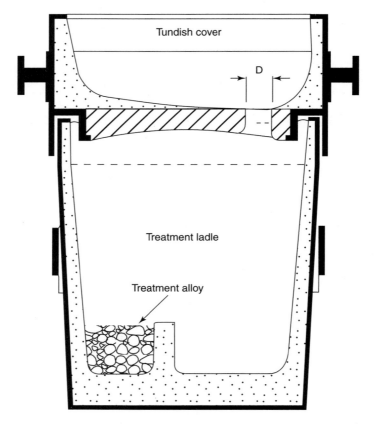

Figure 8.40 Tundish ladle system for magnesium treatment of cast iron. (from Else and Dixon[64]) (courtesy of Institute of British Foundrymen)

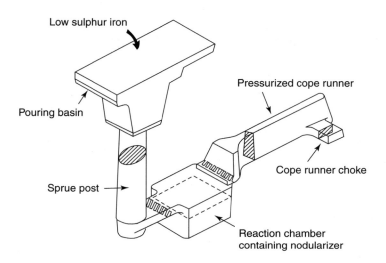

Figure 8.41 Geometry of typical system for magnesium treatment of cast iron within the mould (from Else and Dixon[64]) (courtesy of Institute of British Foundrymen)

system of the individual mould. An example of such a system is shown in Figure 8.41. The active reagent is continuously dispensed into the melt as it enters the mould cavity: the rate of feed is controlled by relating the surface area of the chamber to the flow rate of the iron over the exposed reagent to obtain a "solution factor"[67,68] In a variant of this system the reaction chamber is positioned within a separate treatment box, which can be interposed between furnace and casting ladle, enabling larger batches of metal to be treated.

Inoculation to assist graphite cell and nodule nucleation is often carried out by additions to the ladle stream during filling, but again provision can be made for late treatment. Special equipment is available for automatically regulated powder or wire feed of inoculant directly into the stream entering the mould. Cored wire feed has also found application for the combined desulphurization, nodularization and pre-inoculation of ductile iron[69].

Inoculating agents can be placed in the mould itself, whether in the basin, sprue base or runners, but care is required to ensure uniform absorption by the metal stream. A concise overview of the principal inoculation techniques has been produced by Turner[70].

Flux injection processes. The injection of solid reagents into molten baths achieves rapid dispersion through the melt, enhancing the contact area and shortening diffusion distances for the chemical and physical reactions required. The system has been applied to the desulphurization and purification of ferrous alloys, and to the combined cleaning and degassing of aluminium alloys. Powder dispensing lances are required, through which the reagent is fed into a carrier gas stream and injected below the surface of the bath.

The well-established aluminium alloy flux injection process, reviewed in References 71 and 72, provides for the simultaneous elimination of dissolved gases and inclusions. The flux, in this case based on chemically active alkali halides giving suitable fluidity and wetting characteristics, passes over a calibrated rotary feed table into a mixing chamber, where it is taken up in a nitrogen stream and fed to the injection lance. Oxides and other non-metallics are swept out in a dross, and hydrogen in the emerging gas, producing sound, clean castings. The equipment provides for automatic control of the treatment sequence. Energy savings and environmental benefits arise from this efficient combined process.

The fluxes employed can also incorporate modifiers for the treatment of aluminium-silicon alloys, and phosphorus for hypereutectic alloy refinement. These and other treatments are comprehensively reviewed in References 72 and 73.

The ladle as a reaction vessel. Many of the treatments referred to in the foregoing sections are carried out in the transfer or pouring ladle shortly before casting. Some of the more specialized treatment vessels are themselves essentially modified ladles. Apart from the inherent advantages of late treatment, a further benefit arises from the transfer of part of the refining function out of the prime melter into a secondary unit. This frees the primary vessel for rapid melting and coarse refining, enhancing productivity; in electric melting economies can also be achieved through a high and more uniform pattern of power consumption.

Various refining features can be incorporated in these specialized secondary units: vacuum treatment, injection of gases and solids, controlled stirring by induction or inert gas bubbles, protected alloy additions and, importantly, provision for reheating by induction, arc, plasma torch or exothermic reaction. This may be necessary to maintain the required temperature during and after treatment ready for casting, particularly when dealing with smaller vessels.

Ladle metallurgy has been extensively adopted in the wider steel industry for these functions and the fundamentals were reviewed in Reference 74. In the foundry field the AOD process, first used in furnace and ladle treatments but later in custom designed converter vessels, is widely employed in the production of stainless steel castings. Following primary electric melting and transfer of high carbon melt to the secondary vessel, blowing with argon-oxygen mixtures produces preferential elimination of carbon and protection of the chromium content, recovery of which requires lengthy procedures when using the arc furnace alone. Low cost, high carbon charge materials can thus be employed.

The AOD process has found limited use for other cast steels, given its desulphurizing, degassing and productivity benefits, an aspect reviewed in

Reference 75 in relation to stringent requirements for mechanical properties in carbon and low alloy steel castings.

For smaller volumes of metal, however, the development by SCRATA of the ladle based LMR refining process[76] enabled similar benefits to be obtained by combinations of argon stirring and oxygen lancing in the presence of suitable slags. A key factor in this design was the introduction of the specially developed insulated ladle lining[77] but metal temperature could also be regulated by induced exothermic reactions. Inert gas stirring through porous plugs to achieve the rapid mixing of reactive additions has also been applied to the desulphurization and carbonization of cast irons[78] Powder injection into ladles for desulphurization of cast steels was examined in a further paper concerned with ultra-clean products[79]. Important aspects of ladle metallurgy have been the subject of further contributions by Svoboda[80], Cotchen[81] and Heine[82], including the use of plasma, arc and induction heating in custom designed ladle-furnace hybrids.

In the application of melt treatments of whatever type, close attention is required to the most sensitive production variables, particularly metal temperature and the manner and timing of the additions. Reference has previously been made to the potential for fade following some treatments, hence the emphasis on their application late in the molten metal cycle. A similar principle applies to the introduction of highly reactive constituents of the alloy itself, which would be lost by preferential oxidation if included in the original charge. Such additions may be plunged into the furnace bath or added during tapping into the ladle, avoiding loss by direct combustion. A valuable feature of enclosed process ladles is the facility to make such additions under insert gas protection.

Melt quality control

The condition of the metal during melting can be assessed by various types of test on samples drawn from the furnace. A sample may be used for direct analysis of composition, but quality may also be judged from gas evolution during freezing, fracture appearance and other criteria giving indirect evidence of contamination.

The chemical analysis of bath samples, at one time restricted to a few elements for which rapid methods were available, can be carried out for a wide range of elements using spectrographic and X-ray fluorescence equipment of which highly automatic versions are available. Full bath sample analysis is however usually confined to large and important heats in which composition is being modified by refining; in other cases analysis follows casting.

One of the main aspects of metal quality is the presence of dissolved gases in the melt. Although direct analytical methods can be employed, gas content can be roughly estimated from visual observation of the freezing of a molten sample: the relative tendency of the metal to rise or to pipe is observed. A more sensitive version of this technique is the Straube-Pfeiffer reduced pressure test, in which the solidifying sample is enclosed in an evacuated chamber, the degree of porosity and final shape of the solid being used to indicate the condition of the melt (Figure 8.42)[83,84]. Exact standardization of the test conditions is required. The method is applicable both to light alloys and to copper alloys.

Other melt tests are based on the fracture of solidified samples. Chill tests for assessment of the compositional balance and potential structure of cast iron are based on inspection of fractures of test pieces cast under standard conditions[85–87]. In the widely adopted wedge test (Figure 8.43) the relative extent of the chill structure from the sharp towards the thick end of the sand cast test piece is used as a quality index relating to the carbon equivalent of the iron. Tests using end chilled rectangular test castings are based on a similar principle.

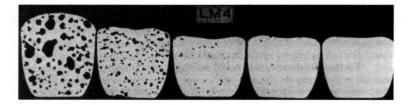

Figure 8.42 Reduced pressure test for gas content in molten metal; series of specimens of varying gas contents in an aluminium alloy (courtesy of Institute of British Foundrymen)

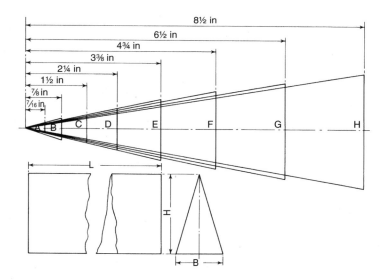

Figure 8.43 Designs of wedge chill test for determination of melt quality of cast Iron (from Reference 86) (courtesy of Institute of British Foundrymen)

Fracture tests can also be employed for some non-ferrous alloys: the colour and texture of fractured samples of certain copper alloys can be used as an index of alloy composition or gas content and thus of the properties attainable in the cast material[88-90]. Such tests require much skill in interpretation, however, as fracture appearance can be affected by many variables.

A number of other melt quality tests have been described in the literature: they include the determination of density and rapid microscopic examination[83,84,88]. Yet another system is the correlation of composition with the thermal arrest temperature of the liquidus[91]: the method can be applied to cast iron, steel and numerous other alloys. The fluidity testing of molten alloys has been fully discussed in Chapter 1.

In the organization and detailed planning of furnace programmes, the sequence of melts can be of vital importance to compositional control. Alloying elements tend to contaminate succeeding melts through skulls left behind on tapping the furnace and through absorption by furnace refractories. Unwanted pickup can also represent an appreciable loss of expensive alloying elements. Although washout heats may be used to clean the furnace, melting sequences are to be preferred which enable predicted alloy transfer to be taken into account in the compounding of each subsequent charge.

Compositional control is improved and melting costs minimized by co-ordination of melting and casting to ensure that metal, once melted, can be discharged from the furnaces without delay. Since heat losses are occurring throughout a furnace campaign it is also desirable to maintain a continuous succession of heats without intermediate shut down.

The pouring stage

The organization of pouring might be regarded as the crucial step in founding, since it is possible to nullify by a single error the accumulated results of all the earlier work. Systematic gating and feeding techniques, moreover, often depend for their full effect upon close control of metal temperature and pouring conditions. Since little time is usually available for casting operations, careful forward planning is needed with respect to mould casting sequence, individual mould requirements and the handling of the metal supply: the operations themselves require firm direction and a capacity for quick decision.

The layout of the casting floor may be based on a fixed casting point to which moulds are fed by conveyor, or moulds may be laid out on the floor, on stationary conveyors or in casting pits. The equipment is relatively simple. For heavy castings crane supported ladles are employed, whilst smaller quantities of metal can be dispensed from underslung hoists or hand shanks. More complex systems use automatic pouring, with synchronized movements of moulds and molten metal supply to maintain a continuous operation. A necessary adjunct is an adequate ladle preheating system, which besides ensuring that the refractory lining is thoroughly dried out, minimizes the temperature drop from furnace to ladle, an increasingly important factor where the volume of metal is small. Various aspects of ladle practice will shortly be further considered.

Metal temperature

Control of pouring temperature is achieved principally through the selection of furnace tapping temperature, which should therefore be governed by casting requirements. The appropriate tapping temperature depends on the expected rate of cooling in the ladle relative to the timing of the casting operations, determined by the number and sizes of moulds to be cast. Prediction of metal temperature at the casting point is assisted by knowledge of the initial temperature drop from furnace to ladle and the subsequent time–temperature relations for the diminishing mass of metal in the ladle. By direct temperature measurement such data can be readily determined and plotted for the average rate of discharge for each size of ladle, after which the time parameter offers a rough means of control.

Direct measurement is normally used for tapping temperature and for some individual pouring temperatures. For the former purpose quick immersion pyrometry is almost universally employed, although optical and radiation pyrometers can provide useful supplementary data. Immersion pyrometry is usually based on platinum–platinum rhodium thermocouples for alloys of melting point up to 1750°C and on chromel–alumel thermocouples below 1300°C. Although readings are in the latter case sometimes obtained by immersion of the hot junction directly in the metal bath, most foundry thermocouples are protected by sheaths. Fused silica is commonly used for liquid steel; other materials include electrographite, cermets, vitreous enamelled steel and heat resisting alloys. In one widely used unit the ends of the metal sheath and the chromel–alumel couple are swaged together to form a robust, quick response immersion probe. Mention must also be made of the expendable hot junction thermocouple developed for immersion pyrometry. A junction of exceptionally fine precious metal wires is embodied in a ceramic assembly, the whole being supported in a cardboard tube carrying compensating lead connections. The assembly survives immersion long enough for a temperature reading to be obtained on a quick response instrument. The disposable junction principle is attractive in its avoidance of possible inaccuracies due to progressive deterioration of normal thermocouple wires.

Of the instruments available for measuring thermocouple e.m.f., portable direct reading deflection indicators of the millivoltmeter type are convenient for general purposes. For more accurate determinations potentiometric measuring systems are preferable. Apart from manually operated portable potentiometers of the quick reading type, electronically amplified signals from a potentiometer can be applied in a direct temperature reading instrument readily visible from the casting floor: fast response potentiometric recorders are also available.

Optical and radiation pyrometers are of limited value for liquid metal temperature measurement owing to the absence of black body conditions and the partial dependence of results on the emissivity of the metal surface. Where conditions can be reasonably standardized, results for a particular alloy can be roughly calibrated against immersion values, but the technique is no substitute for immersion pyrometry where reliable data are needed. When the emissivity and other errors are known, an instrument such as the disappearing filament pyrometer can provide a useful guide to pouring temperature by sighting on the metal stream. Radiation pyrometers can be used to check temperatures of preheated ladle linings before casting.

Fundamentals of temperature measurement and pyrometric instrumentation are comprehensively treated in the British Standards BS1041[92] and BSEN60584[93].

Pouring temperature should be specified as an integral part of the casting method and the tapping temperature and mould sequence determined accordingly. Factors in selection have been emphasised in earlier chapters: the upper limit is established by such factors as hot tearing and the lower by the need for flow in thin sections and the avoidance of skulls in ladles. Within these limits pouring temperature should contribute to the solidification pattern sought by the method of gating and feeding (q.v. Chapter 3). Metallographic structure, particularly grain size and substructure, is a further factor in temperature selection because of the influences of undercooling on nucleation and growth. In general, lower pouring temperatures are possible where heavy, compact castings are being produced; castings with high surface area to volume ratio and short freezing time need higher superheat for satisfactory mould filling.

Equipment and techniques

The principal types of ladle employed in the foundry are illustrated in Figure 8.44. Ladles may be of the lip pouring or teapot type or may be fitted with nozzle and stopper for bottom pouring. The simple lip pouring system is also used for most hand shanks and for crucibles from lift out furnaces; the axially pivotted cylindrical or drum ladle is a variation of the lip pouring system more usually seen in the bucket design of Figure 8.44a. The lip pouring and teapot systems have the advantage that the metal enters the mould with less momentum than in bottom nozzle pouring but the latter gives a high degree of protection against slag inclusions. Although sliding gate and rotary valve systems have proved effective in particular

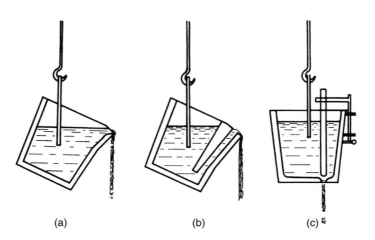

Figure 8.44 Main ladle types: (a) lip pouring, (b) teapot, (c) bottom pouring

circumstances, the nozzle and stopper has remained as the standard bottom pouring arrangement.

The main rules in pouring are to maintain a smooth and uninterrupted flow of metal for the avoidance of cold shuts and to prevent dross or slag from entering the mould: when lip pouring the cover must either be skimmed off or held back from the lip.

Flow rate is governed by the need to minimize turbulence and mould erosion yet to avoid misrun castings or cold laps. The rate is commonly regulated by the mould gating system, in which case pouring is visually controlled to maintain a liquid head in the runner bush without overflow. In tilt pouring this control is readily maintained. In bottom pouring control within limits can be achieved by stopper rod manipulation, but the nozzle diameter is the main factor in flow rate and should preferably be chosen to enable the ladle to operate at or near full aperture. The importance of this aspect of ladle practice was emphatically demonstrated in investigations by Ashton et al[76].

The theoretical rate of flow through an unrestricted ladle nozzle is given by the formula

$$q = a(2gh)^{1/2}$$

where q = volume flow rate,

a = c.s.a. of nozzle,

h = height of surface above the nozzle.

Figure 8.45, derived from this expression, portrays the relationship between flow rate and metallostatic head for various nozzle diameters. From such data the appropriate nozzle diameter can be estimated to match the average capacity of the gating systems but this takes no account of nozzle wear during the emptying of the ladle, nor of frictional losses.

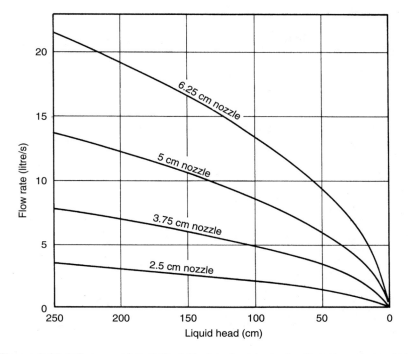

Figure 8.45 Influence of metallostatic head and nozzle diameter on nominal flow rate from bottom pouring ladle

In pouring a large casting, where the level of metal in the ladle and hence the flow rate changes appreciably during the pour, the approximate pouring time t can be estimated as follows:
Volume delivered in time t

$$Q = A(h_1 - h_2)$$

where A = c.s.a. of ladle,
$\qquad h_1$ = height of metal before pouring,
$\qquad h_2$ = height of metal after pouring.

In a small time interval dt, the volume delivered $dQ = A\, dh$. In the same time interval, the volume passing through the nozzle, also $dQ = a(2gh)^{1/2}\, dt$. Hence,

$$A\, dh = a(2gh)^{1/2}\, dt$$

$$\frac{a}{A} \cdot dt = \frac{1}{(2gh)^{1/2}} \cdot dh$$

Integrating,

$$\frac{a}{A} \int_0^t dt = \frac{1}{(2g)^{1/2}} \int_{h_2}^{h_1} \frac{1}{h^{1/2}} \cdot dh$$

$$\frac{a}{A} \cdot t = \frac{1}{(2g)^{1/2}} \cdot 2(h_1^{1/2} - h_2^{1/2})$$

$$t = \frac{A}{a}\left(\frac{2}{g}\right)^{1/2}(h_1^{1/2} - h_2^{1/2})$$

Should a predetermined rate of delivery be needed with lip pouring, this can be achieved by using a ladle designed to discharge metal at a rate proportional to the change in the angle of tilt. These conditions are obtained with a segment shaped ladle of the type shown in Figure 8.46. In this case

$$q = \pi r^2 l N$$

where q = flow rate,
   r = radius of segment,
   l = length of segment,
   N = speed of rotation.

Such an arrangement is used in special circumstances requiring constant flow rates, as in the centrifugal casting of pipes; the segment ladle has also been utilized in fluidity testing (Chapter 1, Reference 8).

Ladle pouring technique was one of the major aspects of a review by Ashton and Wake, concerned with the dependence of casting quality on

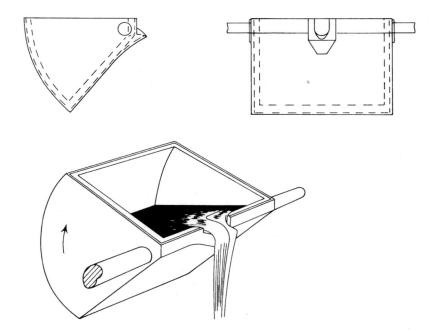

Figure 8.46 Segment ladle for constant rate lip pouring

ladle practice[94]. This was based on industrial surveys and direct investigations of flow from bottom pouring ladles. Nomographs were derived for the prediction of pouring times and the elimination of excessive throttling was again identified as a primary aim in the search for consistency.

Ladle heating. An efficient ladle heating system is crucial to the control of pouring temperature, and helps to avoid solid skulls which reduce product yield and shorten lining life. Much gas can be wasted by the use of ineffectual torches in open vessels. The provision of a cover or hood improves performance but efficient burners are the main requirement and enable lining temperatures exceeding 1000°C to be attained. Self-recuperative and regenerative burners are also available, to increase efficiency by capturing heat from the ladle exhaust gases: the regenerative principle is employed in a system based on alternate firing and exhaust of twin burners. The temperature can be monitored, and in some cases controlled, by suitably positioned thermocouples or infra-red sensors to minimize energy consumption and avoid excessive temperatures; intermittent firing can be used with the same objective. These and other aspects have been fully examined in References 95 and 96.

A relatively modern development in ladle heating is the use of electric power, a highly efficient medium when operated in a fully enclosed ladle. Cylindrical arrays of radiant resistance elements are employed within a protective sleeve or cage. Units rated at 30 kW, attaining ladle temperatures of up to 1000°C, have been successfully employed in iron founding and aluminium die casting[97], although much larger units have been used in the general steel industry. Heavily insulated hoods and temperature controls are essential features of this equipment. The relative fragility of the electrical elements is one factor that has restricted the application of this form of ladle heating.

Ladle insulation. Proprietary low thermal mass disposable liners of preformed refractory insulating board are highly effective in minimizing heat losses, and offer an alternative to high temperature preheating, particularly for smaller ladles, for which single piece moulded versions are available. For larger ladles the joints between flat board segments are sealed with refractory cement, the whole being bedded on a refractory outer lining. The disposable liners are readily removed and renewed, although they can be used in some cases for continuous high temperature runs involving repeated refilling[98,99].

Ladle reheating. Reheating of molten metal in the ladle has already been mentioned in the context of further processing measures, but can also be used simply as an extended control of pouring temperature. One method

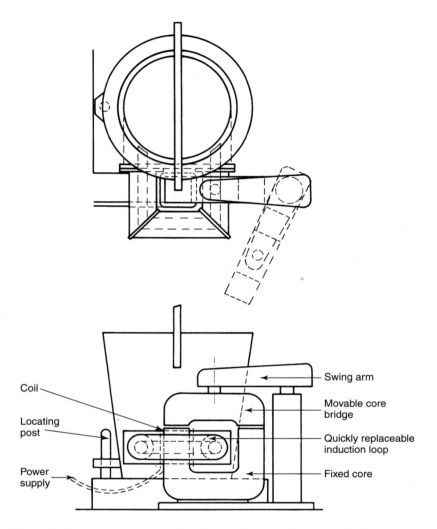

Figure 8.47 Induction loop system for heating metal in ladle. Horizontal loop version, schematic view (Q.R.L. System) (courtesy of Birlec Ltd.)

employs the quickly replaceable loop system, an example of which is shown in Figure 8.47. Induction heating is applied to metal circulated around a horizontal or vertical loop communicating with the main ladle chamber. The loop is readily replaced using detachable flanges, whilst the inductor core is constructed in two sections with a movable bridge piece, enabling the ladle to be readily detached from the heating system as required. Loop heating of ladles as a means of increasing total melt capacity for the production of heavy castings has been reviewed by Smith[97]. Other methods of induction heating for this and more specialized purposes were reviewed in Reference 100. Reaction reheating is also possible in limited circumstances but would normally be an element in a more complicated ladle refining sequence.

Further aspects of the pouring and handling of molten metals were treated in a detailed review paper by Smith[101]. Although relating to ironfoundry practice, most of the general points dealing with ladle practice are applicable across a wide spectrum of casting alloys.

Automatic pouring. In the modern foundry industry, with its extensive use of automatic moulding plant and process controls, there is clearly a need for precise techniques of molten metal delivery into the mould. Despite the difficulties involved, various systems have been developed for incorporation in volume casting operations.

A well designed automatic system has the potential to reduce the variability inherent in manually controlled pouring, and at the same time to improve working conditions in the foundry. Essential features are the capability to position the molten metal nozzle or spout accurately over the mould entry, and to deliver the correct amount of metal at the required rate. In a fully automatic system provision needs to be made to control all the necessary movements with appropriate mechanisms and sensors. A means of heating of the metal in the pouring unit is also desirable if substantial holding times are involved: this also permits retention of molten metal rather than emptying the vessel during interruptions in production. Induction heating on the channel principle is widely used for this purpose.

Automatic pouring units employ varied means of metal transfer, based upon either gravity or pressurized flow. Controlled gravity pouring can be achieved using a tilting vessel: one example of a simple mechanical system is that illustrated in Figure 8.46, designed for constant rate lip pouring. Some fully automatic systems too employ the tilt principle, although the path and shape of the emerging stream may be insufficiently precise in some situations. Other systems have been preferred in much commercial plant, including the stoppered nozzle arrangement as used in conventional bottom pouring ladles, and gas pressurized dispensers.

Repeated operation and continuous immersion of stopper-nozzle systems as required for automatic pouring is made possible by the use of isostatically pressed monobloc units, formed from ceramic bonded aluminia graphite refractories. The normal bottom nozzle system requires allowance for the change in metallostatic head as the vessel becomes depleted, but this effect is reduced in shallow tundish pouring systems and can also be subject to compensating controls.

In the stopper-nozzle system illustrated in Figure 8.48 the metal is contained in a pressure vessel provided with an induction heating loop. The pressure in the vessel is used only to maintain a constant level above the nozzle within the laterally positioned pouring box, the actual pour being controlled by the stopper.

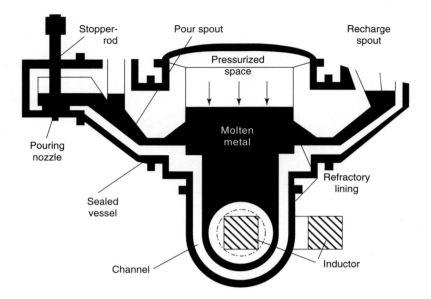

Figure 8.48 Induction heated automatic pouring pressure vessel with stopper rod operation (courtesy of Inductotherm)

The alternative pressure pour system, in an otherwise similar heated vessel, in shown in Figure 8.49. In this case pouring is accomplished by increasing the pressure inside the vessel, to displace the metal though the teapot spout.

The control of automatic pouring can be based on time or weight, the normal method used in the pressure pour system, or may employ sensors to detect the molten metal level as well as to achieve correct alignment with the sprue by detection of the mould top. One type of non-contact optical system, based on laser probes measuring distance, was detailed in Reference 102. A sensor detects the position of the sprue cup when the mould is indexed, and adjusts the ladle into the pouring position. Stopper rod movement is then governed from a probe mounted on the ladle carriage to measure the metal level in the cup. The vessel illustrated in Figure 8.48 is provided with a video camera linked to a computer, which accurately positions the pouring nozzle and operates the stopper rod, to control both pouring rate and amount of metal poured; there is automatic adjustment for changes in metal level or nozzle diameter.

A schematic portrayal of a control system for a pressure pouring vessel with stopper rod closing control is shown in Figure 8.50 (Ref. 103) derived from a major earlier review of the scope of automatic pouring by Powell and Smith[104].

Metal temperature in automatic pouring can be similarly monitored, using remote infrared thermometers to observe the emerging stream[105,106].

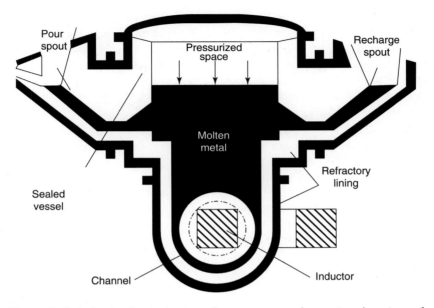

Figure 8.49 Induction heated automatic pressure pouring system (courtesy of Inductotherm)

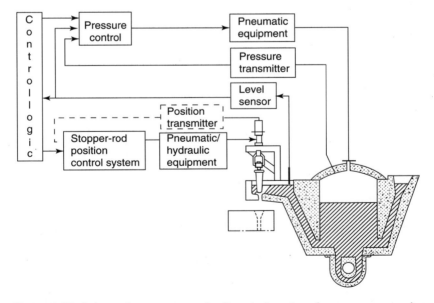

Figure 8.50 Schematic arrangement of control system for pressure-pouring vessel with stopper rod dosing control (ASEA) (from Reference 103 after Powell and Smith[104]) (courtesy of Foundry Trade Journal)

Other features include the use of electromagnetic pumps to propel metered quantities of metal through tubular feed passages, whilst the robot handling of smaller vessels for pouring constitutes yet another automatic system. Finally, mention should be made of the automatic bottom pouring principle, in which small charges for single castings are induction melted in a self-tapping cup, discharging directly into the mould cavity; this was referred to previously and illustrated in Figure 8.38.

Special pouring techniques. The demand for high metallurgical quality and freedom from contamination led to the development of many techniques designed to minimize metal disturbance and atmospheric contact during pouring. In a system described by Campbell and Bartlett[107] and intended for production of premium quality castings in aluminium alloy, the melting crucible is provided with an integral enclosed spout and the whole system is maintained under a nitrogen atmosphere throughout the melting and casting sequence (Figure 8.51). The elongated nozzle of the transport tube has its own nitrogen supply and a special advantage claimed for the system is that the nozzle can be inserted deep into the mould cavity and withdrawn during filling, so combining the advantages of top and bottom gating.

Taylor and Briggs[108] described a system, used for high temperature alloys, in which the furnace was fitted with an extended launder incorporating a stoppered nozzle (Figure 8.52). In this case the nozzle is designed to register in the top of the mould, a seal being formed by a flexible refractory gasket: the nozzle and sprue diameters are identical. A protective slag covers the metal surface during melting and the stopper remains closed during tilting to the pouring position. The mould is separately flushed with argon before pouring.

The SCRATA Closed System Pouring mechanism was designed to achieve direct coupling of a bottom-pouring ladle to the mould sprue, through a nozzle extension tube engaging with the pouring cup. The principle is shown in Figure 8.53 and the operation of the system is fully described in Reference 76. The particular advantage claimed for this technique is the elimination of the ladle throttling often used when pouring relatively small volumes, the flow rate instead being limited by the capacity of the gating system.

During and after pouring various ancillary tasks must be carried out, including the addition of feeding compounds to open heads. Insulating materials or standard feeding compounds must be added without delay but in the case of strongly exothermic materials careful timing of the addition is needed for full effect. Other tasks include lighting off of gas and removal of relief cores.

The interval before knockout is important from the points of view of moulding box utilization and of the temperature of the sand in the system. The knockout temperature of the casting must also be considered in relation to possible metallurgical transformations and to the development of internal

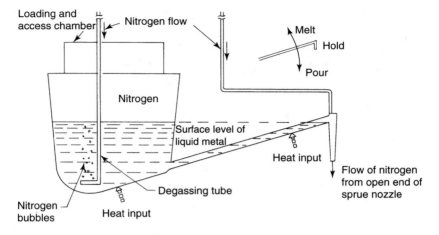

Figure 8.51 Extended spout crucible for inert gas melting and pouring of aluminium alloy (simplified diagram, after Campbell and Bartlett[107]). System shown in holding position, with degassing tube in operation. Transport tube is raised for melting and lowered for pouring (courtesy of American Foundrymen's Society)

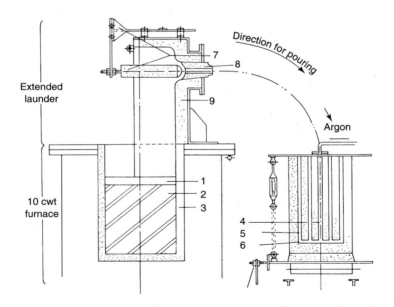

Figure 8.52 Extended spout furnace for protected melting and pouring of high temperature alloys. Furnace in melting position. Key: 1. slag cover; 2. molten metal; 3. furnace lining; 4. downgate; 5. mould cavity; 6. runner system; 7. stopper mechanism; 8. nozzle; and 9. launder lining (from Taylor and Briggs[108]) (courtesy of Foundry Trade Journal)

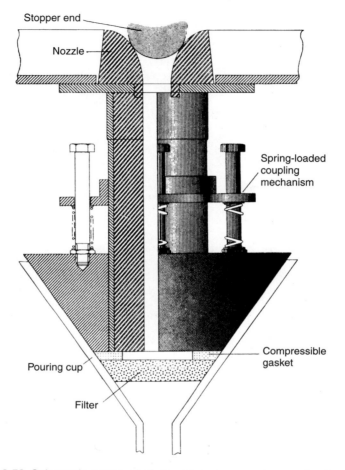

Stopper end

Nozzle

Spring-loaded
coupling
mechanism

Compressible
gasket

Pouring cup

Filter

Figure 8.53 Schematic portrayal of closed-system pouring mechanism (from Ashton et al.[76]) (courtesy of Institute of British Foundrymen)

stresses: this topic was fully considered in Chapter 5. Knockout temperatures can be determined with radiation or contact pyrometers and the information used to establish optimum time intervals.

The knockout operation entails the separation of castings, sand and moulding boxes and can employ various types of mechanical aid. These include punchout rams, vibratory shakeout tables and grids, and rotary drums providing a tumbling motion. Cooling of the castings and sand are achieved in varying degrees, with preliminary screening of the separated sand. Vibratory systems of efficient two-mass designs, with intermediate springs between vibrator and workface, are available and can be applied either to flat decks or in conjunction with rotary drums, whilst high frequency vibration can also be employed, for more difficult sand separation or for fragile castings. In other developments vibration can be applied directly to the individual casting.

Manipulators and remote controls are increasingly available for handling and for variation of operating conditions from cabin or crane cab, separating the operator from the knockout itself, which also has provision for dust and fume extraction. Both extraction and noise control can be facilitated by partial or complete enclosure of the system. Such features can greatly improve working conditions in this exacting operation.

Part 3. Finishing Operations

Much of the remaining work in producing the finished casting is carried out in the fettling shop. Basic requirements are initial cleaning for removal of adhering sand and core residues, separation of feeder heads and runners, and final dressing to eliminate flash and excess metal. Integrated with these operations are heat treatment, rectification and inspection procedures, for which there is no rigid sequence.

The present section will be mainly concerned with the characteristics of fettling equipment and methods, and it should be emphasised at the outset that the task is profoundly influenced by the techniques used in moulding and casting, for example the positioning of gates and feeder heads, and by the degree of surface smoothness and freedom from sand adhesion. The need for fettling can be greatly reduced by close initial attention to two points. Firstly, pattern equipment of high quality and accuracy, embodying careful selection of mould joints and core positions and ensuring close fits, can greatly reduce the incidence of flash and ease its removal. Secondly, feeder head and gate removal should be a major consideration when planning the casting method. The object should be to eliminate contour fettling by placing heads and gates on flat surfaces, permitting accurate, smooth and close cuts: reliance on sculpture by the fettler is then avoided. Where a plane interface is unobtainable templates can be provided to simplify dressing. Gating technique is itself crucial in minimizing burn-on and metal penetration.

Fettling costs represent a major proportion of total production costs, so that preventative measures can make an important contribution to the entire economics of casting manufacture.

Initial cleaning

The main bulk of moulding material is normally removed at the knockout station, but full sand separation and decoring are normally completed in the fettling shop. Rough cleaning with bars or pneumatic tools may be required for confined internal cavities, followed by a general cleaning operation, most commonly shotblasting; alternative treatments include hydraulic and chemical cleaning. These operations may in some cases be performed or repeated at a later stage, for example after head removal, fettling and heat treatment, to give the final finish.

Blast cleaning involves treatment of the surface with erosive jets and can be applied either to individual heavy castings or to batch loads of smaller castings, with varying degrees of manual and automatic working. In some cases the knockout and initial cleaning stages are combined, in plants in which complete moulds, either boxless or even including moulding boxes, are shotblasted; the sand and abrasive are then separated by magnetic and air current sorting.

Shotblasting

Blast cleaning is accomplished by streams of dry abrasive particles energized either by compressed air or high speed centrifugal impellers to impinge on the work. Various designs of pneumatic system can be used, but the highest efficiency for heavy duty operations is achieved by the mechanically energized system. Shot velocities of the order of 80 m/s ($\approx 16\,000$ ft/min) are attained in commercial equipment.

The nature of the abrasive is an important factor both in cost and surface finish. The least expensive material is chilled iron grit but this is subject to loss by fragmentation: a greater rate of replenishment is therefore required than, for example, in the case of the more expensive steel shot. The economics of selection depend also upon the extent of direct loss of unchanged abrasive. Systematic studies reported by Stoch and Lambert[17] revealed the relative merits of various abrasives with respect to cleaning rate, attrition loss and impeller wear (see Table 8.6). Cut wire with its low breakdown and plant wear rates was recommended as most economical where carry-out is negligible, the advantage passing successively to steel shot or grit and to malleable iron abrasives with increasing loss. The cheapest material is most economical where carry-out losses are high. Particle size and other factors were also investigated.

Table 8.6 Relative performance of abrasives*

Abrasive	Cleaning rate	Relative addition rate	Wheel wear	Order of price
Cut wire pellets	1	1	1	1 (high)
Steel shot	1	1.5	1	2
Malleable iron grit	1.5	7.5	1	3
Chilled iron grit	2	15	10	4 (low)

*after Stoch and Lambert (17)

Since abrasive cleaning involves interchange of molecules between the abrasive particles and the abraded surface, non-metallic abrasives such as slag, alumina and silicon carbide need to be used in certain cases to avoid contamination with iron: they are also used to obtain a fine finish on soft alloys and very small components. Compressed air rather than mechanical equipment is used for these purposes.

Numerous designs of shotblast equipment are employed. Heavy castings are normally processed in a fixed chamber, the shot nozzle being either manipulated directly by the operator, clad in a protective suit, or remotely controlled from outside. A similar principle is embodied in the glove box cabinet units used at the opposite end of the scale for the blast cleaning of extremely small castings. Castings of a wide range of intermediate sizes, including machine moulded mass produced components, are processed automatically in either of two basic types of plant. Castings can be supported on a turntable or suspended from a monorail or pendulum conveyor within a fixed chamber or cabinet, the shot stream being traversed for full coverage of the exposed surfaces. Alternatively, the load can be tumbled in the shot stream through continuous rotation or rocking of the chamber itself about a horizontal axis. A special feature of these plants is belt reversal to prevent jamming of the load. Such batch machines can also be provided with automatic feed and discharge conveyors to maintain semi-continuous operation.

Abrasive wear of parts in shotblast plant can incur high costs: in this respect it is particularly important to maintain full loading of the abrasive feed to the discharge nozzle. Some use has been made of rubber facings to replace metal on the most wear-prone surfaces. Provision is required for dust extraction and for cleaning and recirculation of the abrasive: effective separation of sand and fines has been shown to be vital to the economic operation of shotblast plant[17] and reduces the need for complete preliminary cleaning and decoring.

Although much shotblast plant is designed on the batch treatment principle, a close approach to continuous operation can be achieved by conveyor or skip loading and by dual turntable plants enabling loading and unloading to proceed outside the workchamber during blasting. One barrel system embodies four separate chambers for successive loading, blasting, shot evacuation and unloading, with provision to vary cleaning time and speed of rotation in the blast chamber. A fully continuous automatic system is achieved in a three-stage inclined barrel, which integrates an inlet drum, blast unit and discharge drum: the castings tumble and progress through the system on rotation, with feed and discharge via vibratory conveyors.

A substantial review of cleaning with particular reference to shotblasting operations is given in Reference 109; examples of plant are included in References 110 and 111. Specialized functions of shotblast treatment include shot peening, employed to improve fatigue resistance by cold working the surface layers of the material. For this purpose angular grits are avoided in favour of round shot or prerounded cut wire, ensuring smooth indentations rather than pitting of the component surface.

Hydraulic cleaning

In wet blasting systems castings are exposed to high pressure water jets which may also incorporate solid abrasives. Pressures of $16\,MN/m^2$ (≈ 2350 lbf/in^2) or even higher are applied, using systems analogous to those in shot-blasting. Highly effective sand removal can be achieved, but despite early successes the water blast did not significantly displace the dry shotblast system in mainstream production: it has, however been adopted in more limited circumstances for decoring and other tasks.

Chemical cleaning

When compared with blast cleaning there has been little application of chemical cleaning in the foundry industry. The available methods for ferrous castings were reviewed by Sully and Stoch[112]: they utilize baths of molten caustic soda into which additional reagents are introduced to react with and break down the surface oxide layer, thereby loosening adhering sand. In the electrolytic method the same object is achieved by applying current: oxide is reduced by nascent sodium liberated at the casting, which is made the cathode in the cell.

Other methods of cleaning include barrel tumbling and scratch brushing. In the former method, now of declining importance, light castings are rotated with metal pieces to achieve impact abrasion and burnishing of the surfaces: the method has been much applied to cast iron. Wire brushing is primarily used for non-ferrous production, where the normal clean strip often renders a full cleaning operation superfluous.

Runner and feeder head removal

Numerous alternative methods are available for the removal of feeding and gating systems. These extend from simple knocking off or prising away, possible only in some circumstances, through the standard methods of machine cutting by saw, abrasive wheel or parting tool to a highly versatile range of flame cutting processes. Each of these methods fulfils a useful function in the foundry, depending on the type of alloy and dimensions.

Knocking off

This method can be applied to alloys in which the as-cast ductility is low, especially where the area of the junction with the casting is small. To facilitate fracture use is made of various types of restricted neck and knock-off core, both to reduce the metal section and to provide a notch effect; the initial notch may, alternatively, be produced by one of the slitting processes.

Mechanical aids, such as presses and powered wedge devices, are in some cases employed to accomplish the fracture. Similar aids can also be used to remove flash.

Sawing

Although sawing is potentially suitable for most types of cast material it is most widely applied to the softer non-ferrous alloys. Bandsaws, hacksaws and circular saws can all be used, but the bandsaw is the most versatile machine in respect of cut dimensions and working conditions. Bandsaws can be used over a wide range of speeds. The normal saw cutting principle operates at speeds from approximately 25 m/s ($\approx$5000 ft/min) down to 0.25 m/s ($\approx$5 ft/min), the speed being reduced to diminish the heating effect as the length of cut increases. For aluminium and magnesium alloys the higher speeds are suitable.

At still higher speeds the principle of friction cutting comes into effect, sufficient heat being generated to bring the temperature at the root of the cut into the plastic range: incipient melting may even occur and a high rate of stock removal is attained. To achieve adequate heating, speed needs to be increased in relation to the length of cut, velocities as high as 75 m/s ($\approx$15 000 ft/min) being used in some cases. Friction cutting is suitable for ferrous alloys up to approximately 25 mm in section thickness, including hard or work hardening compositions not normally amenable to sawing.

Abrasive wheel slitting

Abrasive cut-off machines provide a rapid means of head removal for sizes within the capacity of the wheel diameters available, being particularly suitable for cast iron and steel, including the hardest alloys. On a full sized machine the wheel can be mounted in a fixed position, the casting being pushed along the workrest by hand, or the casting may be clamped to the table, the wheel mounting being swung into the contact position. Much use is also made of portable machines on which the disc is mounted on a right angled shaft. The wheel thickness used is commonly 3 mm ($\approx\frac{1}{8}$ in), in diameters up to 500 mm ($\approx$20 in): rubber bonding, or laminated construction with fabric reinforcement, provides limited flexibility, but the casting needs to be held firmly to avoid wheel shatter.

For each type of wheel there is an optimum peripheral speed, usually in the range 48–81 m/s (9500–16 000 ft/min) and the machine should therefore incorporate provision for an increase in rotational speed to compensate for wheel wear.

Machining

In certain cases heads can be conveniently removed by machine parting, especially where the process can be combined with other machining operations. The method is particularly suitable for large heads and those where a continuous cut can be achieved, as in the case of axially positioned and annular heads; a narrow parting tool is employed and no further finishing need be carried out.

Flame cutting

Flame cutting or "burning off" techniques are widely employed for the removal of heads from steel castings; cutting is most commonly based on the oxyacetylene system, although propane and other fuel gases can also be used. In cutting ferrous alloys the oxygen–gas reaction is important primarily as a means of preheating the metal to its ignition temperature, after which heat is derived mainly from combustion of the metal itself, with excess oxygen supplied through the torch. This is accordingly provided with preheating jets and a main oxygen orifice. Cutting occurs by a combination of melting and oxidation to slag.

The conventional cutting torch is capable of considerable depths of penetration through carbon and low alloy steels, but difficulty is encountered in the case of alloys with good oxidation resistance, for example stainless and heat resisting steels, in which the progressive action of the flame is impeded by stable oxide films; cast iron also presents some difficulty. Successful cutting of these alloys is achieved by using the powder cutting principle. A stream of iron powder is dispensed through the torch into the flame, providing additional heat of combustion to assist local melting and increasing the fluidity of the oxide products.

Flame cutting is a flexible process in that a skilled operator can follow irregular contours, and heads of a wide range of sizes can be removed. For the longest cuts, which would be too extensive for penetration by normal cutting torch, the thermal lance technique is available. Oxygen is fed through a steel tube of small diameter after separate local preheating of the casting to ignition temperature. The heat for cutting is in this case derived wholly from oxidation of the metal of the casting and consumable lance. The latter can be packed with steel rods to provide a more intense heating effect at the outlet.

Further and related high temperature cutting processes use electric arc or plasma torch heating to achieve local melting of the cast metal. Arc cutting can employ either carbon or metal electrodes and the process is suitable for oxidation resistant alloys; compressed air is usually employed to blow the molten product clear of the cutting zone. Air-arc or oxy-arc torches are particularly suitable for the removal of large feeder heads. Plasma cutting too involves the generation of an electric arc, which is in this case used to transfer heat in the form of an ionized gas stream with the general character of a flame.

Care is required in any high temperature cutting process, including those where intense frictional heat is generated, to avoid adverse metallurgical effects from local heating: martensitic transformations in high carbon and alloy steels, for example, may lead to cracking. As in welding operations this danger can be diminished by preheating.

Dressing and finishing

The last stage of fettling involves dressing for the removal of excess metal and residual adhering sand. Material to be removed includes flash, pads and the stumps of feeder heads and ingates, which need to be dressed flush with the surface. The object is to ensure that the shape, surface finish and dimensions of the black casting conform to design requirements. The principal techniques are chipping and grinding but hot fettling techniques with specially designed torches have also been developed. Apart from surface dressing it may also be necessary to gouge out surface defects prior to rectification: this topic was treated in detail in Chapter 5.

Chipping is carried out to prise away flash and adhering sand. Apart from limited use of hand hammers and chisels, the principal tool is the pneumatic chisel operating at the normal mains pressure in the range 550–690 kPa ($\approx$80–100 lbf/in^2). Clipping, using mechanical aids, can also be used for this purpose.

Grinding operations are carried out on swing frame, pedestal or portable machines. The first of these is suitable for the removal of stumps of feeder heads and other heavy duty grinding on large castings. Fixed head pedestal or bench grinders, using wheels up to approximately 750 mm (30 in) diameter, are suitable for small castings capable of being individually handled.

The most universal tool for general dressing is the portable grinder, using a wide variety of wheels for manual operations on castings of all sizes. Most portable machines are electrically powered: for heavy duty grinding high frequency motors (200 Hz or higher) are preferred. The commonest design uses an axially mounted wheel but angled shafts and flexible drives are also used. Many shapes of wheel are available but most fettling is carried out using plain wheels, for example the $150 \times 25 \times 16$ mm ($6 \times 1 \times 5/8$ in) size. Small high speed pneumatic or electric machines are used for delicate shaping and polishing with mounted points and tungsten carbide burrs.

Grinding technique involves the choice of wheel characteristics, peripheral speed and operating pressure to give minimum overall cost per unit weight of metal removed. The initial step is selection of the appropriate wheel. Wheel quality is defined by numerous characteristics of which the most important are the type and size of abrasive grit, the bond and the degree of porosity. The abrasives used are silicon carbide, alumina and zirconia, coarse grades being used for rapid stock removal and fine for polishing. Pore size must be adequate to leave clearance for the abraded chippings. Bonding materials can be of either the vitrified or synthetic resin type. The former is cheaper but the latter, being more robust, is preferred for high speeds, rapid stock removal and rugged treatment involving uneven pressure; glass fibre reinforcement is employed to increase wheel toughness for the highest speeds. Wheels are also made in various grades of hardness, of which there is an optimum value for a given metal quality and work pressure, representing the most economic combination of cutting rate and resistance to wheel wear. The harder the wheel the greater its wear resistance but the higher the pressure required to achieve an equivalent rate of metal abrasion. The characterization and selection of wheels for particular grinding operations and materials featured in a major report on the dressing of castings[113], which also reviewed aspects of layout, safety and the care and maintenance of equipment; the elimination and removal of flash was the subject of a separate report from the same working group[114].

Most proprietary grinding wheels are designed to operate at recommended peripheral speeds in the range 33–81 m/s ($\approx$6500–16 000 ft/min) and the ideal machine should incorporate variable drive with compensating adjustments to enable the recommended speed to be maintained irrespective of diameter. High speeds are generally associated with lower unit costs of stock removal, but it is vital from the safety aspect that no wheel should be run above its maximum design speed. The other important operating variable is work pressure, which should be sufficiently high to utilize the full available power of the motor and, in conjunction with wheel selection, to give the best combination of stock removal rate and wheel life. To achieve higher contact pressures than normally available, pedestal grinders can be provided with pressure bars enabling leverage to be exerted: in this case wear tends to increase, so that a harder wheel is required to maintain the optimum wear–cutting rate performance. Grinding tests reported in a detailed review by Roebuck[115] demonstrated that increasing the work pressure can greatly reduce unit costs. Although abrasive costs were almost doubled when using a high grinding force, the reduction in grinding time due to increased cutting rate halved the total grinding cost (Table 8.7). Additional factors in selection of wheel type and operating conditions were reviewed by Gray[116]. Irrespective of the particular practice, wheel performance can be enhanced by the use of heavy duty wheel dressers to clear debris and expose fresh cutting edges on the abrasive particles.

A further important grinding system for the dressing of castings employs abrasive belt machines, which are capable of very high rates of stock removal given sufficient power[117]. The fabric backed belts are operated at speeds around 23 m/s ($\approx$4500 ft/min) and the work rate can be enhanced with pressure loading without danger of wheel failure; abrasive belts are noted for their relatively cool cutting action. Workholding devices can be used to avoid the need for direct operator contact. Belt machines are particularly suited to automatic working, since rigid metal backings provide support to the abrasive face and no allowance is required for wear as when

Table 8.7 Summary of foundry grinding test results*

Average force applied (lbf)	18	77
Average metal removal rate (lb/h)	5.2	16.2
Average wheel loss (lb)	2.4	11.6
Relative cost of abrasive per unit of metal removed	1	1.65
Relative cost of time	6.75	2.25
Relative total grinding cost	7.8	3.9

*After Roebuck 115.
1 lbf = 0.454 kgf = 4.45 N

using wheels. Similar abrasive facings are used in disc form, with flexible backings for the dressing of curved surfaces.

Other methods of surface abrasion include filing and wire brushing: these are frequently used for the softer non-ferrous metals, in which fettling problems are generally less severe.

Hot fettling processes. As adopted for ferrous alloys, these have the particular advantage of avoiding dust generation. In powder washing, iron powder is dispensed through a specially designed torch into an oxy-gas flame, which can be played across the casting surface to remove superfluous metal and adhering sand; air-arc and plasma torches can be similarly used. These are essentially variants of the respective cutting processes, available as alternatives to heavy grinding[120].

Automatic fettling and the use of robots

Fettling has traditionally represented the most labour intensive sector of metal founding, with wide use of manual techniques and of batch rather than continuous production. Mechanization in fettling shops mainly entailed systems for handling castings between stages, either individually by crane or conveyor, or as batches on pallets or in stillages.

The introduction of new mechanical and automatic devices, including robots, brought considerable changes in the traditional pattern, mainly in cases where individual operations are subject to frequent repetition. This position came to be expressed in the neat concept of fixed and variable fettling[118,119]. The fixed element, for example the grinding down of feeder head stubs, a standard procedure for all castings of one type, may be seen as suitable for an automatic operation, whilst removal of extraneous material, such as random finning or metal penetration, can remain as a matter for manual attention.

Industrial surveys also identified stub removal as the most costly procedure, followed by general dressing and head removal[120]. This focused interest on the cutting operation itself, and an accurate system was developed for head removal on large steel castings, using a torch traversed by a remote controlled servo-arm manipulator[121]. The closeness of the cut as compared with manual operation greatly reduced the subsequent stub dressing requirement. One application of this system reported cost savings in the range 10–50%, the quality of the cut eliminating the variations encountered with manual cutting[122].

Flat grinding, especially with abrasive belts, typically lends itself to automatic fettling systems in which small castings can be fed successively to the grinding face. Conversely the grinding tool can be applied to larger castings, using remote controlled manipulators as in the cutting operation. Heavy duty grinding heads are applied in this way, with performances exceeding those of swing frame machines.

In many of these applications involving advanced mechanical equipment, usually with some automatic devices, the principal controls are largely retained by the operator whilst the manipulation and heavy work are performed by the machine. The labour content is thus reduced with relatively modest increase in capital costs. Robots, by contrast, involve high capital investment, most commonly justified by frequent repetition, as in the processing of large batches of individual components.

A robot is defined as a reprogrammable device to manipulate and transport parts, tools or specialized implements through variable programmed motions for the performance of specific tasks. Simple robots perform programmed movements of varying degrees of complexity, but more advanced types employ sensors for measurement and for feedback and program modification. The cutting system previously referred to was developed to embody an infra-red laser sensing device mounted on an X–Y positioner, to achieve precise location of the cut for a particular feeder head, amending the basic programme for the casting type[123].

The introduction of advanced robots enabled fully automatic fettling to be envisaged, a topic extensively examined by Godding[119] and McCormack[124], including possibilities of deflection under reactive loading situations, and the high capital costs which need to be offset by continuous and intensive use. Minimal setting up time is also essential in all automatic fettling systems.

A typical sequence is seen in the quoted example of a fettling cell operation, employing a multi-axis robot to pick up castings from a conveyor system for presentation to a grinding machine, followed by positioning on an exit conveyor. An intermediate turnover device is also provided for double sided grinding, the robot performing the successive transfer movements. Automatic adjustments for grinding wheel wear are determined by an optical sensor.

Robotic controls have also been successfully applied to the surface grinding of elaborately contoured castings, including marine propeller and turbine blades. Individual grinding cells embody automatic measurements to detect variations in excess material over the specified shape, the design data being fed directly into the cell. The necessary rough and finish grinding are automatically programmed, followed by final measurement and record generation. The system is instructed through a computer menu and replaced multi-stage operations requiring much longer process times. This and other developments were reviewed in Reference 111.

The dust hazard

The problem of dust generation exists throughout the production of castings, but especially where dry operations are performed, as at the knockout and in the fettling shop. Early concern and past studies of the behaviour of dust[112,125,126] contributed greatly to the development of modern fettling equipment and techniques, to minimize both dust and noise, and to embody effective controls.

The main health hazard arises from silica particles in the size range below 2 μm, which are active in inducing pneumoconiosis. Airborne foundry dust consists predominantly of particles below 2.5 μm with a median size of about 1 μm: a particle of this size has a free falling velocity of only 18 mm/min and so stays airborne indefinitely in normal environments.

Apart from the use of methods and materials which avoid production of harmful dust particles, suppression can be assisted by a number of alternative or complementary measures, namely:

1. general ventilation of the workspace;
2. local extraction in the immediate region of the work;
3. local extraction at the point of dust creation.

These basic approaches are schematically illustrated in Figure 8.54.

General ventilation has been commonly sought through upward movement of air, brought about by high level extraction fans, but an alternative approach employs downdraught ventilation with high level inlets, airflow being created by extraction close to the points of dust formation: in this case local exhaust is used as an aid to general shop ventilation.

Local exhaust systems are based on two different concepts. Extraction may be localized to the work region, either by the use of hoods and booths, positioned in the path of the main dust stream, or by downdraught grids or exhausted fettling benches, designed to maintain a positive air current past the work and away from the operator (Figure 8.54b). This type of system is paralleled by the hood and downdraught systems employed at knockouts. Various custom designed booths have been tailored to individual operations[118,126,127]. One such system encloses the manual air-arc cutting process within a partially sealed cabinet, in which observation is maintained through a glazed panel, with handling through a small opening; noise is also reduced by the use of acoustic linings. Booths can also contain turntables

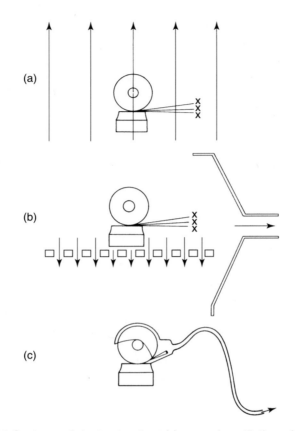

Figure 8.54 Systems of dust extraction: (a) general ventilation of workspace, (b) local extraction adjacent to work, (c) local extraction at tool

for easy manipulation of the casting. Larger castings can be fettled within a chamber fitted with integral dust extraction, the operator in this case also being provided with personal protection. Alternatively the casting may be positioned beside a fixed extractor hood, or mobile hoods with flexible ducting can be deployed close to the workpoint.

In the second type of local system extraction is more intensely localized at the machine or tool, using integral suction attachments providing closely confined high velocity air streams (Figure 8.54c). Such systems are available for each type of fettling equipment. For a pedestal or bench grinding machine the dust is extracted through a perforated work rest and wheel cowling, whilst a swing frame grinder is exhausted through the wheel hood. A portable grinder is provided with an extraction shoe or guard exhausted through a flexible lead; pneumatic chisels are fitted with a concentric rubber suction sleeve. In all these high velocity, low volume systems the attachments must be carefully designed not only to trap the primary dust stream but to avoid escape of the secondary stream which closely follows the rotating wheel.

A particular weakness of this system in relation to portable tools is the cumbersome nature of the connecting hoses required to maintain adequate airflows. This and other points emerging from controlled tests were discussed in Reference 128.

Although masks incorporating filters and more positive breathing aids can be of help in limited circumstances, they do not provide an acceptable solution to the general dust problem.

The foregoing discussions of cleaning and finishing need to be seen against a background of major improvements in casting finish achieved over the period since the advent of modern moulding material compositions and compaction techniques. The heavy incrustations of metal and sand, once commonly encountered, have largely disappeared. Other encouraging developments have included the diminishing need for heavy dressing of feeder head stubs through enhancement of the cutting operation: the closeness and quality of burned finish has even been such as to dispense altogether with subsequent dressing[129].

Despite such progress, fettling operations as a whole retain a high manual content as compared with other stages of casting production. Automatic fettling can change the picture but does entail heavy capital expenditure, which normally calls for extensive use with long production runs. This has led in some circumstances to the development of contract fettling shops: these are able to spread the costs of advanced plants over the products of numerous foundries. On the same principle a single advanced facility can serve several foundries within one large group.

Further operations

Several other measures can be involved in the concluding stages of casting manufacture. Heat treatment, the application of special surface treatments and coatings, and inspection are coordinated in various sequences with fettling shop operations; rough machining is also carried out in specified cases. Inspection was fully discussed in Chapter 6 and the ensuing chapter referred to surface treatments used for castings: these may be applied either at the foundry or subsequently by the purchaser.

Heat treatment

Heat treatment of castings has two broad purposes: the relief of stresses created in cooling, head removal, repair welding or machining, and the development of structure sensitive properties by metallurgical changes. The former aspect was examined in Chapter 5: stress relief is applicable to most types of casting alloy, although often achieved incidentally to other heat treatments, eliminating the need for a separate process. Treatment for the control of structure sensitive properties has many variations according to alloy. Some treatments commonly applied to castings are listed in Table 8.8: in most cases these are not peculiar to cast alloys and are fully detailed in standard metallurgical texts. A comprehensive examination of heat treatment will not therefore be undertaken here but its general position in relation to the casting process can be briefly considered.

Many types of cast alloy owe their properties mainly to the structures formed during solidification: heat treatment plays no part in these cases unless in the role of stress relief or possibly homogenization. In other alloys the as-cast structure is unsuitable and heat treatment is essential to the development of the specified properties. The object in most cases is to strengthen and toughen, by changing the grain size, by solution and controlled precipitation of phases, or by crystallographic changes such as the martensitic transformations. Of similar significance are the structures produced by such specialized treatments as maraging and austempering in steels and cast irons. In other cases the main purpose is to improve corrosion resistance, for example by solution of phases. The essentials of such treatments are the same as for other metal products, although there are special aspects peculiar to castings: these mainly concern mass effect and members of unequal cross-section.

Many castings are of heavy section and section thickness in cast metals has double significance. The normal heat treatment mass effect is present, governing the heating and soaking times required to achieve temperature uniformity and affecting the attainable cooling rate. In addition, however, process times are affected by the original as-cast structure, itself related to mass via the cooling rate obtained during solidification (q.v. Chapter 2). The influence of the as-cast microstructure on homogenization time was examined in Chapter 5: coarse dendritic structures as encountered in heavy sections require much longer times for diffusion of solutes. Similarly, initially coarse microconstituents, such as second phase particles, resulting from slow freezing and large primary grain size, require longer process times in solution heat treatment. This link between casting and heat treatment conditions may be summarized thus:

Casting section $\rightarrow$ freezing rate $\rightarrow$ structural spacing $\rightarrow$ treatment time

$$D \qquad\qquad R \qquad\qquad\qquad d \qquad\qquad\qquad t$$

Applying some known quantitative relationships between these variables:

$$R = k_1 D^{-2} \quad d = k_2 R^n \quad t = k_3 d^2$$

$$\text{whence } t = k D^{-4n}$$

Thus when $n = -0.5, t = kD^2$
and when $n = -0.33, t = kD^{1.33}$

Table 8.8 Heat treatments applied to castings

Treatment	Main purposes and effects
Stress relief	Removal of internal stress by plastic flow: stabilization of dimensions and elimination of risk of fracture in further processing or in service. (Low temperature: no appreciable changes in structure or properties)
Homogeni-zation	Elimination of dendritic and microsegregation by diffusion. (High temperature: time depends on spacing of microstructure). Term sometimes used to mean soaking for temperature uniformity
Annealing	Softening; stress relief; grain refinement and redistribution of dispersed phases, e.g. by solution and recrystallization. (Reheating and furnace cooling). Term also employed in broader sense for homogenization, solution treatment, malleabilising etc.
Normalizing	As in annealing but more pronounced refinement of grain and substructure by air cooling: improved toughness and machinability. (Reheating and air cooling: rate governed by mass. Residual stresses may be introduced)
Malleablizing	Production of malleable cast iron by graphitization of white cast iron: graphite precipitated by high temperature decomposition of cementite (prolonged heating in neutral (blackheart) or oxidizing (whiteheart) atmosphere)
Solution treatment	Solution of second phase constituents to produce single phase structure either permanently (e.g. for corrosion resistance) or as a step in producing controlled dispersion for strengthening. (Reheating to single phase region followed by quenching or rapid cooling)
Precipitation treatment or ageing	Production of controlled second phase precipitates, usually for strengthening (Low temperature, time dependent changes). Softening can also occur q.v. tempering
Quench hardening	Strengthening by non-equilibrium changes, notably martensitic transformations (reheating and quenching: sensitive to cooling rate and thus dependent on nature and temperature of medium and on mass. Residual stresses introduced.)
Tempering	Softening and toughening of non-equilibrium structures by precipitation after hardening treatment; secondary hardening; stress relief. (Low temperature, time dependent changes)
Isothermal treatments	Strengthening by controlled time dependent phase transformations, usually at elevated temperatures; avoidance of quenching stresses (reheating, stepped quench and other sequences)
Austempering	Strengthening treatment sequence applied to ductile cast irons and alloy steels (High temperature austenitization, rapid cooling to intermediate temperature for timed isothermal transformation, followed by air cooling)
Surface treatments	Flame, induction and laser hardening (see also Table 7.5)

It is seen, therefore, that heating and soaking times must be designed to suit the particular structure in the section as well as to achieve the required temperature. The importance accorded to fine initial cast structures explains the extensive use of chills in the manufacture of high duty premium castings (q.v. Chapter 6). These heat treatment issues are fully explored in Reference 130.

The other aspect of heat treatment relating particularly to castings is that of unequal sections and geometric complexity. This can create difficulties in developing the required properties in thick and thin sections simultaneously, whilst stresses tend to be produced by differential expansion or contraction resulting from uneven heating or cooling. Thus the treatment must be a compromise between the requirements of the various sections and must involve slow and uniform heating to minimize stress.

In these respects the careful positioning of castings in relation to the heating or cooling medium and to the furnace load as a whole can sometimes be used to positive effect, for example by placing the heavier sections or heavier castings in the outermost positions in the direct path of radiation, and by spacing castings to assist convective circulation of the furnace atmosphere. Local insulation with sand can be used to protect light members from excessive temperature differences during heating or cooling.

Transfer of heat to or from a casting or load of castings is related to shape in a manner analogous to that encountered in solidification (q.v. Chapter 3). The heating or cooling time will depend on a shape factor or modulus according to the general relation

$$t = k \left(\frac{V}{A} \right)^2$$

Thus the time varies with the square of the ruling section thickness of the individual casting or the load. The aspect of furnace loading of castings was originally examined by Paschkis[131], who stressed the possibility, in furnaces where heat transfer is mainly or partly by radiation, of dividing a multiple furnace load into smaller units for separate and successive treatments, yet increasing the overall furnace throughput by making a disproportionate saving in heating time.

Fundamental aspects of furnace design and operation have been treated in the specialized literature including References 132 and 133.

High temperature heat treatments are frequently followed by a cleaning process for scale removal prior to final inspection.

Hot isostatic pressing

Casting quality can be greatly enhanced by hot isostatic pressing, or HIPping, so as to close residual microporosity in the final casting. Small volume fractions of voids exercise disproportionate effects on mechanical properties, particularly those influencing fatigue performance. Isostatic

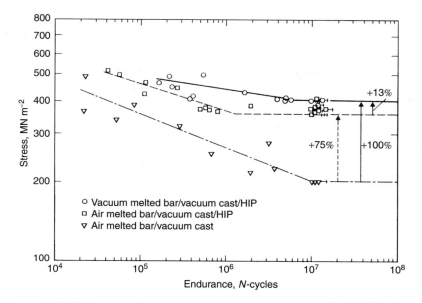

Figure 8.55 S–N curves for investment cast precipitation hardening stainless steel 17/4 PH (from Reference 136) (courtesy of BICTA)

pressure treatment, applied in an inert gas atmosphere at high temperature, brings about plastic deformation and diffusion bonding, so closing and healing internal pores and producing levels of soundness analogous to those of wrought material. External dimensions remain essentially unchanged.

Argon pressures in the range 100–200 MPa ($\approx$15 000–30 000 lbf/in^2) are commonly employed, in tanks with provision for heating to the process temperature and for rapid cooling. The benefit of the process is restricted to internal voids: there is no effect on pores communicating with the surface, since these are subjected to the same gas pressure as that operating on the outer surface of the casting.

Temperatures employed lie in the hot working range: these and other process parameters are detailed in the literature[134]: the aim is to avoid incipient melting or phase degradation whilst maximizing plasticity. The importance of high gas purity is also stressed.

The process has been applied to a wide range of cast materials, including aluminium, copper, nickel, and titanium base alloys, and alloy steels. The general picture emerging is one of substantial increases in tensile strength, elongation and above all fatigue resistance. A typical example relating to gunmetal castings was quoted in Reference 135. Inherent porosity characteristic of this material was practically eliminated after pressure applied in the temperature range 680°–820°C: tensile and elongation increases of 37 and 100% respectively were achieved. Enhancement of fatigue properties in aluminium alloy sand and die castings was also reported.

The typical effect on the fatigue properties of stainless steel investment castings is shown in Figure 8.55, representing the upgrading of an already exceptional cast product; similar considerations apply with respect to cast aluminium airframe components, where fatigue life has been enhanced by a factor of 3 and ductility by 29%[137].

In the field of superalloy and titanium castings HIP treatment is frequently specified as an integral part of the production sequence. In the former case substantial increases in high temperature stress rupture life and elongation have been demonstrated following treatments at over 1300°C ($\approx$2400°F)[134]. Titanium castings treated in the temperature range 800–1000°C can be reduced to zero porosity given the easy diffusion bonding in the alloys at these temperatures[138].

The specialized nature of the HIP operation and associated capital costs have led to the development of contract services, enabling castings to be processed without the need for plant installation by the individual producer[139].

References

1 Campbell, J., *Castings*, Butterworth-Heinemann, Oxford (1991)
2 Davies, W., *Foundry Sand Control*, United Steel Companies, Sheffield (1950)
3 Stoch, C. M. and Lambert, P., *Br. Foundrym.*, **56**, 181 (1963)
4 Nicholas, K. E. L. and Roberts, W. R., *Br. Foundrym.*, **56**, 125 (1963)
5 Stoch, C. M. and Bownes, F. F., *Br. Foundrym.*, **54**, 428 (1961)
6 Bodsworth, T. J. *et al.*, *Br. Foundrym.*, **53**, 137 (1960)
7 Bodsworth, T. J. *et al.*, *Trans. Am. Fndrym. Soc.*, **67**, 47 (1959)
8 Dietert, H. W. and Graham, A. L., *Trans. Am. Fndrym. Soc.*, **61**, 490 (1953)
9 Howell, R. C., *Trans. Am. Fndrym. Soc.*, **71**, 607 (1963)
10 Kremnev, L. A. *et al.*, *Russ. Cast. Prod.*, 101 (1966)
11 Heine, R. W., King, E. H. and Schumacher, J. S., *Trans. Am. Fndrym. Soc.*, **71**, 508 (1963)
12 Nicholas, K. E. L., *Br. Foundrym.*, **60**, 127 (1967)
13 *Foundry International*, **21**, 3, 12 (1998)
14 Boenisch, D. and Lorenz, V., *Trans. Am. Fndrym. Soc.*, **95**, 7 (1987)
15 McCombe, C., *Fndry Trade J. Int.*, **14**, 4, 146 (1991)
16 Sixth Report of Working Group T30, Mould and Core Production, *Foundrym.*, **89**, 3 (1996)
17 Stoch, C. M. and Lambert, P., *Br. Foundrym.*, **56**, 181 (1963).
18 First Report of Sub-committee T.S. 39., *Br. Foundrym.*, **50**, 169 (1957)
19 Second Report of Sub-committee T.S. 39., *Br. Foundrym.*, **53**, 41 (1960)
20 Gardner, G. H. D., *Br. Foundrym.*, **56**, 145 (1963)
21 Fourth Report of Working Group T30. The Vapour Curing Processes, *Foundrym.*, **88**, 269 (1995)

22 Nicholas, K. E. L., The CO_2 silicate process in Foundries, BCIRA, Alvechurch (1972)

23 Sarkar, A. D., *Foundy Core and Mould making by the Carbon Dioxide Process*, Pergamon Press, Oxford (1964)

24 Fifth Report of Working group T30. *Foundrym.*, **88**, 417 (1995)

25 Laing, J. and Rolfe, R. T., *A Manual of Foundry Practice*, 3rd Edn, Chapman and Hall, London (1960)

26 Howard, E. D. (Ed.), *Modern Foundry Practice*, 3rd Edn, Odhams Press, London (1958)

27 Salmon, W. H. and Simons, E. N., *Foundry Practice*, Pitman, London (1957)

28 Dietert, H. W., *Foundry Core Practice*, 3rd Edn, Am. Fndrym. Soc., Des Plaines, Ill. (1966)

29 *Molding Methods and Materials*, Am. Fndrym. Soc., Des Plaines, Ill. (1962)

30 *Smithells Metals Reference Book*, 7th Edn, Brandes, E. A. and Brook, G. B. (Eds.), Butterworth-Heinemann, Oxford (1998)

31 Leyshon, H.J. and Selby, M. J., *Br. Foundrym.*, **65**, 43 (1972)

32 Wilson, C. H. and Driscol, W. J., *Br. Foundrym.*, **69**, 97 (1976)

33 Selby, M. J., *Br. Foundrym.*, **71**, 241 (1978)

34 Taft, R. T., *Foundrym.*, **86**, 241 (1993)

35 First Report of Working Group T10 NF. *Foundrym.*, **88**, 80 (1995)

36 Second Report of Working Group T10 NF. *Foundrym.*, **88**, 158 (1995)

37 Third Report of Working Group T10 NF. *Foundrym.*, **88**, 375 (1995)

38 Ahearne, M., *Foundrym.*, **86**, 137 (1993)

39 McMann, F. C., *Trans. Am. Fndrym. Soc.*, **97**, 661 (1989)

40 Knight, S. J., Manuel, K. and Randle, R. G., *Fndry Trade J.*, **118**, 653 (1965)

41 Paterson, T. M., *Foundrym.*, **89**, 222 (1996)

42 Svoboda, J. M. and Griffith, L. E., *Trans. Am. Fndrym. Soc.*, **105**, 13 (1997)

43 Wells, M. B. and Vonesh, F. A. Jr. *Trans. Am. Fndrym. Soc.*, **95**, 499 (1987)

44 Haywood, I., *Foundrym.*, **85**, 263 (1992)

45 Smith, L., *Foundrym.*, **86**, 131 (1993)

46 Edgerley, C. J., Smith, L. and Wilford, C. F., *Cast Metals* **1**, 4, 216 (1989)

47 Atkins, R. and Crookes, W., *Foundrym.*, **85**, 393, (1992)

48 Drouzy, M. and Mascré, C., *Metall. Rev.*, **14**, No. 131, 25 (1969)

49 Richardson, F. D. and Jeffes, J. H. E., *J. Iron Steel Inst.*, **160**, 261 (1948)

50 Emley, E. F., *J. Inst. Metals*, **75**, 431 (1949)

51 Emley, E. F., *Fndry Trade J.*, **103**, 33 and 63 (1957)

52 Brace, A. W. and Allen, F. A., *Magnesium Casting Technology*, Chapman and Hall, London (1957)

53 Ward, R. G., *An Introduction to the Physical Chemistry of Iron and Steel Making*, Arnold, London (1962)

54 Bodsworth, C., *Physical Chemistry of Iron and Steel Manufacture*, Longmans, London (1963)

55 Report of Sub-committee T.S. 59., *Br. Foundrym.*, **58**, 225 (1965)

56 Report of Sub-committee T.S. 60., *Br. Foundrym.*, **59**, 28 (1966)

57 Report of Sub-committee T.S. 58., *Br. Foundrym.*, **56**, 75 (1963)

58 Campbell, J. S. and Rao, N. J., *Foundry*, **92**, 44 (1964)

59 Report of Sub-committee T.S. 52., *Br. Foundrym.*, **54**, 103 (1961)

60 Howard, J. F., *Fndry Trade J.*, **117**, 284 (1964)

61 Bird, P.J. and Rickards, P. J., Castcon 96, Proc. IBF Annual Conference, Inst. Br. Foundrym. Birmingham, 239 (1996)

62 Hayes, P. J., Tremayne, J. F. and Gibson, R. C., *Foundrym.*, **91**, 23 (1998)

63 Karsay, S. I., *Ductile Iron Production Practices*, Am. Fndrym. Soc. Inc., Des Plaines, Ill. (1975)

64 Else, G. E. and Dixon, R. H. T., *Br. Foundrym.*, **79**, 18 (1986)

65 *Foseco Foundryman's Handbook*, 10th Edn, Brown, J. R., Butterworth-Heinemann (1994)

66 Forrest, R. D., *Br. Foundrym.*, **75**, 41 (1982)

67 Holden, W. W. and Dunks, C. M., *Br. Foundrym.*, **73**, 265 (1980)

68 Weese, S. and Mohla, P. P., *Trans. Am. Fndrym. Soc.*, **103**, 15 (1995)

69 Rotella, J. and Mickelson, R., *Trans. Am. Fndrym. Soc.*, **99**, 519 (1991)

70 Turner, A. M., *Foundrym.*, **93**, 22 (2000)

71 Harriss, R. J. and Cesara, A., *Br. Foundrym.*, **80**, 434 (1987)

72 Cochran, B.P. *et al. Trans. Am. Fndrym. Soc.* **100**, 737 (1992)

73 Gruzleski, J. E. and Closset, B. M., *The Treatment of Liquid Aluminium–Silicon Alloys*. Am. Fndrym. Soc. Inc., Des Plaines, Ill. (1990)

74 Holappa, L. E. K., *Internat. Metals Reviews* **27**, 53 (1982)

75 Dutcher, D. E., *Proc. 1st Int. Steel Foundry Congress*, Chicago Steel Founders' Soc. America, Des Plaines, Ill (1985)

76 Ashton, M. C., Sharman, S. G. and Sims, B. J., *Br. Foundrym.* **76**, 162 (1983)

77 Hearne, B. H., Sharman, S. G., Ashton, M. C. and Clifford, M. J., *Br. Foundrym.*, **76**, 156 (1983)

78 Coates, R. B. and Leyshon, H. J., *Fndry Trade J.*, **119**, 495 (1965)

79 Dainton, A. E., *Trans. Am. Fndrym. Soc.*, **97**, 1 (1989)

80 Svoboda, J. M., *Trans. Am. Fndrym. Soc.*, **98**, 267 (1990)

81 Cotchen, J. K., *Trans. Am. Fndrym. Soc.*, **95**, 315 (1987)

82 Heine, H. G., *Trans. Am. Fndrym. Soc.*, **104**, 805 (1996)

83 Interim Report of Sub-committee T.S. 45., *Br. Foundrym.*, **51**, 91 and 452 (1958)

84 Final Report of Sub-committee T.S. 45., *Br. Foundrym.*, **53**, 120 (1960)

85 Fuller, A. G., *Fndry Trade J.*, **119**, 73 (1965)

86 Report of Sub-committee T.S. 6., *Proc. Inst. Br. Foundrym.*, **39**, A26 (1945–6)

87 Boyes, J. W. and Greenhill, J. M., *Br. Foundrym.*, **58**, 103 (1965)

88 French, A. R., Kondic, V. and Wood, J., *Proc. Inst. Br. Foundrym.*, **48**, B89 (1955)

89 Baker, F. M., Upthegrove, C. and Rote, F. B., *Trans. Am. Fndrym. Soc.*, **58**, 122 (1950)

90 Shelleng, R. D., Upthegrove, C. and Rote, F. B., *Trans. Am. Fndrym. Soc.*, **59**, 67 (1951)

91 Jelley, R. and Humphreys, J. G., *Br. Foundrym.*, **55**, 1 (1962)

92 *B.S. 1041*. Temperature Measurement, Br. Stand. Instn., London

93 BS EN 60584. Thermocouples, Br. Stand. Instn., London

94 Ashton, M. C. and Wake, P., Proc. SCRATA 22nd Annual Conf. Paper 7 (1977)

95 Good Practice Guide No 49. Energy Efficiency Office, Harwell (1992)

96 Taylor, P. B. *et al.*, Metals and Materials, **2**, 8, 495 (1986)
97 Smith, L., *Foundrym.*, **88**, 212 (1995)
98 Hearne, B. H. *et al.*, *Br. Foundrym.*, **76**, 156 (1983)
99 Cator, L., *Trans. Am. Fndrym. Soc.*, **94**, 93 (1986)
100 Kjellberg, B. and Liebman, M., *Trans. Am. Fndrym. Soc.*, **95**, 723 (1987)
101 Smith, J. R., *Foundrym.*, **82**, 260 (1989)
102 Miller, R. and Sjodal, E., *Foundrym.*, **82**, 158 (1989)
103 Fndry Trade J, 165, March (1987)
104 Powell, J. and Smith, J. R., Foundry Technology for the 80's. Proceedings of International Conference, BCIRA (1979)
105 Bargh, F. G., *Foundrym.*, **91**, 289 (1998)
106 Bargh, F. G. and Barnes, G. R., *Trans. Am. Fndrym. Soc.*, **104**, 439 (1996)
107 Campbell, J. and Bartlett, R., *Trans. Am. Fndrym. Soc.*, **73**, 433 (1965)
108 Taylor, L. S. and Briggs, R., *Fndry Trade J.*, **123**, 139 (1967)
109 Second Report of Working Group, P10 *Br. Foundrym.*, **76**, 193 (1983)
110 Little, S. D., *Br. Foundrym.*, **79**, 317 (1986)
111 Stevenson, M., *Foundry International*, **16**, 2, 242 (1993)
112 Sully, A. H. and Stoch, C. M., *Br. Foundrym.*, **52**, 193 (1959)
113 Third Report of Working Group P10. *Br. Foundrym.*, **80**, 82 (1987)
114 First Report of Working Group P10. *Br. Foundrym.*, **70**, 98 (1977)
115 Roebuck, E., *Fndry Trade J.*, **121**, 337 (1966)
116 Gray, S. M., *Br. Foundrym.*, **54**, 239 (1961)
117 Morgan, J. H., *Fndry Trade J.*, **115**, 35, Special Issue, (May, 1966)
118 McCormack, W., Foundry Technology for the 80's. International Conference as Warwick. BCIRA, Paper 15 (1979)
119 Godding, R. G., *Br. Foundrym.*, **76**, 228 (1983)
120 Wallis, R. *et al.*, Proceedings of 24th Annual Conference, SCRATA, Sheffield, Paper 8 (1980)
121 Sims, B. J. and Wallis, R., *Br. Foundrym.*, **78**, 410 (1985)
122 Huntington, K., *Trans. Am. Fndrym. Soc.*, **95**, 339 (1987)
123 *Br. Foundrym.*, **81**, 119 (1988)
124 McCormack, W., *Br. Foundrym.*, **78**, 267 (1985)
125 *Foundry Ventilation and Dust Control.* Proceedings of Conference in Harrogate, 1955, B.C.I.R.A
126 Proceedings of 23rd Annual Conference, SCRATA, Sheffield (1978). Papers by several authors
127 Clayton, A., Stott, M. D. and Sims, R. J., Proceedings of 26th Annual Conference, SCRATA, Sheffield. Paper 5 (1982)
128 Taylor, D., *Br. Foundrym.*, **77**, 295 (1984)
129 Schneider, M. D. and Petersen, R. R., *Trans. Am. Fndrym. Soc.*, **94**, 913 (1986)
130 Flemings, M. C., *Solidification Processing.* McGraw-Hill, New York (1974)
131 Paschkis, V., *Trans. Am. Fndrym. Ass.*, **54**, 287 (1946)
132 Thring, M. W., *The Science of Flames and Furnaces*, 2nd Edn, Chapman and Hall, London (1962)
133 Dryden, I. G., Efficient use of Energy. IPC Business Press (1975)
134 Eridon, J. M., *Metals Handbook*, 9th Edn, Vol. 15, Casting, 538 ASM International, Metals Park, Ohio (1988)
135 Siefert, D. A. and Hanes, H. D., *Trans. Am. Fndrym. Soc.*, **84**, 571 (1976)

136 Investment Castings for the 90s. *Br. Investment Casting Trade Ass.*, (1991)
137 Mocarski, S. J., Scarich, G. V. and Wu, K. C., *Trans. Am. Fndrym. Soc.*, **99**, 77 (1991)
138 Eylon, D., Newman, J. R. and Thorne, J. K., *Metals Handbook*, 10th Edn, Vol. 2, 634, ASM International Metals Park, Ohio (1990)
139 Anon. *Br. Foundrym.*, **79**, 12, 474 (1986)

9

Production Techniques 2 Shell, investment and die casting processes

In the previous chapter the production sequence in founding was considered principally in relation to the process of sand casting. The purpose in this and the following chapter is to examine processes which differ in important respects from conventional sand founding but which make, through the special qualities of their products, a major contribution to the design potential of castings.

Special casting processes may be adopted in the search for greater dimensional accuracy, higher metallurgical quality or in certain cases lower production cost than can be achieved by conventional sand casting techniques. In most cases, however, both equipment and production costs are higher and must be offset by the technical advantages of the product; quality production is also usually required to justify the relatively high initial outlay on tooling.

A common feature of many of these processes, and a strong influence in selection, is the quality of precision: this will be further examined before detailed consideration of the processes themselves. These will be treated in two groups: the present chapter will be concerned with shell, investment and die casting, whilst centrifugal casting and a number of unorthodox casting techniques will be examined in Chapter 10.

Precision in casting

Precision of form is one of the basic attributes of a casting. Since the fundamental purpose of founding is to produce forms at or close to finished size, the success of the casting process might be measured largely by this criterion. The quality of precision encompasses not only the accuracy of individual dimensions but surface finish and general appearance. An ideal casting would not only fall within specified dimensional tolerances but its outlines would be crisp and clear cut, its surface smooth. With precision is coupled the quality of intricacy, seen in the reproduction of elaborate features and fine detail.

Economic aspects. Measures to increase precision can include modifications to standard sand casting procedures as well as the use of separate processes. In most cases, however, additional costs are incurred. For economic reasons the vast majority of castings are not made to the maximum attainable standards of precision. The mainstream of foundry production is of castings which are functionally adequate, the need for high accuracy and smooth surfaces being met by local or general machining.

Increased accuracy may, on the other hand, reduce *total* costs by the reduction or elimination of machining, especially where difficult contours, awkward access or tough alloys are involved. As additional assets, more elaborate design may be feasible or the improvement in appearance may be a decisive element in sales appeal. Such advantages must again be weighed against the extra costs incurred. There remain those cases in which increased precision offers the only technically feasible solution to a particular design problem and in which costs are no longer decisive in selection.

Factors influencing dimensional accuracy. Factors contributing to the dimensional accuracy of castings were examined in some detail in Chapter 7; before considering measures for increasing precision it will be useful to summarize these factors. They are:

1. accuracy of the pattern or die equipment,
2. the accuracy with which the mould reproduces the pattern shape: important sources of error include pattern rap, distortion by sagging, and inexact mutual fit of mould parts,
3. the accuracy with which the casting conforms to the original shape of the mould, depending on the degree of filling and dimensional stability of the mould cavity on pouring,
4. the contraction factor, which determines the predictability of dimensional changes during cooling,
5. finishing operations.

Increased precision must be based upon improvement in one or more of these fields.

Limitations and possibilities in sand casting. Normal sand castings are subject to appreciable dimensional errors from all the foregoing causes. Production is, for example, frequently based on the use of low cost wooden patterns of miscellaneous origins, stored for prolonged periods and showing varying degrees of wear. Limitations to accuracy arise from uncertain clearances in coreprints, dowels and boxpart location systems. Sand mixtures are susceptible to deformation in the green state and the normal finish on sand castings, especially in alloys of high melting point, is far from smooth.

Major errors, however, are not inherent in the process and can be pared down without radical departure from the sand casting principle. The availability of special processes should not obscure what can be accomplished, often at lower cost, by additional precautions or modifications to conventional practice. Such measures include:

1. The use of precise pattern and moulding equipment. Metal patterns and accurate location systems, for example close tolerance box pins and corebox dowels, can reduce mismatch to low levels.
2. Techniques giving more accurate mould impressions. Machine moulding techniques, with their automatic pattern draw mechanisms, are preferable to hand moulding: this is especially true of high pressure squeeze moulding with its output of dense, dimensionally stable mould parts. Manual patching and rubbing need to be eliminated by careful maintenance of equipment.

    Of special value are those techniques enabling mould parts to be partly or completely hardened in contact with the pattern or corebox, using the materials and reactions discussed in earlier chapters. Core assembly moulding techniques are particularly useful since they embody rigid and accurate mould cavities and integrally moulded location systems.
3. Measures to improve surface finish. The use of fine sands or refractory facings can minimize metal penetration and reduce the mean surface roughness of the casting. With improved as-cast surfaces additional care is needed in the selection of finishing processes, such as shotblasting, which also affect surface finish.

With such measures the division between 'conventional' and 'special' casting processes becomes much less distinct. Hot box and shell coremaking, for example, have many similarities, whilst the cold setting processes have features in common with permanent pattern investment casting. In general, however the accuracy of sand castings still falls short of that attained in the more specialized processes.

Special processes of precision casting. Shell, investment and die castings derive their qualities in the first instance from the production of highly accurate moulds; this is achieved partly through precise pattern, die and moulding equipment and partly by the use of moulding materials and techniques designed to give smooth mould surfaces. Secondly, casting conditions are designed to ensure that the liquid metal conforms exactly to the surface of the mould cavity. In the case of pressure die casting pressure injection is the key feature; in shell moulding the exceptional permeability of the shell, in investment casting the use of hot moulds plays an important part.

Each of the processes offers special advantages over a particular range of products: examples of the qualities and limitations influencing process selection will be given in the separate sections. Comparisons of a number of design characteristics such as surface finish and minimum wall thickness have already been made in Chapter 7 (see Tables 7.13 and 7.14 and Figure 7.6). Other comparative data were included in References 1 and 2. Richards[3] critically reviewed and ranked specific features of several processes, stressing opportunities for the use of combinations to optimize individual advantages. Clegg[4], in his work on precision casting processes, provided systematic tabulated comparisons of major characteristics of these and numerous other processes.

Shell Moulding

The essential feature of the resin shell or Croning process is the use of thin walled moulds, in which the external surface follows the contour of the mould cavity. The process thus provided a break with concept of a mould as a cavity within a solid block of moulding material, of a size often determined by that of the nearest available moulding box. The shell mould is extremely light and easily handled and the volume of moulding material is only a small fraction of that required in conventional practice.

Basic operations

The original production sequence is illustrated in Figure 9.1. Plate mounted metal patterns are required, for example of cast iron or light alloy, fitted with integral gating and feeding systems. Each plate is provided with spring loaded ejector pins, the heads of which normally lie flush with its upper surface. The pins are distributed round the periphery of the pattern and are in some cases joined to a common backing plate. A typical pattern assembly is illustrated in Figure 9.2. In other systems springless ejector mechanisms are employed.

The complete assembly is preheated to 180–200°C and its surface treated with a parting medium, commonly a silicone emulsion, at which stage it is clamped against the open end of a dump box containing the moulding mixture. This consists of a fine, dry sand, for example silica or zircon, with 2–7% of a solid thermosetting resin binder. The resins commonly used are phenol novolaks containing hexamine as a hardener: these are designed to set permanently by cross-linking after a brief softening stage. Resins of low nitrogen content were also developed to counter specific surface defects. The binder may be present as a powder mixed with the base sand, but precoated sands are usually preferred for their cleanliness, uniformity and economy in resin consumption. Resins were developed for either hot or solvent coating of the sand grains at the foundry, but commercially precoated sands also find wide application.

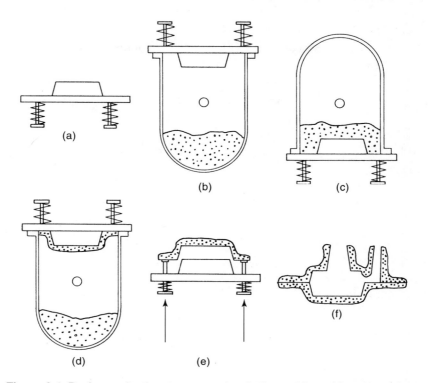

Figure 9.1 Basic production sequence in shell moulding: (a) pattern plate, (b) pattern plate mounted on dump box, (c)–(d) investment, (e) ejection, (f) shell assembly

The dump box is inverted to allow the moulding mixture to fall into contact with the attached pattern plate (Figure 9.1c). The heat from the pattern assembly softens the resin to a depth depending on the dwell time and produces bonding in the heat affected layer. This investment operation is usually completed within 30 s, at which stage the dump box is returned to its original position, leaving the bonded layer adhering to the pattern assembly (Figure 9.1d). This layer is normally between 5–10 mm ($\approx \frac{3}{16} - \frac{3}{8}$ in) in thickness.

The behaviour of the binder on heating is characterized in Figure 9.3. An initial melting stage produces a viscous liquid allowing flow and adhesion between the sand grains: the depth to which this occurs depends on the temperature gradient in the moulding material and determines the thickness of the shell. This is followed by curing, when the full strength of the shell is developed. The properties of the resin should give rapid softening to a consistency just sufficient for flow and adhesion, followed by rapid curing without too great a further fall in viscosity. This minimizes the danger of peelback, where part of the soft shell falls away with the excess

Figure 9.2 Pattern assembly for shell moulding (courtesy of Catton & Co. Ltd.)

moulding material on reversal of the pattern plate. The circumstances associated with varying resin characteristics are illustrated by the three curves in Figure 9.3. The chemistry of shell moulding binders was more fully examined in Reference 5.

Full and rapid curing requires a higher temperature than that used for investment. The pattern assembly carrying the embryo shell is transferred to an oven held at a temperature of 300–450°C, or exposed to radiant heat from a hood carrying gas burners or electrical elements. Hardening proceeds from both surfaces of the shell and is normally completed within 2 min. The shell can at this stage be pushed away from the pattern assembly by actuation of the ejector system (Figure 9.1e). The cured shell is ready for assembly with other shells and cores to form a complete mould (Figure 9.4); a range of more intricate shells, for the production of automobile castings, is shown in Figure 9.5.

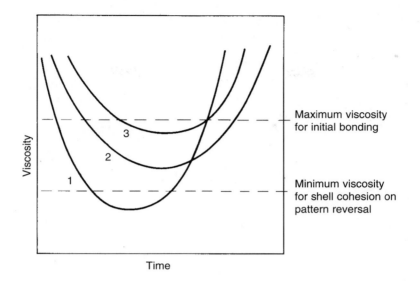

Figure 9.3 Characteristics of shell moulding resins on heating: 1. rapid softening but peel back due to low viscosity, 2. satisfactory characteristics, 3. slow
softening and weak shells due to inadequate resin flow (after Lemon and
Leserve[5])

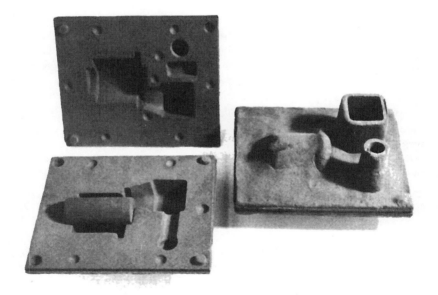

Figure 9.4 Set of shells forming shell mould

Figure 9.5 Range of shell moulds used to produce automobile castings (courtesy of Hepworth Minerals and Chemicals Ltd.) (see also References 19 and 20)

The latitude in resin properties with time and temperature permits an operating equilibrium to be reached in respect of pattern plate temperature; the pattern assembly can thus be returned immediately to the dump box for the investment of a further shell. Pattern plate temperature can be monitored by surface contact pyrometer.

An analogous process is followed in the production of shell cores. Moulding mixture is dumped or blown into the preheated metal corebox, surplus mixture being extracted after the appropriate investment time. Where blowing is employed the use of precoated sand avoids resin segregation. The accuracy and self-venting character of the hollow cores led to their extensive use in fields outside that of shell moulding: a sectioned core was illustrated in Figure 8.32.

In closing, the shells are mutually aligned either by integrally moulded male and female dowels or by loose metal pegs registering in moulded recesses in both shells. Assemblies are held together for casting by spring clips, bolts or adhesive. Alternatively, moulds may simply be weighted or held in clamping fixtures. They may be oriented either horizontally or vertically for casting.

In the process as originally developed, moulds were packed in steel shot for casting to reinforce the thin walls against distortion. It was later found that many castings could be produced without this complication, although heavier castings are then exposed to the danger of swelling. Thicker shells and local reinforcement with metal strips can reduce this tendency, whilst some foundries use sand as a packing medium. In this case additional support can be provided by the application of a vacuum to the container, a feature of the SCRATA Vacustract® system[6], which has the further advantage of removing fume evolved on casting.

Despite the low breakdown temperature of the resin binder, the cast shape is established before the mould disintegrates; cleaning operations are relatively simple. Thermal reclamation of the base sand can be achieved by burning off unchanged resin: this becomes economically important where zircon or other special base sands are employed.

Production systems

The outstanding production characteristic is the simplicity and speed of the basic moulding sequence coupled with the ease of handling and storage of shells. This eliminates the need for skilled operations and makes the process adaptable to widely varying degrees of mechanization. For small production runs from a large variety of pattern plates, manual operations can be based on a single dump box station and oven or can be integrated into a pattern flow system such as that in Figure 9.6, using multistation manual working and oven conveyors. The next stage is the incorporation of dump boxes in simple hand operated machines which coordinate the production sequences for two pattern plates and eliminate heavy lifting in the production of larger shells. A more refined and semi-automatic two station machine is illustrated in Figure 9.7: this is controlled by manually operated valves but is also available in a fully automatic version.

Fully automatic machines can be based either on the continuous working of one or two large pattern plates or on a multistation principle in which numerous plates are indexed through successive operations and can be readily

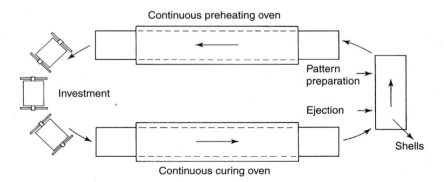

Figure 9.6 Shell moulding layout based on pattern flow for small batch manual production

Figure 9.7 Semi-automatic two station shell moulding machine. Pattern plates rotate to engage with investment bin apertures; central investment bin rotates alternately to twin investment–cure–ejection stations (courtesy of Polygram Casting Co. Ltd.)

interchanged with others without interrupting the production cycle. Heat for investment and curing is provided by gas fired or electric ovens or hoods and in some cases by heating elements beneath the pattern plates. Coremaking can be similarly mechanized: a two station machine for the production of large cores is illustrated in Figure 9.8 and embodies a fully automatic sequence of clamping, shooting, rocking, curing and ejection stages.

Most shell moulds are produced as two part moulds as previously described. In one variation of the technique, however, small components are mass produced on the stack moulding principle, using double sided shells obtained by pressing the moulding material between two heated platens. Some modern machines are designed for either mould or core production, enabling single or double sided shells as well as cores to be produced by blowing. In this case the moulding mixture is blown into the space between the heated pattern plate and a matching contoured blowhead mounted on the machine. The pattern plate with its adhering layer is then pressed against a heated contoured head. Using this type of technique, shells can be provided with ribbed backs if required for strengthening purposes.

The same principle of a moulded back was adopted for the D process[7,8], in which, however, a straightforward coreblowing operation was used in conjunction with a heat curing oil or resin bonded sand mixture. The general arrangement for this is illustrated in Figure 9.9: a preheated contoured carrier shaped the shell back to give a sand thickness of around 10 mm, the mould being blown through the pattern plate mounted on the sand

Figure 9.8 Automatic two station machine for production of large shell cores (courtesy of Polygram Casting Co. Ltd.)

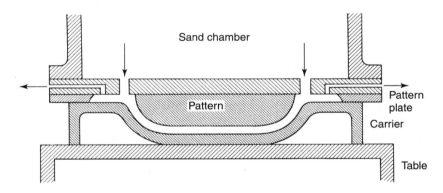

Figure 9.9 Principle employed in D process shell mould production

chamber. The carrier, complete with moulded shell, was then briefly stoved for thermal curing, followed by ejection.

Dry resin based sands also came to be used, reducing the distinction from orthodox shell, and the process was adaptable to CO_2 process moulding. Analogous techniques were described in the Russian literature[9]. In some cases a shaped support flask was retained in position for casting. This concept of a metal backing with a thin refractory coating still offers attractive features, combining some advantages of both sand and die casting.

These numerous variations of the original shell technique were developed mainly in the era of thermally cured moulding systems, including hot box coremaking, but pointed the way to similar developments exploiting the highly versatile range of cold box bonding materials as these became available, offering their own special advantages.

Cold shell moulding

A major advance came with the introduction of the Colshel machine[10], based on the use of gas-hardening resin bonded sand. The back of the shell is in this case formed by a rubber diaphragm, so dispensing with the custom shaped rigid contoured back previously employed when using conventional "wet" sand mixtures.

A uniform layer of the sand mixture is laid on the flat flexible diaphragm within a shallow moulding frame. This is then raised and pressed against the downward facing pattern plate, variations in depth being accommodated by stretching of the diaphragm. Air pressure is applied to the diaphragm and vibration assists further in the compaction of the layer. The pressure is held whilst the hardening gas is passed through from the edge of the vented moulding frame, followed by purging. The frame is then withdrawn and the cured shell removed from the pattern. All the successive machine operations are automatic and the original cycle time was 45 seconds, but rates of 90 shells per hour were subsequently achieved. The shell moulds range in thickness from around 50 mm downwards to that of a normal hot shell, depending on size and on the depth of frame selected.

Amine cured isocyanate, SO_2 hardening, and other cold bonding systems can be employed, and the enclosed design of the machine assists in preventing escape of gases into the atmosphere. The use of cold setting sands enables wood or resin patterns to be used, making the system economically viable for shorter runs than normally required for shell moulded castings. Metal patterns remain preferable, however, for quantity production and maximum accuracy.

A notable feature of the shell moulding process and its derivatives is the high output obtainable within a small floor area and involving minimum ancillary equipment, including moulding tackle and handling facilities. Weights of moulding material and binder consumption are also minimal when compared with other casting processes.

Characteristics of shell moulded castings

Close control can be exercised over the few production variables in shell moulding, principally the temperature and time cycles in investment and curing; the human factor is thus minimized and a high standard of reproducibility attained.

Castings are notable for the intricate detail which can be reproduced. This arises from the sharp precision of the shells themselves and their extreme permeability to displaced air, derived from the small thickness of moulding material. The absence of air cushioning, combined with the smoothness and low heat diffusivity of the shells, facilitates flow through thin sections and sharp delineation of surface detail, even in metals of moderate fluidity. This is seen in the clarity of raised lettering. Application to thin sections is typified in cast iron holloware, in which general wall thicknesses in the range 1.9–2.3 mm ($\approx$0.075–0.09 in) have been attained[11]. The castings in Figure 9.10 demonstrate the ability to form thin fins and other detail susceptible to misrun in sand casting. Holes of diameter as little as 3 mm ($\approx\frac{1}{8}$ in) can be cast if required.

Not only is the sharpness of the shell cavities transmitted to the castings, (see Figure 9.11.), but accuracy across the joint lines is ensured by precisely moulded location systems and coreprints. General values for dimensional tolerances should, however, be treated with caution for the reasons discussed in Chapter 7. Values in the region 0.3–0.7% of the dimension are frequently quoted, and can certainly be held on small dimensions reproduced from a single element of the pattern, but cumulative tolerances exceeding $\pm$0.38 mm ($\approx$0.015 in) are more realistic for most dimensions,

Figure 9.10 Shell moulds and cores, with grid and finned cylinder castings also shown (courtesy of Hepworth Minerals and Chemicals Ltd.)

Figure 9.11 Bracket for electronic unit, in aluminium alloy, produced by shell moulding (courtesy of Sterling Metals Ltd.)

depending on the design of the casting and the magnitude of the dimension. Few systematic studies have been reported. One such study, of pearlitic malleable iron castings[12], showed process capability values of $\pm0.46\,$mm ($\approx0.018\,$in) and $\pm0.61\,$mm ($\approx0.024\,$in) on 75 mm and 150 mm in dimensions within single shells and $\pm0.76\,$mm ($\approx0.030\,$in) on a 50 mm dimension across a parting line. A minimum of draft taper is required, less than $\frac{1}{2}^{\circ}$ being usual and as little as 1 in 1000 attainable if necessary.

The enhanced accuracy and finish as compared with sand castings may eliminate the need for machining in certain cases, but reduced machining allowances or their replacement by small grinding allowances are the most usual benefits from a change to shell moulding. The process is most adaptable to castings which are relatively shallow in a direction normal to the parting line.

The process is suited to ferrous and non-ferrous alloys over weight ranges normally associated with light machine moulding, most output being of castings in the range 0.1 to 10 kg; much heavier castings have been produced in individual cases. Certain alloys, however, are susceptible to surface defects, especially carbon and low alloy steels, in which "orange

Figure 9.12 High definition ferrous castings produced by the shell moulding process (courtesy of Hepworth Minerals and Chemicals Ltd.)

peel" or pinholing can be encountered. Middleton and Cauwood[13] attributed such problems to nitrogen originating in the hexamine hardener of the resin, and nitrogen-free hardeners were developed by suppliers to counter gas porosity. Several researchers evolved measures for the suppression of surface defects[14–16].

Figure 9.13 Shell moulded castings in non-ferrous alloys (courtesy of Hepworth Minerals and Chemicals Ltd.)

Stainless and other high alloy steels are less prone to surface defects, although surface carbon absorption has been detected and could diminish corrosion resistance in some circumstances[17].

In assessing the process, a significant factor is that production costs are only marginally higher than those in sand casting, whilst fixed costs of plant and handling installations can be considerably lower. Initial pattern costs are, however, high and method alterations during the development stage costly: the process is therefore unsuitable for small quantities of castings unless there is no practical alternative. One possible solution to this difficulty, however, lies in the availability of a rapid prototyping route directly relevant to shell moulded castings, in which resin bonded shells can be produced directly from a CAD file. The Direct Croning Laser Sintering Process is discussed, with other advanced prototyping techniques, in Chapter 7. The high cost tooling problem is also greatly reduced with the use of the cold shell process route.

Shell moulding is suited to the quantity production of relatively intricate small castings where the cost of investment casting cannot be justified. Notable applications include automotive rocker arms and crankshafts, whilst much attention was attracted by the early use of the process for the manufacture of crankshafts in spheroidal graphite cast iron[18].

Numerous products are illustrated in Figures 9.12 and 9.13 Further progress and applications of the shell moulding process were reviewed in References 19 and 20, including developments in resin formulation and the use of additives for the prevention of surface defects and shell cracking.

Investment casting

The term investment casting is used to describe a group of processes in which moulds are produced from liquid refractory slurries. These, containing finely divided materials, give the moulds a fine surface texture which is subsequently transmitted to the castings.

The processes fall into two distinct categories, based on expendable and permanent patterns. The difference is a fundamental one and the term precision investment casting is widely recognized as applying to those systems in which jointless moulds are made possible by the use of expendable patterns. In the permanent pattern processes the advantages of a smooth mould are still obtained, but the degree of precision is reduced by the need to assemble the moulds in separate parts.

Investment casting gives high standards of dimensional accuracy, surface finish and design flexibility and, unlike pressure die casting, is applicable to alloys of virtually any composition.

Expendable pattern processes

These have in common the provision of permanent tooling in the form of dies rather than patterns: the latter are produced from fusible or destructible materials as a part of the production sequence for each casting. The essential steps in the manufacture of a precision investment casting are illustrated in Figure 9.14 and may be summarized as follows:

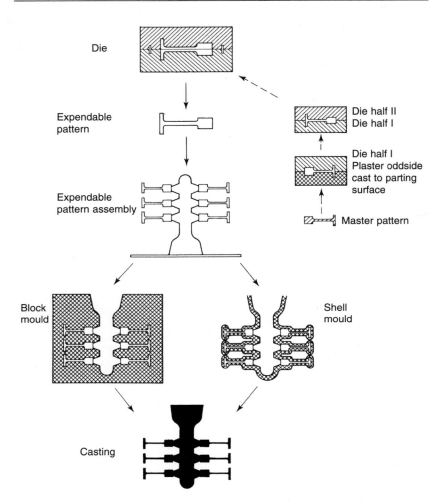

Figure 9.14 Production sequence in investment casting (expendable pattern technique) showing casting production (centre) and steps in fusible metal die manufacture (top right)

1. Construction of a die containing an impression of the casting;
2. Production and assembly of expendable patterns;
3. Investment of the patterns to form a one piece refractory mould;
4. Pattern elimination and high temperature firing;
5. Casting and finishing.

Dies

The die contains an accurate impression of the casting and is used to form the expendable patterns. Dies can be manufactured either in strong alloys such as steel and Duralumin or in softer fusible alloys, whilst epoxy resins or vulcanized rubbers are suitable where accuracy and permanence are less important. Plaster or 'stone' dies can be used for short life purposes.

Steel dies are most satisfactory for long production runs and are machined from the solid by die sinking and assembly in the tool room: this technique gives the highest standard of accuracy and is also used for aluminium tooling. Dies in fusible alloys are formed by casting and require the preliminary production of a master pattern or metal replica of the final casting. This master requires an allowance for subsequent contractions of pattern and metal, partly offset by expansion of the mould: the net allowance used may be up to 2%. The master is embedded to a selected joint line in plaster of Paris and the fusible alloy cast directly against the combined surface after drying (Figure 9.14). Pressure casting can be employed for accurate reproduction of detail. The second half die is produced by casting directly against the first after treatment with a parting agent. The master is then extracted and the die halves are sent for any necessary machining or drilling. Cast dies are more economical than sunk dies for short production runs. Resin dies too are produced by casting. Spray metal dies, built up from zinc alloy, are formed by deposition but following an essentially similar sequence.

To produce flexible rubber dies liquid rubber can be cast completely around the master pattern in a single operation, the parting interface being produced after vulcanizing by cutting to extract the master. In such cases undercut features may be produced without additional joints in the die.

Many dies can be produced in two sections but more complex designs require multiple construction and loose segments to enable the pattern to be extracted by dismantling. Gates for injection of the pattern material are usually machined separately. During production die faces are coated with a release agent or lubricant, for example a silicone based fluid, to assist pattern extraction.

Pattern production

The most common pattern material is wax, the term 'lost wax' being widely used to describe this version of the process. Pattern waxes of low ash content are compounded from natural and synthetic waxes and resins to produce the required combination of strength, contraction characteristics and dimensional stability: the materials usually melt within the range 55–90°C to form low viscosity liquids. Since reproducible contraction values are essential a single blend of waxes is normally maintained; operating temperatures too must be closely controlled because of the high coefficient of expansion of the pattern material (see Figure 9.15).

Patterns are commonly produced using wax injection machines. These embody levels of automation appropriate to the production facility and the wax can be injected in liquid, pasty or solid state. Although paste injection offers great potential for accuracy and rapid production, liquid injection

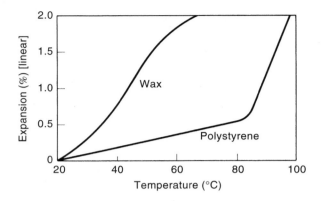

Figure 9.15 Typical thermal expansion curves for wax and polystyrene

is widely preferred and the high solidification shrinkage is reduced by the incorporation of solid fillers: organic materials such as polystyrene are employed to ensure complete burnout on firing. The machine incorporates a clamping system to maintain the required closing force on the die during injection. The pressure is maintained for a short dwell time to minimize shrinkage before the pattern is extracted.

Constant and controlled temperature throughout the machine from wax reservoir to nozzle is desirable and the wax itself is preconditioned by holding at temperature for a period before transfer to the machine.

Thermosetting plastics such as polystyrene can also be used as expendable pattern materials. Such patterns are less fragile than wax and a further advantage is their lower coefficient of expansion, pattern dimensions being consequently less sensitive to temperature variation. Figure 9.15 illustrates the difference in thermal expansion of the two types of material. Polystyrene patterns are produced on automatic hopper fed injection machines in which successive charges of the plastic are melted and injected into the die in a manner analogous to pressure die casting. Polystyrene behaves as a viscous liquid in the temperature range 200–350°C and requires higher injection pressures than those used for wax, up to 140 MPa ($\approx$20 000 lbf/in^2) being normal. The more exacting conditions require the use of steel dies and polystyrene is principally employed for mass produced components or to form exceptionally thin cast features, in some cases in composite patterns with wax.

The production of expendable patterns is attended by many problems similar to those in metal casting. In the case of wax the problem of solidification shrinkage cavities can be similarly countered by the use of internal wax chills. As in die casting, injection speed must be designed to avoid flow marks or misrun patterns on the one hand and entrapped air on the other. The possibility of distortion from residual stress is another similarity.

The next stage is the assembly of the separate patterns into a spray or cluster around a gating and feeding system of the same material so that

Figure 9.16 Wax pattern assembly and corresponding spray of castings (courtesy of B.S.A. Precision Castings Ltd.)

the whole forms a replica of the as-cast assembly (Figure 9.16). Gates and feeder heads are designed on the principle of directional solidification, but enclosed systems are normally used to facilitate investment, the main sprue affording the only access to the outside. The patterns are packed as closely as possible for maximum output.

Wax patterns can be assembled using a heated spatula or soldering iron, whilst polystyrene components are glued with plastic cement or by moistening with a solvent. Accurate positioning can be achieved with jigs and fixtures. Laser welding has also been used in conjunction with automation of these operations.

Investment

Investment casting was originally developed using block moulds produced by casting a grog or slurry into a container surrounding the patterns, followed by consolidation and pattern removal. This method is now only employed in a limited field, for small moulds as used in dental and jewellery casting, and has long been superseded in mainstream production by the ceramic shell system. The shell moulds are produced by alternate dip coating of the pattern assembly in refractory slurry, and stucco finishing with coarse particles. This operation is adaptable to mass production, with short drying times and wide use of dip automation.

Table 9.1 Some refractories used in investment casting

Zirconia, ZrO_2		
Zircon, $ZrO_2 \cdot SiO_2$	Fireclay	⎫
Alumina, Al_2O_3	Chamotte	⎪ calcined
Magnesia, MgO	Mullite	⎬ alumino-
Quartz; flint, SiO_2	Sillimanite	⎪ silicates
Cristobalite, SiO_2	Molochite	⎭
Fused quartz, SiO_2		

Investments consist of suspensions of refractory powders in liquid binder solutions, together with granular refractory material to form the stucco layers. Some of the important refractories employed are listed in Table 9.1. The binders are normally based on soluble silicates which undergo controlled reactions to form bonds of solid silica, themselves highly refractory. The binder solutions can be either water or alcohol based. The alcohol based system predominated throughout much of the period of growth in investment casting and employs ethyl silicate as the binder. Bonding is initiated by formation of a silica gel in a hydrolyzed solution of the ester, as approximately represented by the equation:

$$Si(OC_2H_5)_4 + 4H_2O = 4C_2H_5OH + Si(OH)_4$$

Tetra ethyl   Water   Ethyl alcohol   Silicic acid
silicate

$$\xrightarrow{\text{gel}}$$

This equation applies in the presence of a large excess of water. Since ethyl silicate is insoluble in water, ethyl alcohol is employed as a mutual solvent in the initial binder solution. Hydrolysis is catalyzed and the solution held stable by the presence of a small amount of hydrochloric acid: permanent gelation can be induced by further addition of water and by neutralizing with an alkali to pH > 3.

When using the original block mould system gelation was timed by close control of the pH value over the range 3–6 to permit pouring of the slurry, settling of solids and separation of excess binder solution. In the shell process the gelation of the liquid layer occurs on drying and is accelerated by exposure to ammonia gas after application of the stucco, raising the pH value to ~7. The chemistry of ethyl silicate binders is more fully treated in the literature[21–23].

The water based binder system employs alkaline colloidal silica sols which gel with increasing concentration on air drying, assisted by air flow under controlled humidity. The times required are longer than those involved when using the alcohol based binder but this system has been increasingly preferred on both environmental and cost grounds.

In both systems the complete evaporation of residual solvent and decomposition of intermediate products is accomplished during subsequent heating stages, and the final bond consists of lenses of silica between the refractory grains.

As well as the silica sol and ethyl silicate systems, hybrid binders are used, consisting of mixtures of the two, in which hydrolysis of the ethyl silicate is brought about by the sol itself without additional water. This produces a more stable binder and a greater effective concentration of silica in the alcohol based system, giving more rapid drying and strong shells: considering environmental factors also, such binders are seen to represent a compromise between the previous water and alcohol based types.

Other bonding systems too are available, based on the formation of alternative refractory oxides such as alumina or zirconia, exploiting reactions analogous to the formation of silica from the silicate binders. As one example, colloidal aqueous aluminium hydroxide can be produced by treatment of aluminium nitrate. Setting of the sol is then effected by treatment with an alkali such as ammonia, and decomposition on firing produces the all-alumina bond[24,25].

With melting points well above 2000°C such refractory oxide binders provide a basis for use with equally refractory powders and granular stucco materials, to form moulds which will remain inert under the most rigorous casting conditions. The casting of refractory and reactive alloys is clearly a relevant field, but one area of special interest is the directional solidification of superalloy components, still to be discussed, in which the ceramic shell is subjected to prolonged exposure of the interface to temperatures exceeding 1500°C, requiring preheat temperatures of 1550°C or higher. Mills[26] has extensively discussed effects on refractories under these conditions, including the behaviour of the zirconia-silica refactories used in the ceramic cores. Work in such areas clearly represents one of the frontiers of the entire casting process.

For the production of shells the pattern assembly is submerged is a stirred bath of ceramic slurry containing a fine (<200 mesh) suspension of refractory powder in binder solution. Excess slurry is drained off to leave a uniform surface layer and the granular stucco (>100 mesh) is applied whilst the coating is still tacky: this is accomplished by rotation in a raining machine or immersion in a fludized bed. The coating is allowed to set before repeating the sequence to build a multi-layer mould of the required thickness. The intermediate sets are accelerated by forced air drying; ethyl silicate bonded layers treated with ammonia are sufficiently hardened in minutes.

The shell build process lends itself particularly well to automation, using programmed robot handling of the pattern assemblies through the systematic motions required for dip, drainage, stucco application and drying. Many such plants are in service and an example is shown in Figure 9.17 (see colour plate section).

The initial or primary coat, commonly water based, determines the surface finish on the casting and may also be used as a vehicle for the introduction of inoculants, designed to produce finer and more reproducible microstructures in the castings; zircon is widely used as the powder filler and zircon sand is in some cases employed as a relatively fine stucco material to complete the initial layer. Wetting and avoidance of bubbles are particularly important for the primary coat: initial rinse cleaning of the pattern assembly to remove release agent, and the inclusion of wetting additives in the primary slurry, are used to minimize this problem. Shrinkage of the gelled binders during drying tends to produce cracking of the coating, a problem minimized by close control of temperature and humidity as well as by maintaining consistency in the composition and grading of both slurry and stucco materials: the binders are sensitive to impurities and to changes occurring in storage, for example from solution of carbon dioxide or contact with other contaminants. Drying times of one hour or more are required for this coat before proceeding with the main sequence.

Different slurry and stucco materials, particularly alumino-silicates, are commonly selected for subsequent layers, with coarser stucco gradings, depending on alloy, cost and other factors. In the case of block moulds the main investment incorporates coarser grog materials in a thick graded suspension, which is poured and vibrated round the precoated pattern assembly for consolidation and elimination of air bubbles. Excess liquid is removed, first by decanting and subsequently by evaporation and drying, a much longer process than that required for shells; initial setting and air drying below 45°C for up to 24 hr is followed by gentle heating below 120°C before raising to the firing temperature. Complete shell mould assemblies are illustrated in Figure 9.18.

The complex physical phenomena occurring in the production of silicate bonded investment casting moulds, including shrinkage and the cracking tendency in drying, have been examined in the detailed treatment by Mills[26]; this also covers subsequent changes in the dewaxing and firing stages.

Although most investment casting production employs siliceous binders of the above types, usually in conjunction with the ceramic shell process, alloys of lower melting point can also be cast in block moulds employing plaster of Paris as the bonding agent, blended with fine refractory fillers such as silica or molochite and suspended in water. The plaster is manufactured by partial dehydration of the mineral gypsum (calcium sulphate, $CaSO_4.2H_2O$) to form the hemihydrate $(CaSO_4)_2.H_2O$: the subsequent setting reaction is brought about by reabsorption of water of crystallization to form a coherent network.

Plaster–silica investments containing up to 25% of plaster have been widely used in dental applications and in the manufacture of jewellery, being suitable for pouring temperatures of up to 1100°C, although firing

Figure 9.18 Ceramic shell moulds for investment casting (courtesy of B.S.A. Precision Castings Ltd.)

and preheat temperatures need to be kept below 700°C to avoid premature decomposition of the plaster. In dental work, where cast replicas derived directly from mouth impressions are required, contraction allowances cannot be introduced. The contraction of the casting can, however, be offset by controlled expansion of the mould investment using the different expansions of quartz and cristobalite, and of the plaster: the relative proportions of these constituents can be adjusted to achieve the required behaviour. Expansion curves for three forms of silica and of these investments are shown in Figure 9.19.

Shaping of internal cavities. One of the main advantages of precision investment casting is the ability to produce internal cavities and undercut features as integral parts of the mould: this avoids core location and maintains the principle of a jointless mould. Several alternative techniques are available for forming such cavities and are illustrated in Figure 9.20.

The first three methods involve hollow patterns, into which the mould slurry flows on investment and forms its own core. In some cases the hollow pattern can be produced directly from the die assembly by dismantling and the use of loose pieces (Figure 9.20a). Where the design is more complicated the problem can be simplified by producing the pattern in separate sections which can be welded together to enclose the cavity (Figure 9.20b).

A further technique for the production of hollow patterns employs separately produced water soluble wax inserts, located in coreprints in the die

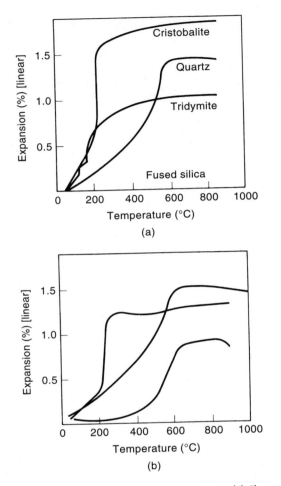

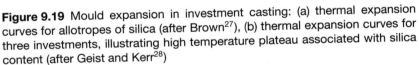

Figure 9.19 Mould expansion in investment casting: (a) thermal expansion curves for allotropes of silica (after Brown[27]), (b) thermal expansion curves for three investments, illustrating high temperature plateau associated with silica content (after Geist and Kerr[28])

before injection of the main pattern material (Figure 9.20c). The inserts are dissolved out before investment in the moulding material, which once again leaves its own core. A suitable wax is polyethylene glycol, which has a higher melting point than the pattern waxes and so remains solid when the pattern is injected round it. Using this technique complex cavities can be produced.

The fourth method of producing internal cavities uses preformed refractory core inserts, which are again located in the die before injection of the

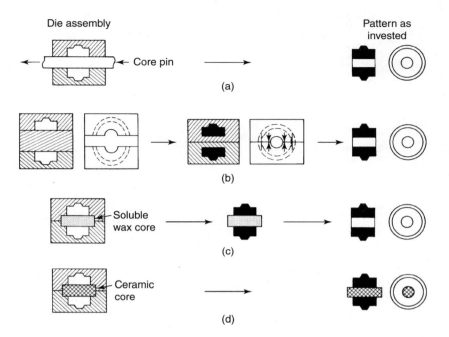

Figure 9.20 Production of internal cavities in investment casting: (a) hollow pattern formed by loose die member, (b) hollow pattern formed by booking and welding, (c) hollow pattern formed by soluble inserts, (d) insertion of preformed ceramic core

pattern material (Figure 9.20d). On investment the core becomes locked into the main body of the refractory, where it remains as an integral part of the fired mould. This method is suitable for very thin or intricate passages into which the investment slurry would only flow with difficulty; the technique is widely used for the production of hollow blades for gas turbines. The ceramic cores, usually made from fused silica, need to be robust enough to withstand forces imposed in wax pattern injection, mould firing and casting, and can be formed by a number of alternative techniques. These include injection moulding of pastes of silica powder bonded with waxes, resins and silicones into steel dies, followed by progressive firing to remove the binder and sinter the refractory. Slip casting of slurries, powder pressing and other systems are also used. Some idea of the capability of such cores is shown in the blade example in Figure 9.21. For some of the most intricate cavities minute platinum wire chaplets are incorporated to provide additional support for the core during casting.

Ceramic core manufacture is commonly undertaken by specialist external suppliers of technical ceramic products.

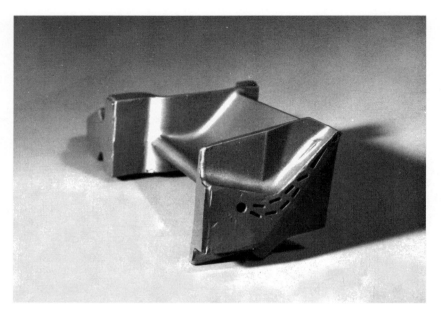

Figure 9.21 Example of confined passages through turbine blade casting, formed by ceramic cores. Ten separate holes through blade section 34 mm width, 7 mm maximum thickness approx (courtesy of Rolls-Royce plc)

Pattern removal and firing

Following the initial setting and air drying stages, the next operation is pattern elimination and consolidation of the bond by evaporation of the remaining volatile constituents. The importance of the early drying stages in avoiding shrinkage cracking in the primary coating has already been stressed; a further factor in cracking is expansion of the pattern on heating, which can generate significant stresses in the ceramic shells with their much lower coefficients of expansion. This problem can be surmounted by rapid heating to develop a steep temperature gradient though the thickness of the ceramic shell and into the wax. Superficial melting of the pattern surface is then achieved before the main bulk of the wax undergoes applicable expansion. The molten film is partly absorbed by the adjacent shell, enabling the remainder of the pattern to expand freely.

Two principal techniques employed to achieve this condition are steam autoclave treatment and flash dewaxing. The autoclave system is most widely adopted: a steam pressure of several atmospheres is almost instantaneously developed in the loaded chamber and dewaxing is completed in a few minutes at temperatures below 200°C, with high wax recovery. In flash dewaxing the temperature is raised rapidly to approximately 1000°C by placing the mould directly into the firing furnace. Wax is drained though the furnace base but recovery is lower owing to volatilization and partial combustion.

This method is also suitable for plastic patterns with their higher melting points, for which the normal autoclave treatment is ineffective.

Other techniques include straightforward heating to a temperature slightly above the melting point of the wax, allowing drainage of the main bulk, a system principally used for block moulds but unsuitable for shells; organic solvent treatments have also been employed but these too have been superseded by the autoclave process.

Mould firing is commonly carried out at temperatures around 1000°C in either batch or continuous furnaces. The process completes the volatilization and combustion of pattern residues and organic mould constituents and the dehydration of the binder, including elimination of combined water. Sintering contributes to the full development of strength and stability in the ceramic. The firing process thus ensures fully inert conditions for contact with the molten alloy. A further purpose in many cases is preheating for casting to facilitate metal flow and the reproduction of intricate features, although this may be carried out as a separate operation. Much higher firing and preheat temperatures are required for moulds to be subjected to directional solidification.

Casting

Melting for investment casting is commonly carried out in coreless induction furnaces, with some use of resistance furnaces for aluminium, and of gas fired crucible units, including immersion heated types, for aluminium and other non-ferrous castings. The furnaces are of standard designs as used in other metal melting operations. Moulds may be gravity cast or may employ pressure or vacuum assistance; centrifugal casting is also used in certain applications, whilst full vacuum melting and casting is an essential feature in some of the most important product sectors. Mould preheating to high temperatures is widely although not universally practiced.

Gravity casting gives satisfactory results in many circumstances, especially when used in conjunction with hot moulds, but assisted pouring further enhances capacity for the running of thin sections and definition of intricate detail. Rollover induction furnaces, designed specifically for investment casting, provide one form of assisted pouring: the mould is in this case inverted and clamped over the furnace mouth, the whole assembly being rapidly rotated through 180° for casting, generating additional momentum for filling. The rollover principle had been a feature of the original Merrick indirect arc furnace, which used additional air pressure to enhance filling. In some more modern systems metal is displaced upwards into the inverted mould by pressure or vacuum suction: these distinctive counter-gravity techniques are variously applied in investment, sand and die casting and are further examined in Chapter 10.

Centrifugal casting is usually carried out on vertical axis machines, with the mould cavities spaced around an axial sprue and provision for clamping the mould assembly to the turntable. Very small moulds as used in dental casting are cast on special machines rotating about either a horizontal or a vertical axis: these provide for melting small charges, which are then centrifuged directly from the crucible into the mould. Some of these machines also employ direct vacuum assistance to the filling process.

Vacuum melting and casting have important applications in the production of high temperature alloys, where such conditions are essential to the protection of reactive alloying elements and the exclusion of atmospheric contaminants. The running of thin sections is also assisted by elimination of backpressure from air in the mould cavity. Various types of specialized vacuum furnace have been developed for investment casting. Features include the provision of vacuum locks through which preheated moulds can be successively transferred into and out of the main vacuum chamber for casting, obviating the need for frequent pumping down from atmospheric pressure. The self tapping high energy induction furnace melting single charge slugs, previously mentioned and illustrated in Figure 8.38, is incorporated in custom built vacuum units for the casting of gas turbine blades. A further special unit is the directional solidification furnace, embodying controlled withdrawal of the mould unit from the heating coil, enabling columnar or single crystal structures to be developed by growth from a water cooled chill surface forming the mould base.

Finishing

Investment castings pass through knockout, cleaning and finishing stages largely analogous to those used in sand casting although necessarily more delicate given the character of the products. Power saws and abrasive slitting wheels are used to separate the castings; effects of subsequent dressing on local dimensions must be considered in positioning points of attachment.

Blast cleaning is used for the removal of residual mould material and selection from a wide range of available abrasives is crucial in its effect on the texture and appearance of already smooth cast surfaces; water blasting is also used. Some general aspects of cleaning and dressing processes are reviewed in Chapter 8 but a detailed illustrated account by Bidwell of various processes as applied to investment castings will be found in Reference 29.

One aspect of special significance is the requirement for core removal. Many dense, intricate cores forming confined cavities are inaccessible to the standard blast cleaning operations and require chemical methods. Such cores, with their high silica content, can be dissolved out using reactions with caustic alkalis or dilute hydrofluoric acid. The former are most commonly used, either as hot aqueous solutions or as molten salts, for which enclosed saltbath plants are available.

Leaching in aqueous solutions can be accelerated by enclosure in high pressure autoclaves, which enable the temperature to be increased beyond the normal boiling point of the liquid, although boiling itself also assists dissolution by local agitation. Final neutralization and rinsing are required to avoid subsequent corrosion from chemical residues.

Other finishing processes include hot isostatic pressing, for which investment casting is a particularly important field: the process is examined in Chapter 8.

Considering the production process as a whole, the position with respect to mechanization and automation in the investment casting industry was the subject of a broad survey by Barnett[30], which confirmed the general picture of extensive robotization in the mould build operation, often using purpose-built units of the kind already illustrated, with the main emphasis on mechanization of individual operations in the patternmaking and finishing areas.

The availability of robotic manipulation has been a major factor in the extension of the process to larger and heavier moulds with envelope sizes of around one metre cube; loads in the range 20 kg to 2 tonnes are feasible.

Concerning pattern production, fully automatic wax injection machines have been seen as suitable mainly for long production runs or for standard gating components and test pieces; investment in semi-automatic tooling and rapid die changing facilities is considered necessary for economic use of automatic plant for short casting runs. The most labour intensive operation remains that of pattern assembly, although such developments as laser welding[31] offer potential for reducing the manual content of this stage in the process.

For the finishing stages automatic cut-off machines are available, including fully programmed computer controlled units, and robot automatic grinders too have found increasing use.

Factors influencing casting quality

The production of precision castings requires an accurate knowledge of dimensional behaviour at every stage of production from master pattern or die to casting, so that allowances can be made where appropriate. Wood and von Ludwig in their comprehensive early work on investment casting[32], listed some thirty variables which affect casting dimensions through their influences upon die contraction, pattern contraction, mould expansion and casting contraction. The overall effect of these factors can, however, be established within narrow limits by experience, as can the individual behaviour of particular designs of casting. As in any casting process dimensions are subject to initial uncertainty, but this can be catered for in some cases by progressive die modification.

The concept of size controlled by variations of investment composition has been mentioned earlier in relation to special cases and was the subject

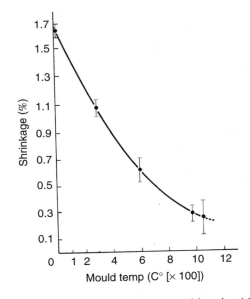

Figure 9.22 Example of relationship between mould preheat temperature and casting dimensions. Co–Cr alloy (from Carter[35])

of systematic studies[33,34]. Preheat temperature too has a strong influence on casting dimensions, as shown in Figure 9.22. Whilst such effects may be used to adjust dimensions, they mainly emphasize the importance of close control of production variables once the initial allowances have been established. Factors which need to be standardized include pattern and investment compositions, and the times used in patternmaking, mould drying and firing. Most important is the control of temperature at all stages of production. A consistent time/temperature cycle for pattern injection should be followed by assembly and investment of patterns in a temperature controlled and preferably air conditioned atmosphere. The need for consistent drying, firing and casting temperatures has already been illustrated.

Characteristics and applications of investment castings

The high accuracy of investment castings arises from the smooth and inert mould surfaces and from the elimination of joints by the use of one-piece moulds. The only alignment operation is that of the die assembly, for which the metal components can be made to fit with close precision. During pouring, ready flow in the preheated mould gives intricate detail and fine finish. The process thus offers a degree of precision unrivalled except in pressure die casting, even though matched in some cases in the narrower dimensional sense.

As in other processes, attainable tolerances depend partly upon design and the circumstances of the particular dimension. General values are widely quoted, however, whilst more detailed treatments and other design recommendations are given in the literature, particularly in References 36–41. Tolerances of around ±0.15 mm are commonly accepted as feasible for linear dimensions up to 25 mm, extending to ±1.5 mm for sizes around 250 mm, although these figures may be halved in favourable circumstances, which include high grade tooling with opportunity for modifications, lower melting point alloys and more rigorous process controls. This situation is usefully portrayed in Figure 9.23, from Reference 37. Tolerances should be specified on a functional basis, with the most exacting requirements confined to critical dimensions.

Quoted values for surface finish lie in the range 0.8–3.2 µm, the lower values being associated with aluminium alloys and the higher for steels.

The implications of the recommended tolerances and surface finish are that, whilst loose fits are attainable between as-cast components, machining allowances are normally advisable for bearing surfaces, precision threads and other close tolerance requirements. Allowances in the range 0.5–2.5 mm are recommended depending on the size of the casting. The importance of liaison on the matter of datum planes was emphasized by Albutt[39]: that subject has been previously examined in Chapter 7.

The minimum metal thickness that can be run depends partly upon the area of the section concerned. Although thicknesses of 0.75 mm or even less can be achieved over small areas of casting wall, 1–1.5 mm is seen as a reasonable minimum for large areas, with aluminium at the lower end. Holes of this size are also feasible: the use of ceramic cores

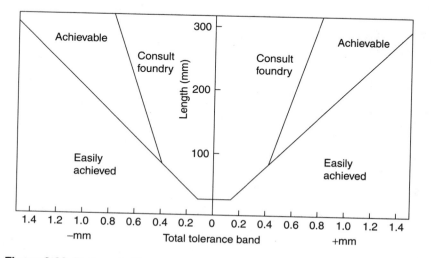

Figure 9.23 Portrayal of linear tolerance regimes for investment castings (from Reference 37) (courtesy of BICTA)

permits the production of cast holes down to a diameter of 0.5 mm and of substantial length, but 3 mm is the normally recommended minimum for holes to be produced by direct investment, with lengths representing up to three times the diameter; this limitation becomes more relaxed at larger diameters.

The general design capability of the process is high, with few shape restrictions. Mould joint considerations are absent and components can be produced without draft taper. Using the wide range of techniques previously described, holes and internal cavities of virtually any shape can be formed.

Design considerations in respect of casting and cooling are broadly similar to those involved in other processes and discussed at length in Chapter 7. The main objectives are smooth flow of metal, directional solidification, and minimum stress generation on cooling. As in other cases these aims are furthered by uniformity or progressive change of section thickness and by the use of corner fillets and streamlined contours. To exploit the accuracy of the process, however, maximum design detail should be incorporated in the casting, leaving machining to be undertaken only where the required precision exceeds the process capability.

Apart from dimensional errors, investment castings are subject to a similar range of defects to those encountered in other processes, for example misrun castings and non-metallic inclusions. In the latter category are defects due to oxidation and mould reaction which are the counterpart of the exogenous inclusions encountered in sand casting: in the case of heat resisting alloys this type of surface defect is known as 'spotted dick' and can be combatted by reducing conditions in casting[42]. Defects with causes peculiar to investment casting include blowholes due to incomplete removal of wax residues and surface defects resulting from local failure of the primary coating.

The application of investment casting is principally in the field of relatively small components, which are usually produced in assemblies around a common sprue. Since dimensional accuracy depends partly on the magnitude of the dimensions, the advantage over other casting processes diminishes with increasing size. Most castings are below 5 kg in weight, with emphasis on intricate parts weighing less than 0.5 kg. Castings exceeding 50 kg are nevertheless produced, although it is in this field of heavier castings that the Replicast® process, employing foam patterns, is particularly suitable. A special strength of the investment casting process lies rather in the production of large box-like castings of very thin wall section as exemplified in Figure 9.24: these exploit the capacity to fill extensive yet confined mould passages, and often to combine, in a single component, elaborate features which might normally require several castings or a fabricated construction. An envelope bounded by a one-metre cube represents the feasible size accepted as being attainable for this type

Figure 9.24 Thin walled box configuration investment castings: largest items 570 mm in height with panels down to 1.2 mm in thickness (courtesy of P.I. Castings Ltd.)

of casting: costs for such items might be relatively high in terms of unit weights, but not when perceived in the overall context of the finished article.

Virtually every type of cast alloy is available, including light alloys, copper base alloys, cast irons, steels and corrosion and heat resisting alloys based on nickel and cobalt. Of special interest amongst the steels are the types employed for applications requiring very high strength. Prominent amongst these is the 17% Cr–4% Ni precipitation hardening composition containing copper and other alloying elements: a component made in this material is shown in Figure 9.25. Typical fatigue properties of such a steel were previously illustrated in Chapter 8 Figure 8.55.

Titanium castings, including very large gas turbine casings, are also produced. The main groups of alloy are reviewed in detail in Reference 36, including those British Standards specific to investment castings; they are also covered, with others, in a guide to alloy selection[43], which interrelates UK and overseas specifications.

The high technical capabilities of the process are paralleled by high manufacturing costs, but the unprecedented growth in the production of investment castings reflects the overall economies to be achieved in end product cost through the elimination of machining, sub-assembly and similar operations. One example of a complex cast component produced as a single unit, so avoiding the need for sub-assemblies, was shown previously in Figure 7.17. Other highly favourable applications are those involving complex contours, such as aerofoil surfaces where forging is not a feasible alternative because of metallurgical factors, or parts in hard, wear resisting alloys which are inherently difficult to machine.

Figure 9.25 Investment cast load bearing aerospace component in 17/4 high strength precipitation hardening stainless steel (courtesy of P.I. Castings Ltd.)

The adoption of investment casting usually requires appreciable quantity production for amortization of die costs, although these are generally much lower than in die casting. Tooling costs are also usually less than for forgings or sintered compacts, for which the hardest and least machinable die steels are required. For the mass production of large numbers, however, investment casting is not price competitive with these processes, despite progress in automation of some of the manufacturing operations.

Any reference to individual applications must give great weight to the superalloy gas turbine blade, with which investment casting has been closely associated from the beginning of its engineering history. This remarkable cast product contributed to a succession of advances in engine design and performance, which called in turn for renewed expansion in the capability of the casting itself. The achievement paved the way for other engineering applications for investment castings. A brief account will be given here: more detailed treatments will be found in References 44 and 45.

The first investment castings employed in gas turbines were stator blades and segments, since wrought structures were considered essential for the more highly stressed rotor blades. With continued development of the nickel base superalloys, however, the temperature range available for forging became progressively more restricted. This was due to the use of increasing volume fractions of the dispersed gamma prime phase $Ni_3(TiAl)$, formed by titanium and aluminium additions, to enhance high temperature creep resistance. The effect was to raise the process temperature required to dissolve the phase and permit deformation, whilst at the same time lowering the solidus temperature, so that forging became impracticable. The adoption of cast alloys allowed the use of much larger volume fractions of the dispersed phase, and vacuum melting and casting prevented loss of the readily oxidized alloying elements whilst ensuring high quality in the cast products.

Progress in gas turbine design also introduced blade cooling by flow of gas through passages within the component. This was facilitated by an advance in investment casting, the use of preformed cores as previously described, enabling higher engine temperatures to be employed within the temperature capability of a particular blade alloy.

Important further advances followed the introduction of controlled directional solidification as discussed in Chapter 2. An early example of an all-columnar cast structure in a gas turbine blade is shown in Figure 9.26. The removal of most transverse grain boundaries in such structures enhanced high temperature ductility and stress rupture life, an improvement later extended by techniques for the production of single crystal blades, using competitive elimination of growing crystals by special designs of mould passage: one such system is shown in Figure 9.27. Once the grain boundaries have been eliminated, subsequent alloy development can focus on the

Figure 9.26 Unidirectionally solidified gas turbine blade showing columnar structure (courtesy of Pratt & Whitney Aircraft, Division of United Aircraft Corporation)

further strengthening of the crystal matrix, at the expense of elements previously required for grain boundary strengthening. Advanced automatic NDT techniques are then also required to maintain full orientation control and to ensure the absence of unwanted boundaries. The progress achieved by these successive developments is summarized in Figure 9.28.

The need for the quantity production of directionally solidified blades led to the design of specialized furnaces of the type exemplified in Figure 9.29. This combines a self tapping induction melting unit with the facility for controlled mould withdrawal from the hot zone, the whole being maintained under vacuum throughout melting and solidification.

Apart from the continuing importance of the aerospace sector, investment castings are used throughout the commercial field. Notable applications include impellers and other pump and valve components in stainless steel and non-ferrous alloys, waveguides, and parts for gun mechanisms. Milling cutters and golf club heads indicate the diversity of the product areas. Further components are shown in Figures 9.30 and 9.31; fuller reviews of applications in general engineering and the dental-medical and jewellery fields are contained in Reference 45.

Figure 9.27 Example of cast metal geometry used for production of a single crystal turbine blade casting, embodying a spiral mould constriction (courtesy of Rolls Royce plc)

Investment casting from permanent patterns

The advantages of liquid investments, with their close conformity to the pattern and smooth mould surfaces, can be utilized in a simpler and less costly sequence involving direct moulding from patterns of metal, plaster or wood. Various processes use this principle and all produce multipart rather than one piece moulds. The castings have a high standard of accuracy and finish, although the degree of precision is second to that of castings

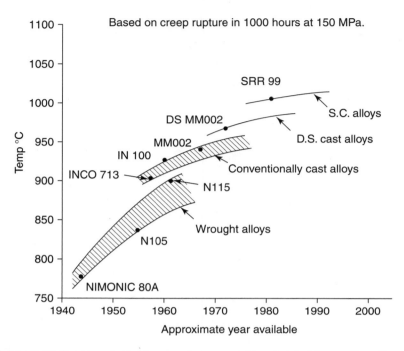

Figure 9.28 Progress in the temperature capability of nickel-base alloys (from Alexander[46]) (courtesy of Rolls Royce plc)

produced from expendable patterns because of the need to align the separate mould parts and cores. The errors are nevertheless small because of the precise mould components. A further feature of some investments is a degree of flexibility after initial setting, eliminating any need for moulding taper and enabling vertical walls or even slight undercuts to be produced.

Plaster of Paris investments have long been established for the production of non-ferrous castings, especially in aluminium and copper base alloys. Although orthodox pattern materials are used, flexible rubber patterns can also be employed to facilitate the production of undercut features in intricate components. The aqueous plaster suspension is poured and allowed to set, after which the pattern is removed and the mould part stove dried at temperatures from 200°C to 700°C or even higher depending on the particular process. Various techniques have been used to develop strength and permeability in the dried mould. Strength can be enhanced by the addition of fillers such as sand, cement, talc and fibreglass to the plaster mix, and permeability by porous fillers or by the use of foaming agents and agitation

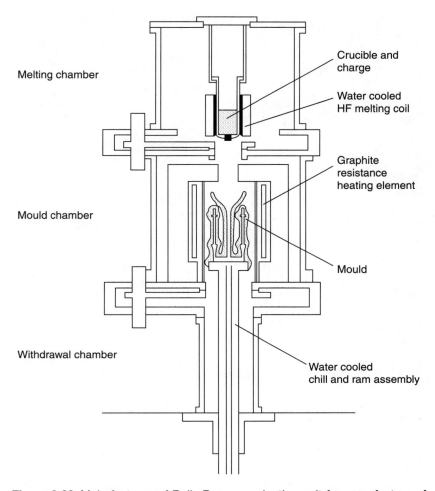

Melting chamber

Crucible and
charge

Water cooled
HF melting coil

Graphite
resistance
heating element

Mould chamber

Mould

Withdrawal chamber

Water cooled
chill and ram assembly

Figure 9.29 Main features of Rolls Royce production unit for manufacture of directionally solidified gas turbine blades (courtesy of Rolls Royce plc)

during mixing. In the Antioch process[47,48] which uses a mix of silica sand, plaster and additives, permeability is obtained by controlled recrystallization of the plaster following steam pressure treatment in an autoclave, producing intergranular air passages. Fuller details of this and other plaster mould processes will be found in Reference 49. Examples of intricate components produced by plaster moulding are shown in Figure 9.32.

In the Shaw or ceramic mould process[50–52], block moulds or contoured shells are produced from a highly refractory investment utilizing the

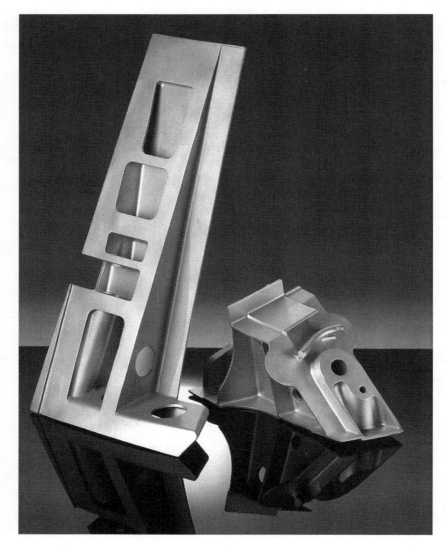

Figure 9.30 Aluminium alloy aircraft engine nacelle castings destined for Airbus (courtesy of P.I. Castings Ltd.)

previously described gelation of ethyl silicate to bond a graded refractory such as sillimanite, mullite or zircon. The hydrolized suspension is poured round the pattern immediately after the addition of a catalyst, for example ammonium carbonate, designed to produce rapid gelation without settling of the solids. The mould part sets to a rubbery consistency, at which stage

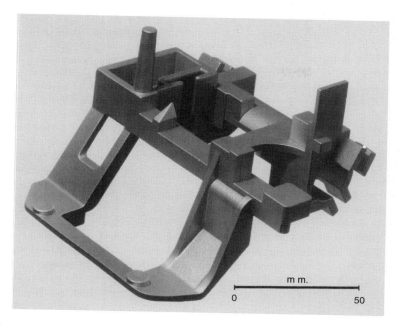

Figure 9.31 Complex investment cast computer component in aluminium alloy (courtesy of P.I. Castings Ltd.)

(a)   (b)

Figure 9.32 Aluminium alloy castings produced by plaster moulding: (a) torque converter component, (b) cathode ray tube housing (courtesy of Sterling Metals Ltd.)

the pattern can be readily withdrawn. Variations in moulding technique include the production of shell or contour back moulds by pouring the slurry between pattern and contour plates and the use of cheaper backing materials or permanent backs of metal or refractory. In the case of contour shells carrier drying is required.

With the evaporation of alcohol on drying, volume shrinkage of the investment produces fine craze cracking, which confers permeability and resistance to thermal shock whilst limiting overall contraction. To assist drying, the alcohol vapour is ignited with gas burners and the mould is subsequently fired at 1000–1100°C for evaporation of final traces of water and alcohol. Moulds may be cast whilst hot if required to assist the casting of thin sections and intricate features. Fired moulds can, however, be cast without preheating.

Although ethyl silicate investments are relatively costly, the moulds are sufficiently refractory for castings to be produced in carbon and alloy steels and in heat and corrosion resisting alloys. Castings weighing several hundred kilogrammes can be produced and the process is particularly suitable for the manufacture of steel dies and die inserts for forging, drawing, extrusion, die casting and glass making; large turbine blades are also produced. Examples of castings are illustrated in Figure 9.33.

Another permanent pattern process with similar features is the Unicast process, in which the ceramic moulds undergo an additional treatment with a surface hardening catalyst, a volatile liquid applied as a spray or by immersion, immediately after pattern removal, to stabilize the ceramic. The excess volatiles are removed by local surface heating before full furnace or torch firing. The process is described in Reference 53, with emphasis on use for quantity production of pump and valve castings; cast to size tooling is again an important field of application.

References to further modern developments in ethyl silicate investments for this type of process have discussed the use of resin additions to replace

Figure 9.33 Cast dies manufactured by the Shaw process, for forming and die casting (courtesy of Avnet-Shaw & Co. Ltd.)

part of the alcohol content of the slurries, giving very high hardened strengths, especially suitable for thin impeller cores and thin-wall moulds[54].

Permanent pattern investment castings are capable of accuracies within the range ±0.5–1.5% of the dimension according to magnitude and position. The processes are adaptable to loose as well as to mounted patterns, even wooden patterns being suitable provided that they possess a finish compatible with the investment: synthetic resin finishes are required to resist solution in ethyl silicate. Loose patterns are accommodated using oddside practice to form the parting surface. Mutual location of block moulds is achieved with integrally moulded tapered locators or registers.

The adaptability to various forms of pattern equipment enables medium precision to be combined with low pattern cost, making the processes suitable for short runs as well as for quantity production. In the latter case high outputs can be achieved in mechanized systems with facilities for automatic mould production.

Die casting

The essential feature of die casting is the use of permanent metal moulds, into which molten alloy is either poured directly or injected under pressure, giving rise to the separate processes of gravity and pressure die casting. Permanent moulds offer obvious advantages in terms of simplicity of production for large quantities of parts, but are subject to limitations yet to be discussed.

Gravity die casting, also referred to as permanent mould casting, has many features in common with other processes of metal founding including sand casting, both in production technique and in type of product. There is little such similarity in the case of pressure die casting, a highly distinctive process which might be held to have more in common with injection moulding than with other casting processes. The use of complex casting machines and the specialized nature of die construction combine to give the pressure process a strong engineering emphasis. A third process, low pressure die casting, has much in common with the gravity process and it too will be briefly reviewed.

Arising from their use of permanent moulds, these processes are noted for low production costs combined with high initial tooling costs: this applies particularly to the pressure process, where the dies can be of great complexity and where heavy capital costs are also incurred on machinery. The processes are thus ideally suited to mass production and unsuitable for jobbing and small quantity requirements, for which other accurate methods are available.

Die casting is in a different category from other precision casting techniques in that its use is confined, for practical purposes, to relatively few alloys. This restriction is imposed by the limited capacity of die materials to withstand intermittent contact with molten metals of high melting point; in addition, the casting alloy itself must solidify and contract without cracking under conditions of hindrance from the die.

Subject to these limitations the degree of accuracy and definition, coupled with the low cost of quantity production, places the die casting processes in a highly competitive position relative to other forming techniques.

Gravity die casting

The gravity die casting process is notable for the very large output of castings in aluminium alloys, for which the process is predominant as a mass production technique, but there is also substantial production of copper base alloys and cast iron, particularly in relatively simple shapes, and a limited output of magnesium alloy. The process is only suitable for fluid alloys owing to the high freezing rates obtained in metal moulds.

Dies for gravity casting are often of comparatively simple construction. A typical two part die is split along a vertical joint line passing through the die cavity, the running, feeding, and venting system being disposed in the same plane (Figure 9.34). Internal cavities may be produced by using separate metal cores, which can in many cases be directly retracted from the casting;

Figure 9.34 Two part hinged die for gravity die casting, showing casting in position; choke gate system embodies continuous slot connecting sprue to riser and two gates connecting riser to casting (courtesy of H. J. Maybrey Ltd.)

in other cases collapsible core construction is required to enable loose core-pieces to be withdrawn sideways from undercut sections. Cavities are frequently produced with the aid of shell, sand or plaster cores, permitting greater flexibility of design. There has also been limited use of soluble salt cores[55]. Dies are provided with various arrangements of pin locators, clamping devices and ejection systems for casting removal.

To facilitate assembly and mutual location the die halves may be hinged. Modern practice is also characterized by extensive die mechanization similar to that used in pressure die casting, with provision of advanced equipment for automatic opening and closing, locking, and actuation of core movement and ejection systems. Air and hydraulically operated mechanisms are employed. A production cycle of less than 1 min is feasible using these systems.

The dies, of high quality cast iron or in special cases die steel, are preheated and coated with insulating refractory dressings before use. During the production cycle the die is brought to an equilibrium operating temperature, usually between 300 and 400°C, by maintaining a regular time cycle for casting and ejection. The cycle can be shortened and production increased by provision of cooling fins, by forced air cooling or by incorporation of water cooling systems in the die blocks. After pouring, the casting is ejected from the die as rapidly as possible to avoid hindrance to contraction on further cooling: careful timing is thus required. Extraction is assisted by generous taper to the joint line: this reduces wear and tear of the die surfaces. Using such precautions a die life of many thousands of castings can be achieved.

For good surface quality, great care is required in pouring to avoid turbulence and dross formation: tilt pouring can be used for steady mould filling. For gating and feeding, similar principles apply to those in sand casting and were reviewed by Glick[56]: the general object is directional solidification. The die cooling system can be used to develop the required temperature gradients in the casting; other techniques include local die heating and varying the thickness of the die dressing to produce differential insulation.

Apart from the advantages of gravity die casting in producing high surface quality and accuracy, mechanical properties are improved by the chilling effect of the metal mould and refinement of the microstructure: specified values for tensile strength and ductility are thus considerably higher than those for sand castings of equivalent composition.

The process is applicable to castings in the light and medium weight range, roughly equivalent to those produced as repetition sand castings by machine moulding. Although the process is inherently capable of producing heavier pieces, most output is of castings in the range 0.5–50 kg. Techniques of the process, particularly die design, were comprehensively treated in Reference 57.

Low pressure die casting

Closely related to gravity die casting in its aims and field of application is the process of low pressure die casting, in which the metal is displaced upwards into the die cavity by pressure exerted on the surface of the liquid in a sealed reservoir. This counter-gravity casting principle is examined further in Chapter 10 in relation to a number of other important process developments. The system employed in low pressure die casting is illustrated in Figure 9.35. In this unit air pressures in the range 20–100 kPa ($\approx$3–15 lbf/in^2) are used for casting aluminium alloys[58-60]. On applying the pressure, metal passes up the coated cast iron tube or 'stalk' and through the gating orifice at a rate enabling air to escape through joints and vents in the die; advanced high temperature ceramics have also been developed for

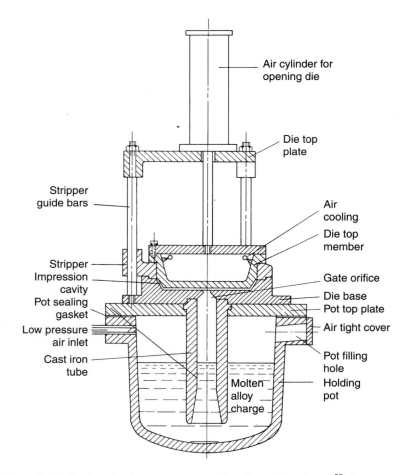

Figure 9.35 System for low pressure casting (from Woodward[58]) (courtesy of Institute of British Foundrymen)

the stalk application. Upward feeding is then maintained from the molten column and reservoir. On releasing the pressure the surplus liquid returns to the bath and all that remains is a small projection on the casting; more than one stalk or gating orifice can be employed if required. Such a system involves no runners or feeder heads in the accepted sense, so that casting yields approaching 100% and very low remelt ratios can be achieved. The operational cycle is necessarily slower than that for gravity die casting owing to the need to maintain the pressure until solidification of each successive casting is complete.

The castings are of a high standard of surface quality and soundness. Sharp definition of detail is obtained, since positive pressure is available to force the metal into the die recesses throughout its height: thin fins and wall thicknesses as low as 1.5 mm ($\approx \frac{1}{16}$ in) are satisfactorily formed. The standard of definition is demonstrated in Figure 9.36. Sand or shell cores can be used to form internal cavities. Other thin walled applications include beer casks and most importantly, motor car wheels.

The low pressure die casting process was examined in some detail in Reference 61 in the context of die casting processes as a whole: some further studies and applications are included in References 62 and 63.

Pressure die casting

In pressure die casting the metal is injected into the die at high velocity and solidifies under externally applied pressure. These conditions give a unique capacity for the production of intricate components at relatively low cost, the process thus being of great engineering importance.

The great majority of pressure die castings are produced in two types of alloy, based on zinc and aluminium. The process was originally established

Figure 9.36 Die casting of automobile cylinder block produced by low pressure process (courtesy of Alumasc Ltd.)

using fusible alloys based on lead, tin and zinc, of which the ductile zinc alloys were later widely adopted for their useful mechanical properties and their excellent response to the conditions of the process. If the intrinsic accuracy of the process is to be fully exploited, dimensional stability of the products is obviously necessary. In some of the early zinc alloys a form of intergranular corrosion resulting from grain boundary segregation of impurities was liable to produce growth and ultimately disintegration of the castings: this led to the adoption of a high purity standard* for the zinc used in die castings and modern castings are both strong and dimensionally stable.

There was subsequent rapid growth of pressure die casting in aluminium alloys, particularly the very fluid silicon containing alloys. The size of castings also increased following the introduction of heavier machines. Smaller but significant quantities of castings are produced in magnesium alloys, in fusible alloys based on lead and tin, and in copper alloys, especially 60–40 brass. Application of the process to steels and other alloys of higher melting point, although proved to be technically feasible using refractory metal dies and die inserts, has undergone only limited development.

Pressure die casting is the subject of a wide body of literature, and specific references mentioned within the following sections include a number of the major general works. References 61 and 64–69 are in this category.

Die casting machines

The main features of a machine are the system for injecting the metal and the means of clamping the die and actuating its movements during the production and ejection of a casting. Two basic types of unit are used for injection, these being referred to as the hot and cold chamber systems. Using either system castings can be produced at rates of one every few seconds.

In the *hot chamber* or 'gooseneck' type of machine, illustrated in Figure 9.37, a reservoir of molten metal is maintained at a temperature well above the melting point. For casting, metal from the chamber is forced through the gooseneck into the die, which is held in direct contact with the exit orifice. In the machine illustrated in Figure 9.37(a) this is accomplished by raising the air pressure in the chamber, the latter being replenished by tilting to bring the mouth of the gooseneck below the level of the metal in the surrounding vessel. This direct air system is however, restricted to pressures in the region of 3.5 MPa ($\approx$500 lbf/in^2), unlike the system illustrated in Figure 9.37b, in which injection is accomplished by the operation of a plunger or piston in a submerged cylinder within the metal bath; in this case the cylinder is replenished through an inlet when the plunger is withdrawn

---

* Total impurity content (Pb, Cd, Sn, Fe, Cu) below 0.01%

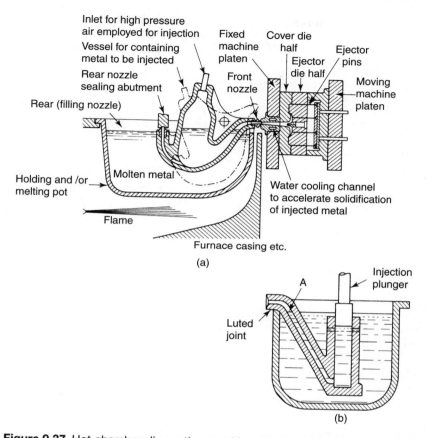

Figure 9.37 Hot chamber die casting machine: (a) direct air operated goose-neck design (from Reed[70]) (courtesy of Institute of British Foundrymen); (b) submerged plunger design (from Barton[67]) (courtesy of Institute of Materials)

after each injection. Pressures exceeding 20 MPa ($\approx$2900 lbf/in^2) can be achieved if required.

Hot chamber machines are extensively used for the casting of zinc base alloys and to a lesser degree for magnesium alloys. They are, however, unsuitable for aluminium and other alloys of higher melting point, since they give rise to contamination with iron due to prolonged contact of the molten alloy with the walls of the chamber. There is, moreover, appreciable entrainment of air in the metal during injection, although this can be diminished by close control of injection speed.

The *cold chamber* machine is thus preferred for casting alloys of higher melting point, particularly the aluminium alloys. The general principle of such machines is illustrated in Figure 9.38. Metal for a single shot is loaded into a cylindrical chamber through a pouring aperture; a piston

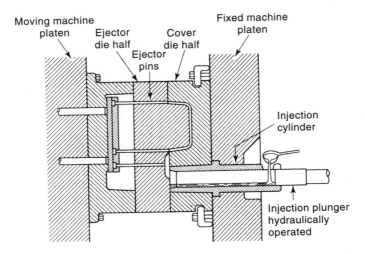

Figure 9.38 Cold chamber pressure die casting machine (from Reed[70]) (courtesy of Institute of British Foundrymen)

then forces the metal into the die, the entire operation being completed in a few seconds so that iron contamination is virtually eliminated. Using this technique much higher injection pressures in the range 70–140 MPa ($\approx$10 000–20 000 lbf/in^2) or even 200 MPa ($\approx$28 000 lbf/in^2) are feasible, enabling lower metal temperatures to be employed and greater intricacy achieved. The castings are also less prone to entrapped air and a higher standard of soundness ensues from the smaller amount of liquid and solidification shrinkage occurring within the die. The most widely used type of cold chamber machine embodies a horizontal chamber as illustrated in Figure 9.38, but a vertical system can also be used. Figure 9.39 illustrates an injection system in which the entry to the die is kept closed by a spring loaded counter plunger. The main plunger stroke compresses the spring sufficiently to force the metal into the die and the return stroke of the counter plunger shears the residual slug of metal from the gating system.

In cold chamber operations the molten metal is usually maintained at constant temperature in an adjacent holding furnace of the bale out type; transfer of successive shots to the machine chamber can be accomplished either by hand or by automatic pumping devices, including the use of vacuum suction, air displacement, and mechanical and electromagnetic pumping systems. One arrangement for metal supply utilizes a single melting furnace feeding holding units at several die casting machines through a system of heated launders. Holding furnaces may be of the electrically heated or fuel fired crucible type, including those operating on the immersion tube and immersed crucible principles.

A crucial feature of any die casting machine is the locking force available to keep the die closed against the injection pressure. The required locking

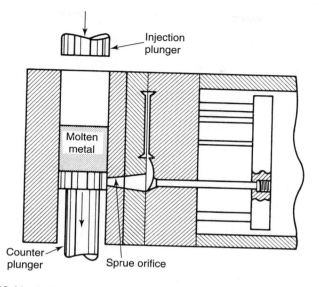

Figure 9.39 Vertical injection system with spring operated counter plunger (from Barton[67]) (courtesy of Institute of Materials)

force is determined by the projected area of the casting in the plane normal to the direction of closing; very high forces are required for some of the larger die castings now produced, several hundred tonnes being common-place. This accounts for the heavy construction of modern machines; rigidity also ensures continued accuracy of die alignment.

The basic principle of a typical machine is illustrated in Figure 9.40. The die halves are mounted on a fixed and a moving platen, the latter actuated by a hydraulic ram. The whole structure is held together by tie bars or can be made as a single cast framework for greater rigidity. Locking of the die assembly can be accomplished by direct action of the hydraulic ram, but hydraulic power is commonly combined with a mechanically acting toggle

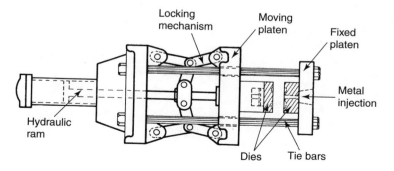

Figure 9.40 Basic principle of pressure die casting machine

link system to reduce the pressure required and to safeguard the system in the event of hydraulic failure. Wedge locking is claimed to offer a number of advantages[71].

A modern die casting machine of advanced design is illustrated in Figure 9.41. This unit is provided with a standard cylinder and pressure intensifier suitable for most liquid die casting applications, but alternative systems with larger cylinders, with or without intensifiers, can be fitted for the casting of very thin walled components, or for use in semi-solid and squeeze casting. Versions of this machine with capacities of up to 3200 tonnes locking force are produced. Modern machines are provided with automatic spray lubrication of the die assembly and plunger, and full sequence control of all the movements and operational phases. The machine shown in Figure 9.41 is provided with a comprehensive process monitoring system, enabling all the functions to be optimized to the requirements of individual components: some of these features will be the subject of further discussion.

Dies

A major step in the production of a die cast component is the engineering of the die with its numerous built-in features. In many cases the main impression is sunk in solid steel die blocks, fixed core inserts being used to form narrow projections running in a direction parallel to the stroke. For very large dies multiple construction is preferred, with die inserts carrying the casting impression and supported in heavy blocks or bolsters. To achieve maximum production of small castings, multiple impression dies are used, enabling several castings to be obtained from a single shot of metal through a common sprue. The die cavity is designed with taper to the joint face, the amount being especially critical on internal surfaces. A taper of 1 in 2000 or 0.05% would be adequate under the most favourable conditions,

Figure 9.41 530 tonne cold chamber pressure die casting machine (courtesy of Buhler Ltd.)

but minimum values in the range 0.2% to 2.0% are recommended for most purposes for outer surfaces; greater allowances are required for core surfaces and for shallow walls. Draft minimizes damage to the die surface during ejection and is thus an important factor in die life.

In production, one half of the die is fixed to the machine and the other mounted on the moving platen connected to the closing mechanism: mutual location is achieved by a system of fixed guide pins registering in holes.

Cavities in the casting are formed by metal cores which can be retracted as part of the ejection sequence. Core movement is accomplished by cam, rack and pinion or piston mechanisms actuated either by the opening stroke of the machine or as a separate movement. The die is also provided with an ejector pin or stripper plate system to extract the casting on opening. This system is normally a part of the moving die block assembly and can be designed to operate during the opening stroke of the machine. The general arrangement of the die blocks during a casting sequence is illustrated in Figure 9.42. For such a system to be effective it is necessary for the draft

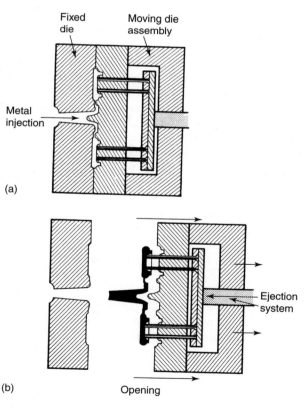

Figure 9.42 Arrangement of die blocks in casting: (a) die closed for metal shot, (b) opening stroke and ejection

taper, and the disposal of the casting impression and cores between the dies, to be designed to retain the casting in the moving die on opening, i.e. that provided with the ejection system.

Water cooling channels may be incorporated in the die blocks to prevent local overheating in the region of the gating system, where the normal temperature gradient becomes accentuated by frictional heating effects. For satisfactory working and die life, the die temperature should be maintained within an optimum operating range depending on the metal concerned; preheating is required at the beginning of a production run. Die temperatures of the order of 200°C for zinc base and 300°C for aluminium base alloys are usual. The die temperature and rate of production are subsequently interdependent. One of the important functions of die face lubrication is to prevent the surface temperature from rising much above the optimum range. The moving parts of the die assembly also require frequent lubrication: a thick suspension of graphite in oil is often used. Die design and engineering are extensively treated in the general reference works previously listed.

Die life is determined primarily by resistance to craze cracking induced by thermal shock from the rapid fluctuation of temperature at the immediate surface. Dies are usually produced in carbon or special alloy steels, using techniques of machining alone, rough machining followed by spark erosion, and casting to form by investment casting. Whilst established die steels give prolonged life in the casting of zinc, aluminium and magnesium alloys, the difficulties increase sharply with alloys of higher melting point. Die and chamber materials are the limiting factor in the extension of pressure die casting to other groups of alloys, although there is considerable production of copper alloy castings.

Using conventional die materials, die life in the casting of the principal types of alloy can be roughly indicated for comparison purposes:

Zinc base alloys:	>100 000 castings
Aluminium or magnesium based alloys:	20 000–100 000 castings
Copper base alloys:	<10 000 castings

These values can only be taken as a general guide since much longer life may be attained in individual cases. Reed, for example[70] quoted a die life exceeding 500 000 aluminium alloy castings when using die steel of composition 5% Cr, 1% Si, 1.3% Mo, 1% V, 0.37% C, hardened to 425–460 HB, indicating the importance of die steel composition and heat treatment. Some benefit to die life has also been found from light carburising treatments, designed to counteract the progressive decarburization encountered after prolonged exposure to aluminium alloy casting conditions.

For alloys of high melting point use is made of special cast irons, heat resisting tool steels, and, where long runs justify the cost incurred, nickel and cobalt base alloys. Such materials give economic die lives even in the production of complex castings in copper alloys. Refractory metal dies based on molybdenum and tungsten offer further possibilities for the most exacting applications.

Casting techniques

The design capability of the process and the characteristics of the product are derived from the rapid flow induced by the high injection pressure, die filling being accomplished in times of the order of 0.05–0.15 s. This enables castings of much higher surface area to volume ratio to be produced than are possible in other casting processes. Even when casting with little superheat, solid–liquid mixture is forced into the most intricate recesses of the die, a principle extended with the use of *rheocasting* as one variant, a process to be further considered in Chapter 10. To facilitate filling, the die cavity is provided with vents at suitable positions along the die faces. Natural venting also occurs through joints in the die assembly, both through the main parting and through core and ejector pin slides. The problem of air exhaustion has also been subject to various special techniques, including vacuum die casting, particularly in relation to the production of exceptionally thin components.

The gating systems used in conventional pressure die casting are of relatively small mass and are attached to thin rather than thick members of the casting. To obtain complete filling of the gating system, minimising air entrainment, runners should be either parallel walled or should converge slightly towards the casting: they are commonly trapezoidal in section. The gate is of smaller cross-sectional area than the runner and provides a marked constriction at the point of entry to the die cavity: gating ratios are high compared with those employed in sand casting. One system for ensuring complete filling of the runner is to place this parallel to the edge of the casting with the gate at right angles, preferably in the opposite half of the die[68], a system analogous to that sometimes used in sand casting and using the inertia of the metal in the runner to prevent its premature entry into the die cavity.

The behaviour of the molten metal during injection and its repercussions in gating and on casting quality have been the subject of many studies and have been discussed in major publications and reviews dealing with pressure die casting[61,64–67,72]. Arising from the rapid rate of metal entry, the characteristic pattern of flow within the die resembles that shown in Figure 9.43. Jet flow is established soon after the metal in the runner reaches the gate: the jet strikes the die at a point opposite to the gate and the metal then spreads laterally along the die walls and returns towards the gate in the form of a thin surface layer: this stage is of great importance in determining surface quality. Turbulent filling of the cavity then occurs in a general direction back towards the gate. In certain cases, involving very thin sections gated across the full width, the pattern may be different: due to wall friction the passage may fill progressively from the gate to the far

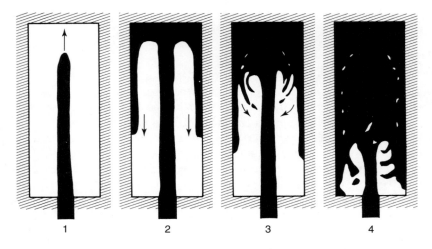

Figure 9.43 Characteristic pattern of metal flow in pressure die casting

extremities of the die. From the pattern of jet flow it is evident that for the most effective expulsion of air from the die cavity, vents need to be positioned where they will remain open throughout filling: i.e. near to the gate as well as at the further ends of the die cavity. A further source of turbulence in horizontal cold chamber casting can be the shot sleeve itself, within which waves generated ahead of the piston can form irregular flow patterns if the rate of piston travel is excessive.

Because of such conditions the presence of some dispersed porosity is regarded as a normal feature of pressure die castings and a major concern is to ensure that this remains clear of the surface. Much effort has, however, been devoted to the improvement of product quality by seeking more rigorous control over casting conditions and by fresh approaches in the design of modern die casting machines.

Early progress followed the introduction of advanced instrumentation of machine and die parameters such as locking force, injection pressure, plunger velocity, temperatures and cycle times, and the development of close controls of injection conditions, including the ability to vary the plunger speed at successive stages of the stroke. An example of data from investigation of such conditions is illustrated in Figure 9.44. A relatively slow speed of travel during filling of the gating system minimizes air entrainment and can give way to high speed during filling of the main cavity. On completion of filling, the pressure builds up to its maximum level, which is maintained during solidification to minimize porosity from shrinkage and trapped air. The effectiveness of this three-stage injection cycle was confirmed in studies of gating using transparent dies[74]. Machines incorporate intensifiers to maximize the pressure available for the final pulse.

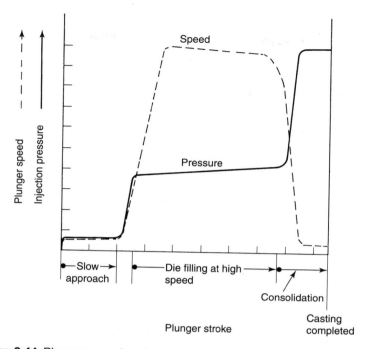

Figure 9.44 Plunger speed and pressure changes during metal injection in pressure die casting (after Gaspar and Munns[73]) (courtesy of Institute of British Foundrymen)

Although the above considerations relating to gating, die filling and inherent porosity still apply widely in die casting practice, various refinements have been introduced in the search for more radical improvements to product quality.

In vacuum die casting, evacuation of the die cavity is aimed at reducing porosity, assisting flow in thin sections and improving surface finish whilst reducing the injection pressure required. One principle is illustrated in Figure 9.45. The die is surrounded by an airtight hood which is evacuated before each injection; in other cases the vacuum may be applied more locally at the die face. In another variant, the Vacural process, the injection chamber too is included in the evacuated space, the vacuum itself being employed to fill the chamber through a suction pipe from the holding furnace below.

In the Acurad process, also developed for aluminium alloy casting and described by McLaren[75], porosity is largely eliminated, and a high degree of soundness achieved, by measures which make the metal flow and solidification phenomena correspond more closely to those occurring in other casting processes. The technique is designed to produce solid front directional filling of the die cavity, followed by directional solidification

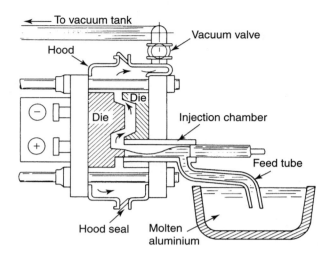

Figure 9.45 Vacuum die casting system for cold chamber machine (from Reed[70]) (courtesy of Institute of British Foundrymen)

under pressure feeding conditions. For this purpose the casting is gated into the heaviest section through a much wider gate than those normally employed. The gate is placed at the bottom so that flow occurs in an upward direction: combined with a low injection rate this produces progressive, non-turbulent filling (filling times in the range 0.30–1.0 s, together with the wider gate, give entry velocities lower by a factor of ≈100 than in conventional practice). Feeding is promoted by graduated water cooling of the die to develop favourable temperature gradients, and by the use of a secondary injection plunger of smaller area than the main plunger: this intensifies the feeding pressure on the liquid within the initially solidified shell.

A similar aim in relation to control of metal flow is pursued in the counter pressure casting technique described by Balewski and Dimov[76]; this principle is discussed in the context of upward fill systems in Chapter 10.

A further refinement, again aimed at the elimination of gas porosity, involves the introduction of a reactive gas, usually oxygen, into the die cavity, gating system and shot sleeve immediately before injection. Atmospheric nitrogen, the main source of porosity, is thus swept away, together with gases from the die lubricant, and the oxygen then reacts with the injected alloy to form a fine dispersion of solid oxide particles instead. This "gettering" action of the molten metal itself has a similar effect to that of a vacuum, producing a slower buildup of back pressure on injection, so resulting in a more rapid fill and readier flow through thin sections. Modified die lubricants are required to avoid rapid oxidation of conventional types containing hydrocarbon based oils.

This "pore-free" system, together with vacuum die casting developments and other advances, have been fully reviewed by Street in Reference 61. Further work in the field is described in Reference 77. The use of oxygen purging has been shown to give considerable increases in mechanical properties with the disappearance of the gas porosity characteristic of the normal product. This also reduces the risk of blistering in heat treatment and surface blemishes in coatings, a crucial requirement in many applications.

Apart from specific process variations of the types outlined above, the general design features of pressure die casting machines have been further developed with the availability of highly sophisticated "real time" computer control systems and rapid response mechanisms. Advanced servo-controlled injection systems have been designed to deal with conditions encountered in the casting of complex and ultra-thin components.

Two advanced machines were the subject of a detailed review by Croom[78]. Very short filling times and high velocities result in hydraulic shock at the end of the injection stroke, which can produce momentary overload sufficient to open the die against the locking force, causing flash and dimensional errors. In the Buhler SC machine all the major variables are closely monitored or regulated and the machine features a servo-controlled shot system. This provides for automatic deceleration of the metal just before the end of the injection stroke. This is achieved by means of a metal sensor mounted in the runner system to detect the advance of the liquid metal: the control system then computes the precise moment for velocity reduction immediately before filling is complete and the velocity control valve is actuated, delaying the final impulse and the danger of flash formation. Turbulence in the shot sleeve is also minimized by close control of the piston travel, to maintain smooth wave motion towards the gate and systematic expulsion of air through the cavity to the vents.

An additional feature of this machine is the provision of controlled hydraulic compression pin actuators for the enhancement of soundness in castings of complex design. The pins are pushed into selected areas of the casting surface at a late stage in the solidification process, so increasing the feed pressure in residual liquid at points susceptible to local shrinkage porosity.

A system specifically designed for the production of fully sound aluminium alloy die castings is the Ube vertical shot indirect squeeze casting machine[78], seen as combining the production characteristics of pressure die casting with the quality of squeeze cast products (q.v. Chapter 10). This employs low metal velocities (0.5 m/s maximum) with the use of large gates favouring directional solidification. The concept is illustrated in Figure 9.46: the vertically oriented injection cylinder is tilted for smooth filling, after which pressure is applied by an upward multi-phase stroke through a servo-controlled system, giving opportunity for air escape ahead of the metal flow into the die cavity. Pressures in the range 600–1000 Bar are developed and compression pin systems are again provided.

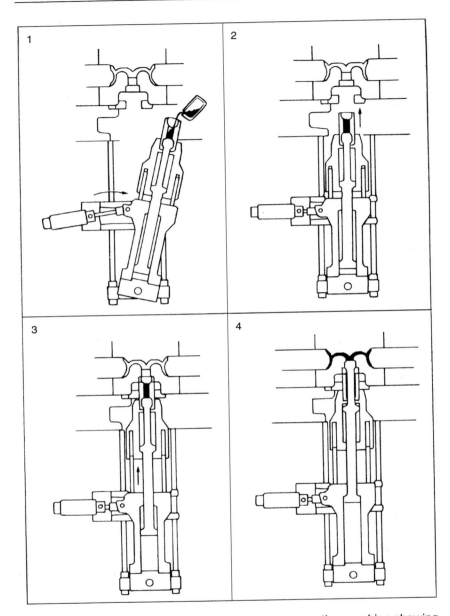

Figure 9.46 The Ube vertical shot indirect squeeze casting machine showing pouring and injection sequence (from Chadwich and Yue[79]) (courtesy of Institute of Materials)

Casting properties superior to those obtained in gravity and low pressure die castings have been achieved, together with hydraulic soundness. It is also possible to incorporate ceramic reinforcing inserts if required, given the low injection velocities employed. It is not, however, possible to produce large thin-walled castings: 4 mm is the effective minimum ruling section under these conditions.

Ejection of die castings is carried out at high temperature, both to minimize production time and to allow free contraction through the greater part of the cooling period: outputs of several castings per minute are then feasible. Castings can be removed from the ejectors either by hand or by automatic devices.

Little fettling in the normal sense is required for die castings. Removal of the gating system and trimming of flash at the joint lines are the main requirements. Flash is removed using trimming dies: these function on the shear principle and may be either hand or press operated. Barrel burnishing, buffing, light local grinding, and surface milling are typical measures used to develop the final finish. Some castings are used with no further processing, but plating, enamelling and other surface treatments are applied in many cases.

As in other fields, the production of pressure die castings has undergone a major increase in the use of automatic as well as mechanical systems at various stages of the process. Apart from the control of the machine variables to optimize injection conditions, these include the previously mentioned automatic metal transfer systems employing pumps, robot ladles and other devices for the control and metering of the supply of molten metal to the casting machine. Robots are also used for insert loading, extraction of ejected castings and for trimming operations; some of the successive steps can be linked in integrated systems controlled by microprocessors.

The rapid repetition and need for precise movements provided fertile ground for these developments, which guarantee reproducibility of conditions. Automatic and programmable die spraying are also provided. Die temperature measurements, cooling and process sequence times are further parameters linked in various forms of automatic control. The overall effect is to reduce product variability and to increase quality, as well as to facilitate maximum machine utilization.

Characteristics of die castings

Pressure die castings are noted for their high accuracy, derived from the unique rigidity of the die and rivalled only by investment castings. This accuracy is combined with smooth surfaces and with design capacity for extremely thin sections and intricate features. The high rate of metal flow makes the process particularly suitable for shapes of high surface area to volume ratio, such as covers, casings and other hollow forms. In applications of this type die castings are frequently in competition with sheet metal products.

Factors governing the dimensional accuracy of castings were reviewed in Chapter 7. In the case of die castings, the attainable tolerances depend principally on design in so far as this affects the disposition of the cavity between the die blocks and the employment of moving core members. Since most variation occurs through misalignment between die elements, very close tolerances are attainable on dimensions located wholly within a single die member: additional allowances are required for any die movements involved. The highest accuracy is attainable in the case of zinc base alloys, progressively wider tolerances being required for alloys of higher melting point. As in other cases the tolerances embody fixed and variable terms, the latter depending on the size of the dimension, and separate values are quoted for normal and critical dimensions.

All the above circumstances are reflected in published recommendations for die casting tolerances, as fully tabulated, for example, in Reference 80, based on the work of the American Die Casting Institute. The most stringent level of $\pm0.076\,$mm ($\pm.003\,$in) relates to critical dimensions below 25 mm, situated within a single die member, for a zinc base alloy casting. This contrasts with a similarly situated 300 mm non-critical dimension on a copper alloy casting, which calls for a tolerance of $\pm1.3\,$mm, larger by a factor of around 17 than the former value.

Further allowances ranging from ±0.1 to $0.9\,$mm need to be added for dimensions crossing the parting plane and/or where separate die movements are involved, so that the total tolerance could exceed $\pm3\,$mm or $0.12\,$in. in extreme cases.

Minimum wall thickness capabilities again vary according to the alloy, but depend also upon the area of the section involved: this aspect is also influenced by the position of the section in relation to the flow of metal in the die. Recommended general minima lie in the ranges 0.6–0.9 mm for small areas according to alloy, extending to 2–3 mm for large areas, although values well below these can be achieved in practice depending on the particular design and machine circumstances.

The pressure die casting process offers reasonable flexibility in design, although there are more restrictions than in sand casting. Many of the general principles discussed in Chapter 7 are applicable: the main exception is in the approach to section thickness. In much pressure die casting practice directional solidification is not the paramount aim and the general principle is to maintain section thickness close to the minimum consistent with running: reinforcing ribs and hollow sections are used for strength. A further principle is to exploit the accuracy of the process by eliminating machining, which tends in any case to expose internal porosity. Should machining be essential on particular surfaces, allowances in the range 0.25–0.64 mm (0.010–0.025 in) in excess of tolerances are normally adequate.

Whilst components of great complexity can be cast, important restrictions exist, for example with respect to some shapes of cored cavity. It is thus essential that die castings be designed for the process from the outset, since the simplification of joint lines, the shaping of cavities for direct core withdrawal and careful control of draft taper minimize production cost and ensure maximum die life. One aid to design flexibility is the soluble salt core, which offers a means of forming undercut cavities without the difficulties attending the retraction of metal corepieces.

The scope for design is extended by the use of preformed inserts. This technique can be adopted to incorporate metals or non-metals of different properties from those of the parent alloy or to eliminate separate machining operations. Threaded bushes or rods can be used to provide durable thread surfaces; bearings can be similarly included. Since no metallic bond is generated, inserts are normally shaped to provide mechanical locking into the parent casting. Provision can be made for automatic loading of inserts into the dies as an integral part of the production cycle.

Detailed treatments of die design are included in the general literature on pressure die casting; the capabilities of the process are illustrated by the examples of castings shown in Figure 9.47 to 9.51 inclusive. A valuable short review of modern progress in die casting techniques and controls, alloy developments, and their effects on product quality and capability, was that by Birch[82].

Limitations to the use of pressure die castings arise principally in respect of size and the restricted range of properties of the available alloys. The size limit is set by the capacity of machines in terms of locking force and by problems in the manufacture and handling of large dies, some of which weigh several tonnes. Locking forces exceeding 3000 tonnes ($\approx$30 MN) have already been used: based on an injection pressure of 70 MPa ($\approx$10 000 lbf/in^2) such a force would enable a casting of projected area >0.4 m^2 ($\approx$600 in^2) to be produced, although much larger castings have been manufactured using lower pressures. Amongst heavier castings produced have been light alloy cylinder heads, door frames and flywheel housings, whilst a large but lightweight component is the magnesium alloy bicycle frame.

Potential defects in die castings were treated in detail by Barton[67], by Elijah for aluminium alloys[83] and by Lewis et al in respect of porosity in zinc based alloy castings[84]; the subject is also discussed in the wider literature. For reasons previously outlined, pressure die castings are susceptible to internal porosity, especially from entrapped air, although careful control can ensure a sound, fine grained surface layer or hardware finish, vital for castings destined for surface treatments such as plating, enamelling and anodising. More radical improvement has, however, become feasible given

(a)

(b)

Figure 9.47 (a) Die cast automotive gearbox unit (b) Die cast automotive gearbox unit: interior detail (courtesy of Buhler Ltd.)

the availability of modern machines incorporating much closer controls on injection variables. With the introduction of hybrid processes embodying fundamentally different approaches to die cavity filling and solidification, the metallurgical quality gap between this and other casting processes should be progressively closed.

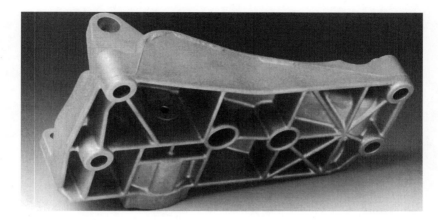

Figure 9.48 Pressure die cast support component (courtesy of Buhler Ltd.)

Figure 9.49 Die cast aluminium alloy node for fabricated frame assembly (courtesy of Buhler Ltd.)

(a)

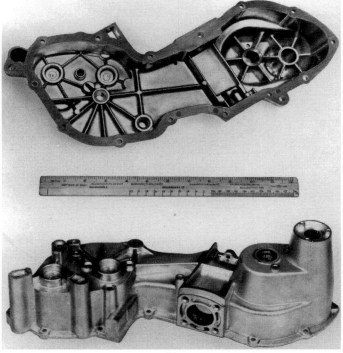

(b)

Figure 9.50 Pressure die castings in zinc base alloy: (a) twin carburetor for heavy motor vehicle (length 180 mm approx.), (b) gear case for washing machine (note ribbed design for strength and rigidity) (courtesy of Fry's Die Castings Ltd. and Dr A. Street)

Figure 9.51 Magnesium alloy die cast chassis and outer casing for portable cellular telephone (from Caton[81]) (courtesy of Foundry Trade Journal International)

Summary: economic comparison of precision casting processes

A common feature of the casting processes examined in the preceding sections is the high quality of the equipment and tooling which determine the contours and mutual fit of the mould sections. Increasingly, however, such advantages are being shared with sand castings manufactured on more conventional production lines. These have evolved away from techniques based on manual skill and requiring individual cutting of gating systems, stopping off, skeleton pattern work, patching, rubbing and sculpture, in favour of methods which exploit the tool room and the precision pattern shop. Reference has already been made in the preceding chapter to high pressure and block moulding and to the significant changes in sand formulations and practice: these greatly reduce the distinction between sand and 'precision' techniques.

The special processes, accordingly, depend for their adoption not on the elimination of gross errors or of crude shaping, but on such positive qualities as fine surface finish, capacity for intricate detail and the reproducibility afforded by exceptionally close control of production variables. A further consequence of these qualities is the high overall yield or conversion efficiency in the utilization of expensive alloys. Reduction or elimination of machining not only lowers shaping costs but reduces metal losses, melting costs, and waste of material in the form of bulky returns. The proportion of scrap castings too may be less, although this must depend on the extent to which the higher quality of the product is accompanied by higher standards for tolerances and finish.

It is thus seen that the cost element in the choice of process should be based on the finished component rather than the black casting: the qualities of precision can then be seen in true perspective against alternative methods of production.

As previously emphasized, the technical capabilities of casting processes show considerable areas of overlap. Thus the final selection of an individual process is frequently determined by the relationship between production and tooling costs for the quantity of parts required. Production cost embraces such items as materials, labour and overheads, including finishing costs, whilst the tooling cost is a fixed total irrespective of the level of production and must therefore be allocated in accordance with the number of castings. This factor in process selection can be expressed as follows:

Process 1 with a low tooling cost but high production cost will give way to Process 2 in which the converse is the case when

$$T_1 + np_1 \geq T_2 + np_2$$
$$\text{or} \quad n \geq (T_2 - T_1)/(p_1 - p_2)$$

where T = tooling cost;
p = production cost per unit;
n = number of castings required.

Even within a single process, alternative cost structures are possible, since a choice can often be made between several forms of tooling according to the quantity of castings required. Examples are the choice between wood and metal patterns in the Shaw process and between plaster, fusible metal and hard metal dies in investment casting. The selection of single or multiple impression dies and of the number of patterns in a pattern assembly are also largely governed by the quantity requirement for castings. This question of mould utilization by multiple casting and the close packing of mould impressions is a theme running through the economics of all casting processes, since no measure has a greater influence on the cost per component. The influence of multiple casting using common running and feeding systems is seen in improved casting yield (q.v. Chapter 3) and in the output of castings from a single moulding operation. This factor is present in machine moulding, shell, investment and die casting production.

The form of analysis required for process assessment was exemplified in the detailed work of the late R. G. Nicholas, with typical comparisons of fixed and variable costs for shell, investment and Shaw processes[85]. These remain broadly valid from the comparative standpoint, although they take no account of special prototyping techniques such as have since become available for the production of small numbers of components without tooling.

For a flat component some 95 mm in length used as an example, the unit production costs, made up of moulding material, labour and overheads, were shown to be lowest for shell moulding by a considerable margin.

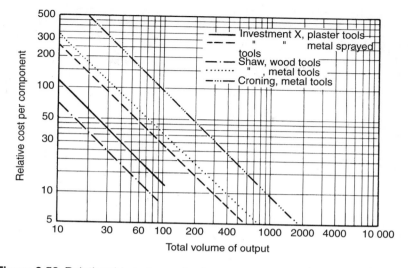

Figure 9.52 Relationship between tooling cost and output for various casting processes (after Nicholas[85]) (courtesy of Institute of British Foundrymen)

The costs of the tooling for each process are represented in Figure 9.52, which clearly demonstrates that this item is dominant for small quantities of castings but becomes insignificant at higher levels of output. A calculation was made using original cost data from the paper in the previously mentioned expression. This showed that the higher tooling costs of shell moulding relative to investment casting would have been offset by the lower production cost when the output reached 840 castings.

Figure 9.53 shows the overall effect of output on the total cost per component, which becomes asymptotic towards the production costs alone. The heavy unit consumption of expensive moulding material in the Shaw process would not, as in the case of investment casting, be reduced by the close packing of numerous castings about a common gating system. This disadvantage would, however, diminish with increasing mass of the individual castings and thus with better utilization of the mould volume.

The adaptability of the Shaw process to the use of wooden patterns also makes it particularly competitive for small quantities of castings and for development work. This advantage, even compared with investment casting using plaster tooling, is clearly shown in Figure 9.53. A calculation using the data for this case in the same formula gives the Shaw process a decisive advantage for quantity requirements below 80 castings.

The position with respect to pressure die castings is that much larger outputs would be required to reduce the die cost element to an acceptable level. Since die costs are an order of magnitude higher than tooling costs for all the other processes, minimum outputs of 5000–10 000 would normally be required to exploit the exceptionally low production costs of this process.

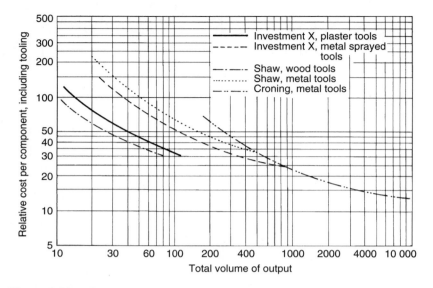

Figure 9.53 Influence of output on total cost per component for various casting processes. (after Nicholas[85]) (courtesy of Institute of British Foundrymen)

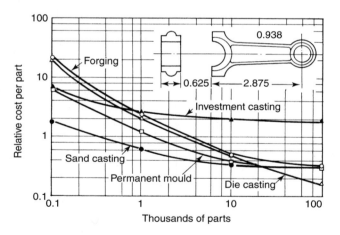

Figure 9.54 Relative costs of producing an aluminium alloy connecting rod by various processes (after Reference 69) (courtesy of American Society for Metals)

This is illustrated in the further example of the aluminium alloy component shown in Figure 9.54. It can be seen that for very large quantities none of the other casting processes, nor even forging (which was included in this comparison), would be economically competitive where pressure die casting would be technically feasible.

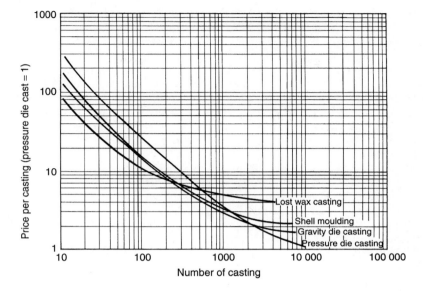

Figure 9.55 Cost comparison for production of small component by four casting processes, including tooling costs (from Davis[86]) (courtesy of Institute of British Foundrymen)

A similar point emerges from the further study illustrated in Figure 9.55, which also (as does the previous example), included gravity die or permanent mould casting in the comparison. Dies for gravity die casting are much less costly, and this process may therefore be economic for as few as 1000 castings where relatively simple shapes are required in suitable alloys. The process is, of course, applicable over a much greater weight range than the other techniques discussed: these characteristics explain its extensive use in the light alloy field.

References

1 Bailey, J. C., *Br. Foundrym.*, **53**, 187 (1960)
2 Short, A., *Br. Foundrym.*, **54**, 400 (1961)
3 Richards, G., *Br. Foundrym.*, **72**, 162 (1979)
4 Clegg, A. J., *Precision Casting Processes*, Pergamon Press, Oxford (1991)
5 Lemon, P. H. R. B. and Leserve, F. L., *Proc. Conf. Shell Moulding in Steelfoundries*, S.C.R.A.T.A., Sheffield (1968)
6 Bish, D. A., Sims, B. J. and Ashton, M. C., Proc. SCRATA 24th annual. Conf. Paper 6 (1979)
7 Dietert, H. W., *Proc. Inst. Br. Foundrym.*, **47**, A104 (1954)
8 *Machinery*, **84**, 501 (1954)
9 Rabinovich, B. B., Platonov, B. P. and Rezinkikh, F. F., *Russ. Cast. Prod.*, 294 (1961)

10 Fallows, J. and McCormack, W., *Fndry Trade J.*, **146**, 458 (1979)

11 Hall, L., *Br. Foundrym.*, **56**, 386 (1963)

12 Truckenmuller, W. C., Baker, C. R. and Bascon, G. H., *Trans. Am. Fndrym. Soc.*, **66**, 81 (1958)

13 Middleton, J. M. and Cauwood, B., *Br. Foundrym.*, **60**, 494 (1967)

14 James, D. B. and Middleton, J. M., *Br. Foundrym.*, **52**, 387 (1959)

15 Thieme, J., *Giessereitechnik*, **6**, 133 and 169 (1960), SCRATA Translation No. 301

16 Powell, R. G. and Taylor, H. F., *Trans. Am. Fndrym. Soc.*, **66**, 403 (1958)

17 Kaplish, B. K. and Protheroe, H. T., *Br. Foundrym.*, **54**, 26 (1961)

18 Grant, H. C., *Trans. Am. Fndrym. Soc.*, **67**, 641 (1959)

19 Terron, C., *Fndry Trade J.*, **12**, Feb, 74 (1987)

20 Curtis, M. W., *Foundrym.*, **83**, 580 (1990)

21 Shaw, C., *Proc. Inst. Br. Foundrym.*, **39**, B99 (1945–6)

22 Emblem, H. G., *Fndry Trade J.*, 379, 386 (1972)

23 Emblem, H. G., *J. Br. Ceram. Soc.*, **74**(6), 223 (1975)

24 Krestovnikov, A. N. *et al.*, *Russ. Cast. Prod.*, 375 (1965)

25 Matusevich, I. S. *et al.*, *Russ. Cast. Prod.*, 149 (1966)

26 Mills, D., in *Investment Casting*, Ed. Beeley, P. R. and Smart, R. F. The Institute of Materials, London (1995)

27 Brown, H., *Proc. Inst. Br. Foundrym.*, **43**, A87 (1950)

28 Geist, K. and Kerr, R. M., (a) *Proc. Inst. Br. Foundrym.*, **40**, A46 (1946–7), (b) *Fndry Trade J.*, **77**, 247, 269 and 291 (1947)

29 Bidwell, H. T., in *Investment Casting*, Ed. Beeley, P. R. and Smart, R. F., Inst. Materials, London (1995)

30 Barnett, S., *Foundry International*, **16**, 1, 226 (1993)

31 Draper, D., Cloninger, T., Hunt, J., Mersereau, T. and Hosler, M., *Trans. Am. Fndrym. Soc.*, **96**, 209 (1988)

32 Wood, R. L. and Von Ludwig, D., *Investment Castings for Engineers*, Reinhold, New York (1952)

33 Leadbetter, M. and Lindop, T. W., *Fndry Trade J.*, **119**, 761 (1965)

34 Lindop, T. W., *Fndry Trade J.*, **123**, 814 (1967)

35 Carter, T. J. and Kidd, J. N., *Br. dent. J.*, **118**(9), 383 (1965)

36 Smart, R. F. and Critchley, D. B., In *Investment Casting*, Ed. Beeley P. R. and Smart, R. F. Inst. Materials, London (1995)

37 Designers' Handbook for Investment Casting. British Investment Casting Trade Ass. Birmingham UK (1990)

38 Horton, R. A., In Metals Handbook 9th Edn. Vol. 15 Casting. ASM International, Metals Park Ohio (1988)

39 Albutt, K. J. Castings Buyer. Oct., 8 (1990)

40 Fourth report of Working Group T20, *Foundrym.*, **90**, 175 (1997)

41 Investment Casting Handbook. Investment Casting Inst. Dallas, Texas (1979)

42 Report of B.I.C.T.A. Metallurgy Committee., *Fndry Trade J.*, **119**, 15 (1965)

43 Guide to Alloy Selection. British Investment Casting Trade Ass. Birmingham UK (1980)

44 Beeley, P. R. and Driver, D., *Metals Forum*, **7**, 146 (1984)

45 Beeley, P. R. and Smart, R. F., Eds. *Investment Casting*, Inst. Materials, London (1995)

46 Alexander, J. D., *Proc. Inst. Mech. Eng.*, **C**, 75 (1983)

47 *Molding Methods and Materials*, Am. Fndrym. Soc., Des Plaines, Ill. (1963)
48 Bean, M., *U.S. Patent 2,220,703*, 5.11.1940
49 Nelson, C. D., *Metals Handbook*, 9th Edn. Vol. 15 Casting, 242, ASM International, Metals Park Ohio (1988)
50 Lubalin, I. and Christensen, R. J., *Trans. Am. Fndrym. Soc.*, **68**, 539 (1960)
51 Dunlop, A., *Metal Ind.*, **83**, 355 and 381 (1953)
52 *Machinery*, **81**, 768 (1952)
53 Greenwood, R. E., *Trans. Am. Fndrym. Soc.*, **84**, 417 (1976)
54 Fischman, J., *Foundry International*, **21**, 4, 14 (1998)
55 Day, R. A., *Fndry Trade J.*, **122**, (April 27, 1967), Diecasting Technology Supplement 9
56 Glick, W. W., *Br. Foundrym.*, **55**, 50 (1962)
57 *The Technology of Gravity Die Casting (Technologie de la Fonderie en Moules Métalliques)*, Translation B. Harocopos, Ed. T. P. Fisher, Butterworths, London (1967)
58 Woodward, R. R., *Br Foundrym.*, **58**, 148 (1965)
59 *Fndry Trade J.*, **122**, 491 (1967)
60 *Fndry Trade J.*, **109**, 163 (1960)
61 Street, A., The Diecasting Book. Portcullis Press, Redhill (1977)
62 Nguyen, T. T. *et al.*, *Trans. Am. Fndrym. Soc.*, **105**, 833 (1997)
63 Müller, W. and Feicus, F. J., *Trans. Am. Fndrym. Soc.*, **104**, 1111 (1996)
64 Kaye, A. and Street, A. C., *Die Casting Metallurgy*, Butterworths, Oxford (1982)
65 Upton, B., *Pressure Die Casting Part 1*, Pergamon Press, Oxford (1982)
66 Allsop, D. F. and Kennedy, D., *Pressure Die Casting Part 2. The Technology of the Casting and the Die*, Pergamon Press, Oxford (1983)
67 Barton, H. K., *Metall. Rev.*, **9**, No. 91, 305 (1964)
68 Barton, H. K., *The Die Casting Process*, Odhams press, London (1956)
69 *Casting Design Handbook*, Am. Soc. Metals, Metals Park, Ohio, 1962
70 Reed, J. C., *Br. Foundrym.*, **53**, 178 (1960)
71 *Metal Ind.*, **103**, 190 (1963)
72 Draper, A. B., *Trans. Am. Fndrym. Soc.*, **59**, 822 (1961)
73 Gasper, E. and Munns, M. G., *Br. Foundrym.*, **54**, 180 (1961)
74 Smith, W. E., *Trans. Am. Fndrym. Soc.*, **71**, 325 (1963)
75 McLaren, J. L., *Die Casting Engineer*, **10**(5), 8 (1966)
76 Balewski, A. T. and Dimov, T., *Br. Foundrym.*, **58**, 280 (1965)
77 Bialobrzeski, A., *Cast Metals*, **3**, 3, 141 (1990)
78 Croom, D. E., *Foundrym.*, **89**, 440 (1996)
79 Chadwick, G. A. and Yue, T. M., *Metals and Materials*, **5**, 1, 6 (1989)
80 Sully, D. L., *Metals Handbook*, 9th Edn. Volume 15, 286, ASM International Metals Park, Ohio (1988)
81 Caton, P. D., *Fndry Trade J. Int.*, March, 14 (1992)
82 Birch, J. M., *Mat. Sci. Tech.*, **4**, 1, 218 (1988)
83 Elijah, E. J., *Trans. Am. Fndrym. Soc.*, **69**, 328 (1961)
84 Lewis, G. P., Craw, D. A. and Bell, R. C., *Trans. Am. Fndrym. Soc.*, **69**, 537 (1961)
85 Nicholas, R. G., *Br. Foundrym.*, **51**, 541 (1958)
86 Davis, F. H., *Br. Foundrym.*, **57**, 166 (1964)

10

Production Techniques 3
Further casting techniques

Centrifugal casting

The essential feature of centrifugal casting is the introduction of molten metal into a mould which is rotated during solidification of the casting. The centrifugal force can thus play a part in shaping and in feeding according to the variation of the process employed.

The principle of centrifugal casting is long established, dating originally from a patent taken out by A. G. Eckhardt in 1809. Following early development during the nineteenth century, the process began after 1920 to be used for the manufacture of cast iron pipes on a large scale and has since been extended to a much wider range of shapes and alloys.

The centrifugal force produced by rotation is large compared with normal hydrostatic forces and is utilized in two ways. The first of these is seen in pouring, where the force can be used to distribute liquid metal over the outer surfaces of a mould. This provides a means of forming hollow cylinders and other annular shapes. The second is the development of high pressure in the casting during freezing. This, in conjunction with directional solidification, assists feeding and accelerates the separation of non-metallic inclusions and precipitated gases. The advantages of the process are therefore twofold: suitability for casting cylindrical forms and high metallurgical quality of the product.

The casting of a plain pipe or tube is accomplished by rotation of a mould about its own axis, the bore shape being produced by centrifugal force alone and the wall thickness determined by the volume of metal introduced. This practice is widely referred to as *true centrifugal casting*. In the case of a component of varying internal diameter or irregular wall thickness a central core may be used to form the internal contours, feeder heads then being introduced to compensate for solidification shrinkage. A further step away from the original concept is the spacing of separate shaped castings about a central downrunner which forms the axis of rotation. These variations are referred to respectively as *semicentrifugal casting* and *centrifuging* or *pressure casting*: in both cases, since the castings are shaped wholly by the mould and cores, centrifugal force is used primarily as a source of pressure for feeding.

The rotational axis may be horizontal, vertical or inclined and important variations exist in respect of mould material and the method of introduction of the molten metal.

Fundamental principles

Centrifugal force

The centrifugal force acting upon a rotating body is proportional to the radius of rotation and to the square of the velocity:

$$F_c = mr\omega^2 = \frac{mv^2}{r}$$

where F_c = centrifugal force (N; pdl)
$\quad m$ = mass $\qquad$ (kg; lb)
$\quad r$ = radius $\qquad$ (m; ft)
$\quad \omega$ = angular velocity (rad/s)
$\quad v$ = peripheral speed (m/s; ft/s)

The gravitational force on the same mass would be given by:

$$F_g = mg$$

where g = acceleration due to gravity (m/s²; ft/s²).
Hence the factor by which the normal force of gravity is multiplied during rotation is given by:

$$G \text{ factor} = \frac{F_c}{F_g} = \frac{r\omega^2}{g}$$

Expressed in the more convenient speed units of revolutions per minute, N, the expression becomes:

$$G \text{ factor} = \frac{r}{g} \left(\frac{\pi}{30}\right)^2 N^2 = \frac{0.011rN^2}{g}$$

More usefully,

$$N = \left(\frac{G \text{ factor} \times g}{0.011r}\right)^{1/2} = 29.9 \left(\frac{G \text{ factor}}{r}\right)^{1/2}$$

$$= 42.3 \left(\frac{G \text{ factor}}{D}\right)^{1/2}$$

where D = rotational diameter (m).

Alternatively, $N = 265 \left(\dfrac{G \text{ factor}}{D}\right)^{1/2}$

where D = rotational diameter (in).

These relationships between rotational speed, diameter and centrifugal force are illustrated graphically in Figure 10.1: this and similar charts or nomograms are normally used to select the speed in accordance with the magnitude of centrifugal force required.

There is no standard criterion for selection of the required force. In true or open bore casting, circumferential velocity is imparted from mould to metal by frictional forces at the mould surface and within the liquid. In horizontal axis casting, the metal entering the mould must rapidly acquire sufficient velocity to prevent instability and 'raining' as it passes over the upper half of its circular path: because of slip, the generation of the necessary minimum force of $1G$ in the metal requires a much greater peripheral mould velocity than would be the case if metal and mould were moving together. One investigation placed the minimum limit in the region 3–4.5G[1] but much greater force is required in practice for full advantage to be taken of the

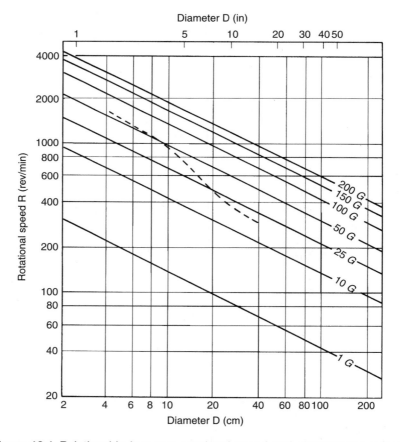

Figure 10.1 Relationship between rotational speed and diameter for various magnitudes of centrifugal force

process. Based on practical observations, Cumberland[2] reported a range of minimum speeds required to avoid ejection of metal. These are represented by the dotted line superimposed on Figure 10.1 and approximate more closely to a constant peripheral velocity of 5 m/s ($\approx$1000 ft/min) than to a fixed magnitude of centrifugal force: the required force diminishes with increasing diameter.

In vertical axis casting the permissible minimum speed is established by the tendency to form a parabolic bore owing to the gravitational component of the force acting on the metal. In this case the effect is defined by the expression:

$$\omega^2 = \frac{2gL}{r_t^2 - r_b^2}$$

where L = axial length   (m; ft)
    r_t = top radius of bore   (m; ft)
    r_b = bottom radius of bore (m; ft)

Hence $N = \dfrac{30(2g)^{1/2}}{\pi} \cdot \left(\dfrac{L}{r_t^2 - r_b^2}\right)^{1/2} = 42.3 \left(\dfrac{L}{r_t^2 - r_b^2}\right)^{1/2}$

where the linear dimensions are expressed in metres.

Alternatively, $N = 265 \left(\dfrac{L}{r_t^2 - r_b^2}\right)^{1/2}$

where the linear dimensions are expressed in inches.

This is roughly in agreement with Carrington[3], who developed further and similar expressions for rotation about an inclined axis.

Other relevant factors in speed selection are the mechanical capabilities of the equipment and the hoop stresses generated in the solidifying casting, which govern the maximum safe operating speed.

Although centrifugal forces exceeding 200G are attained in some cases, most practice is empirically based within the range 10–150G, the highest values being used for open bore cylindrical components of small diameter and the lowest for semicentrifugal and pressure castings. Speeds generating forces of 60–80G are most commonly quoted for true centrifugal castings. As previously emphasized, however, the optimum value of centrifugal force diminishes with increasing diameter.

Cumberland quoted values of 33G for a wide range of plain vertical axis castings and 15G for semi-centrifugal castings in sand moulds, for which lower speeds suffice since there is no longer dependence on centrifugal force to shape the casting. Thornton[4] gave values of 50–100G for die cast and 25–50G for sand cast pots and shaped castings.

Solidification and feeding

The effectiveness of centrifugal force in promoting a high standard of soundness and metallurgical quality depends above all on achieving a controlled pattern of solidification, this being governed by the process used and by the shape and dimensions of the casting. High feeding pressure is no substitute for directional freezing, which remains a primary aim of casting technique.

Considering firstly the casting of a plain cylinder, conditions can be seen to be highly favourable to directional solidification owing to the marked radial temperature gradient extending from the mould wall. Under these conditions the central mass of liquid metal, under high pressure, has ready access to the zone of crystallization and fulfils the function of the feeder head used in static casting. The steepest gradients and the best conditions of all occur in the outermost zone of the casting, especially when a metal mould is employed. Another important factor is the length to diameter ratio of the casting, a high ratio minimizing heat losses from the bore through radiation and convection. Under these conditions, heat is dissipated almost entirely through the mould wall and freezing is virtually unidirectional until the casting is completely solid: the wall of the casting is then sound throughout. This ideal is closely approached when producing long, thin walled tubes in metal dies and an effective casting yield of 100% is sometimes attainable: this type of product can frequently be used in the as-cast form.

As the wall thickness increases or as the ratio of length to diameter is reduced, however, radial temperature gradients become less pronounced. Heat loss from the bore surface eventually attains a level at which the temperature gradient is locally reversed, initiating some freezing from this surface. Under these conditions a zone of internal porosity is associated with the last liquid to freeze, being normally confined within a band of metal close to the bore. To achieve a wholly sound end product this porosity needs to be removed by machining, an operation analogous to feeder head removal in static casting. For this purpose an appropriate allowance must be made on the cast dimensions.

The magnitude of the allowance depends upon the alloy concerned and the shape of the casting. Cumberland[2] demonstrated how empirical observations, in this case for castings in metal dies, can be collected and rationalized into a system for determining feeding allowances on the basis of the shape factor (Figure 10.2). Short pots can be seen to require much greater allowances than long tubes.

When the length to diameter ratio exceeds 15, machining may be unnecessary for many applications. The shape factor is, however, less critical for thin walled castings which solidify under the steep temperature gradients induced by the initial cooling effect of the mould: such components are often used in the as-cast state even when the ratio is below 15.

Thick and heavily shaped components in which the mould cavity is wholly formed by refractories present a different problem. In semicentrifugal castings and in shaped castings spun about an external axis there is

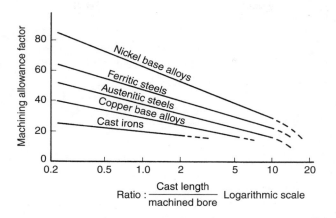

Figure 10.2 Relationship of internal machining allowance to casting geometry. Allowance factor is depth of cut expressed as percentage of thickness from cast o.d. to finished bore (from Cumberland[2]) (courtesy of Institute of British Foundrymen)

no predisposition to directional freezing. If full advantage is to be taken of centrifugal force for feeding, therefore, the mould cavity must be shaped or selectively chilled in accordance with the normal principles of directional solidification, to produce freezing from the periphery towards the centre. Feeder heads are positioned in the latter region to exploit the centrifugal force and need to be sufficiently large to induce favourable temperature gradients throughout freezing. The feeding pressure P arising from rotation is given by the expression

$$P = \frac{\rho\omega^2(r_1^2 - r_2^2)}{2}$$

where ρ = density;
ω = angular velocity;
r_1 = radius of rotation at the given point in the liquid;
r_2 = bore radius.

Centrifugal separation of inclusions

A further function of centrifugal force is seen in the tendency for non-metallic inclusions to segregate towards the axis of rotation. In static casting the rate of separation, expressed in Stokes' Law and previously discussed in Chapter 5, depends on the net gravitational force acting on each suspended particle. In centrifugal casting this force is increased by the factor $r\omega^2/g$, the previously mentioned G factor for which values of up to 200 may be attained in practice. In the casting of annular components this effect, essentially that of a centrifuge, is combined with marked directional freezing and with a short path to the free surface in the bore, so that a high standard of freedom from inclusions is achieved.

Similarly, owing to the pressure gradient within the casting, nucleation of dissolved gases to form bubbles will occur if at all in the bore region, where the gas can escape readily from the casting.

Production techniques

Variations in practice concern the axis of rotation, type of mould and manner of introduction of molten metal: these features govern the design of machines and equipment used. Practice will be considered, however, primarily in relation to the type of product.

In general, horizontal axis machines are preferred for castings of high length to diameter ratio, whilst castings of diameter exceeding the length are more commonly spun about a vertical axis, although certain exceptions exist in both cases.

Mould material is determined partly by shape and partly by the number of castings required. For simple shapes, metal moulds are usually preferred on economic grounds as well as for their effect in inducing directional freezing. These can either be wholly of steel or of cast iron supported in a steel shell. They are used with various types of coating for protection and ease of stripping: coatings range from refractory slurries to dry powdered dressings containing graphite, ferrosilicon and other substances, which serve both to insulate and inoculate. For more complex shapes or for parts required in small quantities, refractory moulds are normally used: these may be based on high strength conventional moulding mixtures or on investments. Graphite moulds have also been used for special applications.

True centrifugal castings

The widest application of centrifugal casting is for the production of components which are essentially cylindrical and therefore suitable for casting by the open bore or 'true' process. Such castings are utilized in either of two ways. Many tubes are used in the as-cast state. These can embody limited taper or simple external features, although the latter must, when casting in metal dies, permit extraction of the casting. Sand linings and core inserts give added scope for the shaping of flanges, ribs and apertures.

True centrifugal casting is also used for the manufacture of plain cylindrical blanks or 'pots' for the production of rings, bushes and other annular components by machining. Pots are usually cast in metal dies, the full range of required sizes being covered by maintaining a stepped die series enabling any combination of machined dimensions to be achieved without excessive stock removal. The cast dimensions are designed to include the previously mentioned shrinkage allowance in the bore; a typical centrifugally cast pot is illustrated in Figure 10.3.

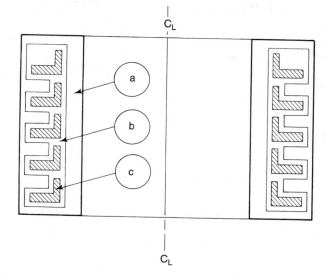

Figure 10.3 Layout of centrifugally cast pot for multiple production of machined annular components: (a) cast pot, (b) rough machined gashed pot as supplied to purchaser, (c) finished component

Long tube production. Most long tubes for either of the above purposes are produced on horizontal axis machines, using metal or sand lined moulds. One of the main problems in the production of long tubes is to achieve uniform distribution of metal over the large surface area of mould. Pouring conditions should produce progressive rather than sporadic spread over the surface: laps, shot and prematurely chilled stringers of metal are then avoided.

With fluid alloys these conditions can be achieved by progressive deposition of metal along the length of the rotating mould, a principle used in the De Lavaud process for the production of socketed cast iron pipes in metal moulds. In this process, illustrated in Figure 10.4, a long pouring spout, supplied from an automatic ladle, is initially inserted to the far extremity

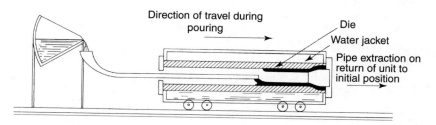

Figure 10.4 Essentials of de Lavaud pipe casting machine

of the mould, a forged steel die enclosed in a water cooling jacket. As pouring proceeds, the rotating mould is withdrawn over the spout so that the metal is laid progressively along the length of the mould wall, control being achieved by synchronizing the rates of pouring, mould travel and mould rotation. The casting is extracted as the mould returns to its original position and its accuracy is checked by weighing. High rates of output are achieved, each machine producing a casting every two minutes.

A more modern plant using the De Lavaud process embodies a high degree of automation to produce an entire range of ductile iron pipes, of diameters up to 1.6 m, on a single casting machine, all the functions being computer controlled. After extraction the pipes are tilted to the vertical for indexation through a rotary heat treatment furnace, before return to the horizontal for the extensive finishing sequence, beginning with eddy current testing for matrix assessment and concluding with cement lining and bitumen coating. The plant was described in Reference 5 and its operation was the subject of a comprehensive review by Else[6]. A vast tonnage of centifugally cast pipes is accounted for by the De Lavaud process throughout the world: a leading plant in the USA[7] itself accounts for some 1500 tonnes per day of such pipes and ancillary fittings.

Perhaps the most rapid output of centrifugal castings has been achieved in the production of rainwater pipes, using a simplified variation of pipe casting, involving circulation of several moulds through a number of stations for coating, casting and extraction: rates of over 100 castings per hour were reported[8].

With the slower rates of cooling obtained in sand moulds, fluid alloys can be poured without the retractable spout system. The Moore system for production of cast iron pipes employs a rammed and dried sand lining in conjunction with end pouring: the process is especially suitable for large pipes required in small quantities. Sand linings are also applied in the production of very heavy bronze cylinders used in the papermaking industry.

For production of tubular castings in a wider range of alloys, metal distribution is achieved by use of a metal die provided with a refractory and insulating coating of rough texture[9]. The coating is applied to the heated bore surface of the rotating die through a spray head at the end of a lance mounted on a reciprocating carriage: the coating shows a controlled roughness when dried (see Figure 10.5).

The prepared die is transferred to a free roller casting machine of the type illustrated in Figure 10.6, where pouring is carried out through a short spout which deposits the metal just inside the end cover of the die. The frictional resistance of the die coating restrains longitudinal flow whilst assisting circumferential pickup. A thick belt of liquid metal is thus accumulated near the pouring spout and further distribution takes place by the orderly advance of a steep wave front along the die. Premature freezing is avoided and a rough but uniform finish obtained on the outer surface of the casting: the

(a)

(b)

Figure 10.5 Coating of die for production of long tubes: (a) spray lance, (b) surface texture of coating (courtesy of Sheepbridge Alloy Castings Ltd.)

Figure 10.6 Free roller centrifugal casting machine (courtesy of Sheepbridge Alloy Castings Ltd.)

roughness can be removed if required by a small machining cut, although many such tubes are used in the as-cast condition.

This process is suitable both for tubes and for long pots. Since heat loss from the bore is lower than in the case of short pots, a smaller machining allowance suffices to clean up to sound metal and a higher yield can be achieved. Not all long tubes are cast on the horizontal axis, however: an example of a vertical axis operation will be referred to later.

Short pot production. Centrifugally cast pots of shorter length can be produced on either horizontal or vertical axis machines, the die assembly being mounted at the end of the drive shaft: this may carry an axially mounted motor or may be belt driven from a separate motor shaft. One type of horizontal axis machine is illustrated in Figure 10.7: this principle is used in cylinder liner production.

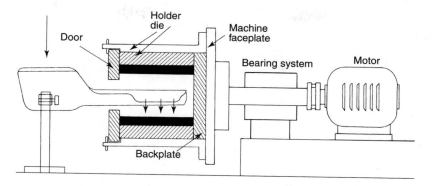

Figure 10.7 Essentials of horizontal axis centrifugal casting machine and die assembly for production of short pots (schematic)

The mould usually consists of an interchangeable metal die supported inside a steel holder; a suitable material for the die itself is haematite cast iron, which shows good resistance to thermal shock and distortion.

Various types of pouring device are used. Horizontal axis machines are commonly fed through an adjustable spout projecting through the end cover of the mould assembly to deposit the metal directly on the die wall. A variation of this practice uses a container or trough provided with a wide pouring lip or weir: on tilting, a ribbon pouring effect gives smooth flow over an appreciable proportion of the die length (Figure 10.8). In pouring horizontal axis castings the metal stream is generally directed against the downward moving side of the mould, allowing maximum opportunity for the metal to acquire angular momentum before passing over the upper part of its path.

For pots of large diameter vertical axis casting is normally preferred. In this case the metal can be introduced through a spout discharging tangentially within the mould cavity (Figure 10.9), or can be poured directly, the stream striking the base before being distributed to the periphery. The former arrangement is attractive in that the downward velocity of the metal can be used to provide initial impetus in the circumferential direction. Ignoring frictional losses the exit velocity attained in such a system would be given by:

$$V^2 = 2gh$$

where V is the velocity at the base of the tube and h the effective height of the system.

Thus, for a height of 1 m the velocity at the discharge end of the system would be approximately 4.4 m/s: after allowing for losses the speed would still represent a useful fraction of the typical peripheral die speed, of the order of 2.5–10 m/s.

A further pouring system, due to Pearce and described by McIntyre[10], was designed to reduce turbulence in vertical axis casting and to provide an

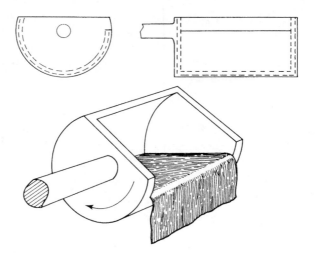

Figure 10.8 Weir type trough for ribbon pouring in horizontal axis centrifugal casting

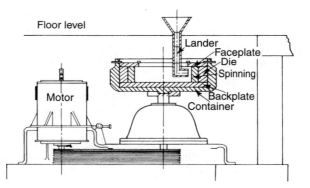

Figure 10.9 Vertical axis centrifugal casting machine (from Thornton[4]) (courtesy of Institute of British Foundrymen)

ideal pattern of mould filling. In this system the charge of molten metal to make a casting is held in a well in the base of the stationary mould, which is then rotated until the metal rises and travels smoothly to the periphery; the system is not known to have been adopted on any significant scale.

Apart from the production of tubes and plain pots, true centrifugal casting can be applied to certain components of more complex shape where these are essentially symmetrical about a central axis. If a metal die is used the external profile must change progressively to allow extraction of the casting, although this problem has been surmounted by using two-piece dies split down the longitudinal axis[11]. Since heavy machining would be required in shaping most types of internal contour from a plain bore, this system is suitable for relatively few components.

Semicentrifugal and pressure castings

Heavily shaped components are normally cast in enclosed refractory moulds in which both internal and external surfaces are shaped by the refractory. In these circumstances provision must be made for directional solidification towards a central source of feed metal, using measures similar to those employed in static casting, for example padding and selective chilling.

For castings of small internal diameter a single central feeder head, continuously connected to the casting through a progressively tapered web, is usually employed (Figure 10.10a). Castings of larger diameter require separate segmental or ring heads; depending on dimensions these may communicate with the sprue either directly or through runners analogous to those used in static casting (Figure 10.10b). For components such as spoked wheels the spokes themselves serve as runners dispensing metal to the outer rim: in such cases the casting can be progressively dimensioned for feeding from a head placed at the central hub.

Components too small or irregularly shaped for centrifugal casting about their own axis are pressure cast by spacing about a central sprue. The latter may fulfil the function of a common feeder head or individual heads may be provided for each casting using a large radius of rotation.

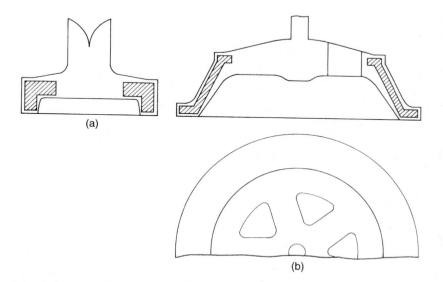

Figure 10.10 Typical feeder head arrangements in semi-centrifugal casting: (a) single central head continuously attached to casting, (b) segment heads as applied to castings of larger diameter

Both semicentrifugal and pressure casting are normally carried out on vertical axis machines to facilitate mould assembly. A typical arrangement, illustrated in Figure 10.11, uses a steel flask consisting of a baseplate, rings and top cover. The mould is either rammed directly in the flask or is built up from high strength cores coated with refractory wash. The complete assembly is attached to the machine turntable by means of a fixture incorporated in the baseplate and is poured down the central sprue.

Since centrifugal feeding requires radial rather than superimposed feeder heads, shallow castings can frequently be stacked for casting from a common sprue, giving a high packing density: high production rates and casting yields are thus attained.

The pressure casting principle is also applied in the precision and dental casting field by rotation of individual investment moulds about an external axis on machines of special design.

Process variables and casting quality

Once the particular process has been established, the main variables controlling casting quality are speed of rotation, pouring temperature, pouring speed and mould temperature. These were examined by Northcott and MacLean[12], Northcott and Dickin[13] and other investigators, principally in the field of true centrifugal casting; their individual significance can be briefly summarized.

Speed of rotation. The main factors influencing speed selection were discussed in relation to the fundamentals of the process. The governing

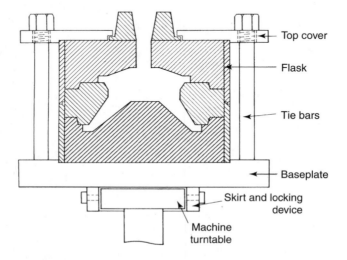

Figure 10.11 Refractory mould assembly for vertical axis centrifugal casting of shaped components

factor in true centrifugal casting is retention of the bore shape against gravity whilst avoiding longitudinal tearing through excessive hoop stress; in semicentrifugal and pressure casting, feeding pressure is the main criterion. Rotational speed also exerts an influence upon structure, the most common effect of increased speed being to promote refinement, although this can also rise from turbulence induced by instability of the liquid mass at very low speeds. On balance, to secure maximum benefit from centrifugal casting, it is logical to use the highest speed consistent with the avoidance of tearing.

Pouring temperature. Pouring temperature exerts a major influence on the mode of solidification and needs to be determined partly in relation to the type of structure required. Low temperatures are associated with maximum grain refinement and with equiaxed structures, whilst higher temperatures promote columnar growth in many alloys. However, practical considerations limit the available range: the pouring temperature must be sufficiently high to ensure satisfactory metal flow and freedom from cold laps whilst avoiding coarse structures and increased risk of hot tearing due to excessive superheat.

Pouring speed. This is governed primarily by the need to finish casting before the metal becomes sluggish, although too high a rate can cause excessive turbulence and ejection. In practice slow pouring offers a number of advantages: directional solidification and feeding are promoted, whilst the slow development of full centrifugal pressure on the outer solidified skin reduces the risk of tearing.

Mould temperature. The use of a metal die produces marked refinement when compared with sand casting, but mould temperature is only of secondary importance in relation to structure. Its principal significance lies in the degree of expansion of the die with preheating: expansion diminishes the risk of tearing in true centrifugal castings for reasons to be discussed in the ensuing section.

Defects

Centrifugal castings are subject to most types of defect encountered in static castings, although there is less susceptibility to haphazard internal shrinkage, gas porosity and non-metallic inclusions. Such defects as do occur tend to be at the outer or inner surfaces, giving a strong probability of detection without recourse to radiography or ultrasonic testing. This point must be qualified in the case of semicentrifugal and pressure castings, in which conditions are nearer to those in static casting with quality more dependent on individual technique.

Particular hazards in centrifugal casting are laps and tears, whilst the phenomenon of banding is peculiar to the process.

Laps arising from imperfect flow are probably the commonest cause of rejection. They are most prone to occur in alloys of low fluidity, especially those carrying strong oxide surface films. Apart from cold metal, turbulence and excessively slow or uneven pouring are the principal causes.

Hot tears are encountered primarily in true centrifugal castings, for which the highest rotational speeds are employed. Longitudinal tears occur when contraction of the casting, combined with expansion of the mould, generates hoop stresses exceeding the cohesive strength of the metal at temperatures in the solidus region. The stress required to produce tearing is of the order of 0.5 MPa, depending on the alloy: since hoop stress is proportional to the squares of the angular velocity and of the radius of rotation, excessive speed is the factor most directly concerned as a production variable. Pouring temperature, pouring speed and mould temperature are also involved, since they determine the temperature distribution and development of cohesion in the casting in relation to the rate of build-up of hydraulic pressure and hoop stresses in the solid shell.

The stages through which the properties of an alloy pass on cooling from the liquid were reviewed in Chapter 5. Consideration of these stages suggests that tearing may be best avoided by rapid cooling of the outer region of the casting below the brittle temperature range just above the solidus. The metal in this region can then sustain stresses whilst the remainder of the section passes through the critical stage. The best conditions are obtained from a combination of a moderate pouring temperature and slow pouring: the latter serves the dual purpose of maximizing the temperature gradient and delaying the development of hydraulic pressure relative to the growing strength of the outer shell. Tears will be more prone to occur, by contrast, under shallow temperature gradients which produce the brittle condition across much of the radial section at one time.

When casting into metal moulds the other major factor in tearing is the rate of die expansion. This can be reduced, and loss of support for the casting delayed, by increasing the thickness and thermal capacity of the die and by preheating to reduce its further expansion on casting. Temperatures of 300°C and upwards are employed for this purpose.

Circumferential tears in tubular castings and tears in semicentrifugal and pressure castings can result from contraction hindrance by the mould. This is a similar problem to that encountered in static casting and discussed in Chapter 5.

Segregation. Centrifugal castings are subject to various forms of segregation, especially in heavy sections. Centrifugal separation of suspended crystals or insoluble phases can occur within the liquid in either direction, creating a tendency for the less dense constituents to be transported towards the centre. At a more advanced stage of freezing low melting point liquid can flow outwards through passages between the rigid network of interlocking grains.

Thus the particular pattern of segregation in a centrifugal casting depends primarily on the mode of freezing. Assuming for example that differential freezing were to produce rejection of a heavier solute element, this would, under perfect directional freezing, segregate to the bore notwithstanding its greater density. Dispersed crystallization, on the other hand, provides opportunity for relative gravitational movement of solid and liquid: in the same alloy, therefore, the less dense crystals would be centrifuged towards the bore with the opposite effect to that previously encountered.

Examples of segregation include tendencies for phosphide eutectic to gravitate outwards in cast iron[10], for copper enrichment to occur towards the centre in aluminium–copper alloys[13] and for lead to segregate to the outside of leaded gunmetal castings[14].

Banding, a condition encountered in horizontally cast, thick walled cylinders, is a form of structural irregularity in which concentric zones of dissimilar microstructure are associated with segregation of alloy constituents and impurities. Various theories have been advanced as to the cause. These range from the influence of vibration in producing independently nucleated growth bands under conditions of constitutional supercooling, leading to entrapment of solute rich liquid[12,13], to disturbances created by irregular flow of liquid during formation of the casting[2]: in the latter case banding is seen as a result of segregation occurring in the freezing of successively deposited layers of liquid. The condition has been found to occur when some critical level of rotational speed is attained, but has also been associated with very low speeds such as produce sporadic surging of molten metal: both the above mechanisms may therefore be involved.

A further suggested cause of banded structures is freezing of metal at the bore and movement of the solidified layer into the casting under centrifugal force due to its higher density[15]: this does not however explain the association of banding mainly with the horizontal axis process.

Characteristics of centrifugal castings

Structure and properties

The main quality characteristic of centrifugally cast material is the high standard of soundness arising from the conditions of feeding. This factor is predominant in the improvement of properties relative to those of statically cast material, there being little difference when the latter is perfectly fed. To this advantage may be added a degree of structural refinement, affecting grain size and the distribution of microconstituents. The extent of this depends on the particular process. The most important contribution to refinement is rapid cooling in metal moulds, but other factors include physical disturbance in the liquid in pouring and rotation and the ability to achieve satisfactory metal flow using lower pouring temperatures than would be necessary in the absence of centrifugal pressure. Refinement is greatest in true centrifugal castings made in metal dies, whilst pressure castings show little structural difference from static castings of similar shape: the same may be said of the degree of freedom from non-metallic inclusions and random defects.

The macrostructure is subject to similar influences to those governing the structure of static castings, the important factors being alloy constitution, the temperature gradients and cooling rates induced by the thermal properties of metal and mould, and conditions for independent crystallization, in this case additionally affected by the motion of the casting.

Alloys undergoing dendritic crystallization are characterized by regions of columnar and equiaxed growth. Factors giving rise to these forms were examined in Chapter 2. The additional factor in the case of centrifugal casting is the relative movement of liquid by slip during acceleration to the speed of the mould. This has been held in some cases to promote columnar growth by disturbance of the growth barrier of solute rich liquid at the interface[16]. The overall effect of motion on structure is, however, complex, since vibration, diminution of thermal gradients in the liquid, and the possible fragmentation of dendrites can also induce the nucleation of equiaxed grains. A further effect of motion is a tendency for columnar grains to be inclined in the direction of rotation, evidently due to the movement of undercooled liquid towards the dendrite probes. The contribution of motion to banding has been referred to in a preceding section.

The structures encountered in a large number of individual alloys, particularly the zones of columnar and equiaxed grains occurring under a wide range of conditions, were described and explained by Northcott and his colleagues[12–14,17]. However, due to the interaction of several mechanisms it has not so far proved feasible to formulate general rules defining the influence of the main casting variables upon grain structure. In practice the most consistent influence is that of a low pouring temperature in producing grain refinement and equiaxed structures, whilst somewhat higher temperatures tend to promote columnar grains by suppressing nucleation and increasing the radial temperature gradient towards an optimum level. This influence of an increase in pouring temperature in suppressing nucleation has been identified by Ebisu[18] as the dominant factor in the production of all-columnar structures: preheat and mould material were seen as being relatively unimportant. The complexity of structure control in centrifugal castings is indicated in Table 10.1, where the various possible phenomena associated with the major variables are related to structural tendencies.

In cast irons the effect of casting conditions on the structure of true centrifugal castings is seen principally in differences in the numbers and sizes of graphite nodules or flake forms associated with cooling rate. In a metal spun ductile iron pipe, therefore, the outermost zone shows fine graphite nodules with extensive carbides, the proportion of graphite increasing and the nodules coarsening towards the bore. After heat treatment the

Table 10.1 Influences of casting variables upon structure

Variable	Effect	Structural tendency
Increasing pouring temperature	Decreases freezing rate and inhibits nucleation	Columnar; coarse
	Increases temperature gradient to optimum	Columnar
	Decreases temperature gradient beyond optimum	Equiaxed
Decreasing mould temperature	Increases freezing rate and promotes nucleation	Equiaxed; fine
	Increases temperature gradient	Columnar
Decreasing pouring rate	Increases temperature gradient	Columnar
	Increases mechanical disturbance: promotes crystal fragmentation, disturbs undercooled layer	Equiaxed; fine columnar
Increasing rotational speed	Increases mechanical disturbance: promotes crystal fragmentation, disturbs undercooled layer	Equiaxed; fine columnar

structures consist of graphite nodules in a fully ferritic matrix; in pipes of heavier section the nodule size increases progressively across the wall thickness. Structures in these and other centrifugally cast iron components, including cylinder liners and piston rings, are treated in detail in Reference 19.

With respect to properties, centrifugal castings have been frequently compared with forgings, although there is a lack of published data on properties for strictly comparable conditions. Ductility is in general lower than in forgings, but there are many instances where a centrifugal casting has satisfactorily fulfilled the same function. At elevated temperatures the cast structure offers positive advantages with respect to creep strength.

Much direct evidence is available of improved properties compared with the normal run of static castings. Probably the most notable range of information, apart from the previously mentioned investigations of Northcott, was provided by Thornton's exhaustive examinations of high duty aircraft castings in alloy steels[20–22]. Mechanical properties were determined from cut-up tests and a remarkable feature of the results was the low degree of scatter compared with that usually obtained in sand castings: no weak zones were encountered and high elongation values testified to the degree of soundness achieved. The greater refining effect of chill rather than refractory moulds was also demonstrated.

In the values for mechanical properties, little difference was observed between circumferential and longitudinal properties in annular castings. That the properties of such castings are not wholly isotropic, however, was shown by Morley[23], who encountered appreciable ovality in test pieces from columnar regions: this indicates that the radial direction would be associated with significant differences in properties. Studies of the properties of columnar structures produced by unidirectional solidification are relevant to this aspect of centrifugal casting properties.

Applications

True centrifugal casting has been firmly established for many years as the leading method for the production of cast iron pipes and cylinder liners, the former typifying the as-cast product and the latter the component machined from a pot. The process was subsequently extended to tubes and annular components in most types of commercial casting alloy including carbon, low alloy and high alloy steels, and copper and nickel based alloys. The process is especially valuable for the shaping of alloys unsuitable for forging or for wrought tube manufacture. Table 10.2 lists some of the more important applications for cylindrical castings, a selection of which is illustrated in Figure 10.12.

Amongst special products mention should be made of bimetal tubes. Alloys of widely different properties can be combined in a single structure by successive pouring of the two alloys under closely controlled conditions, the time interval before the second pouring operation being critical. Alloy combinations are restricted by the relative melting temperatures and the danger of excessive remelting of the outer by the inner layer, but under suitable conditions a metallurgical bond is achieved without this difficulty. The casting phenomena in the production of bimetal tubes with stainless steel as the outer and cast iron as the inner layer were the subject of a detailed study and review by Choi et al[24]. The general pattern was of initial freezing of the second layer followed by remelting; there was evident interaction since high hardness was a feature of the interface.

An important use of the sequential casting principle is seen in the large scale manufacture of rolls on vertical axis machines. This system is preferred for the avoidance of the rotational bending stresses which can produce hot tears in heavy horizontal axis castings, a problem examined in a review by Carless and Tingle[25]. Figure 10.13 illustrates a vertical machine set up for the production of rolling mill rolls, in which a wear resistant working layer can be combined with a tough central core and journals. Computerized programme control of the essential casting variables is employed to

Table 10.2 Typical applications of centrifugal castings

I.		
True (a) as-cast	Pipes for water, gas and sewage	Cast iron
	Tubing for reformers $\left.\right\}$ Radiant tubes	Heat resisting steel
	Rainwater pipes	Cast iron
(b) machined from pots	Bearing bushes	Copper alloys
	Piston rings $\left.\right\}$ Cylinder liners	Cast iron
	Paper making rollers	Copper alloys
	Gas turbine rings	Heat resisting steel and nickel base alloys
	Runout rollers	Carbon steel
	Rolls	Steel, cast iron, bimetal
II.		
Semi-centrifugal	Nozzle boxes	Heat resisting steel
	Gears and gear blanks	Copper alloys Steel
	Pulleys	Steel
III.		
Pressure	Hinges $\left.\right\}$ Brackets	Steel
	Dental castings	Co–Cr alloy
	Jewellery	

Figure 10.12 Examples of true and semi-centrifugally cast products (courtesy of Sheepbridge Alloy Castings Ltd.)

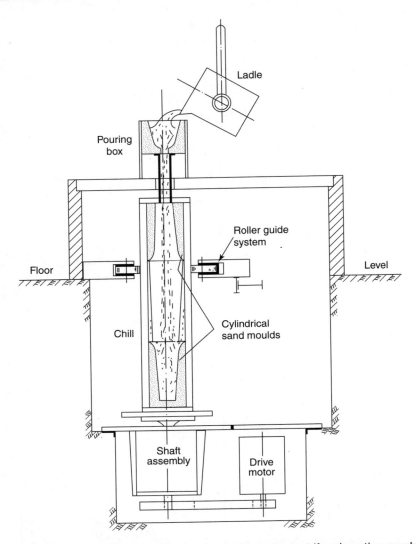

Figure 10.13 General arrangement of a vertical axis centrifugal casting mach-ine for the production of long castings. (from Carless and Tingle[25]) (courtesy of Foundry Trade Journal)

optimize the metallurgical quality of this type of product, an increasing trend in other advanced casting processes.

The Davis wheel was an early example of a multi-composition product: in this case the outermost rim of a steel casting was differentially hardened by the introduction of ferromanganese into the first metal to be poured. Other examples of composite products arise when centrifugal casting is used to line a structural outer shell with an alloy of special properties. A notable case is the production of white metal bearings by pouring the low melting point bearing alloy into tinned outer shells. A further example is the spinning of cast iron friction linings into steel casings for the production of aircraft brake drums.

Centrifugal casting has also been used for the manufacture of blanks for mechanical working. One example is the production of hollow blanks for wrought tube production as an alternative to piercing a solid billet, reducing the total deformation required to reach the finished form. Rings have also been produced by ring rolling and cold expansion of centrifugally cast blanks, with intermediate recrystallization treatment: the final structures are wholly characteristic of wrought material. Crow[26] described the use of centrifugally cast pots for production of flat rolled wide brass strip by slitting, opening out and rolling. Rings have been similarly opened and straightened for bar rolling operations.

Examples of components produced by semicentrifugal and pressure casting are also listed in Table 10.2. Perhaps the most noteworthy are the heavily shaped nozzle boxes and casings for gas turbines and the high strength low alloy steel structural members for aircraft undercarriages and airframes: examples of the former are included in Figure 10.12.

Other special casting techniques

Apart from the widely established processes examined in this and earlier chapters, certain other developments warrant mention for their particular approaches to the problems of shaping by casting. These include unorthodox techniques of shaping, special methods of coring, and systems for the casting of reactive metals.

Unorthodox shaping techniques

In the great majority of casting processes shape is derived through flow of metal into a complete mould impression. Gravity pouring predominates, although important exceptions include pressure and low pressure die casting and the growing use of counter gravity systems to assist flow in other processes. Some of the developments in this area will be reviewed in this chapter. One departure from a fully shaped mould has already been seen in true centrifugal casting, where centrifugal force is employed both to distribute the metal and to form the internal surface of annular castings.

An altogether different and revolutionary approach to shaping from liquid metal is *dip forming*, in which metal is deposited by freezing on to a solid core rod passed continuously upwards through a bath of molten metal at a controlled rate, eliminating the use of any kind of mould or die[27]. This development is outside the scope of the present work, being essentially directed to the production of primary billets and blanks for further working. It is mentioned because it involves reversal of the normal processes of heat flow in casting and takes its place with ingot and continuous casting as a means of forming material of regular cross-section.

In relation to mechanical working, it should be mentioned that shaped sand castings too have been employed in duplex forming processes involving deformation, usually with the object of producing a local improvement in mechanical properties. Watmough[28,29] described the press hardening of steel castings in semi-open and closed forging dies. Very large increases in yield stress and tensile strength were combined with some loss of ductility, although the latter effect was less severe than that encountered in reaching similar strengths by heat treatment. Directionality of properties was reported to be much less than in conventional forgings of similar composition.

In a further special category are the varied developments arising from the structural effects of ultra-rapid cooling from the melt, often referred to as *rapid solidification processing* or RSP. These are typically observed in thin splats formed by the projection of molten metal droplets against a cold surface, previously mentioned in Chapter 2. The metallurgical possibilities opened by rapid solidification have generated a distinctive field of study and experimental work, with an extensive body of literature and some industrial application: only a brief outline is possible here.

A range of available techniques can produce cooling rates up to and even exceeding 10^6 °C/s, as compared with typical values of $\sim 10^{-2}$–10^2 °C/s in normal casting processes. Effects include extreme structural refinement, the eventual formation of amorphous structures or metallic glasses, virtual absence of segregation, extension of ranges of solid solubility beyond equilibrium limits, and creation of metastable non-equilibrium phases.

Many of the processes used for the production of materials under these conditions are based on spray forming or melt spinning, although the melting and freezing of thin surface layers of a prior solid mass can produce similar effects in circumstances requiring enhancement of surface properties.

In *spray forming*, droplets generated by gas or liquid atomization of a molten stream are directed on to a flat or shaped chill surface, using relative movement to overlay the preceding material and build up a solid layer. The nature of this process was analysed by Singer and Evans[30]. The condition of the solid is determined by the rate of deposition and the spray density. *Spray casting* is achieved when droplets arrive whilst the prior splats retain a thin liquid upper layer, so that the respective increments, whilst remaining in position, lose their individual boundaries. Porosity levels of 1% or less can be achieved given suitable conditions. Oxidation needs to be avoided by inert gas protection and in some cases by the use of vacuum or inert gas in subsequent processing.

If the prior splats are already fully solid when further droplets are deposited, the situation resembles that in powder metallurgy and in normal metal spraying operations, whilst if there is a substantial molten pool, lateral flow can occur and the process becomes one of conventional casting.

In *melt spinning* a continuous molten jet is directed on to a moving chill surface to produce a solidified ribbon, which is subsequently peeled away; alternatively the high velocity jet can be continuously cooled in a quenching fluid to produce wire. A typical ribbon casting arrangement using a rotating copper drum as the chill block is illustrated in Figure 10.14.

Products of such processes are suitable for subsequent consolidation by mechanical deformation in hot pressing, rolling and extrusion; in some cases shaped preforms for specific products such as forgings, billets or seamless tubes can be produced, using dies or mandrels to collect the deposited material (Figure 10.15). Direct surface cladding of rolls is also feasible.

The processes can be applied to a wide range of metal compositions, including stainless and high speed steels, nickel base superalloys and aluminium, magnesium, copper and titanium alloys. Metal matrix composites can be produced by co-spraying techniques, and bi-metal products by sequential deposition. The products are variously selected for high strength, elevated temperature properties and special magnetic or electrical characteristics. Numerous commercial applications of spray forming are reviewed in Reference 32.

Major reviews of rapid solidification processing (RSP) are contained in References 33 and 34, whilst notable short reviews were those by Jones[35] and Fleetwood[36].

A further approach to the production of castings with exceptional properties has been seen in the *electroslag casting* of shaped components. This is an adaptation of the method of ingot production in which submerged electroslag melting is combined with the use of water cooled moulds to

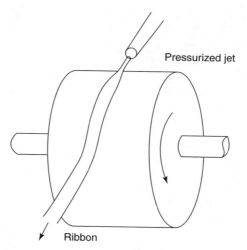

Pressurized jet

Ribbon

Figure 10.14 Principle of system for production of rapidly solidified ribbon by melt spinning on rotating chill

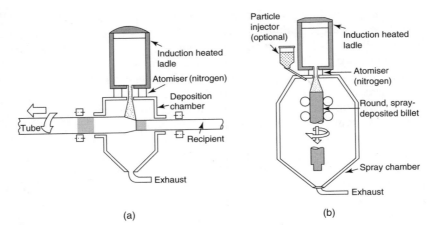

(a)

Figure 10.15 Systems for production of (a) tube and (b) billet by the Osprey process (from Leatham *et al*[31]) (courtesy of Institute of Materials)

Figure 10.16 Electroslag casting of a valve body (a) mould assembly (b) casting (from Mitchell *et al*[37]) (courtesy of Institute of Materials)

form sound structures free from macrosegregation or large inclusions. Steel gear and valve body castings of relatively simple shape, produced in water cooled aluminium moulds, were shown to give isotropic properties equal or superior to those of forgings[37]. Figure 10.16 shows a stainless steel valve body in its partly dismantled mould after casting using this technique.

Other variations of shaping technique are seen in *slush casting* and in the *Cothias Process*. These are essentially forms of die casting used for the production of hollow articles. In the former case the outer surface is formed by a shaped die but the inner surface is produced by decanting excess liquid shortly after pouring, leaving a thin solidified shell. In the Cothias process the metal only partly fills the original die cavity when poured but is then displaced to fill the entire cavity by insertion of a metal core to form the internal contour. Neither process is of engineering importance, applications being primarily for models and statuary in metals of low melting point.

Squeeze casting; extrusion casting

A further variation of die casting is seen in squeeze casting. Metal is poured into a die, of which the halves are initially separated and then brought together to compress the solidifying mass: an element of mechanical deformation is thus introduced and terms such as "liquid forgings" and "liquid stampings" have been applied to the products.

The general principle of squeeze casting is illustrated in Figure 10.17. The liquid metal is accurately metered into the lower half of the preheated steel die, and the top die or "punch" is immediately brought down into the die cavity and maintained under high pressure until solidification is complete. The die motion is controlled to achieve smooth upward metal

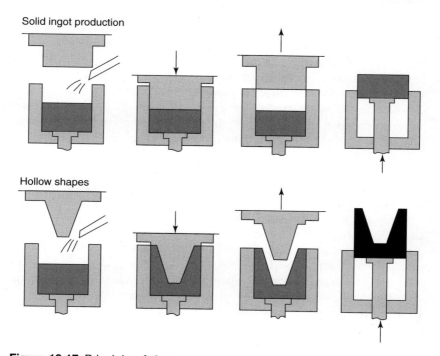

Solid ingot production

Hollow shapes

Figure 10.17 Principle of the squeeze casting process for solid and hollow forms (from Chadwick and Yue[38]) (courtesy of Institute of Materials)

displacement and avoid entrapment of air and oxides. Press loads are much lower than those required for conventional forging, and only a single stage die cavity is required. On completion of freezing the punch die element is raised and the component ejected.

The products are noted for their freedom from internal shrinkage or gas porosity, resulting from the influence of solidification under pressure, and for their excellent surface finish and definition. The principal effect on the cast structure arises from the close contact between metal and die, with the elimination of the air gap as formed in gravity die casting. Solidification rates are thus higher, and the microstructures correspondingly finer, than those obtained in most other casting processes, with corresponding benefits in terms of heat treatment times and mechanical properties. Given the nature of the operation the process is uniquely efficient in terms of product yield, but capital and tooling costs are relatively high. Important process characteristics were reviewed in Reference 39.

In a variant of the process, referred to as indirect squeeze casting, molten metal is injected upwards into the die impression by means of a small diameter piston, though which the subsequent pressure is maintained. This system has, however, more in common with techniques of pressure die casting than with the direct squeeze casting process.

Early work on squeeze casting was carried out in Russia, including the production of large thin-walled panels in aluminium alloy[40], and later work with steel[41]; publications included a substantial monograph by Plyatskii[42]. Most application of the process has been in the aluminium alloy field. Favourable fatigue properties are achieved, and in high strength wrought alloys, squeeze formed material properties were shown to be isotropic and comparable to longitudinal properties in the forgings[43]. An account of varied commercial developments in the field was given by Morton and Barlow[44]. Figure 10.18 shows some of the typical components produced.

Favourable results with magnesium alloys as well as with high strength aluminium alloys were demonstrated by Chadwick and Yue[38], whilst Begg and Clegg have reported similar benefits from microstructural refinement in Zn–Al alloys[45].

Although there were early Russian reports of success with copper alloys[46]; Herrera and Campbell[47] found little property improvement in certain copper alloys when compared with normal cast material, despite evidence of refinement and low porosity, suggesting oxide film or dross formation as a possible reason and emphasizing the need for close control of melting and casting conditions when using the process.

One of the major developments in squeeze casting has been its association with the production of components from metal matrix composites, an application in which the freedom from porosity achieved by solidification under pressure makes a vital contribution to the effectiveness of the reinforcement. This aspect of squeeze casting was the subject of an extensive

Figure 10.18 Typical components produced by squeeze casting (from Morton and Barlow[44]) (courtesy of Institute of British Foundrymen)

treatment by Das *et al*[48] and can be seen against the more general picture of cast composites as outlined below.

Metal matrix composites

Composite materials offer several potential advantages over conventional alloys, including increased strength, stiffness and wear resistance and enhanced high temperature properties. Apart from solid state and spray deposition routes, several different casting techniques can be employed in their production, in which the metal matrix is reinforced by the incorporation of particles or of short or continuous fibres.

The reinforcement can be introduced by the infiltration of bonded preforms placed in the die to achieve uniform or selective distribution as required, a technique suitable for fibrous reinforcements. Alternatively, general distribution of the reinforcement can be obtained by prior mixing into the melt before transfer. This can be suitable for particulate additions, but fibres can undergo degradation during stirring. Feasible volume fractions are limited by increases in viscosity and their effect upon flow. A third approach lies through the formation of in-situ dispersions or aligned phases by reactions in the melt or during solidification: this severely limits the available options in the design of the material and choice of properties.

Concerning casting methods, squeeze casting with its particular advantages was referred to above. A further technique involving infiltration is liquid pressure forming, a variant of low pressure die casting, which employs gas pressure to force the liquid metal upwards through the preform mesh; high die and insert preheat temperatures and melt superheats are required to achieve complete impregnation. This process was described in detail by Mykura[49], including applications based on varied light alloy matrices and reinforcements.

A further notable development is the production of composite investment castings, in which the fibre reinforcement is incorporated in the wax patterns; these are then normally processed so that infiltration occurs on casting, aided by a modest pressure[50]. Investment castings of graphite fibre reinforced magnesium alloys have also been produced[51].

Prior mixing of the reinforcement into the melt can be achieved using mechanical or electromagnetic stirring or gas agitation, although gas bubbles can also cause flotation. A variant of this approach is the use of semi-solid slurry casting or *compocasting*, referred to later in the present chapter: this is mainly used for discontinuous dispersions, being subject to fibre degradation as in other processes involving vigorous stirring. Yet another casting technique, in this case used mainly for the selective deployment of reinforcement to bearing or wear surfaces, is centrifugal casting.

The concepts involved in the production of in-situ cast composites have been examined by Wood et al[52] with special reference to an aluminium alloy reinforced by stable TiB_2 dispersions in volume fractions up to 10%, whilst the production of cast die and tool steels containing smaller fractions of NbC by similar "reactive casting" techniques were reviewed in Reference 53. The dividing lines between these materials, normal composite mixtures and conventional single alloys remain somewhat indistinct. An example of a cast high speed steel microstructure modified and refined by the development of a reaction-formed dispersion of niobium carbide particles in the melt is illustrated in Figure 10.19.

A broad review of solidification processing as applied to metal matrix composites was undertaken by Rohatgi et al[54], with particular emphasis on the importance of wetting to satisfactory distribution, including the use of surface pre-treatments to enhance the process. Fundamentals, techniques and some applications were considered: uses in pistons, rings, cylinder liners and bearings were stressed. A comprehensive monograph on all aspects of metal matrix composites was produced by Terry[55]: this includes references to many individual matrix-reinforcement combinations and their uses.

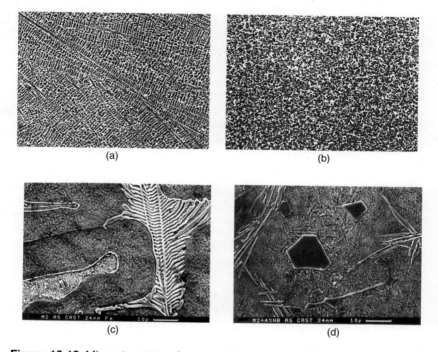

Figure 10.19 Microstructure of a cast M2 high speed steel (a) normal and (b) modified with primary NbC dispersion × 40. (c) and (d) show the respective carbide morphologies (SEM) (courtesy of Dr A. Shokuhfar)

Counter-gravity casting systems

Conventional gravity pouring, however well controlled through ladle practice and gating system design, is inherently prone to the creation of defects through metal turbulence and exposure to surface oxidation. Radical change to the system of metal transfer can be achieved by upward displacement of metal into a superimposed die or mould. The absence of turbulence, coupled with the ability to feed castings in the direction of mould filling, provides favourable conditions for the production of sound castings: the respective advantages of top and bottom gating are combined in essentially the same manner as in total reversal casting (q.v. Chapter 3) yet without the need for mould reversal.

The upward fill principle has been adopted in a number of distinctive applications, some dating back for many years. The Griffin process for the production of railway rolling stock wheels in graphite moulds was an early example[56], operating on a small air pressure difference of about 1.5 bar. The technique has been further applied to ingots and, in conjunction with the slush casting principle, for the manufacture of hollow blanks for tool production. In the latter case the metal is displaced into a cylindrical mould. After a dwell period for growth of a solid annulus of the required thickness, pressure is released and the residual liquid core returned to the bath. The process is used in stainless steel tube production, and the casting of bronze bushes has also been described[57]. The castings are again claimed to be superior to those produced by orthodox means. Results of a systematic experimental investigation of the structure and properties of such tubes in various aluminium and copper base alloys[58] indicated similar levels of tensile strength and surface quality to those produced by conventional metal mould and centrifugal casting. Investigations of several alternative techniques of upward displacement for use in the production of investment castings were reported in Reference 59.

The long established process of low pressure die casting was reviewed in the preceding chapter. Upward filling is also featured in the *counter-pressure* casting technique, previously mentioned in relation to pressure die casting. In one variation of this system[60], the metal is displaced under a pressure differential which can be exactly controlled by a valve system regulating the separate gas pressures in the metal reservoir and the mould cavity.

The counter-gravity principle is employed in further modern processes with the main emphasis on refractory rather than permanent moulds. The Cosworth system uses sand moulds, whilst the CL group of processes were applied firstly to fired ceramic moulds for the production of investment castings, but also to sand and shell mould casting.

Low pressure sand casting: the Cosworth process A highly distinctive development of the upward fill principle is seen in the Cosworth process, which was developed primarily for aluminium alloy casting[61]. Metal transfer is in this case accomplished by low velocity electromagnetic pumping from a zone beneath the surface of the molten bath, so avoiding pickup of dross from the surface and bottom regions. The low turbulence associated with this system ensures high and consistent casting quality: oxide films, inclusions and associated porosity are largely eliminated. The general concept of the process is illustrated in Figure 10.20 and various aspects are examined by Campbell and other authors in References 62–66.

A further feature of this process is its use of precise, resin bonded zircon sand moulds operated on the core assembly principle. Zircon is preferred for its low coefficient of thermal expansion relative to that of silica, which permits more accurate prediction of dimensional behaviour, whilst its similar density to that of aluminium ensures neutral buoyancy behaviour in relation to extended and delicate cores. Thermal reclamation of the high cost base sand is essential to economic operation of the process.

The combination of superior metallurgical quality with enhanced accuracy facilitated the development of thin-walled aluminium alloy castings for automobile engines and other exacting applications. The cylinder head application is considered in detail in Reference 63.

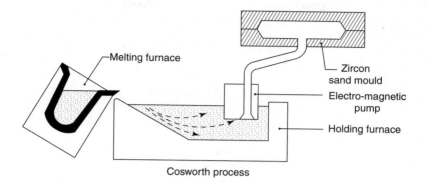

Cosworth process

Figure 10.20 Principle of the Cosworth Process (from Campbell[64]) (courtesy of CDC)

An interesting feature of the further development of the Cosworth process was the reintroduction of the mould reversal principle. Provision of a rollover facility, coupled with a feedmetal supply for the individual castings, eliminates the need to await completion of solidification before disconnection from the metal pump, significantly increasing the output rate for large scale production requirements.

Differences in dimensional behaviour between castings made in silica and zircon sand moulds have been explored in some detail by Campbell[64,67] who has emphasized the special suitability of silica, with its lower rate of heat extraction, for the casting of ultra-thin walls. Since mould temperatures remain low in such cases, the adverse relative expansion factor is much less significant, so that other process options using silica may be preferred in such circumstances.

The CL processes A further and highly original application of the counter-gravity casting concept is seen in the CLA process used for the production of investment castings. Its special feature, also used in the CLV and other variants, is the use of vacuum suction to effect the upward transfer into the mould cavity, providing favourable conditions for the running of castings with extremely thin walls.

Various features of the process and its derivatives have been described in papers by Chandley[68–71] and Taylor[72]: the essentials are illustrated in Figure 10.21. The ceramic shell mould is sealed in a vacuum chamber, from which the sprue projects downwards. Its open end is lowered into the furnace bath and molten metal is drawn upwards into the mould cavity by the action of the vacuum through the shell. On completion of solidification of the castings the vacuum is released and the remaining liquid in the sprue and gates is returned to the furnace bath, giving a high casting yield. A particular advantage is the rapid but progressive fill obtained in multi-level mould cavities.

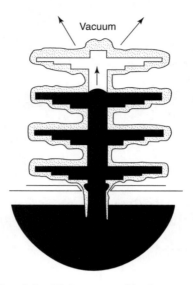

Figure 10.21 Principle of the CLA process for the upward filling of ceramic moulds (from Chandley[69]) (courtesy of CDC)

Smooth transfer is achieved, and the vacuum assistance to metal flow enables pouring temperatures to be greatly reduced from conventional levels, producing refined microstructures and enhanced mechanical properties. The process is used for the production of high grade components in steels and superalloys as well as in aluminium and other alloys of lower melting point. Where the system has been directed to the thin-wall objective, thicknesses of as little as 0.65 mm (0.025″) for some stainless steels and 0.5 mm (0.02″) for cobalt–chromium alloys have been quoted by users of the process[71].

The CLA process employs conventional air melting. In the CLV variant the melting process is itself carried out under vacuum[69,70]. The melting unit is enclosed in a separate vacuum chamber connected to the superimposed mould chamber through a valve, which is kept closed during melting. Before casting the melting chamber is filled with argon: this creates the pressure difference enabling the mould chamber vacuum to draw the metal into the mould cavity when the valve is opened and the sprue submerged in the melt. The process extends the range of products to include the vacuum melted superalloys.

A further variant of the process[69] was designed to introduce conventional chemically bonded sand moulds, incorporating a patented multiple gating system directed towards very high yields, whilst thin shells supported in vacuum-compacted loose sand provide the basis for yet another development[71]. A particular field for application is the production of stainless steel exhaust components, including manifolds, with general wall thicknesses in the range 2–3 mm.

A comprehensive critical treatment of the various factors operating in upward fill and other flow systems available in metal founding was undertaken by Campbell[67], with strong emphasis on the predominant role of the short mould filling stage in determining the quality of castings.

The V process

In the V or vacuum sealed moulding process a vacuum system is employed to compact a mould of dry unbonded sand, held in a moulding box between two plastic films forming the shaped mould face and the flat mould back. The sand is thus supported by atmospheric pressure until casting is complete.

The principle is illustrated in Figure 10.22. To make a mould part the mould face is first created by the suction forming of a heated plastic film against the plate mounted pattern assembly. For this purpose the pattern and plate are extensively vented and mounted on a shallow chamber connected to a vacuum pumping system. A superimposed film, some 0.05 mm in thickness, is softened by radiant heating, draped over the pattern and the vacuum is applied to shape the film closely to the pattern surface. A light refractory coating is applied to the back of the film, dried and the special moulding box placed in position. This has hollow walls with provision for a separate vacuum connection to the interior. The box is filled with dry unbonded silica sand and vibrated for compaction. It is then sealed by a further, thicker plastic sheet and a second vacuum is applied to the box itself, bringing atmospheric pressure to bear on the sand body. A high and uniform mould hardness is achieved throughout the mould cavity.

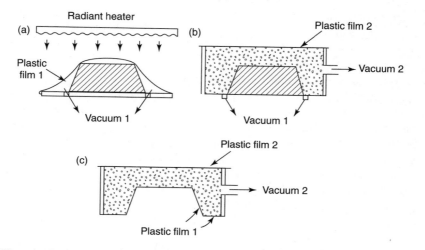

Figure 10.22 Principle of V process. (a) thin plastic film heated and draped over pattern (b) second vacuum applied to box after sand fill and vibration (c) mould part freed for closing by release of initial vacuum

The initial vacuum, that applied through the pattern assembly, is now released and the mould part can be readily lifted off the pattern, whilst maintaining its internal vacuum through a flexible link. Little or no draft taper is required to achieve separation. The second mould part is similarly produced, conventionally bonded cores are inserted if required, and the mould is closed whilst still under vacuum. The small volume of gas produced by vaporization of the film was shown by Kono and Miura[73] to be mainly drawn into the sand through the coating, which continues to prevent metal penetration or inclusions.

To retain the mould shape as the plastic film is melted on entry of the metal, it is necessary to ensure that a satisfactory level of pressure is maintained in the unfilled part of the mould cavity. Risers and vents to outside atmosphere allow the initial escape of heated air and gas, followed by ingress of the air needed to maintain atmospheric pressure for support of the residual mould wall.

After time for solidification the vacuum is released, the casting removed and the sand cooled and returned for re-use. Easy stripping is achieved without the need for mechanical knockout facilities.

The V process is versatile in that conventional machine moulding patterns of any of the normal materials can be employed, subject to the requirement for venting, and a wide range of products have been successfully produced in the main types of casting alloy. Faithful reproduction of pattern detail is achieved, with cast weights ranging from a few kilograms to several tonnes.

A major review of the process was undertaken by Clegg[74] and a number of practical details are included in References 75–77. The vacuum pumps clearly constitute a crucial element in the system and must be capable of coping with dust and fume. A water ring sealed design[76] operates unimpaired under such conditions and even ingested sand is flushed away; large volumes of air can be exhausted.

Although heavy capital investment costs are involved in the V process, the production costs are competitive with sand and die casting processes where the weight ranges overlap. Although layouts are constrained by the cumbersome vacuum connections, custom designed plants offer high productivity: Campbell[64] described one such machine with a carousel arrangement of linear spokes offering advanced automatic handling of all phases of the process. This system is illustrated in Figure 10.23. In a more specialized field, an output of 70 piano frames per hour has been successfully maintained in a Japanese foundry[78].

Some ingenious and varied plants have been designed to apply the V process to products other than shaped castings[79]. These include continuously cast profiled bar and strip, in which all the normal steps in the process are themselves provided on a continuous basis, film being fed from a roll and the mould cavity from a suitably profiled roller. Imprinting from

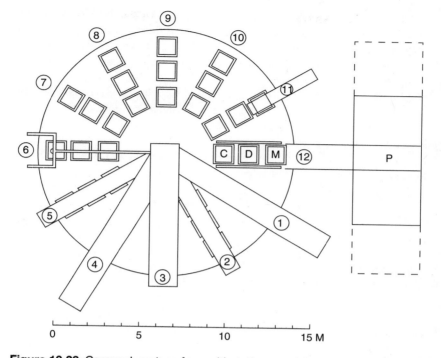

Figure 10.23 Carousel system for multi-station moulding and casting operations using the V-process. 1 film forming; 2 box placement; 3 sand fill and vibrate; 4 top film placement; 5 mould strip and assembly; 6 pouring; 7–10 cooling; 11 mould release; pattern clean/change. (C cope pattern; D drag pattern; M mould position; P pattern change area.) (from Campbell[64]) (courtesy of CDC)

a roller can similarly be used to form cavities for separate shaped castings in a continuous sequence.

The lost foam process

The lost foam or evaporative pattern casting process is based on the expendable pattern principle as used in investment casting, but with the important difference that the pattern assembly is left in the mould to be displaced by the molten metal on pouring.

Low density expanded polystyrene patterns or "preforms" are produced as full replicas of the intended casting, by blowing fine beads of the material into a steam heated multi-part aluminium die, followed by consolidation using steam injection. Patterns can also be fabricated from blocks of the same material by machining, cutting with a blade or hot wire, and adhesive joining. Although suitable for prototypes or for large castings of relatively simple shape this approach is unsuitable for quantity production.

Patterns are assembled, singly or in clusters, with a similarly produced sprue and feeding system, coated with a thin refractory layer and dried; in many cases the feeding system can be produced integrally with the pattern.

In the original *full mould process*[80] conventionally bonded sand was compacted around the pattern assembly in a moulding box, a system still employed for single large heavy castings, but the process subsequently developed with the use of dry unbonded sand. The coated assembly is submerged in loose silica sand in the moulding box and compacted by carefully controlled vibration to fill all the pattern cavities, producing a jointless mould without the need for cores or draft taper. Care is needed to avoid distortion of the delicate patterns.

The most distinctive feature of the lost foam process is the pouring of the molten metal into the mould containing the polystyrene assembly, and the progressive melting and volatilization of the foam as the mould fills. The polystyrene supports a well packed sand, maintaining the shape until replaced by the cast metal. A light vacuum can be applied to the moulding box at this stage to enhance the packing of the sand grains and assist the removal of the evolved vapours through the sand bed.

The nature of the crucial mould filling process has been subject to detailed studies[81,82]: close control is required to avoid retention of gaseous and liquid decomposition products within the casting and the occurrence of porosity and lustrous carbon defects. The permeability of the pattern coating plays an important role in this respect: for aluminium alloys filling is controlled by the rate of elimination of the pattern material and low permeability coatings are preferred, whilst for iron or steel castings more copious gas evolution requires high permeability coatings to maintain rapid and progressive filling, controlled by an ample gating system. Proprietary coating compositions and gradings have been developed to suit the different alloy groups.

After cooling the castings are separated for cleaning and the loose sand is returned to the system for cooling, screening and re-use. The castings are noted for the high degree of shaping freedom and detail reproduction. The surface texture is derived from that of the polystyrene, which has been transmitted to the inner side of the refractory coating: traces of the bead pattern can remain visible on the casting surface.

The lost foam process has been the subject of substantial reviews, in which practical operating techniques and product characteristics have been examined in detail[81,83–87]. Pattern quality is a vital factor, and in some cases the polystyrene preforms are purchased from specialist outside suppliers. Dimensional allowances are required, not only for normal casting contraction but also for the shrinkange of the polystyrene with ageing, which is predictable. Plant and production costs are modest, but high tooling costs require quantity production, although dies once produced last almost indefinitely.

Metallurgical characteristics of lost foam castings have been widely discussed. Steel castings are prone to carbon pickup, which has restricted the use of the process to high carbon compositions, but it has been extensively adopted for aluminium alloys and cast irons. The potential role of gas porosity and surface crevices for lowering fatigue properties needs to be kept in mind[81]. A strong positive factor is the complexity of design that can be achieved by the use of collapsible die construction, as well as by the adhesive joining of separate patterns to form still more elaborate shapes.

A development of the original process was designed to employ magnetization bonding of dry iron particles around the expendable patterns[88]. Although this proved to be feasible it seems that the success of the sand-based system precluded further extensive application.

The Replicast® process

The lost foam principle is taken a stage further in the Replicast® process, in which a more substantial ceramic coating is built up on the polystyrene pattern assembly, using a multiple dip-and-stucco sequence as used in lost wax investment casting. The coated assembly is similarly dried and briefly fired to 1000°C in a prior operation to volatilize the polystyrene, leaving a strong, jointless ceramic mould some 5 mm in thickness, capable of being handled in preparation for pouring. For this the shells are supported in containers in dry sand compacted by vibration; a vacuum is applied to the sand bed to maximize support during casting.

Since the pattern is eliminated beforehand there is no mould gas evolution or potential source of carbon pickup with danger to the quality of the casting. This enables denser polystyrene patterns to be employed than are feasible when using the conventional lost foam process, giving greater pattern strength and smoother castings. Light abrasive blasting suffices to clean the castings, leaving an excellent finish with minimal machining allowances, where these are required at all. Some typical products are shown in Figures 10.24 and 10.25.

Important characteristics of the process and products have been reviewed by Ashton[89] with evidence of accuracy superior to that of sand and resin shell and comparable with investment castings. The process can indeed be regarded as a variant of investment casting but is suitable for the production of much heavier castings than are normally associated with the lost wax process, yet with similar capacity for the production of complex shape and sharp rendering of design detail. The process has been noted for extremely successful application to steel castings: numerous further examples were described and illustrated in Reference 90.

A notable variant of the above system is seen in the subsequently developed Repliwax® process, which employs wax rather than foam patterns, sharing this feature with standard investment casting. The shells, however, are much thinner and are cast cold, being supported in a sand bed consolidated

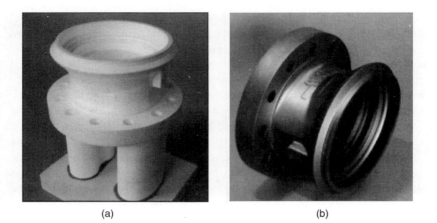

(a)   (b)

Figure 10.24 Alloy steel pressure fitting produced by the Replicast® process. (a) high density expanded polystyrene pattern ready for application of ceramic coating; (b) casting (courtesy of CDC)

Figure 10.25 Drill bit and other special steel castings produced by the Replicast® process (courtesy of Glencast Ltd.)

by vibration; vacuum suction is again applied to the container to enhance support and assist the filling of thin sections. The process has potential for the production of large and accurate castings at competitive cost.

Techniques for intricate coring

The elaborate design made possible by special casting techniques requires the production of more intricate and confined internal passages as part of the cast form.

Using conventional coring techniques, limitations to passage dimensions are imposed by the mechanical strength of the core and the need to exhaust its evolved gases through a limited area of print: these difficulties are aggravated with increasing ratio of length to cross-section. With metals of high melting point there is also the danger of sintering and metal penetration, accentuated with diminishing volume of the core relative to the mass of metal.

Very accurate cored cavities are produced by the normal cold box, shell and hot box techniques, but for closely confined cavities other types of core material can be employed, for example the rammed or slip cast ceramic cores used in investment casting. Based on finely divided refractories, such cores are of low porosity and low volatile content, although they too can be difficult to remove after casting.

For non-ferrous alloys, a number of unorthodox coring techniques have been employed with success for the shaping of narrow and complex passages. The most familiar of these is the use of permanent tube inserts placed in the mould before casting and remaining as linings for the passages so formed. This technique can be used in die casting for cavities too elaborate for the use of retractable metal cores. The preformed insert tubes may be of mild steel, copper or stainless steel and can be utilized in zinc, aluminium or magnesium alloy castings.

Where a duplex structure is unsuitable, several types of removable core may be used. Soluble salt cores are one example of limited potential, but more refractory or stable materials can be similarly eliminated by solution. Dalton[91] described the use in this way of metal tubing sheathed in flexible refractory. In aluminium alloy casting a passage is formed with the aid of copper tubing coated with glass fibre, the whole being set in a sand core embodying prints for location in the mould. The copper is subsequently dissolved out in a mixture of hydrofluoric and nitric acids, the acid producing passivation of the parent alloy.

In the McCannacore process[92], an aluminium alloy casting can be cored out to form intricate passages using preformed heat resistant glass tubing, which is subsequently dissolved out in hydrofluoric acid. Before insertion in the mould the tubing is coated with radiographically dense tungstic oxide wash, enabling X-ray examination to be carried out to detect the complete solution of the glass.

Further techniques of melt processing

Numerous measures aimed at the control of cast structure and at problems of metallurgical quality have been examined in context in earlier chapters. These included vacuum melting and melt treatments, pouring techniques and the control of solidification within the mould. A further requirement for special techniques arises in the foundry processing of reactive metals.

Systems for the casting of reactive metals

With the demand for a widening range of alloys, it became necessary to search for methods of producing shaped castings in alloys which pose special problems on account of their high chemical activity or high melting point. Under such conditions conventional melting furnaces, simple sand casting and the normal techniques of precision casting are all unsuitable because of potential reactions between molten metal and furnace lining, mould and atmosphere. Some indication of the problems encountered can be inferred from Figure 10.26, which indicates the relative melting points and free energies of oxide formation of a number of metals. The latter property does not provide a complete guide to the reactivity, which also depends upon the nature of the oxide layer formed, but elements high and to the right on the diagram might be expected to react with their environment most severely when molten. Aluminium provides a notable exception on account of its stable oxide skin.

One reactive metal in which the problems have been largely overcome is magnesium, for which casting technology is highly developed. Special techniques of melting and fluxing were established to ensure the production of clean metal, whilst metal–mould reaction is suppressed by suitable additions to the moulding material. Major contributions in this field were

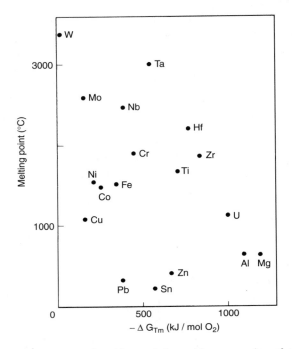

Figure 10.26 Comparison of melting points and free energies of oxide formation for some pure metals

included in the extensive work of Emley[93]. Although magnesium presents further difficulties arising from its low density, which renders it vulnerable to inclusions and blowing defects, its low melting point partly offsets these disadvantages and made possible the production of both sand and die castings of high quality and great complexity.

Problems are much more severe when producing castings in alloys based on metals such as titanium, zirconium, molybdenum and hafnium, but suitable casting systems for these too were eventually developed.

Vacuum or inert gas melting and casting are universal features of such systems. Vacuum melting as applied to alloys containing limited amounts of reactive elements was examined in Chapter 5, but with melts consisting predominantly of such elements metal–crucible reactions prohibit straightforward vacuum induction melting. Although there has been some use of graphite crucibles, the problem of metal–crucible reactions can be avoided by skull melting techniques in which the molten alloy is at all times contained in a solid shell of the same alloy. This can be accomplished by maintaining a steep temperature gradient within the crucible–melt system, using a high conductivity or water cooled metal crucible in combination with direct arc melting. The high temperature of the arc and the elimination of direct contact between melt and container permit the melting and casting of even the most refractory metals.

Various designs of furnace were developed to implement these principles and it proved possible, using water cooled copper crucibles with high power inputs, to achieve molten baths contained within solid skulls representing less than 25% of the total amount of alloy. Tilting crucibles enable such melts to be tapped into moulds enclosed within the melting chamber. Both consumable and permanent electrode systems have been used, under vacuum or in argon or helium atmospheres. The essential concept of one such furnace is shown schematically in Figure 10.27. Turntables or vacuum locks can be provided, to enable series of moulds to be poured in succession, whilst provision for alloy or scrap charge additions is also made in some cases as shown in this example.

There have been further developments in the casting of these metals with the introduction of the induction skull melting furnace. This retains the cold crucible principle, but the power is applied through an induction coil rather than through an electrode-arc arrangement[95,96]. The coil and crucible conditions are designed to push the liquid away from the water cooled crucible, so reducing the contact area and maximizing the size of the melt relative to the total mass of metal. The system also gives greater control over melt temperature, and a more uniform temperature distribution, than are feasible with arc melting. This type of unit has been used for the production of investment castings[95] and a batch weight of over 80 kg has been achieved[97].

A significant example of the use of electromagnetic forces to support molten metal charges out of contact with refractories has been demonstrated

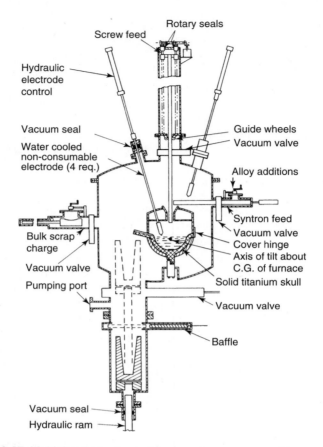

Rotary seals

Screw feed

Hydraulic
electrode
control

Vacuum seal

Water cooled
non-consumable
electrode (4 req.)

Guide wheels
Vacuum valve

Alloy additions

Syntron feed
Vacuum valve
Cover hinge
Axis of tilt about
C.G. of furnace
Solid titanium skull

Bulk scrap
charge

Vacuum valve

Pumping port

Vacuum valve

Baffle

Vacuum seal
Hydraulic ram

Figure 10.27 Multiple electrode skull furnace for vacuum melting and casting reactive metals (from Bunshah[94] after Hamm[59] (courtesy of Reinhold Publishing Corporation)

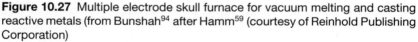

in Reference 98. Experimental melting and casting of aluminium–lithium alloys was carried out in an argon atmosphere, using a special unit designed to maintain the melt in a tall, rounded mass supported on a chill base of minimal area.

Although not directly employed in the production of shaped castings, the related principle of cold hearth refining has relevance to casting, in its potential for the purification of reactive and refractory metals and superalloys[97]. Feedstock is again melted into a water cooled copper vessel, in this case an elongated hearth in which a protective solid skull of the alloy being processed is maintained. As the liquid flows along the hearth from the entry towards the outlet, inclusions sink, float, dissociate or evaporate, and the refined material passes over a lip into a similarly cooled mould for the production of continuously cast ingot of suitable profile.

The melting unit employed in this process can be either an electron beam gun or a plasma torch. The former is preferred where high vacuum conditions are beneficial, as for the removal of gaseous or volatile trace impurities, whilst plasma melting, using helium or argon as the plasma gas stream and capable of operation at around atmospheric pressure, may be selected so as to avoid preferential loss of alloying elements with high vapour pressures. A further advantage over vacuum arc melting is the ability to control the superheat: power increase in the latter process only increases the melting rate. Extensive treatments of electron beam and plasma melting and casting are given in References 99 and 100, and a well illustrated critical review and comparison of the two systems in Reference 101. Although many of these types of development began on a very small scale, tonnage plants have also been put into service.

Apart from the problems of containing melts of these materials, the provision of inert, stable moulds is a further exacting requirement, except in cases where chill moulds can be used for the simplest shapes. It is essential to eliminate volatiles before casting. In the case of titanium alloys both lost wax ceramic shell and rammed refractory moulds have been used for the production of shaped castings. Dunlop[102] described the successful production of titanium investment castings using carbon impregnated ceramic shell moulds based on alumino-silicate refractories bonded with ethyl silicate; skull melting was employed.

Later research with water soluble binders of proprietary composition enabled titanium castings of up to 270 kg in weight to be produced in rammed sand–graphite moulds[103]. Centrifugal casting within the vacuum chamber was also incorporated, to assist mould filling in the "centrifuging" mode. Facilities of this type with a batch weight of 1 tonne were described in Reference 104.

Several examples of modern titanium investment castings, including some of very substantial dimensions, were illustrated in Reference 105.

The commercial production of zirconium castings approaching one tonne in weight, and of a one-piece titanium pump housing weighing over 800 kg, have been reported[106].

A further reactive metal relying upon vacuum melting and casting techniques is uranium, for which either induction or arc furnaces can be employed. Crucibles and moulds may be of refractory coated graphite or a sintered ceramic such as alumina or magnesia. Induction furnaces used for this purpose can be designed for bottom pouring, using either a stopper rod passing through the melt or a water cooled plug of uranium which is allowed to melt out at the appropriate stage. Perhaps most exotic of all is the casting of plutonuim[107], in quantities of up to 5 kg; the only potential weight restriction appears to be the critical mass for nuclear fission.

Although the foregoing techniques have been mentioned primarily in relation to the most highly reactive metals, vacuum melting, inert gas protection and similar measures are applicable in varying degrees to castings in more orthodox alloys, especially where low inclusion and gas contents are at a premium. Special melting and pouring techniques for these types are referred to in other chapters.

In considering problems associated with reactive metals, mention should be made of *levitation melting* as a method of producing small melts for experimental purposes. This technique employs an induction coil of special design in which the balance of electromagnetic forces acting on an enclosed mass of alloy is sufficient to support the alloy without the need for a refractory container. The atmosphere can be controlled and contamination completely avoided, enabling extensive undercooling to be achieved in the absence of heterogeneous nucleation on container walls. Other systems of levitation employ acoustic fields and aerodynamic systems of inert gas layers or cushions, whilst free fall from drop towers has been used for similar studies. These and other techniques of containerless processing of melts have been comprehensively examined in a review by Herlach *et al*[108], but are largely confined to very small masses.

Semi-solid or slurry casting

The processing of alloys by agitation during solidification has been the subject of much modern research and development and some successful application. This followed the discovery by workers at MIT[109] that vigorous stirring of solidifying metal creates a slurry which retains a capacity for flow even when a substantial fraction of the total mass has solidified. In this inter-mediate condition the material exhibits thixotropic behaviour, becoming less viscous with increasing intensity of stirring.

One technique for producing such a slurry is illustrated in Figure 10.28. The solidifying alloy is passed from a liquid reservoir through an annular gap between a cylindrical rotor and a concentric water cooled tube. This produces continuous shear of the growing crystals, preventing dendritic growth, so that the emerging slurry contains small globular solid particles, of a size approximating to the normal dendrite arm spacing, which can remain in relative motion during mould filling. Other forms of agitation, including electromagnetic stirring, are also feasible, and the mobile slurry condition can persist to solid fractions exceeding 0.6.

The semi-solid slurry can be fed directly to the shot sleeve of a pressure die casting machine and injected into a die in the normal way, the process then being referred to as *rheocasting*. Since the temperature is much lower than that of the fully molten alloy, and since much of the latent heat has already been dissipated, die life is extended. The residual shrinkage is also reduced and the high viscosity promotes laminar flow, so that there is reduced porosity such as is often associated with the turbulent flow conditions and shrinkage occurring in normal die casting. Alternatively the material can be cast in the form of a billet and cut into solid slugs for later reheating as single shots. The material reverts to the thixotopic condition,

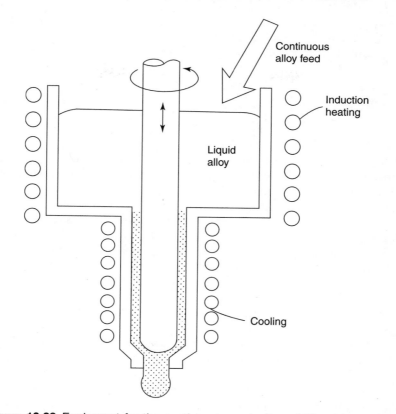

Figure 10.28 Equipment for the continuous production of Rheocast material (after Flemings[110]) (courtesy of Institute of Materials)

again at lower temperatures than those incurred in conventional casting: this is the process of *thixocasting*. The slugs can also be used as billets for *thixoforging*, a variant of the closed die forging process in which very low flow stresses suffice to shape the material from slugs or preforms.

Slurry casting has also proved suitable for the introduction of particulate or fibre reinforcement phases into stirred melt matrices[111]. This development, *compocasting*, provides a production route for composite material which avoids the potential separation of the discontinuous reinforcement. Good wetting is achieved, although some degradation of fibres can result from the stirring effect.

The slurry casting techniques can be employed for the production of near net shape light alloy and copper base components, but have also been shown to be feasible for steels and superalloys[112]. Titanium alloy castings have also been manufactured, using powder pressings as the melting base, giving superior properties to those of conventionally cast material. A wide-ranging review of these and other routes and applications for slurry casting is contained in Reference 113. Some individual light alloy quantity produced components were featured is Reference 114.

A comprehensive review of the entire field of semi-solid processing is that by Kirkwood[115]: a notable feature is the presented evidence of superior ductility and fatigue behaviour as compared with the products of other casting processes.

Integrated casting systems

Any review of the cast metals industry reveals a record of continuous process and product innovation, which has enabled it, at least collectively, to survive strong competition from rival processes. A particular feature has been a change of emphasis towards products with high added value, offsetting a relative decline in the output of low cost tonnage castings. Many examples of advanced process developments have been touched on in the present and earlier chapters, providing solutions to various technical problems. These measures, directed towards the enhancement of casting quality and foundry productivity, have also stimulated efforts to combine separate innovations in more comprehensive systems.

A pioneering approach can be seen in the ambitious integrated casting system concept[116] designed by the then Steel Castings Research and Trade Association, since reconstituted in the Castings Development Centre. Some of the separate features have been described in earlier chapters and were reviewed in a previous paper by Ashton et al[117]. The established main modules constituting the integrated system include a liquid metal refining facility, closed system pouring and Replicast© moulding, although other units would in principle be feasible.

Figure 10.29 (see colour plate section) shows the features and general arrangement: the stated aim is the precise automatic metering of metal of the highest quality into moulds of the highest quality. This system is, however, placed in the further context of an automated layout, within which robots are used in the shell manufacture, handling and fettling stages. An essential feature is the capability of dealing with modest quantities of molten metal by effective insulation and metal reheating facilities. The potential energy saving aspect of this type of system has also been emphasized[118].

Other examples of integrated operations include the coupling of the vacuum melting and casting units in the production of directionally solidified turbine blades, as illustrated in Chapter 9. In such plants differential vacuum pumping can be used to maintain semi-continuous production[119]. Future process developments might be expected to pursue similar linking of successive operations. Analogous connections between design, prototyping, casting production and machine stages have already been touched upon in the context of multi-purpose data files.

References

1 Hall, J. H., *Foundry*, **76**, 76 (1948)
2 Cumberland, J., *Br. Foundrym.*, **56**, 26 (1963)
3 Carrington, F. G., *Trans. Am. Fndrym. Ass.*, **52**, 333 (1944)
4 Thornton, A. E., *Br. Foundrym.*, **51**, 559 (1958)
5 Dowsing, R. J., Metals and Materials, July/August, 27 (1981)
6 Else, G. E., *Br. Foundrym.* **70**, 9, 145 (1983)
7 Fndry Trade J. International **12**, 2, 20 (1989)
8 Morgan, E. and Milnes, H. H. M., *Br. Foundrym.*, **52**, 240 (1959)
9 Samuels, M. L. and Schuh, A. E., *Foundry*, **79**, July, 78; Aug., 84 (1951)
10 McIntyre, J. B., *Fndry Trade J.*, **110**, 459 (1961)
11 Blackwood, R. and Perkins, J., *Trans. Am. Fndrym. Ass.*, **52**, 273 (1944)
12 Northcott, L. and Mclean, D., *J. Iron Steel Inst.*, **151**, 303 (1945)
13 Northcott, L. and Dickin, V., *J. Inst. Metals*, **70**, 301 (1944)
14 Lee, O. R. J. and Northcott, L., *J. Inst. Metals*, **73**, 491 (1947)
15 Tsvetnenko, K. U., *Russ. Cast. Prod.*, 273 (1961)
16 Flemings, M. C., Adams, C. M., Hucke, E. E. and Taylor, H. F., *Trans. Am. Fndrym. Soc.*, **64**, 636 (1956)
17 Northcott, L. and Lee, O. R. J., *J. Inst. Metals*, **71**, 93 (1945)
18 Ebisu, Y., *Trans. Am. Fndrym. Soc.* **85**, 643 (1977)
19 *Typical Microstructures of Cast Metals*, 2nd Edn., Inst. Br. Foundrym., London (1966)
20 Thornton, A. E., *Br. Foundrym.*, **52**, 80 (1959)
21 Thornton, A. E., *Br. Foundrym.*, **56**, 63 (1963)
22 *Symposium on High-Temperature Steels and Alloys for Gas Turbines,* Special Report 43, Iron Steel Inst., London (1952)
23 Morley, J. I., *Ref. 22*, 195
24 Choi, S. H., Kang, C. S. and Loper, C. R., *Trans. Am. Fndrym. Soc.* **97**, 971 (1989)
25 Carless, P. and Tingle, A.B. *Fndry Trade J*, July 2nd 561 (1987)
26 Crow, T. B., *Metallurgia*, **35**, 141 (1947)
27 Carreker, R. P., Jr., *J. Metals*, **15**, 774 (1963)
28 Gouwens, P. R., Watmough, J. and Berry, J. T., *Trans. Am. Fndrym. Soc.*, **67**, 577 (1959)
29 Watmough, J., Berry, J. T. and Gouwens, P. R., *Trans. Am. Fndrym. Soc.*, **69**, 701 (1961)
30 Singer, A. R. E. and Evans, R. W., *Metals Technology* **10**, 61 (1983)
31 Leatham, A., Ogilvy, A., Chesney, P. and Wood, J. V., *Metals and Materials* **5**, 3, 141 (1989)
32 Leatham, A., *Materials World* **4**, 6, 319 (1996)
33 Lavernia, E. J., Ayers, J. D. and Srivatsan, T. S. *Internat. Mat. Rev.* **37**, 1, 1 (1992)
34 Suryanarayana, C., Froes, F. H. and Rowe, R. G. *Internat. Mat. Rev.* **36**, 3, 85 (1991)
35 Jones, H., *Metals and Materials* **7**, 8, 486 (1991)
36 Fleetwood, M. J., *Metals and Materials* **3**, 1, 14 (1987)

37 Mitchell, A., Bala, S., Ballantyne, A. S. and Sidla, G., Solidification in the Foundry and Casthouse. Proceedings of International Conference at Warwick, Book 273, 147, The Metals Society, London (1983)

38 Chadwick, A. and Yue, T. M., *Metals and Materials* **5**, 6 (1989)

39 Second Report of Working Group T20 Casting Processes, *Foundryman* **87**, 386 (1994)

40 Rubtsov, N. N. and Gini, E. CH., *Russ. Cast. Prod.*, 281 (1962)

41 Bobrov, V. I., Batyshev, A. I. and Bidulya, P. N., *Russ. Cast. Prod.*, 153 (1967)

42 Plyatskii, V. M., *Extrusion Casting*, Translation R. E. Hammond and A. Wald, Primary Sources, New York (1965)

43 Williams, G. and Fisher, K. M., Solidification Technology in the Foundry and Casthouse. Proceedings of International Conference at Warwick. Book 273, 137 London, Metals Society (1983)

44 Morton, J. R. and Barlow, J. *Foundrym.* **87**, 23 (1994)

45 Begg, J. and Clegg, A. J., *Trans. Am. Fndrym. Soc.* **104**, 1159 (1996)

46 Batyshev *et al. Russ. Cast. Prod.* 220 (1972)

47 Herrera, A. and Campbell, J., *Trans. Am. Fndrym. Soc.* **105**, 5 (1997)

48 Das, A. A., Yacoub, M. M., Zantout, B. and Clegg, A.J., *Cast Metals*, **1**, 69 (1988)

49 Mykura, N., *Metals and Materials* **7**, 1, 7 (1991)

50 Nolte, M., Neussl, E. and Sahm, P. R., *Trans. Am. Fndrym. Soc.* **102**, 949 (1994)

51 Goddard, D. M., *Trans. Am. Fndrym. Soc.* **94**, 667 (1986)

52 Wood, J. V., McCartney, D. G., Dinsdale, K., Kellie, J. L. F. and Davies, P. *Cast Metals* **8**, 1, 57 (1995)

53 Beeley, P.R., *Br. Foundrym.* **79**, 441 (1986)

54 Rohatgi, P. K., Asthana, R. and Das, S., *Int. Met. Rev.* **31**(3), 115 (1986)

55 Terry, B. with Jones, G., *Metal Matrix Composites*, Elsevier Adv. Tech. Oxford (1990)

56 *Molding Methods and Materials*, Am. Fndrym. Soc., Des Plaines, Ill., 536 (1962)

57 Ryzhikov, A. A. and Timoveev, G. I., *Russ. Cast. Prod.*, 4 (1962)

58 Fukusako, T., Ohnaka, I. and Yamauchi, I., *Br. Foundrym.* **72**, 26 (1979)

59 Lipson, S. and Ripkin, F., *Trans. Am. Fndrym. Soc.*, **73**, 194 (1965)

60 Balewski, A. T. and Dimov, T., *Br. Foundrym.*, **58**, 280 (1965)

61 Campbell, J. *Engineering* **221**, 3, 185 (1981)

62 Campbell, J. and Wilkins, P. S. A., *Br. Foundrym.* **75**, 233 (1982)

63 Wilkins, P. S. A., *Br. Foundrym.* **76**, 72 (1983)

64 Campbell, J., *Cast Metals*, **1**, 3, 156 (1988)

65 Lavington, M. H., *Metals and Materials* **2**, 1, 713 (1986)

66 Smith, R. A. and Wilkins, P. S. A., *Trans. Am. Fndrym. Soc.* **94**, 785 (1986)

67 Campbell, J., *Castings*, Butterworth-Heinemann, Oxford 1991

68 Chandley, G. D., *Modern Casting* **73**, 10, 29 (1983)

69 Chandley, G. D., *Cast Metals*, **2**, 1, 2 (1989)

70 Chandley, G. D. and Cargill, D. L., *Trans. Am. Fndrym. Soc.* **98**, 413 (1990)

71 Chandley, G. D., Mikkola, P. and Shah, R. C., *Trans. Am. Fndrym. Soc.* **104**, 903 (1996)

72 Taylor, I. C., *Castings Buyer* **1**, 3, 12 (1988)

73 Kono, R. and Miura, T., *Br. Foundrym.* **67**, 319 (1974)

74 Clegg A. J. *Fndry Trade J*, **158**, 472 and 486 (1985)

75 Kono, R. and Miura, T., *Br. Foundrym.* **68**, 70 (1975)
76 May, R. P., *Br. Foundrym.* **72**, xxxviii (1979)
77 Anon. *Br. Foundrym.*, **80**, 20 (1987)
78 Kasai, H., *Trans. Am. Fndrym. Soc.* **88**, 535 (1980)
79 *Fndry International*, **16**, 3, 224 (1993)
80 Shroyer, H. F., *Br. Pat.* 850, 331, (3.4.1958)
81 Brown, J. R., *Metals and Materials* **8**, 10, 550 (1992)
82 Walling, R. P. and Dantzig, J. A., *Trans. Am. Fndrym. Soc.* **102**, 849 (1994)
83 Clegg, A. J., *Fndry Trade J. International* **9**, 30, 51–61, 69 (1986)
84 Ballmann, R. B., *Trans. Am. Fndrym. Soc.* **96**, 465 (1988)
85 Griffin, J. C., Patton, K. D. and Bates, C. E., *Trans. Am. Fndrym. Soc.* **99**, 203 (1991)
86 First Report IBF Working Group T20. *Foundrym.* **87**, 6, 223 (1994)
87 Littleton, H. E., *et al. Trans. Am. Fndrym. Soc.* **104**, 335 (1996)
88 Wittmoser, A., *Br. Foundrym.* **65**, 73 (1972)
89 Ashton, M. C., *Metals and Materials* **7**(1), 12 (1991)
90 Grote, R. E., *Trans. Am. Fndrym. Soc.* **94**, 181 (1986)
91 Dalton, R. F., *Trans. Am. Fndrym. Soc.*, **65**, 483 (1957)
92 *Method of Precision Coring in the Casting of Metallic Articles*, British Patent 782,553, Hills-McCanna. (1957)
93 Emley, E. F., *Br. Foundrym.*, **51**, 501 (1958)
94 Bunshah, R. F., *Vacuum Metallurgy*, Reinhold, New York (1958)
95 Reed, D. S. and Jones, M. L., *Trans. Am. Fndrym. Soc.* **99**, 697 (1991)
96 Roberts, R. J., *Trans. Am. Fndrym. Soc.* **104**, 523 (1996)
97 Wightman, P. and Hengsberger, E., *Metals and Materials* **7**, 11, 676 (1991)
98 Reynolds, D. J., Shamsuzzoha, M. and El-Khaddah, N., Solidification Processing 1997. Ed. Beech, J. and Jones, H., Dept of Eng. Materials, University of Sheffield, 45 (1997)
99 Dietrich, W. and Stephan, H., *Metals Handbook*, Vol. 15 Casting, 410, ASM International, Metals Park, Ohio (1988)
100 Pannen, H. and Sick, G., *Metals Handbook*, Vol. 15 Casting, 419, ASM International, Metals Park, Ohio (1988)
101 Apelian, D. and Entrebin, C. H. Jr. *Int. Met. Rev.* **31**, 2, 77 (1986)
102 Dunlop, A., *Fndry Trade J.* **123**, 124 (1967)
103 Dowsing, R. J., *Metals and Materials*, July/August, 26 (1975)
104 Bridges, P. J. and Hauzeur, F., *Cast Metals* **4**, 3, 152 (1991)
105 Beeley, P. R. and Smart, R. F., (Eds) *Investment Casting*. Institute of Materials, London (1995)
106 *Fndry Trade J. Int.* **14**, 4, 145 (1991)
107 Rose R. L. *Trans. Am. Fndrym. Soc.* **81**, 233 (1973)
108 Herlach, D. M., Cochrane, R. F., Egry, I., Fecht, H. J. and Greer, A. L., *Internat. Mat. Rev.* **38**, 6, 273 (1993)
109 Spencer, D. B., Mehrabian, R. and Flemings, M. C., *Metall. Trans.* **3**, 1925 (1972)
110 Flemings, M. C. Solidification and Casting of Metals. Proceedings of 1977 joint conference, Sheffield. Book 192, The Metals Soc, London, 479 (1979)
111 Mehrabian, R., Riek, R. G. and Flemings, M. C., *Metall. Trans.*, **5**, 1899 (1974)

112 Young, K. P., Riek, R. G. and Flemings, M. C., Solidification and Casting of Metals. Proceedings of 1977 Joint Conference, Sheffield. Book 192, The Metals Soc. London, 510 (1979)

113 Kirkwood, D. H. and Kapranos, P., *Metals and Materials*, **5**, 1, 16 (1989)

114 Kapranos, P., *Materials World*, **2**, 9, 465 (1994)

115 Kirkwood, D. H., *Int. Mat. Rev.* **39**(5), 173 (1994)

116 Ashton M. C., Sims B. J. and Sharman S. G. Proceedings of 1st International Steel Foundry Congress at Chicago, Steel Founders' Society of America, Illinois (1985)

117 Ashton, M. C., Sharman, S. G. and Sims, B. J., *Br. Foundrym.* **76**, 162 (1983)

118 Taylor, P. B., Gettings, M., Eyre, N. J. and Bushell, G. C., *Metals and Materials*, **2**, 8, 494 (1986)

119 Goulette, M. J., Spilling, P. D. and Arthey, R. P. Proceedings of 5th International Symposium on Superalloys, Am. Soc. Metals (1984)

11

Environmental protection, health and safety

Foundry operations have significant environmental impacts, both within and beyond the manufacturing plant. Whilst the main focus of foundry technology is upon the economic production of castings, it is increasingly necessary to take account of effects of the processes, not just upon working conditions for foundry employees, but upon the wider environment and its communities.

The foundry industry depends upon a range of natural resources which are wholly or partly consumed during production. These include sands and other minerals, fuel and energy, manufactured chemicals and water, quite apart from the metals and alloys that constitute the commercial product.

Besides the output of castings, waste is generated. Some of this can be recirculated within the plant or recycled in a wider sense, but a proportion is lost to the surroundings, both within the plant and outside. This constitutes pollution, much of which was once accepted as an inevitable by-product of industrial activity. Increasing concern for its damaging effects has resulted in a body of modern legislation, but there has also been a realization of the need to limit the depletion of the world's finite resources, and of the eventual real costs incurred.

Despite these wider concerns the individual manufacturing plant remains the most important focus for action, for two reasons. Firstly the plant as a workplace is most immediately susceptible to the pollution created in the production process, and potentially in its most concentrated forms. Secondly the whole issue of employee health and safety has been brought into sharp relief by stringent modern legislation, through which both employers and employees have legal responsibilities for the consequences.

Environmental protection and employee health and safety are thus seen to be linked together, and can be subject to an integrated approach: the purpose in this chapter will be to review developments in both fields, including the relevant items of legislation and some of the standards designed to assist managers in setting up systems for their implementation.

The nature of pollution

Environmental effects occurring in the foundry industry can be identified under headings employed in Annex A to BS EN ISO 14001[1], a standard to be more fully examined later.

(a) Emissions to air. These include reagents and products of chemical binder reactions, solvent vapours, dust, smoke and fume generated in melting, molten metal treatment and casting, combustion products, grinding dust and other particulates, vapours and gases: the latter include carbon monoxide, chlorine and sulphur dioxide. Much of this airborne contamination passes through local extraction systems or escapes to outside atmosphere through shop ventilation, but provision is required for more stringent control of severe emissions by prior treatment.

(b) Releases to water. These include acids, alkalis, salts, solvents and dissolved organic compounds, and heavy metals, and involve the need for treatment and disposal. Water treatment costs and effluent charges are incurred.

(c) Waste management. This is concerned with the problems of disposal of waste sand, flux residues, dross, slag and similar materials in accordance with statutory procedures, again incurring costs, and emphasizing the need to minimize the problem through economic use and effective recycling.

(d) Contamination of land. This includes the extensive use of landfill sites for the dumping of wastes such as those mentioned above.

(e) Use of raw materials and natural resources. This again relates to the waste issue, and concerns management for maximum process efficiency and product yield, so minimizing consumption of metals, moulding materials, refractories, fuels, energy, and other resources.

(f) Other local environmental and community issues. These include noise and odours, both inside and outside the plant.

Legislation and countermeasures

The key item of legislation relating to problems of pollution is the Environmental Protection Act 1990. Part I of this Act established two separate levels of authority for the control of pollution, depending upon the particular processes and substances concerned.

Integrated pollution control (IPC) applied to Part A processes, those carrying greater potential for pollution, and dealt with discharges to all three media of air, water and land. This was originally enforced by HM Inspectorates of Pollution, but in 1996 HMIP became part of a single new organisation, the Environment Agency.

Local air pollution control (LAPC) applied to Part B processes only, those with lower pollution risks, and is operated by the Local Authorities through their Environmental Health Departments.

Guidance notes have been made available for processes falling in each of the two categories: some relevant metallurgical items are included in Table 11.1. Most practical foundry applications have been found to fall within the Local Authority Air Pollution regime.

Organizations using processes and involving substances prescribed under the Regulations were required to adopt the principle summarized under the acronym BATNEEC, namely to employ the best available techniques not entailing excessive cost, to minimize the release of the harmful pollutants. For those circumstances entailing releases to more than one medium a further principle also applied. BPEO, the best practical environmental option, must be chosen when considering the comprehensive "cradle-to-grave" picture, including raw materials, production process, anti-pollution measures and the treatment and disposal of wastes.

The latter aspect, the proper management of wastes, is also subject to a further requirement of the Environmental Protection Act Part II, namely the Duty of Care, which is aimed at the safe processing and disposal of all wastes, including hazardous substances. This in quite apart from the separate legislation for the protection of employees in the workplace, in

Table 11.1 Guidance notes

Part A Processes (Integrated Pollution Control)	
IPR 2/2	Ferrous foundry processes
IPR 2/3	Processes for electric arc steelmaking, secondary steelmaking and special alloy production
IPR 2/4	Processes for the production of zinc and zinc alloys
IPR 2/8	Processes for the production of aluminium
IPR 2/9	Processes for the production of copper and copper alloys
Part B Processes (Local Air Pollution Control)	
PG 2/3	Electrical and rotary furnaces
PG 2/4	Iron, steel and nonferrous metal foundry processes
PG 2/5	Hot and cold blast cupolas
PG 2/6	Aluminium and aluminium alloy processes
PG 2/7	Zinc and zinc alloy processes
PG 2/8	Copper and copper alloy processes

particular the Control of Substances Hazardous to Health (COSHH) Regulations 1988, which will be fully discussed as a central feature of health and safety within the plant.

Modifications to this overall system have occurred with the introduction of the Pollution Prevention and Control Act 1999, formulated to meet the requirements of an EC Directive 96/61. This is to be implemented by the Pollution Prevention and Control Regulations 2000, designed to achieve integrated prevention and control of pollution, or IPPC.

This has replaced Part I of the 1990 Act, to set up a single regulatory framework covering permissions for discharges under the former IPC and LAPC designations, and there has been some revision of the process categories themselves. The major pollution discharge processes continue to be regulated by the already mentioned Environment Agency, which is also concerned with all discharges to streams under the Water Industries Act 1991. The Local Authorities are responsible for all aspects of pollution control in the less exacting process categories which fall within their sphere.

The guiding principle Best Available Technique, or BAT, is to apply throughout this system. Although more onerous than the previous principle, since more environmental impact items are defined, the cost aspect is still involved, in that economically and technically viable conditions are included as factors to be taken into account in prevention and control measures.

Whilst not in the category of legislation, a further significant development in the environmental field has been the introduction of the international standard ISO 14001, which followed the principles previously contained in BS 7750, dealing with environmental management systems. This will be reviewed after brief consideration of some of the practical measures for the prevention of pollution.

Pollution control begins at source on the principle of prevention being preferable to cure. Good practice to increase product yield and minimize the use of resources brings proportionate benefits. Sand provides the best example, given mounting costs of quarrying, transport and disposal. Maximum use of recycling and reclamation is crucial to this form of economy, whilst optimum binder utilization can make a further significant contribution: this is itself the subject of a Good Practice Guide[2]. An additional aim is to minimize the sand–to–metal ratio by intensive packing of the moulding container volume.

Close attention to plant operation so as to minimize the output of pollutants is the next step, ensuring the efficient combustion of fuels and prevention of leakages.

Specific "end of pipe" measures for the abatement of pollution must be selected to meet all the legal requirements, including quantitative limits, but should also embody provision for the increasingly exacting standards likely to be imposed in the future.

Once the main sources of pollution have been identified, attention to immediate outlet points, and to major discharges at chimneys and drains, allows the emergent streams to be characterized. Sampling and monitoring then enable flow rates and pollutant concentrations to be determined as a basis for equipment selection. Typical conditions such as pH values and temperatures may also be relevant, and a mass balance can be derived from the collected data to provide an overall assessment of inputs and outputs from the plant.

The functions of pollution control equipment are to separate the offending substances from the air or water medium, or alternatively to convert them into relatively innocuous compounds. The emissions or effluents are preferably collected close to their sources, to avoid dilution and so to minimize the volumes requiring treatment.

The selection of equipment to process the collected streams requires comparisons of alternative solutions, on the bases both of technical capability and cost. This requires specialist expertise, whether in-house or from outside sources, including potential suppliers, consultants and advisory bodies. In the latter category the Environmental Technology Best Practice Programme, based at Harwell in the UK, provides much vital information, especially that detailed in References 3 and 4.

Examples of important techniques and their main functions are listed in Table 11.2. Comprehensive tables classifying many such techniques are provided in Reference 3 and include specific details of performance capabilities, with comparative costs, space requirements and any need for supplementary

Table 11.2 Examples of typical pollution treatments

Medium	Equipment/function	Pollutant type
Air	Gravity and cyclone separators / Electrostatic precipitators	Particulate materials
	Scrubbers	Acid gases and particulates
	Thermal and catalytic oxidation / Adsorption	Volatile organic compounds (VOCs)
Water	Gravity separation / Filtration	Suspended solids
	Distillation	Volatile organic compounds, solvents
	Evaporation	Concentration of solute impurities
	Precipitation	Metals
	Adsorption	Solvents
	Electrochemical processing	Metals
	Neutralization	Acids and alkalis

measures before and after passage through the process. Reference to this source is strongly recommended.

The remaining aspect of environmental protection to be considered is the ISO standard 14001, published in the UK as the identical British Standard BS EN ISO 14001:1996 *Environmental management systems – Specification with guidance for use*. This is one of a group of similar standards directed towards providing a general managerial framework for the effective operation of technical measures, in this case for environmental protection. Others relate to quality and to health and safety and are referred to elsewhere. They are non-prescriptive in terms of the technical measures to be employed in the respective fields, but offer a structured approach for the design of systems which will ensure the achievement of legal requirements, and which will meet the needs of interested parties.

The ISO 14001 standard also embodies the equivalent European standard, but there is a further European standard with similar aims, although not identical. This is EMAS, the Eco–Management and Audit Scheme, and the minor differences are explained in Reference 4. A particular feature is this Scheme's requirement for a specific and independently validated public statement of an approved organization's actual environmental performance.

A cardinal feature of BS EN ISO 14001 is the provision for formal certification of compliance to required procedures by an independent external authority, based on a full audit. Integral to such approval is the existence both of a clearly stated policy on environmental issues and of commitments both to legal compliance and to continual improvement. The systematic approach is represented in the basic iterative loop model shown in Figure 11.1.

Five essential steps corresponding to this illustration are detailed in Clause 4 of the specification and guidance to its use is contained in an Annex. The philosophy and some of the practical implications are examined in Reference 4, published in association with the Castings Development Centre. The main features are summarized below:

Policy. This is the documented public statement by top management of commitment to compliance with environmental legislation, and to

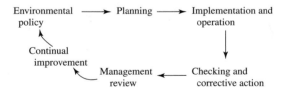

Figure 11.1 Loop model for environmental policy development (adapted from BS EN ISO 14001 by permission of BSI under licence number 2000 SK/0049. The complete standard can be obtained by post from BSI Customer Services, 389 Chiswick High Road, London W4 4AL)

continuous improvement in the attack on pollution. It is intended to be seen in the context of the entire cycle from raw material to outputs, and to permeate the whole organization.

Planning.   This requires the identification and evaluation of all the environmental impacts of the company's activities, and of the relevant legal requirements and other guidelines. Formal registers of these two categories of data are advocated, both as systematic records and as a basis for setting performance objectives, timescales and resource allocation.

Implementation.   The environmental management system requires allocation of authority and responsibilities for all the functions entailed. A particular requirement, however, is the appointment of a management representative with specific responsibility for the system itself, in conformity with the Standard, and for reporting on its performance.

The importance of training at all levels is also stressed, including provision for non-conformance, and emergencies. Communication of information and controlled documentation are further aspects requiring defined procedures, and may form a part of a comprehensive manual covering the whole system.

Checking and corrective action.   Provision is required for documented procedures to monitor and record performance against the objectives, to investigate non-conformance and to take corrective action if required. The need for systematic recording is again stressed. Scheduled internal audits are required to confirm proper operation of the system itself.

Management review.   The completion of the environmental management sequence requires a fully documented and comprehensive review of the system and its performance by the top management of the organization, to be conducted at set intervals. Such a review is strategic in nature and may encompass a need for changes to ensure continued effectiveness and improvement: policy, objectives and procedures may all be subject to revision.

The similarity of the above specification for environmental management systems to that used in the quality management field has already been mentioned. The ISO 9000 series of standards relating to the latter were mentioned in Chapter 6. Although the treatments are not identical in their systems and terminology, they involve a similar approach and procedures with respect to documentation, and some of the plant records may well be common to the two. An annex to BS EN ISO 14001 contains two-way tables identifying the analogous clauses and elements within ISO 14001 and 9001, a continuation of a similar relationship between their predecessors, BS 7750 and 5750 respectively.

This parallel structure suggests that an organisation can achieve benefits and economies from symbiosis between the separate but analogous tasks and management expertise.

Health and safety within the plant

Whilst aspects of environmental pollution and protection do impinge on the health and safety of employees in the workplace, it is the legislation specific to the latter field that has had the greatest impact upon industrial life. The Health and Safety at Work Act came into force in 1974, and bought together various earlier items of legislation and the responsibility for their implementation. The Health and Safety Executive was created as an independent body, combining functions previously performed under several government departments. A separate Health and Safety Commission was also bought into being, made up of representatives of employer and worker organisations and of local authorities, the latter representing the public interest. Its role was to initiate policies and to oversee the work of the Executive.

The Health and Safety Executive is charged with direct enforcement of the relevant laws but also acts as a source of advice and information on all matters relating to occupational health and safety. The long-established functions of the Factory Inspectorate and other bodies were absorbed into the single new organization, which thus became responsible for activities as varied as the guarding of machinery, fire and accident prevention, and the protection of the workforce from exposure to dust, fumes, toxic substances and excessive noise.

A central principle of the Health and Safety at Work Act was the responsibility placed upon employers to protect their employees at work, but also the parallel responsibilities of the employees themselves. The employer must detect hazards and assess whether they expose employees to health and safety risks, and must introduce any necessary protective measures, backed by information and training. Employees must cooperate with these measures and use clothing and equipment provided for their protection.

The most important single development in this field has been the introduction of the Control of Substances Hazardous to Health (COSHH) Regulations 1988, since revised in 1994 and 1999. Although designed primarily to give a legal framework to protection from risks in the workplace, its relevance to wider pollution problems has already been mentioned. An associated development was the introduction of the Noise at Work Regulations 1990, which required the protection of workers from damage to their hearing. These regulations have since undergone revision, in 1992, 1996 and 1997.

The practical implications of these requirements in the foundry industry have been examined in some detail by Hartmann and Johnson in Reference 5. Their full treatment relates particularly to the investment casting process, but the general approach to assessment and action is more widely applicable; reference to this work and its bibliography is recommended.

Under the COSHH Regulations, substances that are or could be damaging to health have been grouped in categories representing different degrees of severity of their potential effects, and specific limits are assigned to each. These, the Maximum Exposure Limits (MEL) and the Occupational Exposure Standard, (OES), are expressed in terms of the concentration of the airborne substance, in parts per million and/or milligrams per cubic metre, over long term (8 hour) or short term (15 minute) exposure reference periods. Values for both limits are published annually by HSE in Guidance Notes EH 40[6], which also explain the precise significance of the two types of limit. Although both relate to inhalation, other means of exposure, particularly by ingestion or through skin contact, must not be overlooked and may require the use of special measures such as protective gloves.

MEL's are assigned to a minority of substances to which exposure by inhalation may have serious health implications, and represent the maximum averaged concentration to which employees must be exposed by inhalation under any circumstances. Such substances include benzene, formaldehyde, isocyanates, crystalline silica and trichlorethylene. OES limits are more numerous and represent the concentration at which no injurious effects would be expected from regular daily exposure to the inhalation. Dusts, apart from silica and other specifically identified substances, are subject to a general 8 hour limit of $10\,\mathrm{mg/m^3}$ for total inhalable dust and $5\,\mathrm{mg/m^3}$ for respirable dust, that falling in the most harmful particle size range.

The published lists provide part of the basis for a risk assessment as required under COSHH Regulation 6. Information from suppliers, including labelling and data sheets, can also contribute to the assessment, but the nature of actual emissions during the casting process is also involved, together with the effects of existing measures to minimize exposure, for example local exhaust ventilation. This supports the view[7] that quantification by environmental sampling is really needed to achieve a satisfactory assessment.

Following the risk assessment, prevention or control measures need to be determined and implemented under COSHH Regulation 7. Prevention may involve a change in the process or in the particular substance used, whilst control may take the form of total enclosure of the process, a local extraction system, the provision of improved general ventilation, or a reduction in the exposure time for the operator. Personal protection, in the form of masks, breathing apparatus and protective clothing, is regarded as suitable only where other measures are insufficient. Once the measures have been implemented, monitoring is required to ensure that that they are continued and remain effective. Further regulations cover this aspect and place obligations upon both employers and employees to ensure that this is the case. In this connection employees must receive proper training and be provided with all the necessary information on the numerous hazards and the appropriate responses to them, including both control measures and emergency procedures: they also share the legal responsibilities involved.

The Noise at Work Regulations impose a general duty on employers to reduce risks of hearing damage, but similarly require assessments to be made in relation to employee exposure to specified levels of noise. Where the daily exposure level reaches 85 dB(A), a recorded quantitative assessment becomes mandatory, and acceptable ear protection must be made available, together with training and information on risks to hearing.

Where the daily exposure reaches 90 dB(A), or where the peak sound pressure reaches or exceeds 200 Pa (140 dB re 20 μPa), the use of the ear protection becomes obligatory on the employee. The areas concerned must be marked out as far as practicable, and exposure to noise must be reduced as far as practicable by means other than ear protectors.

Available measures include the provision of acoustic shielding around noisy equipment, but the most effective long-term approach is the selection of plant and development of practices with inherently lower noise levels. The trend away from jolting as the principal machine moulding compaction technique is one example, whilst the improvement of casting quality to reduce the need for surface grinding and the chipping of flash simultaneously addresses sources of both noise and dust. This needs to be weighed in the balance of costs and benefits accompanying changes in pattern equipment and moulding materials. Where high levels of noise are unavoidable, ear plugs or muffs may be required in line with the above circumstances. This presents no problem where loud noise is intermittent, but continuous use is less amenable and lends more weight to the search for radical solutions.

Risk assessments. The concept of risk assessment has already been referred to in relation to exposure to hazardous substances and noise, but a systematic health and safety risk assessment requires attention to other work activities as well: important areas include mechanical and electrical safety, the guarding of machinery, fire and explosion, slips and falls, manual handling, transport, and exposure to vibration. Advice on these and other aspects is published by HSE and also in annexes to BS 8800, a further standard relevant to health and safety issues.

The sequence followed in the above discussions thus involves the identification of hazards, the assessment and quantification of risks against the background of legal requirements, and the planning, implementation and monitoring of countermeasures, followed by a review. This type of systematic approach provides the theme of BS 8800 1996: Guide to occupational health and safety management systems[8], a standard designed to foster good management in this area and its integration within an overall management system. Its content offers two alternative and slightly different models, one of which is closely related to the previously discussed BS EN ISO 14001 standard for environmental management systems and is based upon it. A major difference, however, is that BS 8800 is not intended for certification purposes, nor to be quoted as a specification: its status is rather that of a management guide. The parallels between these two standards have been illustrated by Lyon[9] in a general review of management integration possibilities in all three areas of quality, health and the environment, for which a model was suggested.

The organization and management of health and safety occupies a role of central importance in modern industry, and HSE itself publishes guidance on the essential principles[10,11]. The diverse requirements have generated a complex matrix of laws and regulations, requiring interpretation and the active participation of managers on all levels. What is required, however is not a bureaucratic structure as much as the promotion of a culture of safety throughout an organization, fostering an awareness of responsibilities and a systematic approach to the problems that need to be overcome.

Whilst the focus of the above discussions has been upon the UK, it is clear that other countries face parallel concerns. A typical examination of the approach to hazardous air pollutants in the USA was detailed by Allen et al[12]. This referred to emissions from sand additives and organic binders, and especially to their decomposition products, in relation to particular American legislation, namely the 1990 Clean Air Act Amendments. The widening from the original emphasis on workplace conditions to emissions to the general atmosphere was stressed.

The foregoing discussions, whether concerning environmental matters or health and safety, relate to an evolving situation in which legislation, regulations, specifications and literature offer a snapshot rather than a permanent picture. Those directly concerned need to keep abreast of a flow of amendments and new developments: reference to the main advisory centres already mentioned in the text will be found useful for this purpose.

References

1 BS EN ISO 14001 Environmental management systems – Specification with guidance for use. Br. Stand. Instn. London
2 Cost-effective management of chemical binders in foundries. Guide GG 104 Environmental Technology Best Practice Programme, Harwell (1998)
3 Choosing cost-effective pollution control. Guide GG 109 Environmental Technology Best Practice Programme, Harwell (1998)
4 Environmental management systems in foundries. Guide GG 43 Environmental Technology Best Practice Programme, Harwell (1998)
5 Hartmann, E. F. and Johnson, P. In *Investment Casting*. Ed. Beeley, P. R. and Smart, R. F., The Institute of Materials, London 212, (1995)
6 EH 40/94 Occupational exposure limits 1994. Health and Safety Executive, HMSO.
7 Drinkall, B. *Foundryman* **84**, 7, (1991)

8 BS 8800 Guide to occupational health and safety management systems. Br. Stand. Instn. London

9 Lyon R. *Foundryman* **90**, 20, (1997)

10 Essentials of health and safety at work. Health and Safety Executive, HMSO

11 Successful health and safety management. Health and Safety Executive, HMSO

12 Allen, G. R., Archibald, J. J. and Keenan, T. *Trans. Am. Fndrym. Soc.* **99**, 585, (1991)

Appendix

Bibliography of some British Standard Specifications relating to the design, manufacture and examination of castings

PART 1. SPECIFICATIONS FOR CASTINGS AND CASTING ALLOYS

Note 1: *Contents*. Chemical compositions are specified except where otherwise stated.

Most specifications give values for mechanical properties including tensile strength and elongation; proof or yield stress, impact, bend, hardness and other test values are given in appropriate cases.

General clauses cover, in varying degree, inspection and sampling, requirements for freedom from defects, rectification practice and dimensional tolerances. Special requirements in certain Specifications include microstructure and machinability.

Note 2: *Other B.S.S. for castings*. In addition to the Specifications listed below, specialized cast products, for example cast iron pipes and rainwater goods and cast steel valves, are covered by separate Specifications*.

Note 3: Numerous British Standards are adopted international standards, prepared under the auspices of the International Organization for Standardization or the European Standardization body CEN. These carry the designations ISO and/or EN after the normal BS prefix. The BSI Standards Catalogue entries also indicate degrees of correspondence between British and related standards.

Note 4: The Appendix is intended as a general guide to relevant British Standards. It is, however, essential to consult the BSI Catalogue and current issues of individual Standards* for definitive user information.

---

* Refer to Subject Index, BSI Standards Catalogue (annual issue)

Alloy group	Specification and subject	Further notes
Aluminium base	BS EN 1706 Aluminium and aluminium alloys – Castings – Chemical composition and mechanical properties. [Replaced BS 1490]	Requirements for composition, properties and other characteristics. Covers sand, chill or permanent mould, pressure die and investment castings. Temper designations; sampling and testing requirements
Copper base	BS EN 1982 Copper and copper alloys – Ingots and castings. [Replaced BS 1400]	Requirements for chemical composition, mechanical properties and other characteristics. Covers sand, permanent mould, centrifugal and pressure die castings, and continuous casting. Sampling and testing requirements
Magnesium base	BS EN 1753 Magnesium and magnesium alloys – Magnesium alloy ingots and castings. [Replaced BS 2970]	Requirements for chemical composition, mechanical properties and other characteristics. Covers sand, permanent mould, pressure die and investment castings. Temper designations, sampling and testing requirements.
Cast iron	BS EN 1560 Founding. Designation system for cast iron. Material symbols and material numbers.	Includes cast iron basic types, with provision for identification of macrostructure and microstructure, including graphite forms.
	BS EN 1561 Founding. Grey cast irons. [Replaced BS 1452]	Designation by specified tensile strength.

Alloy group	Specification and subject	Further notes
		No composition requirement: composition and foundry technique at manufacturer's discretion. Sampling and test bar details. Annexes give mechanical and physical properties and portray relationships between tensile strength and hardness and between tensile strength and wall thickness.
	BS EN 1562 Founding. Malleable cast irons. [Replaced BS 6681]	Designation by tensile strength/elongation combination. White heart (decarburized) and black heart (non-decarburized) types. Mechanical properties specified. Composition only specified for maximum phosphorus content to optimize impact resistance.
	BS EN 1563 Founding. Spheroidal graphite cast ions [Replaced BS 2789]	Designated by tensile strength/elongation combination: properties dependent on structure. Manufacturing process, chemical composition and heat treatment at manufacturer's discretion. Test bars specified
	BS EN 1564 Founding. Austempered ductile cast ions [Replaced BS 2789]	Designated by tensile strength/elongation combination.

(continued overleaf)

Alloy group	Specification and subject	Further notes
		Mechanical properties include Charpy notched impact energy. Composition and manufacture at producer's discretion
	BS 1591 Corrosion resisting high silicon iron castings.	Requirements for chemical composition, heat treatment, manufacture and freedom from defects. Grades for use in chemical engineering and for cathodic protection anodes.
	BS 3468 Austenitic cast iron.	Grades based on mechanical properties and chemical composition. 2 groups, for general engineering and special purposes
Steel	BS 3100 Steel castings for general engineering purposes	Requirements for chemical composition, heat treatment and mechanical properties of cast steels. Non-destructive examination; sampling and testing. Traceability aspects and delivery conditions
	BS EN 10213 (1–4) Technical delivery conditions for steel castings for pressure purposes	1. General: inspection, NDT, documentation, compositional accuracy etc. Informative annex on physical and creep properties 2. Steel grades for use at room temperature and at elevated temperature. 3. Steels for use at low temperatures

Alloy group	Specification and subject	Further notes
		4. Austenitic and austenitic-ferritic grades (Compositions, heat treatments and mechanical properties specified)
	BS 3146 Parts 1 and 2 Specification for investment castings in metal	Part 1 Carbon and low alloy steels Part 2 Corrosion and heat resisting steels, nickel and cobalt alloys. Requirements for chemical composition, mechanical properties, heat treatment, non-destructive testing and rectification of defects.
Nickel base	BS 3071 Nickel–copper alloy castings	Requirements for chemical composition, mechanical properties and other characteristics. Main alloy content 30% Cu with 1–4% Si
	BS 3146 Part 2 – see under Steel	
Zinc base	BS EN 12844 Zinc and zinc alloys. Castings specifications [Replaced BS 1004]	Designations. Chemical composition and other requirements Annex (informative) on properties of zinc alloy die castings at 20°C. Relationship of alloy compositions to other and earlier standards including BS 1004

(*continued overleaf*)

Alloy group	Specification and subject	Further notes
	BS EN 1774 Zinc and zinc alloys – Alloys for foundry purposes–Ingot and liquid [partially replaced BS 1004]	
Several alloy types	BS EN 1559 Foundry. Technical conditions of delivery.	1. General 3. Additional requirements for iron castings. 4. Additional requirements for aluminium alloy castings 5. Additional requirements for magnesium alloy castings 6. Additional requirements for zinc alloy castings

PART 2. FURTHER SPECIFICATIONS RELATING TO THE EXAMINATION OF CASTINGS

Field	Specification and subject	Further notes
Non-destructive examination	BS EN 1330 Non-destructive testing. Terminology [Progressive replacement of BS 3683]	1. List of general terms 2. Terms common to the non-destructive testing methods 3. Terms used in industrial radiographic testing 4. Terms used in ultrasonic testing 5. Terms used in eddy current testing 8. Terms used in leak tightness testing

Field	Specification and subject	Further notes
		9. Terms used in acoustic emission testing
	BS 3683 Glossary of terms used in non-destructive testing [Progressive replacement by BS 3683]	Part 1 Penetrant flaw detection Part 2 Magnetic particle flaw detection Part 4 Ultrasonic flaw detection
	BS EN 444 Non-destructive testing. General principles for radiographic examination of metallic materials by X- and gamma rays	
	BS EN 462 1–5 Image quality of radiographs [Replaced BS 3971]	1. Indicators: wire type 2. Indicators: step / hole type 3. Image quality classes for ferrous metals 4. Experimental evaluation of image quality values 5. Indicators (duplex wire type), determination of image unsharpness value.
	BS 2737 Terminology of internal defects in castings as revealed by radiography	Includes diffraction mottling – each term defined with description of radiographic appearance of defects, illustrated by 65 reproductions of radiographs or micrographs.

(continued overleaf)

Field	Specification and subject	Further notes
	BS EN 1369 Founding. Magnetic particle inspection [Replaced BS 4080-1]	
	BS 4069 Magnetic flaw detection inks and powders	Fluorescent and non-fluorescent inks, concentrates and powders: tests for acceptability
	BS EN 1371 Founding. Liquid penetrant inspection [Replaced BS 4080–2]	Sand, gravity die and low pressure die castings
	BS 6208 Method for ultrasonic, testing of ferritic steel castings including quality levels. [Replaced BS 4080]	
	BS 12223 Non-destructive testing – Ultrasonic examination – Specification for calibration block No 1	
	BS EN 27963 Calibration block No 2 for ultrasonic examination of welds	
	BS EN 1370 Founding. Surface roughness inspection by visual tactile comparators	

Field	Specification and subject	Further notes
	BS 6615 Specification for dimensional tolerances for metal and metal alloy castings.	The adopted international standard ISO 8062
Mechanical testing	BS EN 10002 (1–5) Tensile testing of metallic materials [Replaced BS 18]	1. Method of test at ambient temperature. Defines the mechanical properties 2. Verification of the force measuring system of the machine 3. Calibration of the force proving instruments for verification of the machine 4. Verification of extensometers 5. Method of test at elevated temperatures. Defines the mechanical properties.
	BS 131 Notched bar tests	Part 1 Izod impact tests on metals Part 4 Calibration of Izod pendulum impact testing machine Part 5 Determination of crystallinity Part 6 Method for precision determination of the Charpy V notch impact energy for metals Part 7 Specification for verification of the test machine used for the latter.

(continued overleaf)

Field	Specification and subject	Further notes
	BS EN 10045 Charpy impact test on metallic materials [Replaced parts of BS 131]	1. Test method (V- and U-notches) 2. Method for verification of impact testing machines
	BS 1639 Methods for bend testing of metals	General principles and factors influencing test; single and reverse bend tests. Test pieces.
	BS EN 10003 Brinell hardness test [Replaced BS 240]	1. Test method 2. Verification of Brinell testing machine 3. Calibration of standardized blocks.
	BS EN 10109 Metallic materials. Hardness test (covers the Rockwell test) [Replaces BS 891 and BS 4175]	1. Rockwell test (scales A, B, C, D, E, F, G, H, K,) and Rockwell superficial test (scales 15N, 45N, 15T, 30T and 45T) 2. Verification of Rockwell testing machines 3. Calibration of standardized blocks
	BS EN ISO 6507 Metallic materials. Vickers hardness test [Replaced BS 427]	1. Test method 2. Verification of testing machines 3. Calibration of reference blocks
	BS 3500 Methods for creep and rupture testing of metals	1. Tensile rupture testing 2. Tensile creep testing 5. Production acceptance tests

Field	Specification and subject	Further notes
		6. Tensile stress relaxation testing
	BS 3518 Methods of fatigue testing	1. Guide to general principles. Recommends tests 2. Rotating bending fatigue tests. Procedure where the test piece and not the load is rotated
		3. Direct stress fatigue tests. Without deliberate stress concentrations 5. Guide to the application of statistics Determination of S–N curves, fatigue limit, life etc; analytical methods and examples.
	BS 7448 Fracture mechanics toughness tests	Parts 1, 2 and 4: Methods for determination of critical fracture toughness, crack tip opening displacement and crack extension resistance using fatigue pre-cracked specimens
	BS EN ISO 12737 Metallic materials. Determination of plane–strain fracture toughness [Partly replaces BS 7748 Part 1]	

(continued overleaf)

Field	Specification and subject	Further notes
Chemical analysis	BS 1728 Methods for the analysis of aluminium and aluminium alloys	20 parts treating the determination of individual elements
	BS 1748 Methods for the analysis of copper alloys	11 parts treating the determination of individual elements

Index